Resource Service Management in Manufacturing Grid System

Scrivener Publishing
3 Winter Street, Suite 3
Salem, MA 01970

Publishers at Scrivener
Martin Scrivener (martin@scrivenerpublishing.com)
Phillip Carmical (pcarmical@scrivenerpublishing.com)

Resource Service Management in Manufacturing Grid System

Fei Tao, Lin Zhang and Yefa Hu

Scrivener

WILEY

Co-published by John Wiley & Sons, Inc. Hoboken, New Jersey, and Scrivener Publishing LLC, Salem, Massachusetts.
Published simultaneously in Canada.

Cover design by Russell Richardson

Library of Congress Cataloging-in-Publication Data:

ISBN 978-1-118-12231-0

Printed in the United States of America

10 9 8 7 6 5 4 3 2 1

"To our families for their continuous love, encouragement, and support."

Contents

Acknowledgements **xxi**
Preface **xxiii**
Abbreviations **xxix**

1 Introduction to Manufacturing Grid **1**
1.1 Introduction 1
1.2 Proposal of Manufacturing Grid 2
1.2.1 Several Issues of Manufacturing 2
1.2.2 Proposal of MGrid 4
1.2.3 Technological Driving Forces of MGrid 6
1.3 Concept of MGrid 8
1.3.1 A Brief Outline of Grid and its Applications 8
1.3.2 Concept of MGrid 10
1.4 Basic Features of MGrid 13
1.5 The Connotation of MGrid 14
1.6 Comparison between MGrid and Networked Manufacturing 16
1.7 Comparison between MGrid and Computing Grid 19
1.8 Key Research Contents and Technologies of MGrid 20
1.8.1 General Technologies 20
1.8.2 Supporting Technologies 22
1.8.3 Key Enabling Technologies 23
1.8.4 Application Technologies 25
1.9 Summary 26

2 Resource Service Optimal-Allocation System in MGrid 27
1.1 Introduction 27
2.2 The Architecture of MGrid 28
2.2.1 MGrid Resource Layer 28
2.2.2 MGrid Core Middleware Layer 29
2.2.3 MGrid User Service Middleware Layer 30
2.2.4 MGrid User Portal Layer 30
2.2.5 MGrid User Layer 30
2.3 MGrid Collaborative Manufacturing Platform 31
2.3.1 Conceptual Model of MGrid Collaborative Manufacturing Platform 31
2.3.2 Resource Service Publication 33
2.3.3 The Resource Service Exchange between RSP and RSD 35
2.4 MGrid Resource Service Optimal-Allocation System (MGRSOAS) 35
2.5 The Key Issues and Technologies for Realizing RSOAS 38
2.5.1 Modeling and Digital Description of Resource Service (DDoRS) 39
2.5.2 Resource Service Match and Search (RSMS) 39
2.5.3 QoS Modeling and Evaluation of Resource Service 39
2.5.4 MGrid Resource Service Optimal-Selection and Composition (RSOSC) 40
2.5.5 Resource Service Composition and Network Modeling and Dynamic Characteristics 40
2.5.6 Failure Tolerance Management 41
2.6 Summary 41

3 Digital Description of MGrid Resource Service 43
3.1 Introduction 43
3.2 Classification of MGrid Resource Service and Its Application 44
3.2.1 Classification of MGrid Resource and Resource Service 44

3.2.2 Application Case: Resource Service Design for Magnetic Bearing System 45
3.3 Requirements of DDoRS in MGrid 53
3.4 MGrid and Ontology 54
3.5 Establishing the Method of MGrid-Ontology 55
3.5.1 Step 1: Define the Scope and Requirements of MGrid-Ontology 56
3.5.2 Step 2: Determine Essential Concepts, Reusing Existing Ontologies if Possible 57
3.5.3 Step 3: Analyses and Design of MGrid-Ontology 57
3.5.4 Step 4: Representation of MGrid-Ontology 58
3.5.5 Step 5: Verification and Validation of MGrid-Ontology 58
3.6 Selection of Describing Language for MGrid-Ontology 59
3.7 MGrid Ontology 62
3.7.1 OWL-S 62
3.7.2 MGrid-Ontology 63
3.8 DDoRS Based on MGrid-Ontology 68
3.8.1 Description of ***Agent*** 68
3.8.2 Description of ***MGSP*** 68
3.8.3 Description of ***MGSM*** 71
3.9 Application Case: MGrid-Ontology Based MGrid Resource Service Discovery 71
3.10 Summary 74

4 MGrid Resource Service Match and Search 75
4.1 Introduction 75
4.2 Related Works 77
4.2.1 Service Discovery in Traditional Distributed System 77
4.2.2 Service Match and Discovery in Distributed Manufacturing System 78
4.3 Framework of Resource Service Match and Search in MGrid 79
4.4 SMAs: Similarity Matching Algorithms (SMAs) 82
4.4.1 Word Matching Algorithms (WMAs) 83

4.4.2 Sentence Matching Algorithms (SeMAs) 88
4.4.3 Number Matching Algorithms (NMAs) 89
4.4.4 Entity Class Matching Algorithms (ECMAs) 93
4.5 RS-Matcher: Resource Service Matcher 94
4.5.1 Basic-matching 97
4.5.2 I/O-matching 98
4.5.3 QoS-matching 100
4.5.4 Integrated-matching 101
4.6 Case Study 104
4.6.1 Step 1: Basic-matching 105
4.6.2 Step 2: I/O-matching 106
4.6.3 Step 3: QoS-matching 108
4.6.4 Step 4: Integrated-matching 108
4.7 Performance Results and Discussion 109
4.7.1 Accuracy 109
4.7.2 Efficiency 111
4.8 Summary 112

5 Resource Service QoS Modeling and Evaluation 115
5.1 Introduction 115
5.2 Related Works 116
5.3 Evaluation Indices System of MGrid Resource Service 118
5.4 Evaluation of SEIs and IEIs 120
5.4.1 Evaluation Framework of SEIs and IEIs 120
5.4.2 Structure Model of Agent 124
5.4.3 Evaluation Process of SEIs and IEIs 127
5.4.3.1 Setting of Manufacturing Resources Evaluation Indexes Set U 127
5.4.3.2 Setting the Evaluation Grade Values Set P 128
5.4.3.3 Establishment of the Comprehensive Information Matrix R 128
5.3.3.4 Establishment of the Weight Coefficients Set Q 129
5.4.3.5 Establishment of the Comprehensive Evaluation Matrix M 129

5.4.3.6 Establishment of the Increased Weight Set Matrix B 129
5.4.3.7 The Result of Comprehensive Evaluation Quantitative Value V Based on Matrix M and Matrix B 129
5.5 Classification and Modeling of MGrid QoS 130
5.5.1 QoS Modeling from the Whole-lifecycle Management of MGrid Resource Service QoS 130
5.5.2 QoS Modeling from MGrid Architecture Views 131
5.5.3 MGrid QoS Attribute Parameters Modeling 132
5.6 Evaluation of MGrid QoS Attribute Parameter 133
5.6.1 Time 133
5.6.2 Cost 136
5.6.3 Reliability 136
5.6.4 Maintainability 136
5.6.5 Trust-QoS 137
5.6.6 Function Similarity 137
5.7 Application Case: QoS-based MGrid Resource Service Management 138
5.8 Summary 141

6 Resource Service Trust-QoS Evaluation 143
6.1 Introduction 143
6.2 Related Works 146
6.2.1 MGrid Resource Scheduling Based on QoS 146
6.2.2 QoS and Trust-QoS 147
6.3 Resource Management and Trust Relationship Management in MGrid 151
6.4 MGrid Resource Service Trust-QoS Relationship Model 153
6.5 MGrid Resource Service Trust-QoS Evaluation Model 156
6.5.1 Related Conceptions and Definitions 156
6.5.2 Intra-domain Trust-QoS Evaluation Model of Resource Service 161

6.5.3 Inter-domain Trust-QoS Evaluation Model of Resources Service 161
6.6 Data Structure Design 162
6.7 Trust-QoS Evaluating and Updating Algorithms 166
6.7.1 Direct Trust-QoS Evaluating Algorithm 166
6.7.2 Recommended Trust-QoS Evaluating Algorithm 166
6.7.3 Comprehensive Combining Recommended Trust-QoS Evaluating Algorithm 169
6.8 Real-time and Dynamical Updating Algorithm of Trust-QoS Degree 170
6.9 Trust-QoS Evaluation Case Study 171
6.10 Application Case: Trust-QoS Based MGrid Resource Service Scheduling 175
6.10.1 Requirements of Trust-QoS Based MGrid Resource Service Scheduling 175
6.10.2 Trust-QoS Based MGrid Resource Service Scheduling Framework 176
6.10.3 Trust-QoS Based MGrid Resource Service Scheduling Algorithms 180
6.11 Summary 184

7 Resource Service Optimal-selection and Composition Framework 185
7.1 Introduction 185
7.2 MGrid-RSOSCF: MGrid Resource Service Optimal-selection and Composition 188
7.2.1 Architecture of MGrid-RSOSCF 188
7.2.2 Life-cycle of MGrid Resource Service Composition 190
7.3 T-Layer: Task Layer 191
7.3.1 Basic Models of MGrid Task 191
7.3.2 Main Functions and Services in T-Layer 195
7.4 S-Layer: Resource Service Match and Search Layer 196
7.5 Q-Layer: Resource Service QoS Synthetically Processing Layer 196
7.5.1 Main Functions and Services in Q-Layer 197
7.5.2 QoS Extraction 197
7.5.3 QoS Comparison 200

7.6 O-Layer: Resource Service Optimal-selection Layer 201
7.6.1 Simple Resource Service Optimal-selection Method 201
7.6.2 The Other Resource Service Optimal-selection Methods 204
7.7 C-Layer: Resource Service Composition Layer 204
7.7.1 RSC-Engine: Resource Service Composition Engine 204
7.7.2 RSCEP-Generator 207
7.7.3 RSCEP-Selector 208
7.7.4 RSCE-Controller: Resource Service Composition Executing Controller 210
7.7.5 RSC-Monitor: Resource Service Composition Monitor 210
7.7.6 RSC-Coordinator: Resource Service Composition Coordinator 211
7.8 Summary 212

8 Resource Service Optimal-selection Based on Intuitionistic Fuzzy Set and Non-functionality QoS 215
8.1 Introduction 215
8.2 Framework of Resource Service Selection 220
8.3 Resource Service Optimal-selection Based on IFS in MGrid 222
8.3.1 Symbols and Notations 222
8.3.2 Preliminaries on Intuitionistic Fuzzy Set (IFS) 223
8.3.3 RSOS Based on IFS 225
8.3.3.1 Evaluating c_j of RS^m at Time Periods t_k 225
8.3.3.2 Calculating the Synthetic Evaluation of C_j of RS^m form t_0 to t_c 226
8.3.3.3 Determining the Weights of QoS Criteria 228
8.3.3.4 Evaluating Fuzzy Synthetic of Each CRS 230
8.3.3.5 Calculating the Closeness Coefficient of Each CRS 231

8.3.3.6 Ranking the Order of Candidate Resource Services (CRSs) 232
8.3.4 Data Structure Design 232
8.4 Case Study 233
8.4.1 Step 1: Extracting the Related Data of RS^1, RS^2, RS^3, RS^4, and RS^5 233
8.4.2 Step 2: Evaluating the QoS Criteria of Each Candidate Resource Service at Different Time Periods 238
8.4.3 Step 3: Calculating the Synthetic QoS Criteria of Each Candidate Resource Service 238
8.4.4 Step 4: Determining the Weights of Each QoS Criterion 238
8.4.5 Step 5: Calculating the Closeness Coefficient of Each CRS 238
8.4.6 Step 6: Ranking the Order of CRSs 245
8.5 Performance Analysis and Discussion 246
8.5.1 Scalability and Efficiency 246
8.5.2 Effectiveness 246
8.5.3 Comparison with the Method of Bedi (Bedi *et al.*, 2007) 248
8.6 Summary 251

9 Correlation Relationship Management in Resource Services Composition 253
9.1 Introduction 253
9.2 Related Works 254
9.3 Motivation 255
9.4 Correlation Relationship among Resource Services 258
9.4.1 Combinable Correlation 258
9.4.1.1 Definition and Model 258
9.4.1.2 Combinable Correlation Degree 259
9.4.2 Business Entity Correlation 263
9.4.3 Statistical Cooperate Correlation 264
9.5 QoS Computation Model of Correlation-aware Resource Services Composition 265
9.6 Case Study: Correlation-aware Resource Services Composition 268
9.7 Summary 273

10 Resource Service Composition Optimal-selection 275
10.1 Introduction 275
10.2 Problem Formulation and Review 279
10.2.1 Motivating Example 279
10.2.2 Basic Models of Resource Service Composition and their QoS Evaluation 281
10.2.2.1 QoS Properties and QoS Description Mode Supporting Service Correlation 281
10.2.2.2 Basic Composite Models of CRS and their QoS Evaluation 284
10.2.3 Formulation of MO-MRSCOS Problem 286
10.2.4 Pareto Solution of MO-MRSCOS 290
10.3 Review of Standard PSO 293
10.4 Improved PSO for MO-MRSCOS Problem 295
10.4.1 Representation 295
10.4.2 Particle Movement (update) 296
10.4.2.1 Selection of Inertia Weight 297
10.4.2.2 Selection of Acceleration Coefficients 298
10.4.2.3 Biggest Personal Position Value 299
10.4.2.4 Biggest Velocity Value 299
10.4.3 Generation of Personal Best (*pbest*) 300
10.4.4 Generation of Global Best (*gbest*) 300
10.4.5 Population Trimming in PSO 303
10.4.6 The Procedures of the Proposed PSO Algorithm for MO-MRSCOS Problem 304
10.5 Performance Analysis and Discussion 305
10.6 Summary 309

11 Resource Services Composition Flexibility 311
11.1 Introduction 311
11.2 Related Works 313
11.3 The Analysis, Definition and Classification of RSC Flexibility 315
11.3.1 The Factors Influencing RSC 315
11.3.2 The Definition and Classification of RSC Flexibility 317
11.4 The Measurement of RSC Flexibility 322

11.4.1 Task Flexibility 323
11.4.2 Resource Service Flexibility 327
11.4.3 QoS Flexibility 330
11.4.4 Correlation Flexibility 335
11.4.5 Flow Flexibility 341
11.5 Case Study and Experiment Results 346
11.5.1 The Change of Tasks 347
11.5.2 The Change of Resource Services 348
11.5.3 The Change of QoS of Resource Services 350
11.5.4 The Change of Correlations among Resource Services 352
11.5.5 The Change of Flows of RSC 353
11.6 Summary 355

12 Resource Services Composition Network 357
12.1 Introduction 357
12.2 Scale-free Network (SFN) 362
12.3 The Theoretical Hypothesis: Composition Service Network is a Scale-free Network 364
12.4 Concepts and Definitions in CoRCS-Net 365
12.4.1 Web Service, Composition Service, and Composition Service Network 365
12.4.2 Combinable Relationship (CoR) 368
12.4.3 Decision Rule of Combinable Relationship 373
12.4.4 Combinable Strength 374
12.4.5 Definition and Establishment of CoRCS-Net 375
12.5 The Evolving Behavior of CoRCS-Net 377
12.5.1 The Elementary Evolving Operators of CoRCS-Net 377
12.5.2 The Influence of the Evolving Operators on the Combinable Strength in CoRCS-Net 379
12.6 Theoretical Proof of the Scale-free Characteristics of CoRCS-Net 392
12.6.1 The Variation Rate of Combinable Strength for Each Service Vertexes in the CoRCS-Net 392
12.6.2 The Proof of Scale-free Characteristic of CoRCS-Net 396

12.6.3 Significance 399
12.7 Summary 399

13 Failure Detection and Recovery in Resource Service Optimal-Allocation 403
13.1 Introduction 403
13.2 Related Works 404
13.2.1 Failure Management in Manufacturing System 404
13.2.2 Failure Management in Distributed System 406
13.3 Define and Classification of MGrid Failure 407
13.3.1 Virtual Link Related Failure 408
13.3.2 Resource Service Related Failure 408
13.3.3 Task Related Failure 411
13.3.4 Application Related Failure 412
13.4 Architecture of MGrid Failure Management System 413
13.5 Detection of MGrid Failure 415
13.5.1 Detection of Virtual Link Related Failures 415
13.5.2 Detection of Resource Service Related Failures 416
13.5.3 Detection of Task Related Failures 419
13.5.4 Detection of Application Related Failures 421
13.6 MGrid Failure Recovery Based on ECA Rules 422
13.7 Implementation and Simulation 429
13.7.1 Comparison of Success Rate 430
13.7.2 Comparison of Execution Time 431
13.8 Conclusion 434

14 Summary of the Application of Grid Technology in Manufacturing 435
14.1 Introduction 435
14.2 Review of MGrid Theories 435
14.2.1 Architecture of MGrid 445
14.2.2 Resource Management System (RMS) 447
14.2.3 Modeling and Encapsulating of MGrid Resource Services 448
14.2.4 Resource Service Discovery and Scheduling 449

14.2.5 QoS Management 450
14.2.6 Service Composition 451
14.2.7 Workflow Management 452
14.2.8 Job Management 453
14.2.9 Reliability Management 454
14.2.10 Security and Trust Management 455
14.2.11 Others 456
14.3 Investigation of Application Research on MGrid 456
14.4 Key Future Research Issues 463
14.4.1 Basic Protocols, Standards, and Criteria 464
14.4.2 Service-oriented MGrid Architecture with Semantic Support 464
14.4.3 MGrid Ontology 464
14.4.4 Quantitative Research on MGrid Resource Service 464
14.4.5 Resource Service Composition and Its Optimal-selection 465
14.4.6 Cost and Price Management, and Electronic Payment of MGrid 465
14.4.7 Performance Measurement and Evaluation of an MGrid System 466
14.4.8 Commercial Operation Model of MGrid 466
14.4.9 Embedded Connect Technology for Physical Equipment 467
14.5 Summary 467

15 Cloud Manufacturing: Development and Commerce Realization of MGrid 469
15.1 Introduction 469
15.2 Concept and Architecture of Cloud Manufacturing 470
15.2.1 Concept of Cloud Manufacturing 470
15.2.2 Architecture of Cloud Manufacturing 472
15.3 Core Enabling Technologies for Cloud Manufacturing 474
15.4 Typical Characteristics of Cloud Manufacturing 475
15.5 Difference and Relationship between Cloud Computing and Cloud Manufacturing 476

15.6 Classification of Cloud Manufacturing Service Platform 477
15.7 Key Technologies and Main Research Contents of Cloud Manufacturing 479
15.8 Key Advantages and Challenges of Cloud Manufacturing 479
15.9 Summary 481

Bibliography **483**

Index **509**

Acknowledgements

This book is a summary of Dr. Tao's research on resource service management in manufacturing grid (MGrid) and cloud manufacturing (CMfg) system during his study and work from September 2005 to August 2011 in Wuhan University of Technology (WHUT), the University of Michigan-Dearborn (UMD), and Beihang University (BUAA). Therefore, Dr. Tao would like to send special thanks to Professors YF Hu, ZD Zhou, MZ Yang, BY Shen, and YF Ding in WHUT, Professor D Zhao in UMD, and Prof. L Zhang in BUAA.

The authors would like to express their special thanks to China Machine Press and journal publishers. This book is an English-language version of the authors' Chinese book (*Theory and Practice: Optimal Resource Service Allocation in Manufacturing Grid* by F Tao, YF Hu and L Zhang, published by China Machine Press in 2010) but with more than over 40% new content. Some of the material has been published in *IEEE Transactions on Industrial Informatics*(TII) , *International Journal of Production Research* (IJPR), *International Journal of Advanced Manufacturing* (IJAMT), *European Journal of Operational Research* (EJOR), *International Journal of Computer Integrated Manufacturing* (IJCIM), *International Journal of Manufacturing Technology and Management* (IJMTM), *Enterprise Information Systems*(EIS), *Knowledge and Information Systems* (KIS), *Proceedings of the Institution of Mechanical Engineers, Part B, Journal of Engineering Manufacture.* (JEM), *Chinese Mechanical Engineering*, and some conferences such as IEEM'2009 and INDIN'2006.

Dr. Tao would also like to express his special thanks to Professor A. Y. C. Nee for his encouragement, invaluable help and advice over the years and to his suggestion of writing this book.

Some of the book's research and writing were made possible with the financial support of the following research projects: Excellent Doctoral Dissertation Fund from WHUT (Wuhan University of Technology), Nature Science Foundation of China (No.51005012, No.61074144 and NO.50335020), Hubei Digital Manufacturing Key Laboratory Opening Fund (No.SZ0621), Aeronautic Bairen (One Hundred Outstanding People) Plan and Weishi Youth Foundation (No.YWF-10-02-007) from Beihang University.

Thanks for Dr. H. Guo's contribution to the Chapters 9 and 11. Dr. Tao would like to give special thanks to his master students for proofing some chapters: Y. Cheng, Y.L. Liu, and L. LV.

Thanks for the help from Martin D. Scrivener, the President of the Scrivener Publishing LLC., as well as the hard and efficient work by the other people at the publishers.

Fei Tao, Lin Zhang, Yefa Hu
Beijing, September 2011

Preface

In order to realize the goals of TQCSEFK (fastest Time-to-market, highest Quality, lowest Cost, best Service, cleanest Environment, greatest Flexibility, and high Knowledge), many advanced manufacturing technologies and models such as computer-integrated manufacturing (CIM), networked manufacturing (NM) virtual manufacturing (VM), intelligent manufacturing (IM), green manufacturing (GM), agile manufacturing (AM), concurrent engineering (CE) have been proposed. These technologies or models have played very important roles in manufacturing related fields, and have made great contributions to the development of digital manufacturing. Manufacturing grid (MGrid) is a new manufacturing model combining grid technology with the supports of some common and unified architecture, standards, and criteria computing (e.g., web service, WSDL, UDDI, SOAP, WSRF, OGSA, OGSI).

The aim of this book is to make advanced computing technologies (service computing, grid computing, and cloud computing) to be fully used in manufacturing so as to enhance the utilization and sharing of manufacturing resources, and to speed up the transformation from production-oriented manufacturing to service-oriented manufacturing. It achieves this by constructing a concrete theory for MGrid and by detailing implementation methods for MGrid resources and services. Specifically, the book:

(1) Breaks through the application research field of grid technology, turning grid technology from traditional large-scale science computing to application in manufacturing.

(2) Establishes the theoretical foundation for MGrid Resource Service Optimal Allocation (RSOA). From the perspectives of manufacturing science, systematic, ontology, fuzzy mathematics, information theory, set theory, and social psychology, this book explains the basic theory and the implementation method for addressing MGrid RSOA in many aspects such as implementation framework of RSOA, digital description, match and search, Quality of Service modeling and evaluation, optimal-selection, composition of resource service, and resource service combination network, failure detection and recovery of RSOA.
(3) Provides relevant theories and methods of MGRSOA, such as resource service match and search, QoS evaluation, optimal selection, composition, failure-tolerance management. The theories can be used not only in addressing the RSOA problem in MGrid, but also in resolving the RSOA problem in other related SOA-based distributed system.

Outline of This Book

This book consists of 15 chapters.

In Chapter 1, the motivations and driving forces of MGrid are introduced. The connotation of MGrid, including the concept, basic features, and differences between MGrid, networked manufacturing, and computing grid are investigated. The key research contents and technologies of MGrid, including its four commonly known categories and thirty one items, are also studied.

In Chapter 2, the service-oriented architecture of MGrid is proposed, as well as an MGrid collaborative manufacturing prototype platform. The resource service optimal-allocation system which supports the running of the MGrid collaborative manufacturing platform is investigated. The key functions and components for the system are described, as well as its key implementation technologies.

In Chapter 3, the issue of digital description of resource service (DDoRS) in MGrid is investigated. A method for establishing an MGrid-Ontology is presented and an MGrid-Ontology is built. A method for DDoRS based on established MGrid-Ontology is proposed. An MGrid-Ontology based MGrid resource service discovery framework is proposed.

In Chapter 4, the resource services match and search (RSMS) is studied. The describing information of resource services are classified into four categories: word concept information, sentence information, number information, and entity class (or data structure) information. The similarly matching algorithms (SMAs) for the four kinds of basic describing information are presented, including word matching algorithms (WMAs), sentence matching algorithms (SeMAs), number matching algorithms (NMAs), and entity class matching algorithms (ECMAs).Under the supports of the proposed SMAs, the process of resource services match and search are divided into four steps: basic-matching, I/O-matching, QoS-matching, and integrated-matching.

Chapter 5 studies the evaluation indices system of resource service including special evaluation indices (SEIs), individual evaluation indices (IEIs), and general quality of service (QoS) evaluation indices. The evaluation framework and method for SEIs and IEIs are studied. The modeling of MGrid QoS from the points of QoS whole-lifecycle management, MGrid architecture views, and QoS attributes parameters are investigated. A QoS-based MGrid resource service management framework is proposed.

In Chapter 6, the concept of resource service trust-QoS is presented and introduced in order to enhance the validity and success rate of MGrid resource scheduling, and provide high credible resource service abilities and results to user. The trust problems existing in the resource service transaction are put forward. The trust-QoS relationship model which is capable of capturing a comprehensive range of trust relationships exist in MGrid system is put forward. A two-layer resource service trust-QoS evaluation model (intra-domain trust-QoS and inter-domain trust-QoS evaluation models) are put forward. The quantitative evaluating algorithms of trust-QoS degree value are proposed, as well as the real-time and dynamic updating algorithms of trust-QoS degree value. A trust-QoS based

MGrid resource service scheduling framework and associated realizing algorithms are proposed to illustrate the application.

In Chapter 7, an MGrid resource service optimal-selection and composition framework (MGrid-RSOSCF) is investigated. The process of resource service optimal-selection and composition is divided into five steps in MGrid-RSOSCF and the five key problems to realize MGrid-RSOSC are analyzed. The proposed MGrid-RSOSCF consists of five layers and each layer provides the corresponding necessary services and algorithms to address one problem. The five layers are: (1) *T-Layer* is responsible for MGrid task decomposition, (2) *S-Layer* is responsible for resource service match and search, (3) *Q –Layer* is responsible for QoS processing, (4) *O-Layer* is responsible for evaluating and ranking the candidate resource service, and (5) *C-Layer* is responsible for resource service composition and optimal-selection.

In Chapter 8, user's feeling is taken into account in resource service optimal selection (RSOS) in MGrid system. The non-functionality QoS evaluation of resource services is based on users' feeling and transaction experiences using intuitionistic fuzzy set (IFS).The dynamics of non-functionality QoS is considered, and a time-decay function is introduced into non-functionality QoS evaluation. A method is proposed for resource service optimal-selection based on IFS and non-functionality QoS. The performance and advantage of the proposed method are discussed.

In Chapter 9, the issue of correlation-aware composite resource service in MGrid is considered. Three kinds of correlations relationship (i.e., combinable correlation, business entity correlation, and statistical cooperate correlation) in services composition are investigated. The impact of each kinds of correlation relationship on the whole quality of resource service composition is investigated, and QoS computation model based on the three correlations is proposed. The case study indicates that the higher quality of services composition can be achieved when considering the correlations in resource services composition.
In Chapter 10, the multi-objective MGrid resource service composition and optimal-selection (MO-MRSCOS) problem is studied. The formulation is presented for an MO-MRSCOS problem with the given multi-objective (e.g., time minimization, cost minimization and reliability maximization) and multi-constraints. The basic resource service composite modes (RSCM) for composite resource

service are described, and the principles for translating a complicated RSCM into a simple sequence RSCM are presented for simplifying the resolving process and complexity of MO-MRSCOS problem. A method based on the principles of particle swarm optimization (PSO), is proposed for solving MO-MRSCOS problem. Unlike previous works: (a) the proposed PSO algorithms combine the non-dominated sorting technique to achieve the selection of global best position and private best position; (b) the parameters of particle updating formulation in PSO are dynamical generated in order to make a compromise between the global exploration and local exploitation abilities of PSO; and (c) To maintain diversity of solutions in population, *permutation-based* and *objective-based* population trimming operators are applied in PSO.

In Chapter 11, the concept and the classification of resource service composition (RSC) flexibility are presented, and the measurement method of RSC flexibility is investigated to achieve the optimal-selection of RSC based on flexibility.

In Chapter 12, the resource service composition network based on complex network theory is investigated. The principles for establishing and modeling combinable relationship-based composition service network (CoRCS-Net) are studied, and nine combinable relationships among services in CoRCS-Net were investigated and fourteen elementary evolving operators for CoRCS-Net dynamic evolution are designed.

In Chapter 13, the potential failures that would generate during MGrid resource service scheduling are investigated. Thirteen failures are defined in detail, which are classified into four classifications: (a) virtual link related failures, (b) resource service related failures, (c) task related failures, and (d) application related failures. A failure management system is proposed, which provides failure-tolerance service in MGrid resource service scheduling. Corresponding detection mechanisms and methods to each defined failure are presented in detail, associated with the corresponding failure recovery methods.

In Chapter 14, the related works on the application of grid technology in manufacturing are investigated, including research on manufacturing grid (MGrid) theories and applications, and then several key future research issues of MGrid are discussed.

In Chapter 15, combing the new technologies and existing theories and technologies of current enterprise information, a computing and service-oriented manufacturing model, i.e., cloud manufacturing (CMfg), which is the future commercial realization of MGrid, is discussed based on the previous work of this book. The concept, architecture, core enabling technologies, and typical characteristics of CMfg are abstractly studied. Four typical CMfg service platforms, i.e., public, private, community, and hybrid CMfg service platforms are investigated. The key advantages and challenges for implementing CMfg are analyzed, as well as the key technologies and main research contents.

Abbreviations

AM: Agile manufacturing
App_AccessRight_Failure: Accessing right failures
App_DesignCode_Failure: Application design or coding failures
ASR: Application System Resource
Bandwidth_Failure: Bandwidth failure
BuC: Business entity correlation
CEAgent: Chief Evaluation Agent
CG: Computing grid
CIM: Computer-integrated manufacturing
C-Layer: Resource service composition layer
CMfg: Cloud manufacturing
CoR: Combinable relationship
CoRCS-Net: Combinable relationship based composition service network
CR: Computational Resource
CRS: Composite resource service
CRSS: Candidate resource service set
DDoRS: Digital description of resource service
Dep-phase: Deploy phase
Des-phase: Design phase
E&M-phase: Execution and monitor phase
E-Agent : Evaluation agent
ECA: Event–condition–action
ECMAs: Entity class matching algorithms
EEAgent: Evaluation expert agent
eiCoR: Equivalent input combinable relationship
EIS-Agent: Evaluation indices set agent
eoCoR: Equivalent output combinable relationship
eqCoR: Equivalent or competition combinable relationship

ERP: Enterprise resource planning
exCoR: Exact combinable relationship
FD: Failure detector
FR: Failure recovery
GA: Genetic algorithms
GRAM: Grid resource allocation management
IaaS: Infrastructure as a service
IEIs: Individual evaluation indices
IFS: Intuitionistic fuzzy set
IM: Intelligent manufacturing
IOPE: Inputs, outputs, preconditions, and effects
IoT: Internet of thing
irCoR: Input replaceable combinable relationship
MatchEngine: Resource service match engine
MCRS: Mixed composite resource service
MCS: Manufacturing cloud service
MCSs: Manufacturing cloud services
MDS: Miscomputing Discovery Service
MGJMS: MGrid job management system
MGrid: Manufacturing grid
MGrid-Ontology: MGrid ontology
MGrid-RSOSCF: MGrid resource service optimal-selection and composition framework
MGRS: MGrid resource service
MGRSOA : MGrid resource service optimal allocation
MO-MRSCOS: Multi-objectives MGrid resource service composition and optimal-selection
MRCM: MGrid resource conceptual model
MROM: MGrid resource objective model
MRSCOS: MGrid resource service composition and optimal-selection
MRSOAS: MGrid resource service optimal-allocation system
MRSRTask :Multi-resource service request task
NIS: Negative idea solution
NM: Networked manufacturing
NMAs: Number matching algorithms
OGSA: Open grid service architecture
OGSI: Open grid service infrastructure
O-Layer: Optimal-selection layer
orCoR: Output replaceable combinable relationship
OWL-S: Ontology web language for services

PaaS: Platform as a service
PIS: Positive idea solution
ppCoR: Partial pre-order combinable relationship
prCoR: Partial replaceable combinable relationship
psCoR: Partial successor combinable relationship
PSO: Ppartial swarm optimization
Q-Layer: QoS synthetically processing layer
QoS: Quality of Service
QoS-RSM: QoS-based resource service management
RMS: Resources Management System
RS_AbilityChange_Failure: Resource service ability changed failure
RS_Composition_Failure: Resource service composition failure
RS_Overload_Failure: Resource service overload or saturation failure
RS_Quit_Failure: Resource service quit failure
RSC-Coordinator: Resource service coordinator
RSCE-Controller: Resource service executing controller
RSC-Engine: Resource service composition engine
RSCEP: Resource service composition executing path
RSCEP-Generator: Resource service composition executing paths generator
RSCEP-Selector: Resource service composition executing paths selector
RSC-Monitor: Resource service monitor
RSD: User enterprise or resource service demander
RSIC: Rresource service information center
RS-Matcher: Resource service matcher
RSMS: Resource service match and search
RSOS: Resource service optimal-selection
RSOSC: Resource service optimal-selection and composition
RSP: Resource enterprise or resource service provider
SaaS: Software as a service
SCM: Supply chain management
SCRS: Sequence composite resource service
SEIs: Special evaluation indices
SeMAs: Sentence matching algorithms
SFN: Scale-free network
SLA: Service level agreement
S-Layer: Resource service match and search layer
SMAs: Similarity matching algorithms
SMEs: Small and medium-sized enterprises

SOAP: Simple object access protocol
SRSRTask :Single resource service request task
StC: Statistical cooperate correlation
Task_Cancel_Failure: Task cancellation failure
Task_Require_Change_Failure: Changed task requirements failure
Task_RS_Mimatch_Failure: Mismatch failure between a task and resource service
Task_Suspension_Failure: Task suspension failure
T-Layer: Task layer
TQCSEFK: Fastest Time-to-market, highest Quality, lowest Cost, best Service, cleanest Environment, greatest Flexibility, and high Knowledge
TriFNMAs: Triangular fuzzy numbers matching algorithms
UDDI: Universal description, discovery and integration
VL: Virtual link
VL_Disconnect_Failure: Virtual link disconnected failure
VM: Virtual manufacturing
VO: Virtual organization
WMAs: Word matching algorithms
WSDL: Web service description language
WSRF: Web service resource framework

1

Introduction to Manufacturing Grid

1.1 Introduction

Grid technology has been recognized as a promising paradigm for the next generation of manufacturing systems. Researchers have attempted to apply grid technology to product design, manufacturing enterprise resource integration and sharing, enterprise management, enterprise collaboration, optimal manufacturing resource allocation and scheduling, and to enable the digitalization of enterprise information as an implementation methodology. It has been already close to a decade since the first appearance of the concept of manufacturing grid (MGrid). Many projects on the application of grid technology in manufacturing have been carried out, and a large number of research papers on MGrid have been published. In order for MGrid to be better known and received by researchers in the manufacturing community, this chapter will cover (a) the motivations and driving forces of MGrid, (b) a detailed description of the connotation of MGrid, such as its concept and basic features and (c) the differences between MGrid, networked manufacturing, and computing grid. The key research contents and technologies of

MGrid, including its four commonly known categories and thirty one items, will also be investigated (Tao *et al.*, 2011d).

1.2 Proposal of Manufacturing Grid

1.2.1 Several Issues of Manufacturing

Modern manufacturing has changed significantly due to intense global competition, economic globalization, resource globalization, and the rapid development of advanced manufacturing, information, computer, and management technologies. The key mission of manufacturing has changed from enlarging production scale in the 1960s, reducing production cost in the 1970s, promoting product quality in the 1980s, rapidly responding to markets in the 1990s, to emphasizing knowledge and service in the present decade. The introduction of computer and information technologies in manufacturing, and the rapid development and application of Internet technologies have sped up the development of manufacturing. Manufacturing is now moving towards the direction of achieving globalization, digitalization, integration, and intelligence. Some of the current issues that have emerged are as follows (Tao *et al.*, 2010a; Tao *et al.*, 2011d; Tao *et al.*, 2008a):

1. *There is a contradiction between the scarcity and redundancy of manufacturing resources.*

Currently, on the one hand, some manufacturing enterprises, especially small and medium-sized enterprises (SMEs), are unable to accomplish their manufacturing orders because of their lack of advanced high precision equipment. On the other hand, some enterprises own this equipment, but it may have relatively little manufacturing mission and utility.

2. *The design process of a new product has become more complex.*

The entire life-cycle of a product must be considered during its development phase. High performance computing, complex simulation, multi-disciplinary optimization and cooperation are often involved in a new product development. It is impossible for an enterprise to own all of the manufacturing capabilities and resources. Therefore, an enterprise will not be able to complete the entire development process of a new product relying on its in-house resources alone.

3. *There is an inadequacy in the traditional resource and capability sharing mode.*

Sharing of resources, such as document and picture transmission, is inadequate to meet the requirements of global competition, and resource and cooperation globalization. Deep sharing such as full connectivity, remote access, and interoperability of computing resources (e.g., high performance computing devices), data and information resources (e.g., parts library), application and service (e.g., online simulation, design, and analysis), equipment (e.g., CNC, manufacturing cell), application systems (e.g., large-scale and application-specific software) are required. Novel technologies have to be introduced in order to meet the new requirements of resource sharing.

4. *There is a dependency on information and software systems.*

Modern manufacturing enterprises use many information and software systems such as ERP (enterprise resource planning), PDM (product data management), SCM (supply chain management), CRM (customer relationship management), CAD (computer-aided design), MIS (manufacturing information system), and CAPP (computer-aided process planning). Undoubtedly, these systems are required and are very useful for an enterprise, but they are very expensive if an enterprise needs to own all of them. In addition, it is very difficult for an enterprise to integrate and share the data among different information systems. Therefore, new methodologies are needed to enable each enterprise to benefit from these information systems at low cost, and in a convenient way.

5. *High subcontracting cost hinder the development of SMEs.*

The cost of subcontracting can represent a significant portion in the entire cost of a new product. Reducing subcontracting cost (e.g., fees on renting manufacturing resources from other enterprises) and improving the competitive power of an enterprise are some of the important issues to be considered.

Taking into account the above issues, it is apparent that it will be very difficult for an enterprise to remain competitive using its in-house resources and capabilities alone. In order to survive and grow in a competitive market, it has to cooperate with other enterprises, and make use of the global resources in its production methods (Ding *et al.*, 2005a). In other words, it has to realize the general sharing of global manufacturing resources in order to achieve the goal of TQCSEFK (fastest Time-to-market, highest Quality, lowest Cost, best Service, cleanest Environment, greatest Flexibility, and high Knowledge) (Tao *et al.* 2010a; Tao *et al.* 2006b).

1.2.2 Proposal of MGrid

In order to realize the goals of TQCSEFK, many advanced manufacturing technologies and models such as computer-integrated manufacturing (CIM) (Barash 1980, Allen and Bryan 1986, Sabbaghi and Montazemi 2004), networked manufacturing (NM) (D'Amours *et al.*, 1999, Akkermans and Van der Horst 2002, Poler *et al.*, 2008) virtual manufacturing (VM) (Onosato and Iwata 1993, Wadhwa *et al.*, 2009), intelligent manufacturing (IM) (Márkus 1987, Gholamian *et al.*, 2007, Oztemela and Tekezb 2009), green manufacturing (GM) (Dickinson *et al.*, 1995, Jiang *et al.*, 2008a), agile manufacturing (AM) (Frayret *et al.*, 2001, Elkins *et al.*, 2004), and concurrent engineering (CE) (Thomas 1996, Koufteros *et al.*, 2001) have been proposed. These technologies or models have played very important roles in manufacturing-related fields, and have made great contributions to the development of digital manufacturing.

Although each of the above-mentioned advanced manufacturing technologies or models has its own emphasis, they all center on network, cooperative work, and resource sharing. Without the support of common and unified architecture, standards, and criteria, the technologies would lack expandability, openness, and flexibility, thereby limiting their implementation. In addition, resource sharing is the bottleneck that further hinders their implementation. Therefore, it is imperative to find a unified architecture based on international standards, and to establish a virtual manufacturing platform based on network and information technologies.

At the same time, after the proliferation of Internet and web technology, information technology is now experiencing a third wave, i.e., grid technology. The grid concept was first proposed in the 1990s, and was introduced based on the concept of a power grid. A grid is a group of emerging technologies based on the Internet, and is known as the next generation Internet.

Ian Foster, widely known as the "Father of the Grid," and his co-authors stated in their article "The Anatomy of the Grid" (Foster *et al.* 2001a), that grid computing is concerned with "*Coordinated resource sharing and problem solving in dynamic, multi-institutional virtual organizations*". The key concept is the ability to negotiate resource-sharing arrangements among a set of participating parties (providers and consumers) in order to use the resulting resource pool for some specific purposes. In his subsequent article, "What is the Grid? A Three Point Checklist" (Foster 2002a), he suggested that the essence of the definition of grid computing can be captured in the following three point checklist (Foster 2002a):

- Coordinate resources that are not subject to centralized control,
- Use standard, open, general-purpose protocols and interfaces, and
- Deliver nontrivial qualities of service.

Apparently, the key concepts of grid are *resource sharing* and *coordinated problem solving in a virtual organization*. However, resource sharing and coordinated work are also the two bottlenecks that hinder the development of the current modes of manufacturing. This is where grid technology can be effectively used to solve these bottlenecks.

Grid technology can enable full connectivity of distributed and heterogeneous resources located in disparate places, enterprises, organizations, and yet realize resource sharing and enjoy interoperability. In a grid system, users can obtain most of the services they need for establishing a professional application system according to specific manufacturing requirements. Furthermore, the grid has the characteristics of being distributed, dynamic, autonomous, and transparent, which are useful features in the realization of the sharing of manufacturing resources. Therefore, VM organizations based on grid technologies have gradually evolved into a new manufacturing paradigm under a network-centric environment. MGrid has been put forward under this scenario, and it has been named the next generation manufacturing model (Qiu 2002).

Furthermore, the rapid development and prevalent application of international standards and protocols such as web service, grid service, WSDL (Web Service Description Language) (WSDL 2009), UDDI (Universal Description, Discovery and Integration) (UDDI 2009), SOAP (Simple Object Access Protocol) (SOAP 2009), WSRF (Web Service Resource Framework) (WSRF 2009), OGSA (Open Grid Service Architecture) (OGSA 2009),

OGSI (Open Grid Service Infrastructure) (OGSI 2009), Globus Toolkit (Globus 2009), etc., have provided the basis for implementing MGrid.

Since the concept of MGrid was first introduced in 2003, many projects applying grid technology in manufacturing have been carried out, and a large number of research papers on MGrid have been published. In order for MGrid to be better understood and accepted by researchers in the field, this chapter will aim to describe the connotations of MGrid by investigating the existing research pertaining to it.

1.2.3 Technological Driving Forces of MGrid

There are primarily five Technological Driving Forces (TDFs) of MGrid (Tao *et al.*, 2008a):

1. The first TDF of MGrid is the rapid development of grid technology. The appearance and development of grid technology has established a reliable and effective foundation for a general resource sharing system (Foster *et al.*, 2004a). This makes the common sharing and mutual operation of all kinds of resources possible, including computational, memory, communication, data, and knowledge resources, etc., and eliminates the resource and information island (Czajkowski *et al.*, 2001; Foster *et al.*, 2001a).
2. The second TDF of MGrid is the requirement for continuous change and innovation of the product design and manufacture model. In addition to changes from the initial independent design and manufacture, the current integrated design and manufacture, the collaborative design and manufacture, etc., the requirements for high performance computation, complicated simulation, collaborative working in heterogeneous environments and multi-fields is extremely urgent. MGrid presents an opportunity to overcome any insufficiencies in innovation and low technological value added in the process of a new product's design and development.
3. The third TDF of MGrid are changes to the resource sharing model and its ability in modern enterprise.

Competitive globalization, resource globalization and cooperation globalization have already brought a huge change to enterprises regarding their resource sharing model and its ability. Large amounts of information have already brought about a great challenge for networked manufacturers. In the MGrid system, resources' multiplicity, complexity, and dynamic property go far beyond the current Internet (Ye and Gu *et al.*, 2004). The modern manufacturing enterprise's resource sharing is not only simple document and picture transmission, but is also the comprehensive sharing of all kinds of heterogeneous resources geographically distributed and owned by different organizers, including computational resource sharing (e.g., high performance calculating device), data and information sharing (e.g., network components storehouse), application and service sharing (e.g., on-line analysis processing), equipment sharing (e.g., senior processing center), application system sharing (e.g., large-scale expensive software), etc., in order to solve the problems faced due to collaboration.

4. The fourth TDF is the expansion of information technology (IT) application and the difficulties of enterprise informatization. The application of IT in enterprises has become more extensive, and has been applied from the original departments to the enterprise's interior integration, leading to today's integration among enterprises. From a depth perception, it has gone from original information integration and process integration, to today's advanced manufacturing technology application integration and knowledge integration. All these changes have brought about higher costs, longer cycles, poor controllability, difficult integration, poor adaptability, and many other problems to an enterprise's informatization. Grid technology can fully make use of the idle resources of software and hardware in an enterprise's IT system, and can actualize the integration of all existing information systems in the form of grid service, thereby reducing the investment of an enterprise's informatization.

5. The fifth TDF is the rapid development and application of related technologies and criteria such as Web Service, Grid Service, Web Service Description Language, Universal Description, Discovery and Integration, Simple Object Access Protocol, Web Service Resource Framework, Open Grid Service Architecture, Open Grid Service Infrastructure, Globus Toolkit and so on. These are fundamental for the realization of MGrid.

1.3 Concept of MGrid

1.3.1 A Brief Outline of Grid and its Applications

The traditional Internet has realized the connection of computer hardware, and the Web has realized the connection of different homepages. Grid has made it possible for all resources to connect comprehensively on the Internet. Grid is a group of emerging technologies based on the Internet. It is a computation and data processing virtual system, which integrates many kinds of services provided by distributed resources, and provides synthesized problems a processing environment. It merges the high speed Internet, high performance computer, large-scale database, sensor, long-distance equipment, and so on, into an integrated whole. It also enables the comprehensive sharing of global computation resources, memory resources, data resources, information resources, knowledge resources, expert resources and equipment resources, etc. Hence, it can make it easy for people to use all kinds of resources such as those involving computation and memory. So far, from its emergence to its commercial application, the development of Grid Computation has roughly experienced three stages as shown in Table 1.1.

Grid technology is featured in heterogeneity distribution and sharing, scalability, adaptability and dynamic, structural unpredictability, and multi-level management domain. At present, the mature grid application and research mainly concentrate on distributed supercomputing, distributed instrumentation system, tele-immersion and information integration.

Due to the endless promotion and development of grid applications, researchers all over the world are paying more and more

Table 1.1 The stages of grid technology development (Tao *et al.* 2008a)

Stage	Date	Achievement and Key Research Question
The first stage	The early 1990s–1797	The stage of Grid emergence. The key point was the Meta-computing problem, namely, to provide the ability of high performance computation.
The second stage	1998–2001	Ian Foster described Grid in detail, and proposed five-hour glasses structures grid system structure. The key point was to support the large-scale data and the development of computation middleware, so as to solve the grid platform isomerism, the extension and the compatible question. It also was applied in the scientific research field.
The third stage	Feb. 2002	In the Global Grid Forum, Globus Alliance and IBM issued OGSA, (Open Grid Services Architecture)
	Jun. 2003	Globus Alliance etc issued the bottom standard of OGSA——OGSI (Open Grid Services Infrastructure), connected the Grid and the commercial Web service, and issued GT3.
	Jan. 2004	In the Globus World conference, Globus Alliance, IBM and HP etc issued the new standard draft of Grid Service–WSRF (Web Services Resource Framework), and converted OGSI into WSRF; WSRF defined five series of standards of using WS to visit the condition resources: WS–ResourceProperties, WS-ResourceLifetime, WS-BaseFaults, WS-ServiceGroup, WS-RenewableReferences and WS-Notification.

(Continued)

Table 1.1 (cont.) The stages of grid technology development (Tao *et al.* 2008a)

Stage	Date	Achievement and Key Research Question
The third stage	April, 2005	The platform of Grid–GT4 based WSRF was issued. It was a symbol that Grid entered the commercial application field.
	Until now	Some new research correlative with Grid, such as Knowledge Grid, Intelligent Grid, Semantic Grid, Manufacturing Grid etc are being researched. The application of Grid in each domain is the key research.

attention to grid technology. Many new research areas and new concepts along with grid are emerging, such as computational grid, data grid, knowledge grid, information grid, semantic grid, intelligent grid, service grid, manufacturing grid, etc. Among all of these, manufacturing grid is becoming one of the most important fields of grid technology.

1.3.2 Concept of MGrid

The concept of MGrid was first put forward in 2003. Research on MGrid is still at its infancy and there has been no unified and well accepted definition till now. Following are several related influential concepts of MGrid:

The definition of MGrid by Fan *et al.*, (2003a) from Tsinghua University is: *"MGrid is an integrated supporting environment both for the sharing and integration of resources in enterprises and socially for the cooperating operation and management of the enterprises"*. A nine-layered architecture for the MGrid system was proposed. The nine layers in the architecture include: network, unit and protocol, resource encapsulation, grid middleware, enabling, application, portal and enterprise cooperation, and MGrid operation and management system.

In *2004 "IEEE International Conference on System, Man and Cybernetics (SMC2004),"* Qiu (2004) from Penn State University defined MGrid as the next generation of manufacturing. He pointed out that MGrid was created for low-cost, high productivity production through controlled resource sharing across enterprises just like a computational grid. All participants in MGrid provide designated production service for making different products. A layered MGrid architecture based on five-layered grid architecture was proposed and a simple prototype application of MGrid was presented.

McFarlan from the Institute for Manufacturing (IfM) at the University of Cambridge gave a definition of grid manufacturing in 2004 as: "*The (dynamic) harnessing of significant, disparate manufacturing capabilities and resources in order to satisfy one or more business requirements*" (McFarlan 2004). In this definition, capabilities, connections/linkages, and interoperability are the key elements as shown in Figure 1.1. The corresponding characteristics, driving forces, key issues, and background for grid manufacturing were analyzed.

Based on the above definitions of MGrid and the investigations of related studies on MGrid, the authors came up with the following definition of MGrid (Tao *et al.*, 2011d):

"*MGrid is to utilize grid, information, computer, and advanced management and manufacturing technologies, etc. to overcome the barrier resulting from spatial distances in collaboration among different corporations to allow dispersed manufacturing resources (including design, manufacturing, human, and application system resources) to be fully connected, and it is a manufacturing service pool supporting manufacturing resource sharing, integration, and inter-operability among different enterprises.*"

In MGrid, all the manufacturing resources distributed in heterogeneous systems and locations can offer various manufacturing services to users in a transparent way by encapsulating and integrating them into different corresponding resource services. Users can therefore use all the resources in MGrid as conveniently as they use local resources. MGrid is indeed a realization of sharing various resources.

MGrid is developed from traditional networked manufacturing (NM), and it is a new stage of NM. The differences between MGrid and NM from the aspects of distribution and centralization, maintenance of resources, dynamics of solutions, and standardization have been investigated by Tao *et al.*, (2008b), Zhang (2006a), Liu (2004), and Wen (2005). In addition, some key technologies for implementing MGrid were adopted from computing grid (CG), but

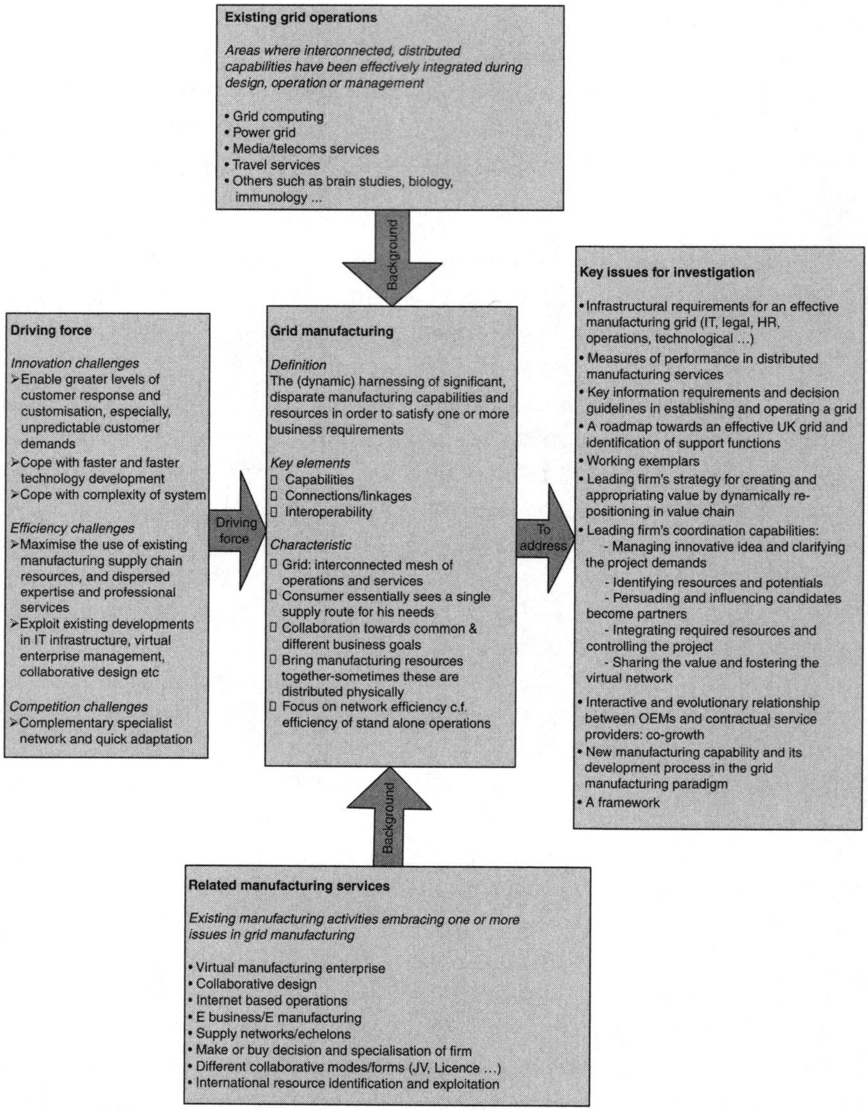

Figure 1.1 Connotation of grid manufacturing defined by the Institute for Manufacturing at University of Cambridge (McFarlan 2004).

compared with CG, MGrid has the characteristics of more flexible interaction, high real-time requirement, multiparty cooperation, resource multiplicity, high reliability requirements, etc. (Tao *et al.*, 2008b).

1.4 Basic Features of MGrid

Combined with research achievements and the features of networked manufacturing and grid technology, the basic features of MGrid are as follows (Tao *et al.*, 2008b):

1. *MGrid is a new advanced manufacturing model* (Qiu 2004). It is a method based on grid technology, information technology, and computing technology, that is employed to organize and manage an enterprise's whole process of design, manufacturing, sales, service, etc., under a networking (a next generation network) environment.
2. *The goal of MGrid is to respond to the market's requirements promptly.* MGrid is a concrete substantial form of Contemporary Integrated Manufacturing (CIM) theory and Agile Manufacturing (AI) system under networking. The goal of MGrid is to achieve rapid design, rapid manufacture, rapid detection, and rapid recombination to promote an enterprise's core competitive power and resolve the TQCSEFK problems.
3. *MGrid has broken through the resources' geography limit and traditional sharing model.* MGrid utilizes grid technologies, information technologies, and computer and advanced management technologies to overcome the barricade resulting from the spatial distance in collaboration among different corporations by enabling all kinds of dispersive manufacturing resources (e.g., design resources, manufacturing resources, human resources, application system resources, etc.) to be fully connected, thereby breaking through the resources' geography limit and traditional sharing model.
4. *MGrid provides a transparent manufacturing service.* In an MGrid system one can only see some units which can provide special services (e.g., computing, storage, foundry, assembly etc.). Perhaps in the future, when someone wants an individual product, what he will need to do is just login MGrid and submit his requirements, and then there will be a group of given units working collaboratively and the MGrid will return the result to the user.

5. *MGrid emphasizes the collaborative work and service among enterprises.* Through the resources sharing and cooperation among enterprises, MGrid will strengthen the interoperation both in resources and techniques. That is to say, through MGrid, any enterprise can cooperate with other enterprises or organizations to develop a new product. At the same time, it can perfect the product's collaborative service among enterprises and the model of manufacturing will develop into an integrated, multi-level distribution system.

1.5 The Connotation of MGrid

The connotations of MG are plentiful, and the theory of MGrid is developed from many relative theories, such as grid theory, collaborative theory, system theory, information theory, etc. MGrid embodies many traits, such as multi-attributed and transparent traits, which are found in sharing, high abstract, self-similitude, autonomy, and management (Tao *et al.*, 2008b)

1. *Resource distribution and share.* On the one hand, the manufacturing resources existing in MGrid are geographically distributed and belong to different systems or organizations, and each enterprise participating in MGrid is different in many aspects and has its own special market orientation and goal. MGrid should resolve distribution problems such as distributed task allocation, distributed scheduling, distributed monitoring, etc. On the other hand, although manufacturing resources in MGrid are distributed, they are also shared completely. That is to say, in MGrid, any resource can be used by any other user or organization. Through a network, MGrid connects all kinds of geographically distributed enterprises and resources, and forms a virtual organization (VO) which is centralized in logic but distributed in physics. In this VO, all resources can be shared and all enterprises can be employed to work collaboratively towards the same target of dealing with a distributed manufacturing task or problem.

2. *Self-similitude.* MGrid has the same dynamic rules and diversity both in large and small scale. In MGrid there is some similitude between the whole and the part. The part usually has some characteristics of the whole, while some characteristics of the whole are also found in the part.
3. *Dynanism and diversity.* In MGrid, the resources or services existing in the current moment may be broken down so they can't be used in the next moment, or may exit from MGrid, and the new resources or services may continuously join into MGrid as time passes. The dynamic nature of the manufacturing grid includes dynamic increases and dynamic decreases. In addition, manufacturing resources in MGrid are diversified. Therefore, MGrid must solve the communication and cooperation problems among these heterogeneous resources.
4. *Autonomy and multiplicity of management.* The resources in MGrid primarily belong to an organization or individual. Therefore, the owners have the highest level of authority over the resources management and should be allowed to have their own management capacity, which is the autonomous character of MGrid. At the same time, grid resources must also be under the unified management of the whole MGrid system, otherwise connecting and sharing among different resources cannot be achieved, so MGrid is not only autonomous but also multiplicitous in regard to management.
5. *Highly abstract and transparent.* MGrid reflects all resources in a highly abstract way in "power wiring boards" which can be seen by users, and other resources are transparent to the users. Manufacturing grid data integration makes the interaction between the various subsystems take place in a transparent manner, which means that all the mutual details are concealed from users so that they accept the various systems as a fully seamless integrated system. Manufacturing grid has the following aspects of transparency:

- Transparency of location: Users in MGrid do not need to know the physical storage location of the resources they use. In MGrid, what the users care about is what service the resources can provide. Just as one do not need to know where the transmission is from, nor what form it is in, when we use electricity.

- Transparency of namespace: In MGrid, all resources or service are described with the same naming rules (or namespace).
- Registration transparency: Once a user is enrolled in MGrid, all the services and functions can be accessed with one logging in.
- Distributed memory transparency: A user can handle any resources distributed in MGrid in the same way they operate their local machines or systems.

1.6 Comparison between MGrid and Networked Manufacturing

MGrid is developed from networked manufacturing (NM). In this section, the differences between MGrid and NM have been analyzed according to the following four aspects: distributed and centralized, maintenance of resources, dynamics of solutions, and standardization (Tao *et al.*, 2008b).

- *Distributed and centralized.* The present mode of NM, or the platform of ASP networked manufacturing, actually is a centralized control mode which embodies the monopolized sharing of services or resources. In MGrid, each node directly corresponds to an enterprise or resource, so the organization of the node chains is the organization of virtual enterprises. But in networked manufacturing, the resources of each enterprise would be assembled together and managed in its own platform, and then the enterprise organizations would be realized according to the above principle, so the maintenance of resources is centralized. The resource management in MGrid is distributed and is centralized in networked manufacturing.
- *The maintenance of resources.* Each enterprise realizes the management of its resources through maintaining its own services in MGrid, but the platform of

networked manufacturing embodies the centralized control of resources and needs the centralized management of resources. The former embodies the distributed mode that provides the advantage of dynamic expansion, not only in the depth but also in the extent of resources, and has more extensibility and dynamic property.

- *Dynamics of solutions*. The application process in MGrid is the process of forming the business chain based on services, thus being mapped into virtual enterprises. Its solution is dynamically generated, namely, firstly generating the business flow and then forming the virtual enterprise. However, networked manufacturing embodies a whole independent system, which provides a fixed amount of resources or an established solution to the customers. The former is more dynamic.
- Standardization and oneness. Networked manufacturing lacks a uniform platform, uniform technologies, uniform standard and criterion at the present time. But there have existed uniform grid platforms (e.g., Globus Toolkit), uniform technologies (e.g., web service, UDDI, SOAP, etc.), uniform standards and criterions (e.g., OGSA, OGSI, WRSF, etc.), and so on, which can be used to realize a real and dynamic MGrid platform.

A further comparison between MGrid and NM is illustrated in Table 1.2.

Table 1.2 Comparison between MGrid and networked manufacturing

	Networked Manufacturing	MGrid
Aim	Resource sharing and cooperative manufacturing	
Resource or member	Enterprise (fixed)	Resource, enterprise, manufacturing ability (enter or leave at any time)

(Continued)

Table 1.2 (cont.) Comparison between MGrid and networked manufacturing

	Networked Manufacturing	MGrid
Organization and management of resource or service	Centralized control and management	Self-organization, distributed control and management
Establishment of virtual organization	Based on partner selection	Based on dynamically service discovery, selection, and composition
Cooperation way	Between enterprises	Between resources or services
Implementing technologies	.NET, J2EE, CORBA, etc. (no uniform technology)	Uniform technology (e.g., Web service, grid Service) and platform (e.g., Globus Toolkit)
Solutions to a problem or a job	Fixed service and solutions	Solutions are dynamically generated based on the services in the form of service chain or virtual enterprise
Access control strategy	Centralized access control	OGSI, Public key infrastructure, etc.
Standardization criteria	No uniform standardization	OGSA, WSRF, OGSI
Expandability	Low (e.g., need huge customized work)	High
Dynamic	Low (e.g., fixed member and solutions)	High (e.g., members can enter or leave dynamically, solutions are generated dynamically)
Application scope	Distributed enterprise, group, regional manufacturing	Global manufacturing

1.7 Comparison between MGrid and Computing Grid

The main differences between MGrid and computing grid (CG) are as follows (Tao *et al.*, 2008b):

1. *More flexible interaction.* MGrid services offer interaction between resources and the user. Although users need to communicate with services in CG, there are not many ways to interact, such as online and offline interaction in MGrid.
2. *High real-time requirement.* In addition to reflecting the connecting state and load state of MGrid resources in real time, the service results and the effects (the finished result of task) of the resources must be reflected in a timely manner. The users' requirements must also be responded to in time.
3. *Long life cycle.* In CG, a service or trading may only exist for several seconds. In general, it is shorter than one day. However, trading in MGrid may last a few hours or even longer than several days.
4. *Multiparty cooperation.* Different from CG, a computing service is assigned to finish a simple computing task. It does not need to complete one task together with other different services, but a task should be completed by the coordinated work of a number of different resources or services in MGrid.
5. *Resource multiplicity.* In CG, resources are limited to computing resources, communication resources, and storage resources. However, in MGrid, in addition to these resources, equipment, software, technical, materials, and public service resources are also involved.
6. *Special knowledge requirement.* In MGrid, manufacturing task decomposition, the combination and discovery of resource services are often based on specific manufacturing knowledge. It is more complex than those in CG.
7. *Functional complexity.* MGrid provides not only the functions of simple computation, transmission, and storage, but also the functions of collaborative design, simulation, fault diagnosis, etc.

8. *Data complexity*. In MGrid, data involve pictures, NC codes, 2D and 3D graphs, etc., which can take many forms and are more complex than those in CG.
9. *Professionalism*. Manufacturing is a professional activity, but not in the case of CG which is rather impersonal. To meet the requirements of business, process, function, service, etc., MGrid offers a good degree of professionalism.
10. *High reliability requirements*. Compared with CG, the services required in MGrid meet not only the time requirements, but also the complex and reliability requirements such as high precision and manufacturability.
11. *Online/offline resources sharing*. In CG, the sharing of resources is primarily done online. In MGrid, the manufacturing equipment, manufacturing cell, and production line to the production workshop, are not part of the network and online resources, but they can still be shared in MGrid offline.
12. *Online/offline result submission*. The way to send jobs and receive results in CG is primarily online. Besides online, there is an offline mode in MGrid, e.g., through entity communication such as transport and purchase.

1.8 Key Research Contents and Technologies of MGrid

In this section, the key research contents and technologies for implementing MGrid are presented, and they can be classified into four categories: *general technologies, supporting technologies, key enabling technologies*, and *application technologies* (Tao *et al.*, 2011d). The key research contents and technologies of MGrid represented in the Venn diagram are shown in Figure 1.2.

1.8.1 General Technologies

Item 1 *Concept and Connotation*: The first task in carrying out a new research problem is to investigate its concept and connotation. As mentioned in Section 1.3.2, there has been no complete and unified definition of MGrid until

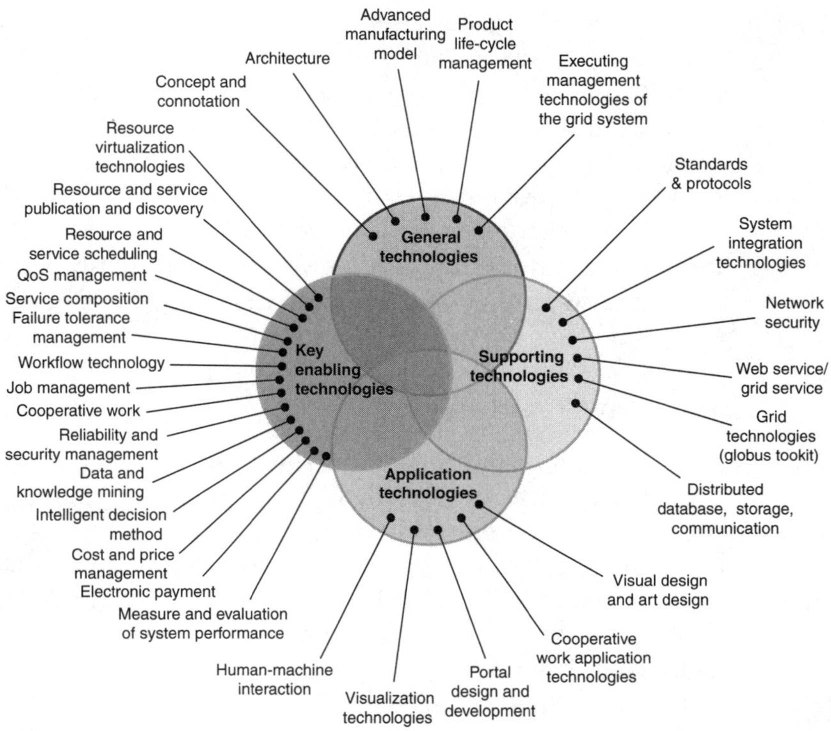

Figure 1.2 Venn diagram of key research content and technologies of MGrid (Tao *et al.*, 2011d).

recently. The connotation of MGrid is evolving with the emergence of new technologies, such as cloud computing and cloud service. Therefore, it is necessary to research the concept and connotation of MGrid under a new environment.

Item 2 *Architecture*: The basis of the architecture that guides the development and implementation of MGrid, including (a) resource and service organization methods, such as the resource management system based on ecology, and (b) service-oriented MGrid architecture with the characteristics of being dynamic, intelligent, self-organizational, and semantics supporting.

Item 3 *Advanced manufacturing model*: The theories and methods which are based on advanced manufacturing models and technologies such as NM, CIM, VM, DA, IM, GM,AM, and CE.

Item 4 *Product life-cycle management*: One of the aims of MGrid is to encapsulate all the resources, manufacturing ability, data, and information involved in the entire life-cycle of a product into services, so as to make them easily sharable by other users. Hence, product life-cycle management technologies are very important elements.

Item 5 *Executing management technologies of the grid system*: This involves constructing, implementing, and maintaining theories and methods for grid and distributed systems.

1.8.2 Supporting Technologies

Item 6 *Standards and protocols*: The standards and protocols technologies include those which support the modeling of services, service encapsulation, communication, access, and failure tolerance, etc. Existing standards and protocols, such as OGSA, OGSI, UDDI, WSDL, SOAP, WRSF, Globus Toolkit, Web service, and grid service are utilized as much as possible.

Item 7 *System integration technologies*: The technologies for system integration, such as enterprise information and application integration technologies.

Item 8 *Network security*: The network security technologies in MGrid include (a) general and standard security architecture and systems, (b) authentication methods and mechanisms for registration of MGrid members, and (c) safety accessing control strategies, the security mechanisms for data storage and communication, such as encryption mechanisms.

Item 9 *Web Service/Grid service*: The development and application technologies on web and grid services, which are used for developing and managing MGrid service.

Item 10 *Grid technologies (Globus Toolkit)*: The application of grid programming technologies, core grid services (such as high performance scheduling, high throughput resource management technologies, performance data collection, analyses and visualization technologies, and security technologies), and Globus Toolkit and its core grid middleware.

Item 11 *Distributed database, storage, communication*: The distributed database, distributed storage, and distributed communication technologies that support MGrid operations.

1.8.3 Key Enabling Technologies

Item 12 *Resource virtualization technologies*: The technologies that encapsulate the distributed and heterogeneous manufacturing resources or capabilities into services, including modeling of resource information, resource virtualization model, semantic description of services, and so on.

Item 13 *Resource and service publication and discovery*: The methods and technologies for resource service publication and discovery, including resource service publication and discovery mechanisms, resource service match algorithms, and resource service search algorithms.

Item 14 *Resource and service scheduling*: The methods and technologies responsible for allocating jobs to corresponding resources or services at specific times under multi-objectives and multi-constraints.

Item 15 *Quality of service management*: The methods and technologies for QoS managements, such as the establishment of QoS index system, description of QoS, modeling of QoS, evaluation of QoS, QoS ontology, life-cycle management of QoS (Tao *et al.*, 2009c), QoS information storage and extraction.

Item 16 *Service composition*: The methods and technologies for assembling several services in some sequence according to the requirements of a complex task, including (a) resource service composition architecture, (b) basic construction model for composite resource service, (c) correlation relationship management of composite resource service, such as the definition of correlation relationship among resource services (e.g., combinatorial correlation, entity or organization correlation, and historical correlation), representation of correlation relationship, and correlation relationship mining, (d) uncertainty management of resource service composition, (e) flexibility of resource service composition, (f) resource service composition optimal selection method, and (g) establishment of composition net and its dynamic characteristics (Tao *et al.*, 2011d; 2008c; 2010e) .

Item 17 *Failure tolerance management*: The methods and technologies for detecting failure and failure recovery, including (a) the definition, classification, and description of failure in MGrid, such as virtual link related failure, resource service related failure, job related failure, application related failure, (b) dynamic failure detection model and method, and (c) dynamic failure recovery strategies.

Item 18 *Workflow technology*: The methods and technologies for workflow management, including (a) concept and architecture of workflow in MGrid, (b) modeling of workflow in MGrid, such as process model, organization model, resource model, and service model, (c) resource service scheduling and composition method based on workflow, and (d) monitoring of workflow.

Item 19 *Job management*: The methods and technologies for managing jobs in MGrid, such as job description, decomposition, scheduling, deployment, and allocation.

Item 20 *Cooperative work*: Technologies for cooperative design, business, and supply chain management in MGrid.

Item 21 *Reliability and security management*: The methods and technologies with regard to reliability and security management, such as (a) definition and classification of reliability (including systems, resources, services, and results), (b) evaluating criteria system of reliability, and evaluation model and algorithms of reliability, (c) dynamic and online collection, statistics and analysis of reliability data, and (d) strategies and methods to improve reliability of MGrid systems.

Item 22 *Data and knowledge mining*: The methods and technologies for mining data and knowledge used in the operations of resource discovery, matching, composition, etc. They are primarily the methods and technologies for (a) manufacturing knowledge mining, such as information extraction of manufacturing businesses, (b) semantic information mining of manufacturing processes, (c) service composition relationship mining, correlation relationship mining of service, and (d) correlation-aware QoS mining and evaluation.

Item 23 *Intelligent decision method*: The intelligent decision supporting system and methods based on intelligent optimization algorithms (e.g., genetic algorithms, evolutionary algorithms), particle intelligence, fuzzy theory, and analytic hierarchy process (AHP) which are used to support the operations for resource service discovery, match, optimal selection, and composition in MGrid.

Item 24 *Cost and price management*: Involves the research on (a) cost factors and its measurement from the perspectives of resource service providers, resource service demanders, and managers of MGrid systems, such as costs for hardware (e.g., cost of new resources, replacement of hardware components), business premises, electricity for computing hardware, software, personal, and data communication (e.g., fee for local area networks and connection to the Internet) (Opitz *et al.*, 2008), and (b) pricing and cost management strategies.

Item 25 *Electronic payment*: The method and technologies for electronic accounting, electronic payment, and rewards and punishment strategies.

Item 26 *Measure and evaluation of system performance*: The establishment of performance index systems, evaluation models and measuring methods, methods for performance monitoring and dynamic measurement.

1.8.4 Application Technologies

Item 27 *Human-machine interaction*: The methods for portal design of human-machine interaction, data collection, and data input and output methods.

Item 28 *Visualization technologies*: These are the technologies for (a) visualization in scientific computing (e.g., simulation and visualization of complex manufacturing environment, extreme manufacturing, biological manufacturing, spatial information, and geographic information), (b) information visualization (e.g., visualization of multi-field coupling, complex data, complex product, and complex manufacturing process), and (c) visualization of data and knowledge mining.

Item 29 *Portal design and development*: Technologies for portal design from different users' perspectives, such as resource and service providers, resource and service demanders, and managers of MGrid.

Item 30 *Cooperative work application technologies*: The application technologies of cooperative design, cooperative control, and cooperative office.

Item 31 *Visual design and art design*: The technologies that make the portal of the MGrid system more friendly and appealing, such as the layout design of the MGrid system, logo, interface, and art design.

Among the above 31 items, resource virtualization and publication, resource service discovery and scheduling, QoS management, service composition, job and workflow management, and cost and price management, failure tolerance, and reliability management are the key problems which need to be addressed for implementing MGrid. Although some of them are also the key problems in other manufacturing systems, such as NM, virtual manufacturing, and e-manufacturing, the detailed environments have changed. MGrid system is a complex system and it is developed from the grid system and the other advanced manufacturing systems, it inherits the existing challenges in both grid and current manufacturing system.

1.9 Summary

In this chapter, the motivations and driving forces of MGrid were analyzed. The connotation of MGrid, including the concept, basic features, and differences between MGrid, networked manufacturing, and computing grid were investigated. The key research contents and technologies of MGrid, including its four commonly known categories and thirty one items, were also studied.

2

Resource Service Optimal-Allocation System in MGrid

1.1 Introduction

Current research on manufacturing grid (MGrid) primarily concentrates on its concept, architecture, application prototype platform, application foreground, etc. But the MGrid resource service digital description method – the foundation for MGrid resource sharing, and MGrid resource service optimal-allocation – the basic theories and key technologies to realize MGrid resource sharing, were not effectively addressed till now. In order to realize MGrid resource service sharing and optimal-allocation, the following problems must be addressed (Tao *et al.*, 2010a):

1. How to publish resource service into MGrid system and realize its digital description?
2. How to find out the qualified candidate resource service according to user or task requirements?
3. What is the quality of the resource service and how to evaluate it?
4. How to realize resource service optimal-selection and composition according to different task requirements?

5. How to provide failure-tolerance (including both failure detection and recovery) service during the process of MGrid resource service sharing and optimal allocation?

It can be concluded that the above problems concern MGrid resource service optimal-allocation. This chapter focuses on these problems and aims to provide the basic theories and key technologies for MGrid resource service optimal allocation (MGRSOA).

In order to realize MGRSOA, after having investigated the connotation of MGrid in Chapter 1, a five-layered service-oriented MGrid architecture will be put forward in this chapter. Dependent on the proposed MGrid architecture, an MGrid collaborative manufacturing prototype platform is proposed. Considering the requirements of resource service optimal-allocation in the MGrid collaborative manufacturing prototype platform, a comprehensive MGrid resource service optimal-allocation system is proposed. The basic theories and key technologies to realize the MGRSOAS are outlined in this chapter, including MGrid resource service digital description, match and search, QoS modeling and evaluation, optimal-selection and composition, failure-tolerance (i.e., failure detection and recovery).

2.2 The Architecture of MGrid

A five-layered service-oriented MGrid system architecture is shown in Figure 2.1. It can be divided into the following five layers:

- MGrid resource layer
- MGrid core middleware
- MGrid user service middleware layer
- MGrid user portal layer
- MGrid user layer.

2.2.1 MGrid Resource Layer

The MGrid resource layer includes all kinds of resources, including (1) Human Resource (HR), (2) Equipment Resource (ER), (3) Material Resource (MR), (4) Application System Resource (ASR), (5) Technology Resource (TR), (6) Public Service Resource (PSR),

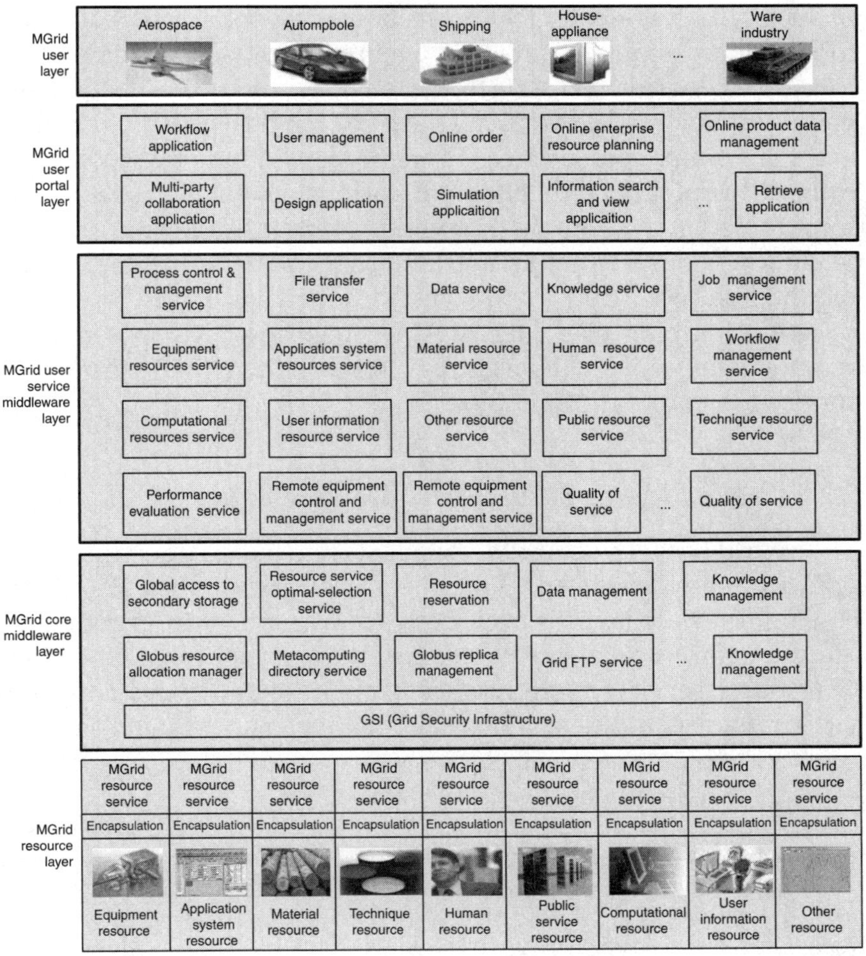

Figure 2.1 The architecture of MGrid (Tao *et al.*, 2008a).

(7) Computational Resource (CR), (8) User Information Resource (UIR), and (9) Other Resource (OR). It is responsible for controlling local resources that are accessible from anywhere on the Internet and offers the access interface and protocols for the upper layer to access these resources.

2.2.2 MGrid Core Middleware Layer

The MGrid core middleware is responsible for providing a series of tools, services and protocols developed with Globus Tookit 3 based on OGSA, OGSI, etc. It includes resource reservation, global access to secondary storage, data management, meta-computing directory service, grid FTP service, registry service, security, QoS service, etc., along with resource service optimal-selection evaluation and QoS evaluation services. Therefore, it can hide the geographically distributed, heterogeneity of various resources in the MGrid resource layer. In addition, it provides a public framework, and different kinds of services can add different contents and implementing methods according to its own characteristics. It also provides a program compiler, class library, API, integrated development environment etc., which will be employed when developing the MGrid service based on grid technologies.

2.2.3 MGrid User Service Middleware Layer

The MGrid user service middleware layer is created upon the development environment and MGrid service provided by the MGrid core middleware layer. It provides various services for meeting the requirements of collaborative manufacturing, including nine kinds of resource services, performance evaluating and monitoring services, remote equipment control and management system, data service, and job management service, etc.

2.2.4 MGrid User Portal Layer

The MGrid user portal layer is the entry toolkit of MGrid in web pattern, which provides friendly and easily collaborative interactive interfaces such as collaborative design, simulation, information search and view, online order, multi-party collaboration application, workflow application, job management, user management, etc.

2.2.5 MGrid User Layer

The MGrid user layer interfaces with each field of manufacturing, e.g., aerospace, automobile, shipping, war industry, etc. Based on the functions of the four layers mentioned above, each enterprise user can submit a manufacturing task and use all the resources and services in MGrid through network and application tools.

2.3 MGrid Collaborative Manufacturing Platform

2.3.1 Conceptual Model of MGrid Collaborative Manufacturing Platform

In accordance with the architecture of MGrid, a collaborative manufacturing prototype platform based on MGrid is designed as shown in Figure 2.2. There are primarily two kinds of user in the MGrid collaborative manufacturing prototype platform:

- Resource enterprise or resource service provider (RSP), and
- User enterprise or resource service demander (RSD).

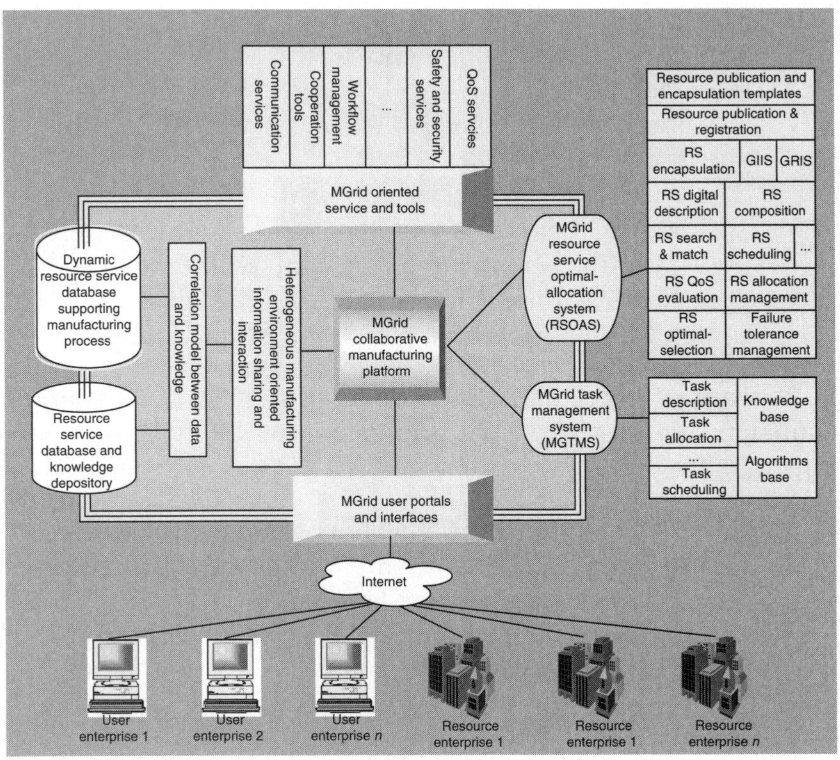

Figure 2.2 Conceptual model of MGrid collaborative manufacturing platform.

The former RSP publishes its idle resource, product, manufacturing capability or service in order to obtain the most profit from its resource. And the later RSD searches the optimal manufacturing resource and service required at the lowest cost, and selects the corresponding partner to establish the collaboration manufacturing net. The partition of resource enterprise (or RSP) and user enterprise (or RSD) are not absolute; RSP may become the RSD, and RSD sometimes plays the role of RSP in the MGrid according to the practical requirements.

The brief running flow of the MGrid collaborative manufacturing prototype platform is described as follows:

1. Resource enterprises (RSPs) publish their idle resource, product, manufacturing capability or service via the resource service publication center, which are then encapsulated into service. The publication process is described in Section 2.3.2.
2. User enterprises (RSDs) submit or define their manufacturing task requirements in the MGrid job management system (MGJMS) via the Internet. These requirements are then parsed into standard requirements documents.
3. The MGrid resource service optimal-allocation system (MRSOAS) then finds out the appropriate resource service for the task and executes it under its supporting core services, such as resource service match and search, optimal-selection, composition, scheduling, etc.

The core of the implementation of the MGrid collaborative manufacturing prototype platform is the MGRSOAS. It is responsible for:

- Searching and finding out all the resource services that accord to the RSP's requirements, and generating the candidate resource services set;
- Evaluating the QoS (Quality of Service) of each candidate resource service, including Performance-QoS and Trust- QoS;
- Providing optimal-selection algorithms and selecting the optimal resource service needed to serve the RSD according to the QoS and optimal-selection evaluation results;

- Assembling several (or many) resource services into a composite resource service to serve the task if the task cannot be implemented by invoking a single resource service; and
- Providing failure tolerance service, and so on.

2.3.2 Resource Service Publication

The MGrid resource service publication process is stated as follows:

1. The resource service provider (the resource enterprise) logs in and submits the resource or service to the *resource service publication center* of MGrid via the human-machine interface through the Internet with the intention of joining MGrid.
2. The MGrid system verifies the identity and validity of the provider and converts the submitted resource information into a standard resource service description (described in Chapter 3) through a parser. It then delivers the request and result to the corresponding domain node (*resource service class model*) responsible for registration.
3. After checking the validity of the resource service, the corresponding domain then feeds back the successful message to the provider and saves the resource service in the resource service information center if the registration is valid, or returns the unsuccessful message if resource registration has failed.
4. The resource publication center returns the GSH to the corresponding resource provider.
5. The resource provider will periodically send out the living messages to the according domain, which is called soft-state registration. At the same time, the specific domain may subscribe with the resource provider for some metadata. When the subscribed metadata changes, the resource provider will notify the domain to register again in time.

The brief sequence diagram of resource service publication in MGrid is shown in Figure 2.3.

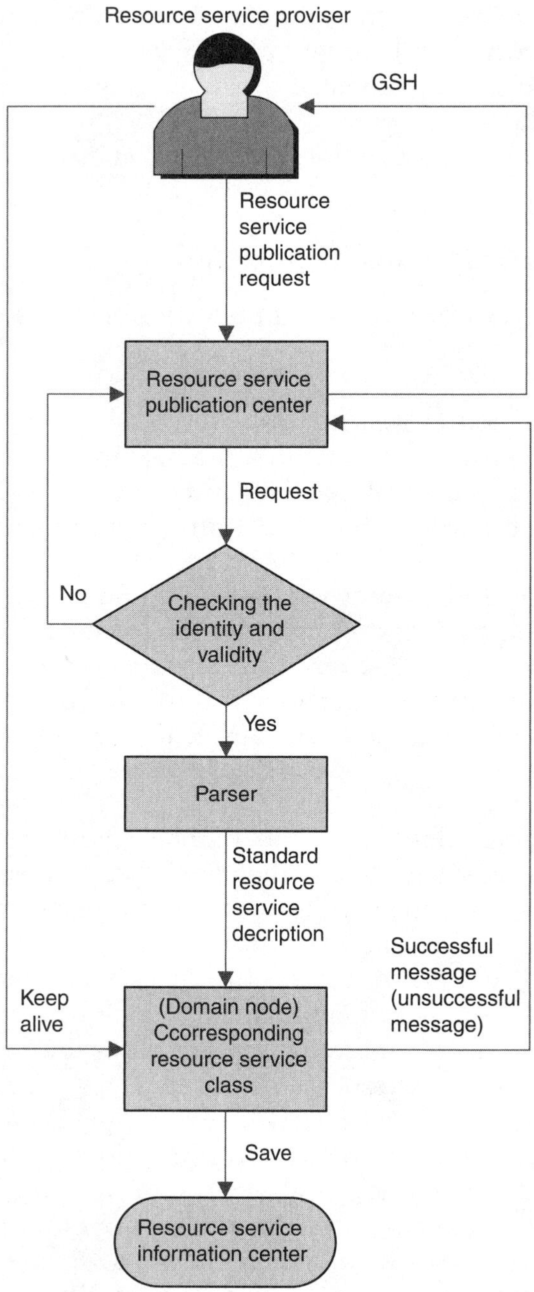

Figure 2.3 The brief sequence diagram of resource service publication.

2.3.3 The Resource Service Exchange between RSP and RSD

The main workflow of the resource service exchange between RSP and RSD in the MGrid is shown in Figure 2.4. Its brief executing process is described as follows (Tao *et al.*, 2008a):

1. RSPs publish resource information through the interface of the *resource service publication center* as described in Section 2.3.2. The published resource is encapsulated into the MGrid resource service and registered in the resource service information center.
2. RSDs submit the manufacturing task to the job management center, and the submitted manufacturing job is decomposed into several subtasks (or resource service requestors), and then the requests are sent to the MGRSOAS.
3. The MGRSOAS searches the eligible resource services according to the requirements of the resource service requestors from the resource service information center, and generates a candidate resource service set. It then selects out the optimal resource services from the set according to the QoS and optimal-selection evaluation results of each candidate resource service.
4. The RSDs request the corresponding resource services according to the result of (3).
5. The RSPs respond to the RSD's request and create the corresponding service instances.
6. The RSPs return the addresses of the resource service instances to the RSD.
7. The RSDs and the service instances are bound.

2.4 MGrid Resource Service Optimal-Allocation System (MGRSOAS)

According to the above function requirements of resource service optimal-allocation, a MGrid resource service optimal-allocation system (MGRSOAS) is designed as shown in Figure 2.5. The proposed MGRSOAS can provide fundamental functions such as

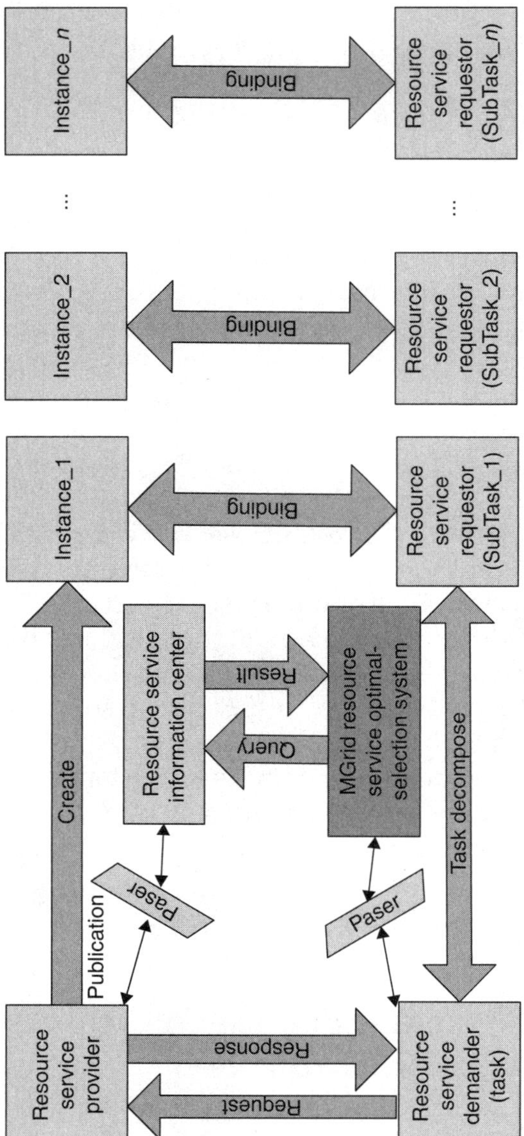

Figure 2.4 The main workflow of resource service exchange in MGrid.

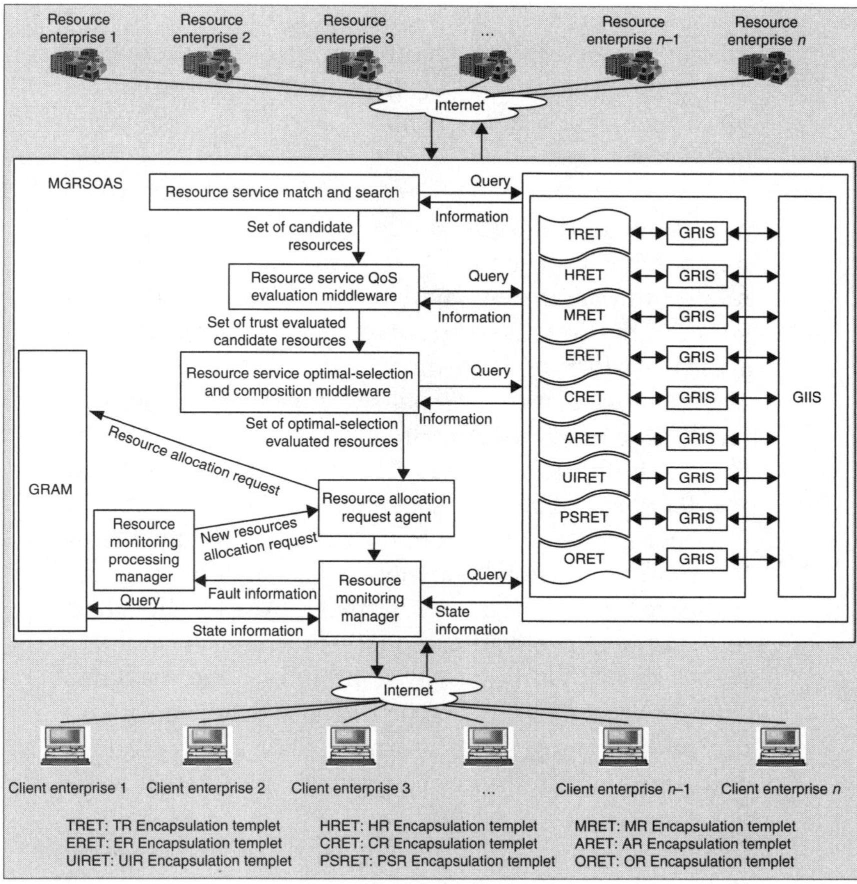

Figure 2.5 Resource Service Optimal-Allocation System (RSOAS) in MGrid (Tao *et al.*, 2008a).

remote resource service publication and encapsulation, resource service discovery, resource service Performance-QoS and Trust-QoS evaluation, resource service optimal-selection evaluation, resource service monitoring, and so on, by integrating MDS and other modules provided by Globus.

The main workflow of RSOAS is as follows:

(a) The user submits a job or a service request to *resource service match and search agent* through the portal of the MGrid system;
(b) *Resource service match and search agent* sends the query message to the corresponding resource services

encapsulation template's GRIS and GIIS to search for all the resources that accord with the requirements, and then generates a set of candidate resource services and sends it to *resource service QoS evaluation middleware;*

(c) The *resource services QoS evaluation middleware* evaluates each candidate resource QoS (including performance-QoS and Trust-QoS) value according to the related information provided by GRIS and GIIS, and submits the results to *resource service optimal-selection and composition middleware;*

(d) The *resource service optimal-selection and composition middleware* provides comprehensive evaluation to the candidate resource service, and sends the ordered results to the *resource service allocation request agent.*

(e) The *resource service allocation request agent* transforms the manufacturing task into a special language for resource allocation and informs GRAM to allocate it to the optimal-selected resources;

(f) *Resource monitoring manager* monitors the resource states, makes decisions regarding any instances of failure by analyzing state information of each resource, and communicates with the *resource monitoring processing manager;*

(g) *Resource monitoring processing manager* displays the state of the resource and the type of failure that occurred to the user, and requests whether the job executed by the resource where failure occurred should be migrated and rescheduled or not.

2.5 The Key Issues and Technologies for Realizing RSOAS

In addition to the functions of GRAM, GRIS, and GIIS which can be realized by using the middleware provided by the Globus Toolkit or other systems, the realization of MGRSOAS also involves the operations of resource service description, match and search, selection, composition, scheduling, failure detection and recovery, and so on. They are the core services provided by MGRSOAS to resource service management. In this book, the basic theories, methods, and technologies for realizing the core services are investigated, so as to establish the basic resource service management theories and methodologies.

2.5.1 Modeling and Digital Description of Resource Service (DDoRS)

Digital description of resource service is the foundation for realizing the MGrid resource service optimal-allocation. It provides basic data and information support for the operations of resource service match and search, optimal-selection, composition, scheduling, failure detection and recovery, and so on, which is examined in Chapter 3.

2.5.2 Resource Service Match and Search (RSMS)

From the user's perspective, the abstract and main function is to find out the appropriate resource service according to the user's (or task's) requirements, and to realize the intelligent and rapid mapping between the user's (or task's) requirements and the available resource service. Therefore, how to search out the appropriate resource service according to the user's (or task's) requirements, and generate the related candidate resource service set, i.e., resource service match and search (RSMS), is the first step for realizing MGRSOA. Chapter 4 will examine the implementing mechanisms and algorithms of RSMS.

2.5.3 QoS Modeling and Evaluation of Resource Service

From the service perspective, MGRSOAs selects the appropriate resource services from all the available and candidate resource services in order to establish the business process or flow. The business requirements are realized by invoking the specific resource services. Therefore, the quality of the service of MGRSOA is decided directly by the quality of the selected resource services, and some quality of service evaluation function is required. In Chapter 5 and Chapter 6, the evaluation criteria of the resource service and its QoS (including trust-QoS), and the evaluation models and algorithms of the QoS of resource service are examined.

2.5.4 MGrid Resource Service Optimal-Selection and Composition (RSOSC)

In the MGrid system, there are primarily two kinds of manufacturing tasks (or resource service requests): (a) single resource service request task (SRSRTask), which can be completed by invoking only one resource service, and (b) multi-resource service request task (MRSRTask), which is completed by invoking several resource services in a certain sequence. For a SRSRTask, the system searches the resource services that are qualified for its function requirements and chooses the optimal one to execute it, namely *resource service optimal-selection (RSOS)*. For an MRSRTask, in addition to the search for all qualified resource services according to each subtask, the system has to select one candidate resource service for each subtask. Then the system generates a new composite resource service (CRS) and selects the optimal resource service composite path from all possible paths to execute the task with the given multi-objective (e.g., time minimization, cost minimization, and reliability maximization) and constraints, namely *resource service composition and optimal-selection (RSCOS)*. RSOS and RSCOS are the core of MGRSOA. In Chapter 7, a comprehensive framework for MGrid resource service optimal-selection and composition is described. Chapter 8 studies the RSOS based on intuitionistic fuzzy set (IFS) and non-functionality QoS. The correlation between relationship management and flexibility management in resource service composition is studied in Chapter 9 and Chapter 11, respectively. The methods and algorithms for realizing resource service composition and optimal-selection (RSCOS) based on improved PSO are introduced in Chapter 10.

2.5.5 Resource Service Composition and Network Modeling and Dynamic Characteristics

During the resource service composition, the following issues have to be answered: (a) How to consider the correlating relationship among the selected resource services in the composition? (b) How to promote the probability that a resource service will be invoked and used?

In addition, the user or enterprise who provides and uses the resource service, must know (a) what kinds of service they should develop or provide, (b) how to enhance their service's utilization, (c) how to realize their service's added value, and (d) how to protect their core services.

In order to answer the above questions, the resource services have to be researched from a macroscopic and systemic point of view, e.g., by researching their dynamic characteristics in a network. Therefore, in Chapter 12, the modeling method of resource service composition network and its key dynamic characteristics are studied.

2.5.6 Failure Tolerance Management

As mentioned before, the realization of MGRSOAS involves the following operrations: resource service description, match, search, selection, composition, scheduling, invoking, transforming, and so on. In the process, various failures would be generated by networks, resources, tasks, and applications. How to detect and recover these failures is very important for improving the service quality of an MGrid system. In Chapter 13, the failure management system is studied, and the detailed failure detection and recovery method and algorithms are also studied.

2.6 Summary

In this Chapter, the service-oriented architecture of MGrid is proposed, as well as an MGrid collaborative manufacturing prototype platform. The resource service optimal-allocation system which supports the running of the MGrid collaborative manufacturing platform is investigated. Key functions and components for the MGRSOA system are described, as well as its key implementation technologies. In addition, the structure and contents of this book about resource service management are outlined.

3

Digital Description of MGrid Resource Service

3.1 Introduction

One of the core ideas of a manufacturing grid (MGrid) is resource service sharing and optimal-allocation (RSSOA), The detailed operations include, (a) match and search of resource service, (b) comprehensive processing of QoS information, (c) resource service optimal-selection and composition, (d) failure detection and recovery, (e) resource service scheduling, etc. (Tao *et al.*, 2011d). One of the foundations for realizing the above operations is digital description of resource service (DDoRS). Without digital describing information of resource service, the operations of RSSOA would have no data and information support. Hence, DDoRS is one of the key issues for addressing RSSOA. The problem of DDoRS in MGrid is examined in this chapter, MGrid-Ontology is established, and a method for DDoRS based on the proposed MGrid-Ontology is presented. The main focus of this chapter will be: (a) to investigate the classification of MGrid resources and resource services, (b) to research the method for establishing an

MGrid-Ontology, and present a method for digital description of resource service based on established MGrid-Ontology, and (c) to research an MGrid-Ontology based MGrid resource service discovery framework to illustrate how the proposed works can be used in the MGrid system.

3.2 Classification of MGrid Resource Service and Its Application

3.2.1 Classification of MGrid Resource and Resource Service

MGrid resource refers to the resource involved in each production action during the whole life-cycle of a product, including materialized resources (MRs) and the ability to resolve a manufacturing problem. The former, MRs, consists of manufacturing equipments, manufacturing cells, product lines, tools, materials, products, software, etc. The later, the ability to resolve a manufacturing problem, includes human resources, manufacturing information, etc. (Shi *et al.*, 2006a; Hu *et al.*, 2009a). Along with the increased application of computer technologies in industry, a lot of advanced manufacturing modes and technologies emerged, such as CIMS (computer integrated manufacturing system), CE (concurrent engineering), VM (virtual manufacturing), AM (agile manufacturing), and e-manufacturing. As a result, the organization, management, planning, and control methods for manufacturing resource are continuously changing. All these changes affect the production plan, product design, process design, and manufacturing control. In addition to the characters in traditional manufacturing models, resources in the MGrid system also have the features such as dynamic, heterogeneous, distributed, and decentralized control (Tao *et al.*, 2008a; 2008b; Hu and Tao *et al.*, 2009a).

Modeling of MGrid resource service aims to encapsulate each kind of resource into a manufacturing resource service in order to provide the user enterprise (Tao *et al.*, 2008a) with different resource services during the transaction. Manufacturing is a multi-principle, multi-layer, multi-constraint, and multi-task process, and the domain knowledge and resource involved in a manufacturing action are not single, but complicated and integrated.

In an MGrid system resources differ greatly from each other in terms of the nature (e.g., physical characteristics, location, dynamicity, sensitivity, functionality, etc.), the demands placed on them (e.g., time, quality, cost, and service), and the ways in which they are engaged (e.g., discovery, brokering, monitoring, diagnosis, and adaptation) (Iamntichi *et al.*, 2000). But one can take resources of similar characteristics and demands as the same category. Therefore, in order to enhance the efficiency of the management and service of MGrid resource, improve the flexibility, and expand the fault tolerance of the MGrid system, the resources in MGrid are classified into 9 categories according to their nature, the demands placed on them, and the ways in which they are engaged. The 9 categories are (Tao *et al.*, 2008b):

1. Human Resource (HR)
2. Equipment Resource (ER)
3. Material Resource (MR)
4. Application System Resource (ASR)
5. Technical Resource (TR)
6. Public Service Resource (PSR)
7. Computational Resource (CR)
8. User Information Resource (UIR)
9. Other Resource (OR)

Then each kind of resource can be encapsulated into a *resource service class model* according to related resource categories, the nine *resource service class models* in MGrid are shown in Table 3.1 (Tao *et al.*, 2008a) and Figure 3.1 (Tao *et al.*, 2008a).

3.2.2 Application Case: Resource Service Design for Magnetic Bearing System

According to the above MGrid resource service classification, 10 related resource services and their partial interfaces have been designed and developed (Tao *et al.*, 2010a), including:

1. Equipment resource partial service class and interface class as shown in Table 3.2.
2. Human resource partial service class and interface class as shown in Table 3.3.

Table 3.1 MGrid resource service abstractive classes

Name of Class	**The definition of Class**	**Instance**
HR-Class	The people that has certain ability required in manufacturing activities, such as operation, manage or technique and so on	designer, manager, technician, etc.
ER-Class	All kinds of equipment that have certain function required in manufacturing activities	Machine tool, CNC, clamp, measure, etc.
MR-Class	Material, semi-manufactured goods, raw and processed materials, fuel, finished – product and so on required in manufacturing activities	Material, raw and processed materials, fuel, product, etc.
ASR-Class	All kinds of software required in manufacturing activities, including design systems, simulation systems, analyzing systems, management systems, graph processing systems, etc.	CAD, CAPP, CAE, PDM, ERP, ANALYSE, Pro/E, PHOTOSHOP, COREDREW, Solidworks, etc.
TR-Class	The set of technical resources and condition required in manufacturing activities	Drawing, design flow, process planning flow, etc.
PSR-Class	Information querying, training, maintenance and so on, provided for resources users	International Standards, national standards, etc.
CR-Class	Computational resources required in manufacturing activities	CPU, storage, etc.
UIR-Class	The information that records the basic information about both resources provider and user, such as credit, etc.	Enterprise information, user information, etc.
OR-Class	Other resources which are not includes in above class, such as log, account information and public information.	Log, account, etc.

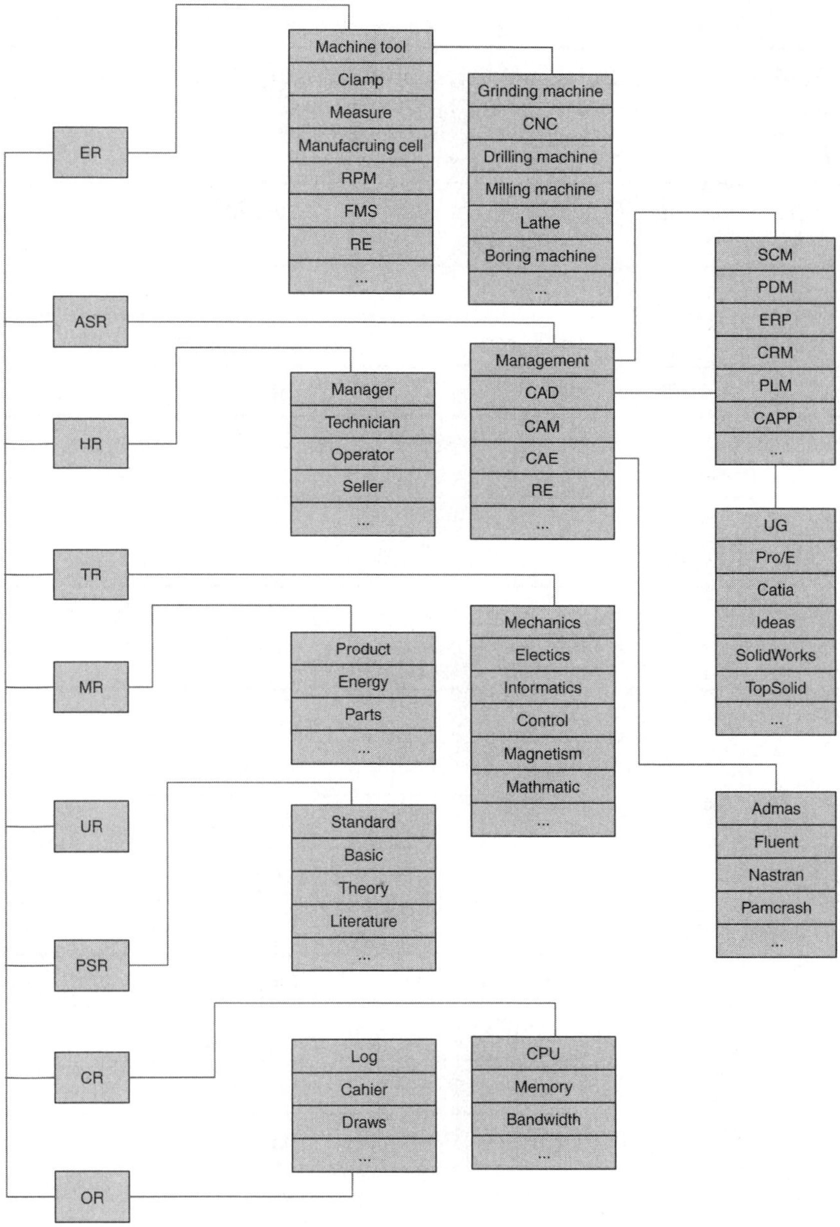

Figure 3.1 Classification of MGrid resource service.

Table 3.2 Equipment resource partial service class and interface class

Equipment Resource	**Equipment Resource Service Class**	**Equipment Resource Service Interface Class**
Machine Tool	MachineToolServ	MachineToolServInt
Rapid molding machine	RapidMoldMacServ	RapidMoldMachineServInt
Cutting tool	CuttingToolServ	CuttingToolServInt
Test-bed	TestbedServ	TestbedServInt
Clamp	ClampServ	ClampServInt
Measuring Equipment	MeasurEquipServ	MeasurEquipServInt

Table 3.3 Human resource partial service class and interface class

Human Resource	**Human Resource Service Class**	**Human Resource Service Interface Class**
Technician	StructDesigServ	StructDesigServInt
	AnalyserServ	AnalyserServInt
	TesterServ	TesterServInt
	MaintManServ	MainManServInt
Manager	FinaManagService	FinaManagServiceInt
	DepotManagerServ	DepotManagerServInt
Marketing	InvestigatorServ	InvestigatorServInt
	SellerServ	SellerServInt
	BagmanServ	BagmanServInt
Worker	MechanicServ	MechanicServInt
	LocksmithServ	LocksmithServInt
	ElectricianServ	ElectricianServInt
	LathemanServ	LathemanServInt

3. Technical resource partial service class and interface class as shown in Table 3.4.
4. Material resource partial service class and interface class as shown in Table 3.5.
5. Application software system partial service class and interface class as shown in Table 3.6.
6. Public service resource partial service class and interface class as shown in Table 3.7.
7. User information resource partial service class and interface class as shown in Table 3.8.
8. Computational resource partial service class and interface class as shown in Table 3.9.

Table 3.4 Technical resource partial service class and interface class

Technical Resource	Technical Resource Service Class	Technical Resource Service Interface Class
Mechanics	EngMecServ	EngMecServInt
	MaterialMecServ	MaterialMecServInt
	StructMecServ	StructMecServInt
	TheorMechanicsServ	TheorMechanicsServInt
Electrics	AnalogCircuitServ	AnalogCircuitServInt
	NumCircuitServ	NumCircuitServInt
	IntegratedCircuitServ	IntegratedCircuitServInt
Magnetism	MagneticFieldServ	MagneticFieldServInt
	MagneticForceServ	MagneticForceServInt
Control Theory	FuzzyContrServ	FuzzyContrServInt
	AutomaticContrServ	AutomaticContrServInt
Informatics	InforSecurityServ	InforSecurityServInt
	InforProcessServ	InforProcessServInt
Mechanics	OptimTheoryServ	OptimTheoryServInt

Table 3.5 Material resource partial service class and interface class

Material Resource	Material Resource Service Class	Material Resource Service Interface Class
Magnetic Material	PermMagMaterServ	PermMagMaterServInt
	ElecMagMaterServ	ElecMagMaterServInt
Standard Parts	BearingServ	BearingServInt
	PowerAmplifServ	PowerAmplifServInt
	ElectricMotorServ	ElectricMotorServInt
	SensorsServ	SensorsServInt
	CircuitBoardServ	CircuitBoardServInt
Energy	FlywheelServ	FlywheelServInt
	BatteryServ	BatteryServInt

Table 3.6 Application software system partial service class and interface class

Application Software System	Application Software System Service Class	Application Software System Service Interface Class
CAD software	UnigraphicsServ	UnigraphicsServInt
	ProEngineerServ	ProEngineerServInt
	CatiaServ	CatiaServInt
	IdeasServ	IdeasServInt
	SolidWorksServ	SolidWorksServInt
	TopSolidServ	TopSolidServInt
CAM software	UnigraphicsServ	UnigraphicsServInt
	MastercamServ	MastercamServInt
	CatiaServ	CatiaServInt

(Continued)

Table 3.6 (cont.) Application software system partial service class and interface class

Application Software System	Application Software System Service Class	Application Software System Service Interface Class
Management software	PDMServ	PDMServInt
	ERPServ	ERPServInt
	SCMServ	SCMServInt
	CRMServ	CRMServInt
	PLMServ	PLMServInt
	CAPPServ	CAPPServInt
	MISServ	MISServInt
CAE software	AnsysServ	AnsysServInt
	AdmasServ	AdmasServInt
	FluentServ	FluentServInt
	NastranServ	NastranServInt
	PamcrashServ	PamcrashServInt

Table 3.7 Public service resource partial service class and interface class

Public Service Resource	Public Service Resource Service Class	Public Service Resource Service Interface Class
Standard	ProcessStandServ	ProcessStandServInt
	InternaStandServ	InternaStandServInt
	IndustrialStandServ	IndustrialStandServInt
Basic Theory	MathematicsServ	MathematicsServInt
	MechanicsServ	MechanicsServInt
	ElectroMagneticsServ	ElectroMagneticsServInt
	InformaticsServ	InformaticsServInt
	CyberneticsServ	CyberneticsServInt

(Continued)

Table 3.7 (cont.) Public service resource partial service class and interface class

Public service resource	Public Service Resource Service Class	Public Service Resource Service Interface Class
Literature	JournalArticleServ	JournalArticleServInt
	NewspaperServ	NewspaperServInt
	ConfArticleServ	ConfArticleServInt
	ReportServ	ReportServInt

Table 3.8 User information resource partial service class and interface class

User Information Resource	User Information Resource Service Class	User Information Resource Service Interface Class
User	RSPInforServ	RSPInforServInt
	RSDInforServ	RSDInforServInt
Product	ProductInforServ	ProductInforServInt
Marketing information	MarketInforServ	MarketInforServInt
Customers information	CustomerInforServ	CustomerInforServInt

Table 3.9 Computational resource partial service class and interface class

Computational Resource	Computational Resource Service Class	Computational Resource Service Interface Class
CPU	CpuServ	CpuServInt
Memory	MemoryServ	MemoryServInt
Bandwidth	BandwidthServ	BandwidthServInt

Table 3.10 Other resource partial service class and interface class

Other Resource	Other Resource Service Class	Other Resource Service Interface Class
Log	LogServInt	LogServInt
Design Draws	DesignDrawsServInt	DesignDrawsServInt
Cahier	CahierServInt	CahierServInt

Table 3.11 Magnetic bearing design partial service class and interface class

Design Service	Design Service Class	Design Service Interface Class
Structure Design	StructureDesignServ	StructureDesignServInt
Control Design	ControlDesigServ	ControlDesigServInt
Static Analysis	StaticAnalyServ	StaticAnalyServInt
Dynamic Analysis	DynamicAnalyServ	DynamicAnalyServInt
Simulation	SimulationServ	SimulationServInt

9. Other resource partial service class and interface class as shown in Table 3.10.
10. Magnetic bearing design partial service class and interface class as shown in Table 3.11.

3.3 Requirements of DDoRS in MGrid

As previously described, resource service sharing and optimal-allocation (RSSOA) in MGrid includes the following five phases: (a) match and search of resource service, (b) comprehensive processing of QoS of resource service, (c) resource service optimal-selection and composition, (d) failure detection and recovery, and (e) resource service scheduling. One of the foundations to realize the above operations of RSSOA is DDoRS. Hence, DDoRS is one of the key issues for addressing RSSOA. Without DDoRS, the operations of RSSOA would have no data and information supports. Taking into consideration the requirements of the above operations involved in RSSOA, the DDoRS in MGrid must satisfy the following requirements (Hu and Tao *et al.*, 2009a):

- It must be machine readable. DDoRS must have the means to be understood and used by different users, application systems, middleware, etc.
- It must be reusable and extensible in order for the describing model of the resource service to be established and extended easily and rapidly.
- It must provide data and information supports to the operations of RSSOA, such as resource service match and search, QoS information extraction and evaluation, optimal-selection and composition, failure detection and recovery, etc.
- It must be consistent. The description of the resource service should be able to express the functionality and characteristics both on a syntactic and semantic level, and should have no conflicts.
- It must have the ability to integrate and represent the domain knowledge of MGrid.

Taking into account the above requirements, in this work ontology technology is adopted to realize the digital description of resource service in MGrid. In the following sections, the method for establishing an MGrid-Ontology and constructing an MGrid-Ontology, are researched then a digital describing method of resource service based on the established MGrid-Ontology is further studied.

3.4 MGrid and Ontology

Ontology is a formal explicit specification of a shared conceptualization. It is a research area of artificial intelligence (AI) and is usually used for knowledge representation in order to build up a knowledge base, knowledge-based system, or knowledge level agent, such as the research of enterprise modeling (Li and Tsai 2005). According to Tao *et al.*, (2008a), it is known that in order to realize RSSOA in MGrid, the effective exchange and sharing of resources and information among different users, organizations, application systems, and enterprises must be realized. Due to different application backgrounds and requirements, the same concept (or term) may have different meanings for different users. These different meanings may result in the blocking of the interoperation and sharing of MGrid resources and services. In order to realize RSSOA in MGrid, an MGrid ontology (MGrid-Ontology) must be established because it is one of the most important systems and methods that allows an explicit representation of knowledge in the field of MGrid. It can explicitly represent the concepts, objects, processes, and rules involved in RSSOA in an uniform way using an ontology describing language even if implicitly each user, organization, unit, or employee interconnected in MGrid uses a personal conceptual schema that could be shared or at least understood among them (Cristani and Cuel 2004). MGrid-Ontology has the ability to describe the knowledge relationship of functions, attributes, and manufacturing abilities of resource services. So MGrid resource services and tasks can be described in an uniform and standard way, which provides the operations of RSSOA with data and information supports. Therefore, developing a good MGrid-Ontology seems to be one of the critical issues in realizing MGrid. The following section will emphasize the establishment of MGrid-Ontology, and research the method for DDoRS based on MGrid-Ontology (Hu and Tao *et al.*, 2009a).

3.5 Establishing the Method of MGrid-Ontology

Figure 3.2 shows the method for establishing an MGrid-Ontology. The establishing process of MGrid-Ontology can be summarized in the following five steps (Hu and Tao *et al.*, 2009a):

1. Define the scope and requirements of MGrid-Ontology
2. Determine essential concepts, reusing existing ontologies if possible
3. Analyses and design of MGrid-Ontology
4. Representation of MGrid-Ontology
5. Verification and validation of MGrid-Ontology (repeating steps 3–4 until requirements are satisfied)

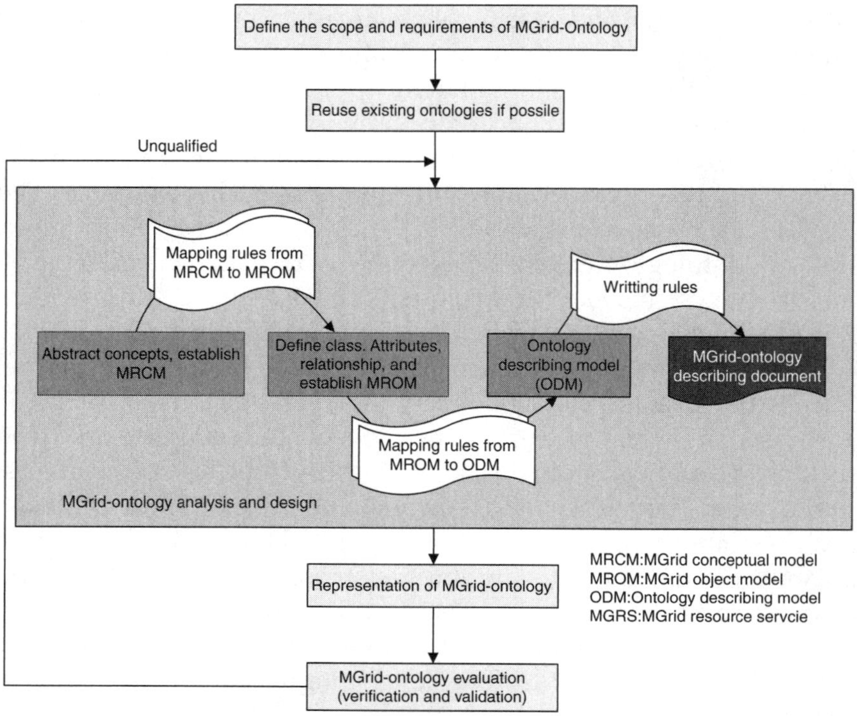

Figure 3.2 Establishing process of MGrid-Ontology.

3.5.1 Step 1: Define the Scope and Requirements of MGrid-Ontology

Similar with the developing processes of software, the first phase for developing MGrid-Ontology is requirement analyses. In this phase, the following issues should be addressed:

(a) The *scope* that MGrid-Ontology should cover,
(b) the *goal* and use of MGrid-Ontology,
(c) the *information* that must be provided by MGrid-Ontology, and
(d) the *users* of MGrid-Ontology.

The *scope* of MGrid-Ontology is focused on the understanding of the main concepts related with the process of RSSOA (including resource service digital description, match and search, QoS evaluation, optimal-selection and composition, and failure detection and recovery) within and between users (including organizations, agents, application systems, etc.).

The main *goal* of MGrid-Ontology is to provide a better understanding of the process of RSSOA, provide a common understanding about the terms used in MGrid, and avoid communication gaps among different users.

The *information* that must be provided by MGrid-Ontology is primarily digital describing information of resource services (such as functional information, manufacturing ability information, QoS information, contact information, state information, etc.) which support the operations of RSSOA.

The main *users* of MGrid-Ontology are MGrid-Ontology engineers, knowledge or service providers, knowledge or service demanders, resource service engineers, etc.

3.5.2 Step 2: Determine Essential Concepts, Reusing Existing Ontologies if Possible

Define the essential concepts in MGrid, search and review the existing ontologies before developing related ontology in regards to a defined concept, and reuse existing ontologies when it is possible. Reuse of ontology means to integrate and use the existing ontology, such as Ontolingua [http://www.ksl.stanford.edu/software/ontolingua/], DAML ontology [http://www.daml.org/ontologies/], etc. Reuse of existing ontologies can reduce the process for establishing upper level ontology of a concept, so that the developer only needs to emphasize establishing domain ontology.

3.5.3 Step 3: Analyses and Design of MGrid-Ontology

This step is the key to develop an MGrid-Ontology. It is responsible for (a) defining the concepts, attributes, and relationships in MGrid; (b) assigning relationships between defined concepts and attributes; (c) constructing a hierarchy of concepts; (d) mapping to an upper lever ontology, such as OWL-S ontology; and (e) building a ontology model (such as OWL-S Model). Taking the above requirements into consideration, the following three substeps are designed to realize the analyses and design of MGrid-Ontology.

1. Abstracting the main concepts of MGrid and establishing MGrid resource conceptual model (MRCM).
2. Defining the related classes, attributes, and their relationship in MRCM, establishing MGrid resource objective model (MROM), i.e., realizing the mapping form MRCM to an object-oriented model (e.g., UML model) according to some of the transforming rules.
3. Realizing the digital description of classes and attributes with an ontology describing language. In other words, realizing the mapping form MROM to an ontology describing model according to some transforming rules.

3.5.4 Step 4: Representation of MGrid-Ontology

Representation of MGrid-Ontology means to realize the describing of MGrid-Ontology using a formalized language. There exist a lot of languages for ontology representation. In Section 3.6, we will compare the existing ontology describing languages and select a language to represent MGrid-Ontology.

3.5.5 Step 5: Verification and Validation of MGrid-Ontology

MGrid-Ontology should be evaluated before being used, including verification and validation (VV). Verification is primarily responsible for checking whether the established MGrid-Ontology is qualified according to the requirements, while validation is responsible for checking whether the established MGrid-Ontology is consistent with the domain model of the real world. In this work, MGrid-Ontology is evaluated from the following three dimensions (Gangemi *et al.*, 2006):

1. *Structural.* The *structural* dimension of MGrid-Ontology is focused on syntax and formal semantics. In this form, the topological, logical and meta-logical properties of ontology can be measured by means of a context-free metric.

2. *Functional.* The *functional* dimension is related to the intended use of a given ontology and its components, i.e., their function is a context. The focus is on the conceptualization specified by ontology.
3. *Usability-profiling.* The *usability-profiling* dimension focuses on the ontology profile (annotations), which typically addresses the communication context of ontology, i.e., its pragmatics.

3.6 Selection of Describing Language for MGrid-Ontology

Ontology describing language plays a key role in the establishment of MGrid-Ontology. A describing language that can realize representation and reuse of MGrid knowledge, and support service operations of RSSOA (e.g., service match and discovery, QoS information extraction and evaluation, failure detection and recovery) in MGrid, must satisfy the following requirements:

1. Has powerful representation and reasoning ability for automatic service discovery, and intelligent match and search of resource service.
2. Has the ability to describe manufacturing capability of resource service in an uniform way.
3. Has powerful supports of the interior logic system.
4. Is compatible with the existing standard of W3C.
5. It is extensible.
6. It has the feature of XML and must be based on the semantic web.

Taking into consideration the above factors and the requirements of resource service sharing and optimal-allocation (MGRSOA) in MGrid, we have selected six different existing ontology describing languages, and compared them from the aspects of representation ability, reasoning ability, etc. The comparison results are shown in Table 3.12.

It is shown by the comparison results that the comprehensive ability of OWL-S (Ontology Web Language for Services) is the best. OWL-S, a standard language based on the semantic web and developed from DAML+OIL, has also been recognized

Table 3.12 Capability comparison of different ontology describing language (Ni 2006a)

	SHOE	**OML/ CKML**	**RDF(S)**	**OIL**	**DAML+OIL**	**OWL-S**
Syntax	**HTML / XML**	**XML**	**XML**	**RDF /XML**	**RDF /XML**	**RDF /XML**
Type constr.	+	+	+	+	+	+
Semantic	+	+	–	+	+	+
Local scope	+	+	+	+	+	+
Global scope	–	+	+	+	+	+
Version edit	+	–	–	–	–	+
Reasoning ability	–	–	+	+	+	+
Subclass of	+	+	+	+	+	+
Class equivalency	+	+	–	–	+	+
Attribute equivalency	+	+	–	–	+	+

Instance equivalency	−	−	−	−	+	+
Distribution define	+	−	+	+	+	+
Extensible of ontology	+	−	+	+	+	+
Describing logic	−	−/+	−	+	+	+
Partition of computing character	−	+	−	+	−	+
Partitions	−	+	−	+	+	+
Document note	+	+	+	+	+	+
Default value	−	−	−	−	−	−
Concept attributes	−	+	−	+	+	+
Instance attributes	+	+	+	+	+	+
Quantity constr.	+	−	−	+	+	+
Interior logic	Logic frame-work	Conceptual graph	Semantic net	Describing logic	Describing logic	Describing logic

"+" denote "Support", "Yes", "Have"; "-" denotes "No", "don't support" and recommended by W3C. Therefore, this chapter has selected OWL-S as the describing language of MGrid-Ontology. Furthermore, in computer science, several tools can be used by users and system developers to create good and effective ontologies, such as OntoSaurus [http://www.isi.edu/isd/ontosaurus.html], Sesame+OMM [http://www.openrdf.org], WebOnto [http://www.aktors.org/technologies/webonto/], Protégé-2000 [http://protege.stanford.edu/], WebODE [http://webode.dia.fi.upm.es/WebODEWeb/index.html], OilEd [http://oiled.man.ac.uk/building/], OntoEdit [http://www.ontoknowledge.org/tools/ontoedit.shtml], COHSE [http://cohse.semanticweb.org/], MnM [http://kmi.open.ac.uk/projects/akt/MnM/], TextToOnto (Maedche and Staab 2008), KAON (Bozsak *et al.* 2002; http://kaon.semanticweb.org) , Ontolingua [http://www.ksl.stanford.edu/software/ontolingua] and Jena [http://jena.sourceforge.net,2007.10], etc. The Protégé-2000 was selected as the tool to develop MGrid-Ontology.

3.7 MGrid Ontology

3.7.1 OWL-S

OWL-S is an ontology for the specification of web services, which combines characteristics of WSDL, OWL semantic markup, and a language for rules specification (e.g., SWRL). OWL-S allows a provider to deploy a complete description of the service's capabilities through three modules (see Figure 3.3):

(a) Service Profile, which describes what a service does,
(b) Service Model, which describes how a service operates, and
(c) Service Grounding, which describes the characteristics of access to a service.

OWL-S is a general describing ontology for web service and it does not consider the specific requirements in an industrial field. Therefore, it cannot be directly used to describe MGrid resource service or task information. Hence, according to the functional requirements of resource service operations of RSSOA (such as resource service match and search, QoS evaluation, optimal-selection and

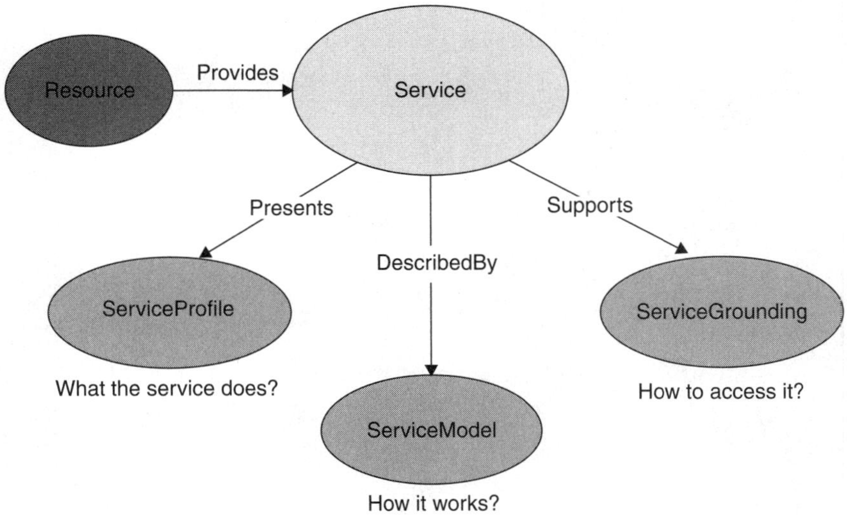

Figure 3.3 OWL-S model (Forte *et al.* 2008).

composition, failure detection and recovery, etc.) and based on the proposed methods for establishing MGrid-Ontology, we established an OWL-S based MGrid-Ontology as shown in Figure 3.4 (Hu and Tao *et al.*, 2009a).

3.7.2 MGrid-Ontology

MGrid-Ontology primarily uses the following fifteen classes to describe an MGrid resource service (MGRS) or task requirements.

1. **MGridService. MGridService** inherits from **OWL-S Service**. It is used to define an MGRS. The specific contents of an MGRS are described primarily by its modules, i.e., **MGridProfile, MGridModel** and **MGridGrounding**.
2. **MGridProfile. MGridProfile** inherits from **OWL-S Profile**. It provides a concise representation of the service provider and of the capabilities of a service (i.e., what the service does) through the advertising of the functionalities description (e.g., *serviceName, serviceDescription, contactInformation*). It also provides attributes for the categorization of the service offered

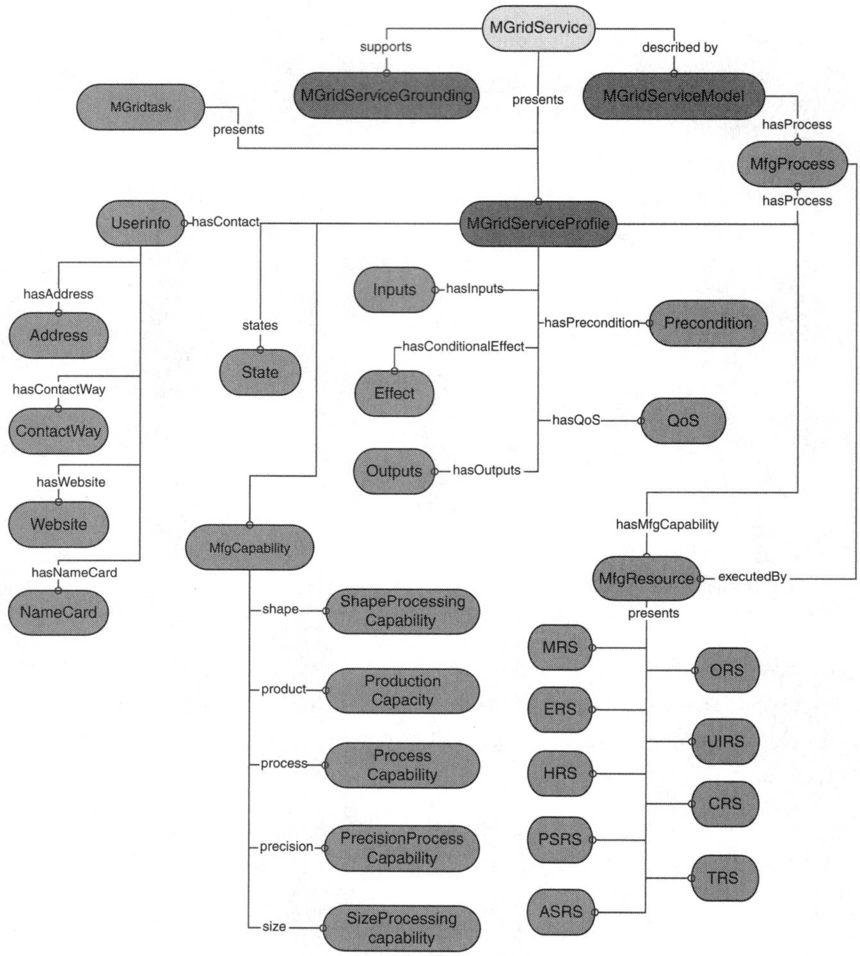

Figure 3.4 MGrid ontology based on OWL-S.

(e.g., *serviceParameter*, *serviceCategory*), and the functional representations of the service in the form of **Inputs**, **Outputs**, **Preconditions**, and **Effects** (IOPE). IOPE is described by the properties *hasParameter*, *hasInput*, *hasOutput*, *hasPrecondition*, and *hasEffect*. In addition to the above functions, **MGridProfile** defines nine subclasses (e.g., **ERS**, **ASRS**, **MRS**, **TRS**, **HRS**, **PSRS**, **CRS**, **UIRS**, and **ORS**) to describe the specific information of the related nine kinds of resource service as mentioned in Section 2.

Furthermore, **MGridProfile** also defines related subclasses to describe QoS evaluation information, state information, contacting information, process information, manufacturing capability information, etc., as well as related properties such as *evaluates, states, contacts, hasProcess, hasMfgCapability*, etc., as shown in Table 3.13(Hu and Tao *et al.*, 2009a).

3. **MGridModel. MGridModel** inherits from **OWL-S Model**. It describes how a service operates. It realizes the description of an MGrid process model using the property of *hasProcess*.
4. **MGridGrounding. MGridGrounding** inherits from **OWL-S ServiceGrounding**. It describes the way another entity (such as soft agent, resource service demander and provider) access a resource service. It describes the specific communication protocol and message used in the service interaction, including related protocol, transmission, orientation, etc. In MGrid, **MGridGrounding** uses the WSDL, SOAP, etc., as the resource service accessing interface protocol.
5. **MGridTask. MGridTask** defines the requirements of an MGrid task. The description of an MGrid task is the same as that of a resource service. **MGridTask** connects with the instances of the classes **MfgProcess**, **MfgCapability**, **MfgResource**, and **QoS** by the property of *presents*, in order to define the process requirements, capability requirements, resource requirements, and QoS requirements of an MGrid task.
6. **MfgProcess. MfgProcess** describes the working way of the invoked resource service, i.e., the serving process of resource service. There are three types of processes: (a) *Atomic Process* supplies abstract specification of the information exchanged with the requester, corresponding to the operations the supplier can execute directly. Each atomic process has to provide related grounding; (b) *Simple Process* cannot be invoked and it does not have grounding. It is usually used as the abstract of an element, such as the view of atomic processes or the simplified representation

Table 3.13 Properties of MGridprofile and their value sets

Classification	Properties	Value class	Classification	Properties	Value class
Functional description information of MGRS	*hasInput*	**Input**	Contact information	*contacts*	**UserInfo**
	hasOutput	**Output**	State information	*states*	**MGridState**
	hasPrecondition	**Precondition**	Process information	*hasProcess*	**MfgProcess**
	hasConditionalEffect	**Effect**	9 kinds of MGRS	*mfgResource*	**ERS, HRS, TRS ASRS, MRS, PSRS, CRS, UIRS, ORS**
QoS information	*evaluates*	**QoS**			
Manufacturing capability information	*hasMfgCapability*	**MfgCapability**			

of a composite process; and (c) *Composite Process is* employed to describe collections of processes organized through some type of flow control structure, e.g., *sequence, parallel, selective, circular*, etc. Each process has corresponding properties, such as *executedBy, processType, connection,* to describe which kinds of resource services execute the process, the type of the process, and the connection with the previous resource service. The value set of *executedBy* is class **MfgResource**.

7. **MfgCapability. MfgCapability** primarily describes the manufacturing capability of a resource service, including the manufacturing capability a resource service can provide and the requirements of an MGrid task. It is primarily used to support semantic-based resource service match and discovery in the aspect of manufacturing capability. It is realized by the property of *hasMfgCapability*.
8. **MfgResource. MfgResource** (i.e., **ERS**, **ASRS**, **MRS**, **TRS**, **HRS**, **PSRS**, **CRS**, **UIRS**, and **ORS**) defines the specific information of the related nine kinds of resource service.
9. **QoS. QoS** is used to describe the non-functional information, i.e., QoS information, which is used to describe the evaluation information of resource service, such as *cost, trust-QoS, reliability, maintainability, functional similarity*, etc.
10. **Inputs. Inputs** define the information that should be provided to invoke a resource service.
11. **Outputs. Outputs** describe the information expected after the interaction with a resource service.
12. **Effect. Effect** is used to define the set of statements that should hold true if the service is invoked successfully.
13. **Precondition. Precondition** is used to define the set of conditions that should hold prior to a resource service being invoked.
14. **MGridState. MGridState** is used to describe the state information of a resource service. It defines the basic concept referring to the state of equipment resource

service such as *loading state, running state,* and *performance state.*

15. **UserInfo. UserInfo** is used to define the basic information of resource service provider and demander, including *address, enterprise card, contact way, website,* etc.

3.8 DDoRS Based on MGrid-Ontology

DDoRS is the information foundation of MGrid resource service optimal-allocation, including service match and discovery, QoS information extraction and evaluation, optimal-selection and composition, failure detection and recovery, etc. MGrid-Ontology has provided conceptual model and logic foundation for DDoRS. The following sections show how to use MGrid-Ontology to realize the digital description of resource service provided by RSP (resource service provider) and task requirements of RSD (resource service demander).

Each MGrid resource service (MGRS) is defined as a 5-tuple based on MGrid-Ontology, i.e., MGRS = (*Agent, MGSP, MGSM, MGSG, Res*), where *Agent* denotes the entity of resource service provider; *MGSP* denotes the service profile of MGRS; *MGSM* denotes the service modeling of MGRS; *MGSG* denotes the service grounding of MGRS; and *Res* denotes the resources owned by the entity (Hu and Tao *et al.*, 2009a).

3.8.1 Description of *Agent*

MGrid resource service ***Agent*** is the resource service entity that can solve a manufacturing problem or a RSP. It has self-control ability in relation to its own resource and its state. The primary describing attributes of *Agent* include *agentName, agentAbility, agentAim, agentType,* and the corresponding value set is class **UserInfo**.

3.8.2 Description of *MGSP*

MGSP provides a concise representation of the capabilities of resource service through the advertising of the functionality description, including general description, functional information,

QoS evaluation information, manufacturing capability information, connection information, state information, process information, specific information of the nine kinds of resources, contact information, etc. (Hu and Tao *et al.*, 2009a).

MGridServiceProfile = (*general, functional, QoS, mfgCapability, connection, state, mfgResource)*

1. *General* denotes the overall information of resource service and helps the user know the main contents of a resource service. *General* = {*serviceName, serviceDescription, contactInformation, serviceCategory, CreateTime, Creator,...*}. *serviceName* describes the name of a resource service; *serviceDescription* gives out the simple description of a resource service; *contactInformation* gives out the contacting way of the corresponding RSP, its value set is class **UserInfo**.
2. *Functional* is the functional representations of a resource service. It describes the functional information transforming (i.e., inputs and outputs) and the state changing (i.e., precondition and effect) of service execution. *Functional* = {*Inputs, Outputs, Precondition, Effect,...*}, they are realized by the following properties:

    - *hasInput* defines the input information required when a corresponding resource service is invoked. E.g., a structural parameter and 3D model. Its value set is the class **Inputs** in MGrid-Ontology and Inputs = $\{i_1, i_2, i_3, \ldots, i_n\}$.
    - *hasOutput* defines the expected output information after the interaction with a resource service. E.g., a 3D model, an engineering drawing, a simulation result, etc. Its value set is the class **Outputs** in MGrid-Ontology.
    - *hasPrecondition* is used to define the set of conditions that should hold prior to a resource service being invoked. For example, pay cost first before using a resource service. Its value set is the class **Precondition** in MGrid-Ontology.
    - *hasEffect* defines the set of statements that should hold true if the service is invoked successfully. Its value set is the class **Effect** in MGrid-Ontology.

3. *QoS* is the representation of QoS information for resource service. It describes non-functional (i.e., QoS) information of a resource service, including both performance QoS (such as *time, reliability, maintainability, satisfaction,* etc.) and description QoS (such as trust-QoS, cost, etc.). It provides data and information supports when it evaluates the quality of a resource service. QoS is an extensible vector and *QoS* = {*time, cost, trust-QoS, reliability, maintainability, functional similarity,...*}. The detailed QoS model associated with its evaluation method can be found in Chapters 5–6.
4. *mfgCapability* describes the manufacturing capability of a resource service, including the manufacturing capability a resource service can provide, and the requirements of an MGrid task. It is primarily used to support semantic-based resource service match and discovery on the aspect of manufacturing capability. *mfgCapability* = {*shapeProcessingCapability, PrecosionProcessCapability, SizeProcessingCapability, ProductionCapability, ProcessCapability,...*}.
5. *state* describes the current working or usability state of a resource service. *State* = {*Idle, UnderMaintain, Load, fullLoad, Overload,...*}.
6. *mfgResource* defines the specific information of the related nine kinds of resource service. They only fit the very kind of the resource service belonging to the same classification. E.g., the breaking-down rate, the requirement of running environment, action precision, processing size and shape, orientation precision, position precision of the equipment resources (ER); the version, the capacity of secondary development, the opening degree of source code, compatibility, dependability of application system resources (ASR); the working experience, technique grade, knowledge, age, the title of a technical post of human resources (HR); the period of validity, the requirements of the employ environment, recycling, and the life-span of material resources (MR). The value sets are the classes **ERS, HRS, TRS, ASRS, MRS, PSRS, CRS, UIRS,** and **ORS.**

3.8.3 Description of *MGSM*

MGSM can be seen as the server process of a resource service. It describes the working or invoking way of a resource service, including *automatic process, simple process* and *composite process. MGSM* has the properties of *hasInput, hasOutput, hasPrecondition,* and *hasEffect*. The corresponding value sets are classes **Input**, **Output**, **Precondition**, and **Effect,** respectively.

- **Inputs and Outputs**. According to the locating criterion and data stream of process, the source of a process's input can be other processes and resource services. In the same way, output of a process may be passed to other process or resource services.
- **Precondition and Effect.** The execution of a process may change the state of an event. There is an obvious difference between process outputs and effects. Precondition defines the set of conditions that should hold prior to a process being exactly invoked, while Effect describes the state of an event and outputs describe the corresponding information.

3.9 Application Case: MGrid-Ontology Based MGrid Resource Service Discovery

In order to illustrate how the previous work can be used in MGrid resource service management, an MGrid-Ontology based resource service discovery is put forward as shown in Figure 3.5. Its brief work flow is described below (Hu and Tao *et al.*, 2009a).

Service publication:

Step 1 The RSP (Resource service provider) publish their resource services to MGrid through corresponding user interface.

Step 2 Published resource services with OWL-S interfaces are then loaded into MGRSR (MGrid resource service repository), in which the elements of resource service such as the *serviceName, serviceDescription, serviceContact,* and properties are registered with MGRSR.

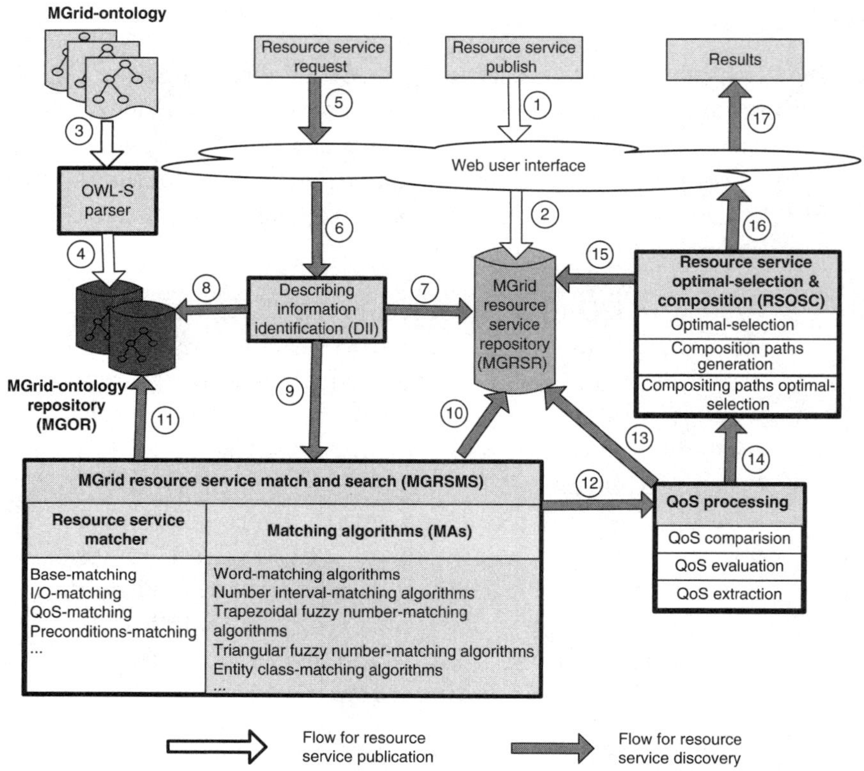

Figure 3.5 MGrid-Ontology based MGrid resource service discovery.

Step 3 While publishing the resource service, the RSP may also publish resource service ontology which can be defined in OWL-S.

Step 4 The ontology is parsed by an OWL-S parser and loaded into the MGrid-Ontology repository (MGOR).The MGOR is used by a reference engine to refer the semantic relationships of describing information when search and match service and query.

Service discovery:

Step 5 The RSD (Resource service demander) submit their resource service requirements to MGrid through corresponding user interface. The requirements are formatted into expected resource service describing information and properties.

Step 6 The requirements are then forwarded to the DII (describing information identification).

Step 7–8 DII accesses the MGRSR to identify and mark the properties of available resource services in MGRSR that are irrelevant to the properties used in the service requirements based on the MGrid-Ontology defined in MGOR.

Step 9 The requirements are then forwarded to MGRSMS (MGrid Resource Service Match and Search),

Step 10–11 MGRSMS invokes its interior components *RS-Matcher* (i.e., Resource Service Matcher) and *MAs* (i.e., Matching Algorithms), which access MGRSR to calculate the similarity of requested services and available resource services in MGRSR based on the MGrid-Ontology defined in MGOR. The detailed match and search mechanism and algorithm of resource services will be described in detail in Chapter 4.

Step 12 After **Step 10–11**, RSD's requirements are matched with their functional aspect. But the quality of the discovered resource service is out of consideration. In addition, the discovered resource service qualified for RSD's functional requirements may be far more than one. How to select the optimal one is another issue that has to be addressed during resource service scheduling. Hence, MGRSMS forwards its results to the component *QoS processing*, which in turn filters functionally matched resource services with nonfunctional aspect (i.e., QoS) after QoS processing, including *QoS Extraction*, *QoS evaluation*, *QoS comparison* and so on. The details will be described in Chapter 5.

Step 13 *QoS processing* accesses MGRSR and extracts related QoS information of the discovered resource services by invoking interior components *QoS extraction*, then evaluates and compares extracted QoS properties by invoking internal services *QoS evaluation* and *QoS comparison*. The detailed algorithms of QoS processing will be described in Chapter 5 and Chapter 6.

Step 14 A list of discovered resource services that are marked with their non-functionally matched degree is forwarded to RSOSC (resource service optimal-selection and composition).

Step 15 In an MGrid system, the requirements (or resource service request) of RSD can be classified into two kinds: (a) single resource service request task (SRSRTask), which can be completed by invoking only one resource service, and (b) multi-resource services request task (MRSRTask), which is completed by invoking several resource services in a certain sequence. For an SRSRTask, the system searches out the resource services that qualified for its function requirements and selects the optimal one to execute. For an MRSRTask, in addition to the search for all qualified resource services according to each subtask, the system has to select one candidate resource service for each subtask and generate a new composite resource service (CRS); the system then has to select the optimal resource service composite path from all possible paths to execute the task with the given multi-objective and multi-constraint. Therefore, RSOSC is designed to address the above issues. The detailed algorithms of RSOSC will be described in Chapter 8 and Chapter 12.

Step 16–17 The final results are then presented to the user via the corresponding web user interface of MGrid.

3.10 Summary

Digital description of resource service (DDoRS) is one of the foundations to realize the operations for service sharing and optimal-allocation. In this chapter, the issue of digital description of resource service (DDoRS) in MGrid is investigated. First a method for establishing an MGrid-Ontology is presented, and an MGrid-Ontology is then built. A new method for digital description of resource service based on established MGrid-Ontology is proposed. An MGrid-Ontology based MGrid resource service discovery framework is proposed, which illustrates how the proposed works can be used in MGrid resource service management.

4

MGrid Resource Service Match and Search

4.1 Introduction

In the MGrid system, there are primarily two kinds of users (Tao *et al.*, 2008a; 2011d): (a) resource enterprise or resource service provider (RSP), and (b) user enterprise or resource service demander (RSD), as shown in Figure 4.1. The former, RSP, publishes its idle resource, product, manufacturing ability, etc., and provides manufacturing resource service to meet the user's requirements. The latter, RSD, searches for the optimal manufacturing resource and service required, and selects the corresponding partner to establish a collaboration manufacturing net.

One of the key technologies to realize resource service exchange in MGrid is resource service match and search (RSMS). Many works on service search and scheduling are carried out in a distributed system such as a computing grid system. Some methods such as QoS-based service search and scheduling, min-min scheduling, and genetic algorithms (GA) based scheduling, have been proposed and studied. But compared with a computing grid system, MGrid resource services have characteristics such as more flexible interaction, real time, long life cycle, multiparty cooperation, resources multiplicity, knowledge, functional complexity, data complexity, higher reliability requirements, and online/offline of resources sharing and task submitting. So RSMS in MGrid cannot simply employ the methods used in grid computing.

Therefore, according to the special characteristics and requirements of MGrid, RSMS in MGrid is studied in this chapter. First, a RSMS framework is proposed, as well as its working flow. Secondly, the digital describing information of resource services are classified into four categories (Tao *et al.*, 2009b): (a) word concept information, (b) sentence information, (c) number information, including number interval and fuzzy number, and (d) entity class (or data structure) information. Then the similarity matching algorithms (SMAs) of each kind of describing information are investigated, including (a) *word matching algorithms (WMAs)*, (b) *sentence matching algorithms (SeMAs)*, (c) *number matching algorithms (NMAs)*, and (d) *entity class matching algorithms (ECMAs)*. Based on the proposed MAs, the match and search processes of MGrid resource services

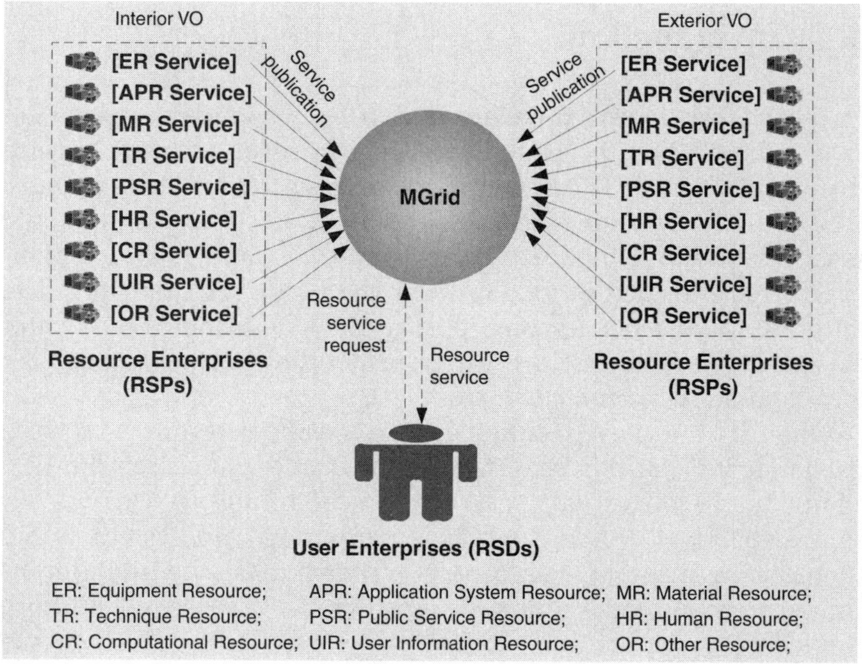

Figure 4.1 The relationship of RSD, RSP and MGrid (Tao *et al.*, 2008a).

are divided into four phases: first, matching the basic information of resource services, such as service name and service description, namely *basic-matching*; second, matching the inputs and outputs information of resource services, namely *I/O-matching*; third, matching the QoS (Quality of Service) information of resource services, namely *QoS-matching*, and; fourth, combining the above three matching results and generating an integrated matching result, namely *integrated-matching*. The matching functions and algorithms of each phase are described in detail. A case study illustrates the application of proposed methods, and the accuracy and efficiency of the proposed method are measured.

4.2 Related Works

4.2.1 Service Discovery in Traditional Distributed System

A variety of papers dealing with service discovery have recently been published. Most existing service discovery techniques aimed at web service discovery are simply based on inputs and outputs of service (Li and Horrock, 2004; Paoluncci *et al.*, 2002; Sycara *et al.*, 2002). Some of them take into account preconditions, effects, and other parameters that describe services. In addition, some researchers investigate service discovery from the perspective of behavior signature (Shen and Su, 2005), service community (Perryea and Chuang, 2006), context of service (Doulkeridis *et al.*, 2007; Balken *et al.*, 2007; Raverdy and Issarny, 2005; Lee and Helal, 2003), user experience (Kokash *et al.*, 2007), role-based interaction model (Alberto *et al.*, 2007), met-interaction (Tomas *et al.*, 2007), market-orientation (Jia *et al.*, 2006a), UML (Zisman and Spanoudakis, 2006), goal-based web service discovery with sophisticated semantic matchmaking (Stollberg *et al.*, 2007), etc.

The above research primarily concentrated on abstract service discovery mechanisms and methods. Although these papers proposed efficient and fast searching schemes for service discovery, their schemes do not consider practical similarity matching algorithms (SMAs) between basic describing information of services. Without the supports of SMAs for basic describing information of services, the upper service discovery cannot be realized. Furthermore, the proposed mechanisms and methods were concerned with the computing resource services in a distributed system such as a computing grid system. Compared with traditional distributed systems, services in MGrid are more complicated and various, and have characteristics such as more flexible interaction, longer life-cycle, being knowledge based, and stronger reliability request. Hence, the existing service discovery methods cannot be simply and directly used in MGrid system. Therefore, in addition to proposing a match and search framework for MGrid resource services, this chapter emphasizes the matching algorithms of basic describing information of resource services (Tao *et al.*, 2009b).

4.2.2 Service Match and Discovery in Distributed Manufacturing System

Bianchini *et al.*, (2005) investigated ontology-based architecture for the service discovery system. They classified services into concrete services, abstract services, and subject categories, and studied five kinds of service matches, including exact match, plug-in match, subsume match, intersection match, and mismatch. Liu *et al.*, (2003) researched MGrid resource scheduling based on QoS, and classified the process of resource scheduling into QoS-based resource search, QoS-based resource scheduling, dynamic negotiation, etc. Deng *et al.*, (2006a) and Chen *et al.*, (2004a) proposed a resource characteristic-based scheduling strategy and applied it in the development of customized artificial joints. Zhang *et al.*, (2006a) proposed a multi-objective optimization mathematical model. Based on the model and combined with user weights, they designed a genetic algorithms (GA) based MGrid resource scheduling process. Lv *et al.*, (2006a) researched marked equilibrium based MGrid resource optimal-allocation methods. The scheduling methods they researched are based on the precondition that the advertising resource services were searched from the database. Tan and Fan, (2005) studied semantic-based service match and composition under a networked manufacturing environment. Zhang *et al.*, (2006b; 2006c) researched an ontology-based approach of automated service chaining for MGrid. In order to supplement the shortage of service discovery middle agents (e.g., matchmaker, broker, yellow page, blackboard, etc.) that do not guarantee efficient and rapid matching results, Lee *et al.*, (2005a) proposed a new matching algorithm based on the marriage matching algorithm of the ATM network to improve the middle agents' performance.

However, the describing information of services considered in their matching methods primarily is limited to the ontology of concepts, inputs, and outputs. In fact, in addition to these, there are a number of describing information, sentence describing information, data structure describing information, all of which play a very important role in MGrid resource service description. Furthermore, the specific matching and searching algorithms are not given out in the above research. In order to supplement the above shortages and realize effective RSMS in MGrid system, similarity matching algorithms (SMAs) for basic describing information of resource services are investigated in these studies. A four-phase method for resource service match and search based on the proposed SMAs will be presented.

4.3 Framework of Resource Service Match and Search in MGrid

Definition 4.1 – **Resource service matching type.** *A* is a requested resource service (or a task requirement) and *B* is an advertising resource service in RSIC (resource servcie information center), Af_i is an arbitrary describing parameter of *A*, and Bf_i is the corresponding describing parameter of *B*. The matching types between *A* and *B* are defined as follows (Tao *et al.*, 2009b):

(a) If $\forall i, \exists j, Af_i \equiv Bf_j$, then the match type between *A* and *B* are defined as *exact match;*

(b) If $\forall i, \exists j, Subs(Bf_j, Af_i)$ (i.e., Bf_i is a super-concept of Af_i), then the match type between *A* and *B* are defined as *plug-in match.* In this condition, *B* can satisfy *A*'s requirements and can be selected to execute the task;

(c) If $\forall i, \exists j, Subs(Af_i, Bf_j)$ (i.e., Bf_j is a sub-concept of Af_i), then the match type between *A* and *B* are defined as *subsume match.* In this condition, *B* can partly satisfy *A*'s requirements;

(d) Otherwise *A* and *B* have no similarity, and *B* cannot satisfy *A*'s requirements at all, defined as a *mismatch.*

In fact, most advertising resources in RSIC cannot accord to the request of a task or a query 100 percent, but they are qualified to execute the task. If the system only selects the *exact match* resource services as candidates, there may be no match result, or some qualified resource services may be excluded. Consequently, the match and search accuracy is reduced. In order to overcome the above shortcomings, a relaxation-matching strategy is used in the process of RSMS in MGrid. When the similarity between two compared resource services (i.e., advertising resource service and requested resource service) equate or exceed a scheduled threshold value, then the advertising resource service is considered as a qualified candidate. The proposed framework of resource service match and search (RSMS) is shown in Figure 4.2. The brief workflow are described as follows (Tao *et al.* 20009b):

(a) A user or RSD (resource service demander) submits its request (i.e., a manufacturing task or a resource service request) to *MGTMS* (MGrid task management system) via corresponding human-machine interface of MGrid.

(b) *MGTMS* decomposes the task into several corresponding subtasks that cannot be decomposed again, and submits them to the system *parser.*

(c) The system *parser* transfers the decomposed task's requirements into standard resource service describing information, including general information, inputs information, outputs information, QoS information, etc.

(d-e) The system selects the corresponding resource services information from RSIC (resource service information center), and the system *parser* transfers them into corresponding general information, inputs information, outputs information, QoS information, etc.

(f) The *RS-Matcher* (resource service matcher) matches the requested resource service information with each advertising resource service extracted from RSIC by invoking the component *SMAs* (similarity matching algorithms).

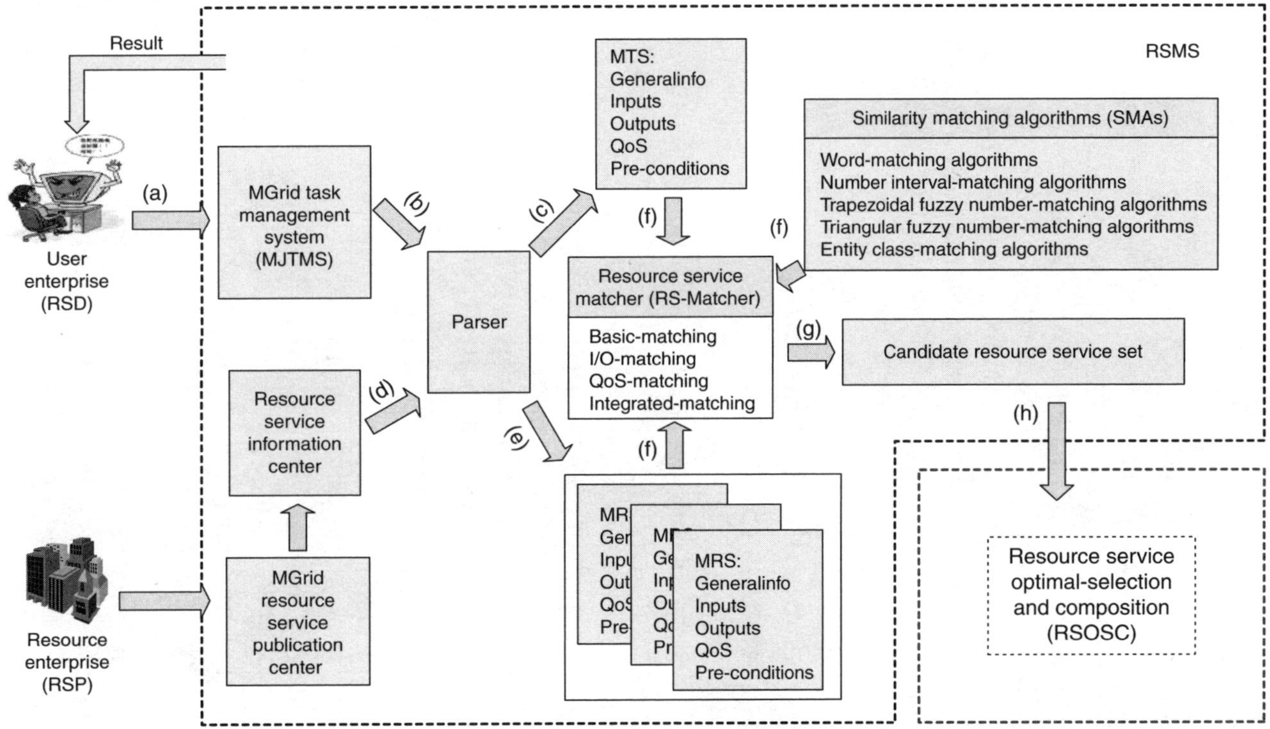

Figure 4.2 Framework of match and search for MGrid resource service and its working flow chart (Tao *et al.* 2009b).

(g-h) The indices of the qualified resource services are recorded in the *candidate resource service set*. Then RSOSC (resource service optimal-selection and composition) selects the optimal resource service.

In the proposed RSMS framework, the key components are *similarity matching algorithms (SMAs)* and *resource service matcher (RS-Matcher)*. *SMAs* provides *RS-Matcher* with similarity matching algorithms of basic describing information about resource service, such as *WMAs, SeMAs, NMAs, ECMAs*. *RS-Matcher* is responsible for matching resource services and tasks, including *basic-matching, I/O-matching, QoS-matching*, and *integrated-matching*. The following sections design and describe the detailed working flow and algorithms of *SMAs* and *RS-Matcher*, respectively.

4.4 SMAs: Similarity Matching Algorithms (SMAs)

Digital describing of resource service (DDoRS) is the foundation for realizing MGrid. It provides data and information supports for the operations involved in the implementation of MGrid, including match and search of resource service, QoS modeling and evaluation of resource service, resource service optimal-selection and composition, failure detection and recovery, etc. In the authors' research, digital describing of resource service is realized based on OWL-S and MGrid-Ontology. According to the digital describing of resource service method in Chapter 3, each MGrid resource service (MGRS) is described as a 5-tuple based on OWL-S and MGrid-Ontology as follows (Tao *et al.*, 2009b):

Definition 4.2 – **Digital description of resource service.** MGRS = (*Agent, MGSP, MGSM, MGSG, Res*)

The detailed description of *Agent, MGSP, MGSM, MGSG*, and the method for establishing MGrid Ontology along with the detailed method for DDoRS are researched in Chapter 3. In order to simplify the process of RSMS and enhance the efficiency, the describing information of MGrid resource service is classified into the following four categories:

- *Word* concept information, such as resource service name, resource service description, etc.

- *Sentence* information, such as resource service description, etc.
- *Number* information, including *fuzzy number* and *interval number*. E.g., cost of service, manufacturing precision, etc.
- *Entity class* (or *data structure*) information.

Different kinds of describing information have different characters and should employ different matching functions when calculating the similarity between them. In the following sections, the similarity matching algorithms (SMAs) of each kind of describing information are investigated respectively, including *word matching algorithms (WMAs)*, *sentence matching algorithms (SeMAs)*, *number matching algorithms (NMAs)*, and *entity matching algorithms (EMAs)*.

4.4.1 Word Matching Algorithms (WMAs)

Provided that Aw_i is a word concept describing information of a requested resource service *A*, and Bw_j is that of an advertising resource service *B*, the matching degree of the corresponding word concepts describing information between *A* and *B* is defined as the similarity between Aw_i and Bw_j (Tao *et al.*, 2009b).

Many researchers have investigated the similarity measurement between word concepts in a virtual system. Resnik (1995) proposed a method for measuring semantic similarity based on information theory. Jiang and Conrath (1997) presented a corpus statistics and lexical taxonomy based approach for measuring word semantic similarity. But research on similarity measurements for word concepts in the manufacturing field is insufficient. Lin *et al.*, (2005a) studied a measurement method for the process knowledge service ontology (PKSO) concepts. They employed the function $S_c(C_1,C_2)=1/(1+p)$ to evaluate the similarity of two different concepts, C_1 and C_2, in PKSO, where *p* denotes the shortest *path length* between C_1 and C_2. However, this method is imprecise if it is applied to calculate the similarity of two manufacturing concepts. For example, in Figure 4.3, the shortest *path length* from *optical-shaft* to *step-shaft*, and from *optical-shaft* to *straight-shaft* both are 2. According to the above method, the two pairs of concepts have the same similarity value 1/3, which is unpractical. This is because they only considered the *path length* between two compared concepts, but the depth, overlap, and density of concepts are ignored. Hence the calculating result is inauthentic.

In order to address this deficiency, more information should be considered in order to adjust the above method. In fact, the more the same words Aw_i and Bw_j contain or share, the stronger the similarity between them. Therefore, the *overlap* of the pair of compared concepts in the hierarchy of ontology should be taken into account while calculating their similarity. Furthermore, concepts at upper layers of the hierarchy have more general semantics and a weaker similarity between them, while concepts at lower layers have more concrete semantics and a stronger similarity. So the *depth* of concepts in the hierarchy is also a decisive fact to similarity. Hence, the similarity between a pair of ontology concepts is decided not only by *path length*, but also by *depth* and *overlap* (Li and Bandar *et al.*, 2003; Tao *et al.*, 2009b).

(a) ***Path length (p) of two compared word concepts.*** The *path length* of two compared word concepts is the number of the border involved in the shortest path connecting the compared words. E.g., in the ISA hierarchy figure of shaft (see Figure 4.3), the number of the border involved in the shortest path between *shaft* and *driven-shaft* is 2, so the *path length* between them is 2. The *path length* is an important factor that affects the similarity between two compared words. In general, the bigger the *path length* is, the weaker the similarity is. Contrarily, the smaller the *path length* is, the stronger the similarity is. Therefore, a monotonically decreasing transfer function, $f_p(p)$ is designed with respect to *path length*, p, and $f_p(p) = e^{-p}\,(0 \le p)$ (Li *et al.*, 2003). The function curve of $f_p(p)$ is shown in Figure 4.4 (a).

(b) ***Overlap (r) of two compared words.*** The *overlap* between two compared word concepts is the number of the same upper concepts they share. The bigger the *overlap* is, the stronger the similarity between them is. E.g., in Figure 4.3, word concepts *rotating-shaft* and *optical-shaft* have 3 same upper concepts in the ISA hierarchy figures, so the overlap of them is 3. While the *overlap* between word concepts *rotating-shaft* and *driven-shaft* is only 2, therefore, the similarity between *rotating-shaft* and *optical-shaft* is stronger than that between *rotating-shaft* and *driven-shaft*. Let $f_r(r)$ denote the transfer function of *overlap* r. Then $f_r(r)$ is a monotonically increasing function with respect to r, and $f_r(r) = 1 - e^{-r} (0 \leq r)$. The function curve of $f_r(r)$ is shown in Figure 4.4 (b).

(c) ***Depth* of two compared words (*h*).** The *depth* of each compared word concept is derived by counting the levels from it to the top of the lexical hierarchy. E.g., in Figure 4.3 the *depth* of *driven-shaft* is 2. When the compared words have the same path length, the similarity between them increases as the sum of their level numbers increases, or decreases as the difference of their level numbers decreases. E.g., in Figure 4.3, *optical-axis* and *step-shaft*, *rotating-shaft* and *driven-shaft* have the same path length, but the sum of the level numbers of the former is bigger than the later, so the similarity of the former is bigger than the later. In the equation, h_i

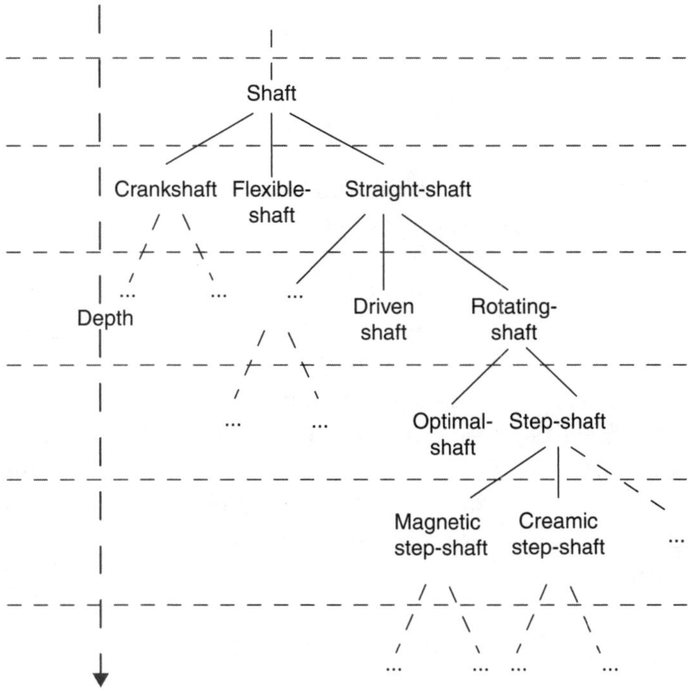

Figure 4.3 Hierarchical relationship of shaft ontology.

and h_j are the depth of two compared word concepts, Aw_i and Bw_j, respectively, and $h = |h_i - h_j|$ and $f_h(h)$ denote the transfer function of *depth;* $f_h(h)$ is a monotonically decreasing function with respect to depth h, and $f_h(h) = e^{-h} = e^{-|h_i - h_j|} (0 \le h_i;\ 0 \le h_j)$. The function curve of $f_h(h)$ is shown in Figure 4.4(c).

Therefore, the similarity, $\text{Match}_w(Aw_i, Bw_j)$, between Aw_i and Bw_j is a function of the attributes p, r and h, as follows:

$$\text{Match}_w(Aw_i, Bw_j) = f(p,r,h) = u_p^{-f_p(p)} \times u_r^{-f_r(r)} \times u_h^{-f_h(h)} \quad (4.1)$$

Where:

$$\begin{cases} f_p(p) = e^{-p} & (0 \le p) \\ f_r(r) = 0 & (r = 0) \\ f_r(r) = 1 - e^{-r} & (0 < r) \\ f_h(h) = 1 & (h = |h_i - h_j| = 0) \\ f_h(h) = e^{-h} = e^{-|h_i - h_j|} & (0 \le h = |h_i - h_j|) \\ u_p, u_r, u_h \in [1, \infty) \end{cases} \quad (4.2)$$

u_p, u_r and u_h are the scaling factors of $f_p(p)$, $f_r(r)$ and $f_h(h)$, respectively. They are the decisive factors to a faster approach to 100% similarity 1. At the same time, u_p, u_r and u_h can be seen as the weights of $f_p(p)$, $f_r(r)$ and $f_h(h)$ respectively, occupy in the whole similarity between Aw_i and Bw_j, or the respective influencing degree to the whole similarity. E.g., the bigger the u_p is, the faster the value of $f_p(p)$ approaching to 1 (as shown in Figure 4.4[d]), and the bigger weight of *path length, p,* occupies in the whole similarity. The same to $f_h(h)$ and *depth, h,* (as shown in Figure 4.4[f]). Contrarily, the bigger the u_r, the slower the value of $f_r(r)$ approaching to 1 (as shown in Figure 4.4[e]), and the smaller weight of overlap occupies in the whole similarity.

4.4.2 Sentence Matching Algorithms (SeMAs)

Provided that At_i is a sentence describing information of a requested resource service, and Bt_j is that of an advertising resource service. The similarity between At_i and Bt_j is decided by *key words, sentence length,* and *word order*. Set T_i and T_j are the word set of sentence At_i and Bt_j

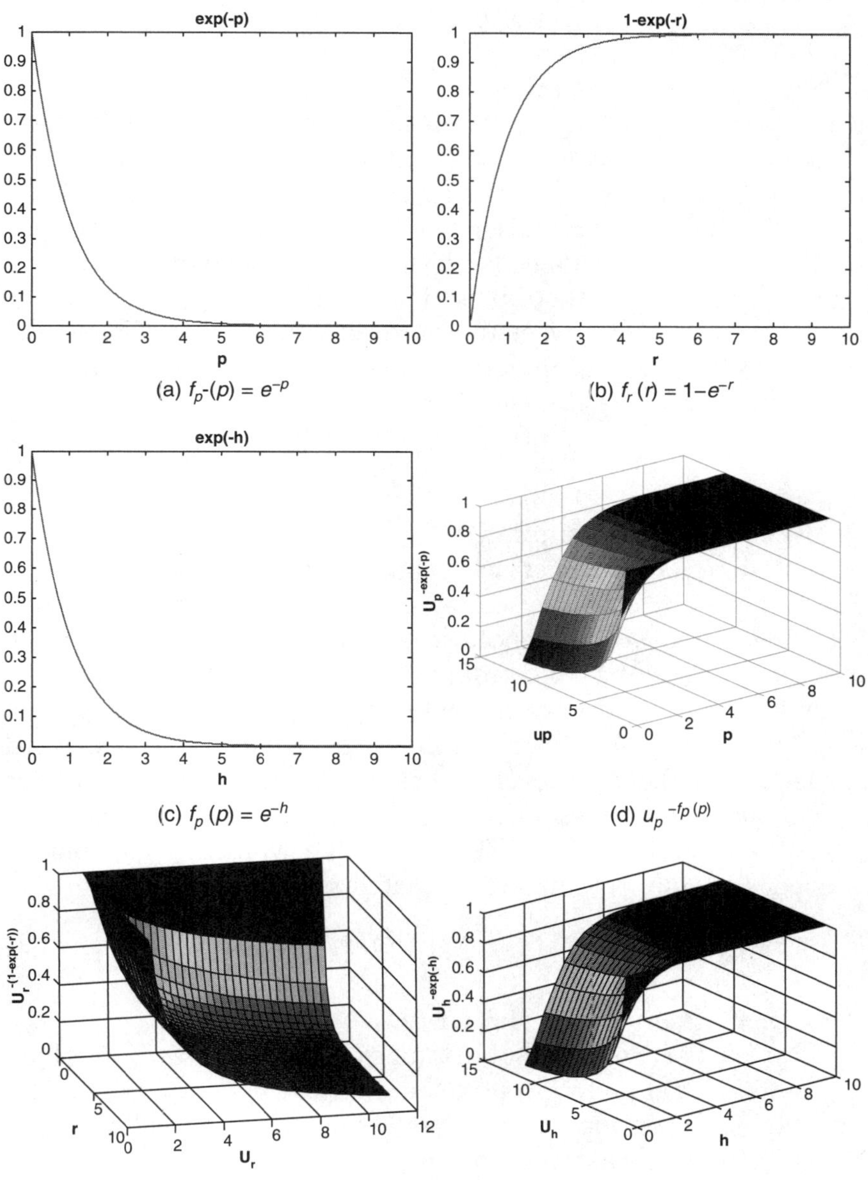

(a) $f_p\text{-}(p) = e^{-p}$

(b) $f_r(r) = 1-e^{-r}$

(c) $f_p(p) = e^{-h}$

(d) $u_p{}^{-f_p(p)}$

(e) $u_r{}^{-f_r(r)}$

(f) $u_h{}^{-f_h(h)}$

Figure 4.4 Function curves of word concept similarity.

respectively,and $W^{\cup} = T_i \cup T_j = \{w_1^{\circ}, w_2^{\circ}, \cdots, w_m^{\circ}\}$ $(m = 1,2,3,\cdots)$ denotes the union set of T_i and T_j, $W^{\cap} = T_i \cap T_j = \{w_1^{*}, w_2^{*}, \cdots, w_n^{*}\}$ $(n = 1,2,3,\cdots)$ denotes the intersection set of T_i and T_j. Obviously, $m \geq n$. The respective similarity measuring methods of *key words, sentence length,* and *word order* are described as follows (Tao *et al.*, 2009b):

(1) ***Key Words* similarity**

Let $s_{i,k}$ and $s_{j,k}$ $(0 \leq s_{i,k}, s_{j,k} \leq 1)$ denote the strongest similarity of each word w_k° $(0 < k \leq m)$ in $W^{\cup}$ within the corresponding words in T_i and T_j, respectively. If w_k° appears in the sentence T_i (or T_j), $s_{i,k}$ (or $s_{j,k}$) is set to 1. Otherwise, if w_k° is not contained in T_i (or T_j), a similarity score is calculated between w_k° and each word in sentence T_i (or T_j) using the method in Section 4.4.1. Then $s_{i,k}$ (or $s_{j,k}$) is the strongest similarity between the words in sentence T_i (or T_j) and w_k°. Therefore, the *key word* similarity, $S_1(At_i, Bt_j)$, between At_i and Bt_j can be formulated as (Li *et al.*, 2006a) :

$$S_1(At_i, Bt_j) = 1 - \left| \sum_{k=1}^{m} s_{i,k} - \sum_{k=1}^{m} s_{j,k} \right| / \left(\sum_{k=1}^{m} s_{i,k} + \sum_{k=1}^{m} s_{j,k} \right) \qquad (4.3)$$

(2) ***Word order* similarity**

A unique index number is assigned for each word in $W^{\cup}$. The index number is the corresponding order number in $W^{\cup}$. Set $O_{rder} = \{1,2,3,\cdots,m\}$ denote the order number vector set. Correspondingly, the word order vector O_{T_i} and O_{T_j} are formulated to denote the word order in sentence T_i and T_j, respectively. Let $|O_{T_i}|$ and $|O_{T_j}|$ denote the absolute-value of the sum of backward sequence numbers of every two neighboring order numbers in O_{T_i} and O_{T_j} , respectively. E.g., if O_{T_i} = {1,2,4,3,3,5}, then $|O_{T_i}| = |1 + 2 + (-1) + 0 + 2| = 0$. Therefore, the word order similarity between At_i and Bt_j can be formulated as:

$$S_2(At_i, Bt_j) = 1 - \left| |O_{T_i}| - |O_{T_j}| \right| / \left(|O_{T_i}| + |O_{T_j}| \right) \qquad (4.4)$$

(3) ***Sentence length* similarity**

Sentence length is defined as the number of the total words involved in a sentence. Let l_i and l_j denote the *sentence length* of At_i and Bt_j, respectively. Then the *sentence length* similarity between At_i and Bt_j can be formulated as $S_3(At_i, Bt_j) = 1 - |l_i - l_j| / (l_i + l_j)$. In general, when the *key words* similarity and *word order* similarity are fixed, the bigger the *sentence length* similarity value is, and the stronger of similarity between two compared sentences is.

During the practical sentence similarity evaluation process, *key words* similarity plays the highest role, and *word order* and *sentence length* play a secondary function. Therefore, different weights should be assigned to *key words* similarity, *word order* similarity, and *sentence length* similarity. Let α, β, δ $(0 \le \alpha, \beta, \delta \le 1, \alpha + \beta + \delta = 1)$ be the corresponding weights respectively. Therefore, the total similarity of sentences At_i and Bt_j is formulated as:

$$\text{Match}_T(At_i, Bt_j) = \alpha \times S_1(At_i, Bt_j) + \beta \times S_2(At_i, Bt_j) + \delta \times S_3(At_i, Bt_j)$$

$$= \alpha\left(1 - \frac{\left|\sum_{k=1}^{m} s_{i,k} - \sum_{k=1}^{m} s_{j,k}\right|}{\sum_{k=1}^{m} s_{i,k} + \sum_{k=1}^{m} s_{j,k}}\right) + \beta\left(1 - \frac{\left|\left|O_{T_i}\right| - \left|O_{T_j}\right|\right|}{\left|O_{T_i}\right| + \left|O_{T_j}\right|}\right) + \delta\left(1 - \frac{\left|l_i - l_j\right|}{l_i + l_j}\right) \quad (4.5)$$

4.4.3 Number Matching Algorithms (NMAs)

When matching two MGrid resource services, some number parameters cannot be avoided. Number describing information of MGrid resource services can be classified into two subclasses: (a) *numerical interval,* and (b) *fuzzy number*. The former, *numerical interval,* is used to describe a specific value or value range, such as cost, and date of delivery. The later, *fuzzy number,* is used to describe some uncertain parameters, such as fuzzy grades, fuzzy evaluations, etc. The specific similarity measuring method for numerical interval parameters and fuzzy number parameters are different.

(1) *Numerical interval* similarity

The task of numerical interval matching is to calculate the matching degree of numerical interval parameters between the requested and advertising resource service. E.g., a resource service demander (RSD) asks for a resource service within a cost from 100 dollars to 200 dollars, while the price of an advertising resource service (RSP) is higher than 120 dollars and lower than 200 dollars. So what is the matching degree between the RSD and RSP? Whether the RSD will accept the RSP to execute its task is decided by the similarity between their costs.

When calculating the similarity of number parameters, number information usually can be transformed into numerical interval. E.g., the above mentioned cost (i.e., from 100 dollars to 200 dollars) asked by the RSD can be denoted as interval [100, 200], and the price (i.e., higher than 120 dollars and lower than 200 dollars) of the RSP can be transformed into interval [120, 200]. Therefore, the number matching is transformed into numerical interval matching.

Provided that Ad_i is a number parameter describing information of a requested resource service *A*, and Bd_j is that of an advertising resource service *B*. Let *Ad* and *Bd* denote the transformed corresponding numerical interval about Ad_i and Bd_j respectively. The similarity between Ad_i and Bd_j is formulated as:

$$\text{Match}_{\text{D}}(Ad_i, Bd_j) = \begin{cases} 1 & (Ad \cap Bd = Ad, or, Ad \cap Bd = Bd) \\ \dfrac{|Ad \cap Bd|}{|Ad|} & (Ad \cap Bd \neq \Phi, and, Ad \cap Bd \neq Ad) \\ 0 & (Ad \cap Bd = \Phi) \end{cases} \tag{4.6}$$

Where '$|\ |$'denotes the length of the corresponding numerical interval. E.g., $|(10,20)| = 10$, and $|(10,\infty)| = \infty$.

(2) *Fuzzy number* similarity

In real manufacturing activities, situations are very often uncertain. Fuzzy numbers and fuzzy expressions are often used to resolve the uncertain problems existing in manufacturing activities, such as linguistic variables, knowledge expression, control system, database, decision making system, failure diagnosis, etc. E.g., fuzzy evaluations to a certain index during manufacturing activities usually are described as {*bad, middle, general, good, excellent*} or {*poor, middle, high*}. All these linguistic variables or parameters cannot be read and understood by machine and computer, so they are often transformed into fuzzy numbers. E.g., the fuzzy evaluation {*poor, middle, high*} can be represented by fuzzy number {0.2, 0.6, 0.8}. In order to enable the description of uncertain factors and parameters to be easily read and understood by the machine in an MGrid system, uncertain factor parameters are represented using fuzzy numbers, including *triangular fuzzy numbers* and *trapezoidal fuzzy numbers*. Hence, when the system searching and matching resource service according to the user's requirements, the problem of fuzzy number matching must be addressed. In this work, *triangular fuzzy numbers* and *trapezoidal fuzzy numbers* are primarily used to describe the uncertain factors and parameters of resource service information. The matching methods and algorithms between fuzzy numbers are investigated in the following sections, including *triangular fuzzy numbers* matching algorithms (TriFNMAs) and *trapezoidal fuzzy numbers* matching algorithms (TraFNMAs).

Before describing the matching algorithms of fuzzy numbers, the concept of *triangular fuzzy numbers* and *trapezoidal fuzzy numbers* will be briefly introduced:

Definition 4.3 – **Triangular fuzzy number.** A triangular fuzzy number is a fuzzy number represented with three points as follows (Lee, 2005b):

$$\tilde{A} = (m, \alpha, \beta)$$

This representation is interpreted as membership functions (Figure 4.5).

$$\mu_{\tilde{A}}(x) = \begin{cases} 0 & , \quad x < m - \alpha \\ (x - m + \alpha)/\alpha & , \quad m - \alpha \le x < m \\ (m + \beta - x)/\beta & , \quad m \le x < m + \beta \\ 0 & , \quad x \ge m + \beta \end{cases} \tag{4.7}$$

Definition 4.4 – **Trapezoidal fuzzy number.** A trapezoidal fuzzy number is defined as (Lee, 2005b):

$$\tilde{A} = (m_1, m_2, \alpha, \beta)$$

The membership function of this fuzzy number is interpreted as follows (Figure 4.6):

$$\mu_{\tilde{A}}(x) = \begin{cases} 0 & , \quad x < m_1 - \alpha \\ (x - m_1 + \alpha)/\alpha & , \quad m_1 - \alpha \le x < m_1 \\ 1 & , \quad m_1 \le x < m_2 \\ (m_2 + \beta - x)/\beta & , \quad m_2 \le x < m_2 + \beta \\ 0 & , \quad x \ge m_2 + \beta \end{cases} \tag{4.8}$$

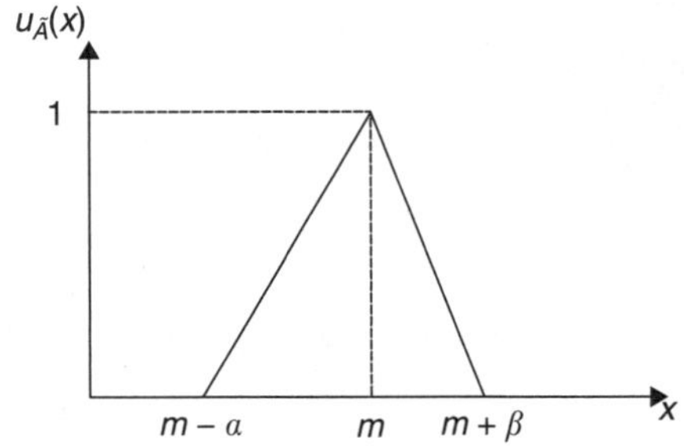

Figure 4.5 Triangular fuzzy number $\tilde{A} = (m, \alpha, \beta)$.

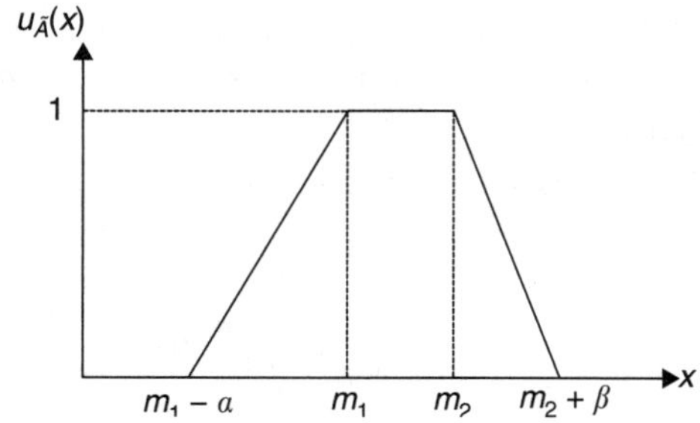

Figure 4.6 Trapezoidal fuzzy number $\tilde{A} = (m_1, m_2, \alpha, \beta)$.

Set Ad_i is a fuzzy number parameter describing information of a requested resource service *A*, and Bd_j is that of an advertising resource service *B*, then the similarity between Ad_i and Bd_j is formulated as (Yang *et al.*, 2005a):

$$\text{Match}_D(Ad_i, Bd_j) = \begin{cases} 1 & (Ad_i = Bd_j) \\ \exp(-d^2(Ad_i, Bd_j)/\sigma) & (Ad_i \neq Bd_j) \end{cases} \tag{4.9}$$

Where σ is a constant bigger than 0, and the corresponding variables in the above function are as follows:

1) If Ad_i and Bd_j are triangular fuzzy numbers, and $Ad_i = (m_i, \alpha_i, \beta_i)$, $Ad_j = (m_j, \alpha_j, \beta_j)$, then

$$\begin{cases} \ell = \int_0^1 L^{-1}(w)dw, \\ r = \int_0^1 R^{-1}(w)dw \\ D_* = |(m_i - \ell\alpha_i) - (m_j - \ell\alpha_j)| \\ D^* = |(m_i + r\beta_i) - (m_j + r\beta_j)| \\ d^2 = (m_i - \alpha_j)^2 + (D_*)^2 + (D^*)^2 \\ \sigma = (D_* + D^*)/2 + (|(m_i - \alpha_i) - (m_j - \alpha_j)| + |(m_i + \beta_i) - (m_j + \beta_j)|)/2^3 \end{cases} \tag{4.10}$$

2) If Ad_i and Bd_j are trapezoidal fuzzy *numbers*, and $Ad_i = (m_{1,i}, m_{2,i}, a_i, \beta_i)$; $Ad_j = (m_{1,j}, m_{2,j}, a_j, \beta_j)$, then

$$\begin{cases} \ell = \int_0^1 L^{-1}(w)dw, \\ r = \int_0^1 R^{-1}(w)dw \\ D_* = \left|(m_{1,i} - \ell\alpha_i) - (m_{1,j} - \ell\alpha_j)\right| \\ D^* = \left|(m_{2,i} + r\beta_{2,i}) - (m_{2,j} + r\beta_{2,j})\right| \\ d^2 = ((m_{1,i} - m_{1,j})^2 + (m_{2,i} - m_{2,j})^2)/2 + (D_*)^2 + (D^*)^2 \\ \sigma = (D_* + D^*)/2 + (\left|(m_{1,i} - \alpha_i) - (m_{1,j} - \alpha_j)\right| + \left|(m_{2,i} + \beta_i) - (m_{2,j} + \beta_j)\right|)/2^4 \end{cases} \quad (4.11)$$

4.4.4 Entity Class Matching Algorithms (ECMAs)

Set Ac_i is an entity class parameter describing information of a requested resource service A, and Bc_j is that of an advertising resource service B. Let Ac and Bc be the corresponding description sets (i.e., synonym sets, sets of distinguishing features, and sets of the entity class in the semantic neighborhood) of Ac_i and Bc_j, respectively. The similarity between Ac_i and Bc_j is calculated according to Tversky's model (Tversky, 1977) and Rodriguez's similarity functions (Rodriguez and Egenhofer, 2003) combined with set-theory.

$$\text{Match}_C(Ac_i, Bc_j) = \frac{|Ac \cap Bc|}{|Ac \cap Bc| + \alpha(Ac_i, Bc_j)|Ac/Bc| + (1 - \alpha(Ac_i, Bc_j)|Bc/Ac|)} \quad (4.12)$$

Where

$$\alpha(Ac_i, Bc_j) = \begin{cases} \dfrac{depth(Ac_i)}{depth(Ac_i) + depth(Bc_j)} & (depth(Ac_i) \le depth(Bc_j)) \\ 1 - \dfrac{depth(Ac_i)}{depth(Ac_i) + depth(Bc_j)} & (depth(Ac_i) > depth(Bc_j)) \end{cases} \quad (4.13)$$

'$\cap$' is the intersection operation and '$/$' is the difference operation of the set (e.g., $Ac/Bc = Ac\text{-}(Ac \cap Bc)$. '$|\ |$'is the cardinality of a set). α is a function that defines the relative importance of the noncommon characteristics. The function *depth()* corresponds to the shortest path from the entity class point (e.g., the point of entity class Ac_i) to the upper intersection point of entity Ac_i and Bc_j (Rodriguez and Egenhofer, 2003). If Ac_i and Bc_j are the same concepts in the ontology, then $\alpha(Ac_i, Bc_j) = 0.5$.

4.5 RS-Matcher: Resource Service Matcher

Definition 4.5 **-Resource service matching model.** Set A is a requested resource service, and B is an advertising resource service, and the matching model between A and B is defined as follows (Tao *et al.*, 2009b):

$$\text{Match}(A,B) = \prod(\text{Match}_{\text{bas}}(A,B), \text{Match}_{\text{i/o}}(A,B), \text{Match}_{\text{QoS}}(A,B)) \quad (4.14)$$

Where:

- $\text{Match}_{\text{bas}}(A,B)$ denotes *basic-matching*. It is primarily responsible for matching the general information between A and B, such as *ServiceName* and *ServiceDescription*. The detailed methods and algorithms of *basic-matching* are shown in Section 4.5.1.
- $\text{Match}_{\text{i/o}}(A,B)$ is the *I/O-matching* (i.e., inputs and outputs matching). It is primarily responsible for matching the inputs and outputs information between A and B. The detailed methods and algorithms of *I/O-matching* are shown in Section 4.5.2.
- $\text{Match}_{\text{QoS}}(A,B)$ is *QoS-matching*. It is primarily responsible for matching the QoS information between A and B, such as *time, cost, reliability, trust, maintainability,* and *satisfaction*. The detailed methods and algorithms of *QoS-matching* are shown in Section 4.5.3.
- $\prod()$ is a comprehensive processing function. It integrates the matching results of *basic- matching, I/O-matching,* and *QoS-matching,* and generates an integrated matching results.

The matching process of MGrid resource service is a stepwise precision process, and it primarily has four steps:

- First, matching the basic information, such as *serviceName* and *serviceDescription*, of A and B according to the *basic-matching* methods and algorithms described in Section 4.5.1. If the result of *basic-matching,* i.e., $\text{Match}_{\text{bas}}(A,B)$, is lower than the *basic threshold value,* $\zeta_{bas} (0 \le \zeta_{bas} \le 1)$, which is set by the user or system, then it illustrated that the advertising resource service cannot satisfy the general requirements of the user. Then the system stops further matching and returns the failed message to the user.

- Second, if *basic-matching* result arrives ζ_{bas} and the user requires further precision matching, then the system carries out *I/O-matching* according to the methods and algorithms described in Section 4.5.2. If the result of *I/O-matching*, i.e., $\text{Match}_{\text{i/o}}(A,B)$, is lower than the *I/O threshold value*, $\zeta_{i/o}(0 \leq \zeta_{i/o} \leq 1)$, which is set by user or system, then it illustrated that the advertising resource service is unqualified for the user's I/O requirements. Then the system stops further matching and returns the failed message to user.
- Third, if the advertising resource service qualified for the user's I/O requirements and further precision matching is required, then the system carries out *QoS-matching* according to the methods and algorithms described in Section 4.5.3. If the result of *QoS-matching*, i.e., $\text{Match}_{\text{QoS}}(A,B)$, is lower than the *QoS threshold value*, $\zeta_{QoS}(0 \leq \zeta_{QoS} \leq 1)$, which is set by user or system, then it illustrated that the advertising resource service is unqualified for the user's *QoS* requirements. Then the system stops further matching and returns the failed message to user.
- Fourth, if the advertising resource service is qualified for all the requirements (i.e., arrived at all the *threshold values* of ζ_{bas}, $\zeta_{i/o}$ and ζ_{QoS}), then the system calculates the final overall matching according to the methods and algorithms described in Section 4.5.4. If the integrated matching result is equal to or over the total *threshold value*, ζ $(0 \leq \zeta \leq 1)$, then it concludes that the advertising resource service satisfied the requirements of the user very well, and it can be selected to execute the requested task.

The matching process flowchart is shown in Figure 4.7. The *basic-matching*, *I/O-matching*, and *QoS-matching* can be used separately or together, which is decided by the practical requirements and conditions of the user or system.

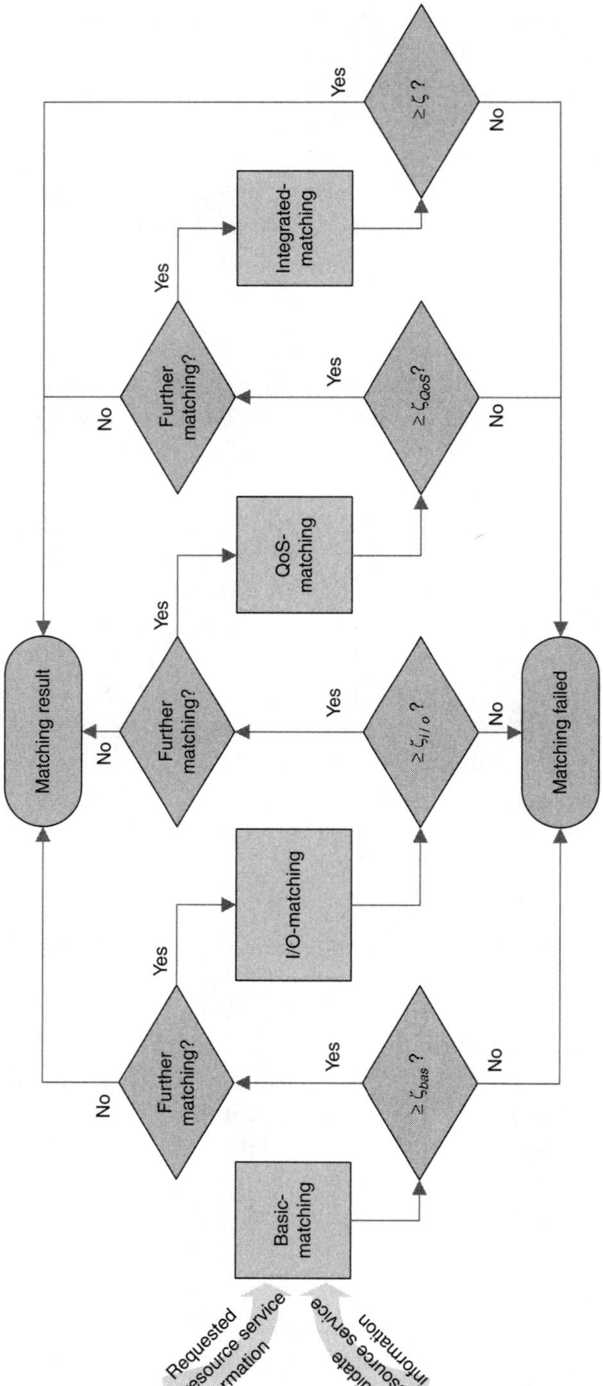

Figure 4.7 Four- steps matching process flowchart of MGrid resource service.

4.5.1 Basic-matching

Basic-matching is primarily responsible for matching the general information between resource services, such as *ServiceName* and *ServiceDescription*. Because *ServiceName* and *ServiceDescription* usually are described with words or sentences, the system primarily employs the proposed *word concept matching algorithms (WMAs)* and *sentence matching algorithms* (SeMAs) to calculate *basic-matching* as follows:

$$\text{Match}_{\text{bas}}(A,B)=\begin{pmatrix}\omega_1\times\text{Match}_{\text{w}}(A.ServiceName,\ B.ServiceName)+\\ \omega_2\times\text{Match}_{\text{T}}(A.ServiceDescription,\ B.ServiceDescription)\end{pmatrix} \quad (4.15)$$

For $0\le\omega_1,\omega_2\le1$ and $\omega_1+\omega_2=1$. ω_1 and ω_2 are the weights of *ServiceName* and *ServiceDescription* respectively.

Set *A. General* is the general describing information set of requested resource service *A*, and *B. General* is that of advertising resource service *B*. Let ζ_{bas} is the *basic-matching* threshold value declared by *A* (i.e., if an advertising resource service wants to execute the task of *A*, the *basic-matching* value between it and *A* must be equal to or over ζ_{bas}, otherwise it will be an also-ran). According to Expression 4.15, the pseudocodes of algorithms for basic matching between *A* and *B* are as follows:

Inputs: ***A*. General ; *B*. Genera;** ζ_{bas}
Outputs: resource service basic-matching results, $\text{Match}_{\text{bas}}(A,B)$

1. Calculating Match_{w} (*A.ServiceName, B. ServiceName*) according to Expression 4.1
2. Calculating Match_{T} (*A.ServiceDescription, B. ServiceDescription*) according to Expression 4.5
3. Calculating $\text{Match}_{\text{bas}}(A,B)$ according to Expression 4.15
4. If ($\text{Match}_{\text{bas}}(A,B)<\zeta_{bas}$)
5. *B* is unqualified for *A*'s general requirements, return failed matching message
6. End if
7. *B* is qualified for *A*, return successful message and matching result $\text{Match}_{\text{bas}}(A,B)$

4.5.2 I/O-matching

As stated in the above section, *I/O matching* is primarily responsible for matching the inputs and outputs information between requested and advertising resource services. It assumes that the parameters set of outputs and inputs of a resource service is $D \cup C \cup W$, and D, C, W denote the subsets of number parameters, entity class parameters, and word concept parameters, respectively. The matching function of *I/O-matching* is defined as follows:

$$\mathrm{Match}_{i/o}(A,B) = \left(\begin{array}{l} \sum_{i=1}^{N_d} \omega_{d_i} \mathrm{Match}_D(A.d_i, B.d_i) + \\ \sum_{j=1}^{N_c} \omega_{c_j} \mathrm{Match}_C(A.c_j, B.c_j) + \\ \sum_{k=1}^{N_w} \omega_{w_k} \mathrm{Match}_W(A.w_k, B.w_k) \end{array} \right) \tag{4.16}$$

Where $d_i (d_i \in D)$, $c_j (c_j \in C)$, $w_k (w_k \in W)$ denotes the $i^{th} (i = 1,2,3,\cdots,N_d)$, $j^{th} (j = 1,2,3,\cdots N_c)$, $k^{th} (k = 1,2,3,\cdots,N_w)$ parameter in D, C, W respectively. $\omega_{d_i}, \omega_{c_j}, \omega_{w_k}$ are the corresponding weights of d_i, c_j, w_k respectively, $0 \le \omega_{d_i}, \omega_{c_j}, \omega_{w_k} \le 1$ and $\sum_{i=1}^{N_d} \omega_{d_i} + \sum_{j=1}^{N_c} \omega_{c_j} + \sum_{k=1}^{N_w} \omega_{w_k} = 1$.

Set *A.Inputs* and *A.Outputs* are the inputs and outputs describing information sets of requested resource service *A*. I/O parameter set of *A* can be classified into three different subsets, W_A, D_A, C_A (the intersection of any two of W_A, D_A, C_A is empty), which denote the subsets of word concept parameters, number parameters and entity class parameters, respectively. Accordingly, Let W_B, D_B, C_B (the intersection of any two of W_B, D_B, C_B is empty too) be the subsets of word concept parameters, number parameters and entity class parameters of *B*'s I/O parameters set. Let $\zeta_{i/o}$ be the I/O matching threshold value declare d by *A*. According to Expression 4.16, the pseudocodes of *I/O-matching* algorithms between *A* and *B* are as follows:

Inputs: W_A, D_A, C_A ; W_B, D_B, C_B ; $\zeta_{i/o}$

Outputs: resource service I/O matching results $\mathrm{Match}_{i/o}(A,B)$

1. If W_B is not empty
2. For each $B.w_k$ in W_B and $A.w_k$ in W_A
3. Calculating $\mathrm{Match}_W(A.w_k, B.w_k)$ according to Expression 4.1
4. End for
5. End if
6. If D_B is not empty
7. For each $B.d_i$ in D_B and corresponding $A.d_i$ in D_A
8. If $B.d_i$ is a *number interval* parameter
9. Calculating $\mathrm{Match}_D(A.d_i, B.d_i)$ according to Expression 4.6
10. End if
11. If $B.d_i$ is a *triangular fuzzy number* parameter
12. Calculating $\mathrm{Match}_D(A.d_i, B.d_i)$ according to Expressions 4.9 and 4.10
13. End if
14. If $B.d_i$ is a *trapezoidal fuzzy number* parameter
15. Calculating $\mathrm{Match}_D(A.d_i, B.d_i)$ according to Expressions 4.9 and 4.11
16. End if
17. End for
18. End if
19. If C_B is not empty
20. For each $B.c_j$ in C_B and corresponding $A.c_i$ in C_A
21. Calculating $\mathrm{Match}_C(A.c_j, B.c_j)$ according to Expression 4.12
22. End for
23. End if
24. Calculating $\mathrm{Match}_{i/o}(A,B)$ according to Expression 4.16
25. If $\mathrm{Match}_{i/o}(A,B) < \zeta_{i/o}$
26. *B* is unqualified for *A*'s I/O requirements, return failed matching message
27. End if
28. *B* is qualified for *A*'s I/O requirements, return successful message and $\mathrm{Match}_{i/o}(A,B)$

4.5.3 QoS-matching

As previously defined, *QoS-matching* is primarily responsible for matching the QoS information between resource services, such as *time, cost, reliability, trust, maintainability,* and *satisfaction*. Because QoS parameters are main number parameters, the *QoS-matching* primarily depends on number matching algorithms (NMAs). Therefore, *QoS matching* can be formulated as follows (Tao *et al.*, 2009b):

$$\mathrm{Match}_{\mathrm{QoS}}(A,B)=\sum_{i=1}^{m}\omega_{\mathrm{QoS}_i}\mathrm{Match}_{\mathrm{D}}(A.\mathrm{QoS}_i,B.\mathrm{QoS}_i) \quad (4.17)$$

For $(i=1,2,3,\cdots,m), 0\le\omega_{\mathrm{QoS}_i}\le 1$ and $\sum_{i=1}^{m}\omega_{\mathrm{QoS}_i}=1$. Where ω_{QoS_i} is the weight of each corresponding QoS parameter and m is the total number of QoS parameters.

Set *A.QoS* is the general describing information set of requested resource service *A*, and *B. QoS* is that of advertising resource service*B*. Let ζ_{QoS} be the *QoS-matching* threshold value declared by *A*. According to Expression 4.17, the pseudocodes of algorithms of *QoS-matching* between *A* and *B* are as follows:

Inputs: $A.QoS, B.QoS, \zeta_{QoS}$
Outputs: resource service I/O matching results $\mathrm{Match}_{QoS}(A,B)$

```
For each QoS parameters A.QoS_i in A.QoS and B.QoS_i in
B.QoS
If A.QoS_i and B.QoS_i are number interval parameters
Calculating Match_D(A.d_i, B.d_i) according to Expression
4.6
End if
If A.QoS_i and B.QoS_i are trapezoidal fuzzy number
parameters
Calculating Match_D(A.d_i, B.d_i) according to Expressions
4.9 and 4.10
End if
If A.QoS_i and B.QoS_i are number interval parameters
Calculating Match_D(A.d_i, B.d_i) according to Expressions
4.9 and 4.11.
End if
```

End for
Calculating $\text{Match}_{QoS}(A,B)$ according to Expression 4.17
If $\text{Match}_{QoS}(A,B) < \zeta_{QoS}$
B is unqualified to *A*'s QoS requirements, return matching failed message
End if
B is qualified for *A*'s QoS requirements, return successful message and $\text{Match}_{QoS}(A,B)$

4.5.4 Integrated-matching

Practically speaking, there is far more than one advertising resource service for a requested resource service or task. The system has to search and match the qualified resource from a mass of potential resource service sets according to user requirements. So the searching and matching process is not a simple comparison between two resource services, but a very complicated process.

Let the requested resource service (or task) describing information be *A*. The declared matching threshold values are ζ_{bas} (i.e., the basic-matching threshold), $\zeta_{i/o}$ (i.e., the I/O-matching threshold), ζ_{QoS} (i.e., the QoS-matching threshold), and ζ (i.e., the integrated-matching threshold). The requested general information, inputs information, outputs information, and QoS information of *A* are A_j*.Gereral*, A_j*.Inputs*, A_j*.Outputs*, and $A.QoS_i$, respectively. W_A, D_A, C_A are the corresponding subsets of word concept parameters, number parameters, and entity class parameters of I/O parameters set. The intersection of any two of W_A, D_A, C_A is empty.

It is assumed that there are $m(m = 1,2,3\cdots)$ potential advertising resource services for selection and the advertising resource service set is $B = \{B_1, B_2, B_3, \cdots, B_m\}$. Let B_j*.Gereral*, B_j*.Inputs*, B_j*.Outputs*, B_j*.QoS* be the corresponding description of an arbitrary resource service, B_j, in *B*. W_B, D_B, C_B are the corresponding subsets of word concept parameters, number parameters, entity class parameters of B_j's I/O parameters set.

According to above matching methods and algorithms, the comprehensive searching and matching algorithms are as follows (Tao *et al.*, 2009b):

Inputs: *A.Gereral, A.Inputs, A.Outputs, A.QoS*, W_A, D_A, C_A, ζ_{bas}, $\zeta_{i/o}$, ζ_{QoS}, ζ, W_B, D_B, C_B, B_j.*Gereral*, B_j.*Inputs*, B_j.*Outputs*, B_j.*QoS*, W_B, D_B, C_B;
Outputs: Qualified resource service set

1. For each B_j in *B*

Basic-matching:

2. Calculating Match_w (*A.ServiceName*, B_j. *ServiceName*)according to Expression 4.1
3. Calculating Match_T (*A.ServiceDescription*) B_j . *ServiceDescription*) according to Expression 4.5
4. Calculating Match_{bas} (A, B_j) according to Expression 4.15
5. If Match_{bas} $(A, B_j) < \zeta_{bas}$)
6. Delete B_j from *B*, return failed matching message and back to Step 1
7. End if
8. Record basic-matching result Match_{bas} (A, B_j), and execute the following I/O-matching

I/O-matching:

9. Classify the I/O parameters set of B_j into three subsets, $W_{B'}$, $D_{B'}$, $C_{B'}$, which denote the corresponding subsets of word concept parameters, number parameters, and entity class parameters, respectively.
10. If W_B is not empty
11. For each $B_j.w_i$ in W_B and corresponding $A.w_i$ in W_A
12. Calculate Match_w $(A.w_i, B_j.w_i)$ according to Expression 4.1
13. End for
14. End if
15. If D_B is not empty
16. For each $B_j.d_i$ in D_B and corresponding $A.d_i$ in D_A
17. If $B_j.d_i$ is a *number interval* parameter
18. Calculate Match_D $(A.d_i, B_j.d_i)$ according to Expression 4.6

19. End if
20. If $B_j.d_i$ is a *triangular fuzzy number* parameter
21. Calculate Match_D $(A.d_i, B_j.d_i)$ according to Expressions 4.9 and 4.10
22. End if
23. If $B_j.d_i$ is a *trapezoidal fuzzy number* parameter
24. Calculate Match_D $(A.d_i, B_j.d_i)$ according to Expressions 4.9 and 4.11
25. End if
26. End for
27. End if
28. If C_B is not empty
29. For each $B_j.c_i$ in C_B and corresponding $A.c_i$ in C_A
30. Calculate Match_c $(A.c_i, B_j.c_i)$ according to Expression 4.12
31. End for
32. End if
33. Calculate $\text{Match}_{i/o}(A, B_j)$ according to Expression 4.16
34. If $\text{Match}_{i/o}(A, B_j) < \zeta_{i/o}$
35. Delete B_j from B, return failed matching message and back to Step 1
36. End if
37. Record the matching result of $\text{Match}_{i/o}(A, B_j)$ and execute the following QoS-matching

QoS-matching:

38. For each QoS parameters $A.QoS_i$ in $A.QoS$ and corresponding $B_j.QoS_i$ in $B_j.QoS$
39. If $A.QoS_i$ and $B_j.QoS_i$ are *number interval* parameters
40. Calculate $\text{Match}_D(A.QoS_i, B_j.QoS_i)$ according to Expression 4.6
41. End if
42. If $A.QoS_i$ and $B_j.QoS_i$ are *trapezoidal fuzzy number* parameters
43. Calculate $\text{Match}_D(A.QoS_i, B_j.QoS_i)$ according to Expressions 4.9 and 4.10

44. End if
45. If $A.QoS_i$ and $B_j.QoS_i$ are *number interval* parameters
46. Calculating $\text{Match}_D(A.QoS_i, B_j.QoS_i)$ according to Expressions 4.9 and 4.11
47. End if
48. End for
49. Calculating $\text{Match}_{QoS}(A, B_j)$ according to Expression 4.17
50. If $\text{Match}_{QoS}(A, B_j) < \zeta_{QoS}$
51. Deleting B_j from B, return matching failed message and back to Step 1
52. End if
53. Memorizing the matching the result $\text{Match}_{QoS}(A, B_j)$ and executing the following integrated matching

Integrated matching:

54. Calculating $\text{Match}(A, B_j)$ according to Expression 4.14
55. If $\text{Match}(A, B_j) < \zeta$
56. Deleting B_j from B, return matching failed message and back to Step 1
57. End if
58. Returning the result $\text{Match}(A, B_j)$ and back to Step 1
59. End for

4.6 Case Study

Suppose the requested describing information (A) for a parameterized radial magnetic design service on our experimental prototype platform, MBRSSP-MGrid (Tao *et al.*, 2008a), is shown on the left in Table 4.1. The declared threshold values are $\zeta_{bas} = 0.85$, $\zeta_{i/o} = 0.75$, $\zeta_{QoS} = 0.80$ and $\zeta = 0.80$. An available advertising resource service's corresponding describing information is shown on the right in Table 4.1.

Table 4.1 Brief Describing Information of Two Compared Resource Services Based on OWL-S: On the left is the requested information by a user and on the right is the corresponding describing information of an advertising resource service. The describing information is processed for the sake of illustration

Requested resource service (*A*)	**Available advertising resource service (*B*)**
A{	B{
General: {*ServiceName*{Radial Magnetic bearing parameterized design Service}, *ServiceDescription*{Providing radial magnetic bearing parameterized design service according to the parameters and QoS requirements submitted by user}}	**General:** {*ServiceName* {Radial Magnetic bearing parameterized design Service'}, *ServiceDescription* {Providing client with parameterized design service for radial magnetic bearing according to the submitted parameters and QoS requirements}}
Inputs: {{0.40,0.30,0.30}, {order}, {ParameterList {magnetic material, conducting wire, air gap, rotor diameter, static bearing capacity, bias, magnetic flux intensity}}, {3D_softwares}, {0.20,030,0.10,0.10}}	**Inputs:** { {0.40,0.10,0.10},{order}, {ParameterList {conducting wire, air gap, rotor diameter, static load capacity, bias, magnetic flux intensity}}, {3D_softwares},{0.55,0.55,0.15,0.15}}
Outputs: {3D picture, 2D picture}	**Outputs:** {3D picture, 2D picture}
QoS: { *C*{140, 185}, *Trust*{0.30,0.10,0.10}, *Ma*{0.25,0.25,0.15,0.15}{},{},{},{ }}	**QoS:** { *C*{165, 215}, *Trust*{0.20,0.10,0.10}, *Ma* {0.20,0.30,0.10,0.10}{},{},{},{}}
Precondition: {pay 50% money}	**Precondition**: {{pay 50% money}}
Operation: {{computing and design}}	**Operation:** {{computing and design}}
Effect: {}	**Effect:** {}
}	}

4.6.1 Step 1: Basic-matching

Step 1.1 Calculate the similarity of *A.ServiceName* and *B.ServiceName*.

According to Expression 4.1 and Expression 4.2, Match_w (*A.ServiceName,B. ServiceName*)=1.0000.

Step 1.2 Calculate the similarity of *A. ServiceDescription* and *B. ServiceDescription.*

According to Table 4.1, it is known that *A.ServiceDescription*={Providing radial magnetic bearing parameterized design service according to the parameters and QoS requirements submitted by user} and *B.ServiceDescription* ={Providing client with parameterized design service for radial magnetic bearing according to the submitted parameters and QoS requirements}.

If set *A. ServiceDescription* = $li = At_i$ and *B. ServiceDescription*= $li = Bt_j$, then the union set (i.e., $W^{\cup}$) of At_i and Bt_j is as follows:

$W^{\cup}$ ={Providing radial magnetic bearing parameterized design service according to the parameters and QoS requirements submitted by user client with for}.

The required variables (i.e., $l_i, l_j, O_{T_i}, O_{T_j}, |O_{T_j}|, |O_{T_i}|, \sum_{k=1}^{m} s_{i,k}, \sum_{k=1}^{m} s_{j,k}$) in Expression 4.5 are as follows:

$li = 17$, $l_j = 18$, $OT_i = \{1,2,3,4,5,6,7,8,9,10,11,12,13,14,15,16,17\}$

$OT_j = \{1,18,19,5,6,7,20,2,3,4,8,9,10,15,11,12,13,14\}$

$$|O_{T_i}| = 16,\ |O_{T_j}| = 13,\ \sum_{k=1}^{m} s_{i,k} = 17,\ \sum_{k=1}^{m} s_{j,k} = 18.$$

Set $\alpha = 0.7$, $\beta = 0.2$, $\delta = 0.1$, respectively. According to Expression 4.5 the final similarity between At_i (i.e., *A.ServiceDescription*) and Bt_i (i.e., *B.ServiceDescription*) is calculated as follows:

$$\begin{aligned}\text{Match}_T(At_i, Bt_j) &= \alpha\,(1-\frac{|17-18|}{17+18}) + \beta(1-\frac{|13-16|}{13+16}) + \delta(1-\frac{|17-18|}{17+18}) \\ &= 0.9564\end{aligned}$$

Step 1.3 Calculate the basic-matching degree of *A* and *B*.

Set $w_1 = 0.1$, $w_1 = 0.9$, according to Expression 4.15.

$\text{Match}_{bas}(A,B) = 0.1^* \text{Match}_w$ (*A.ServiceName,B. ServiceName*)+ 0.9* $\text{Match}_T(At_i, Bt_j) = 0.9555$.

4.6.2 Step 2: I/O-matching

Step 2.1 Classify the parameters of I/O into D, C, W

Let D_A, C_A, W_A be the subsets of number parameters, entity class parameters, and word concept parameters of the I/O parameters of *A*, and D_B, C_B, W_B be that of *B*. Then

D_A = {{0.40,0.30,0.30},{0.20,030,0.10,0.10}}.

W_A = {{order}, {3D_softwares},{3D picture},{ 2D picture}}.

C_A = {{*ParameterList*}}={{magnetic material, conducting wire, air gap, rotor diameter, static bearing capacity, bias, magnetic flux intensity}}.

D_B = {{0.40, 0.10, 0.10}, {0.55, 0.55, 0.15, 0.15}}.

W_B={{order}, {3D_softwares},{3D picture},{ 2D picture}}.

C_B={{*ParameterList*}}={{magnetic material, conducting wire, air gap, rotor diameter, static bearing capacity, bias, magnetic flux intensity}}

Step 2.2 Calculate the similarity of the word concept parameters of I/O.

Let W_A={{order}, {3D_softwares},{3D picture},{ 2D picture}}={ $A.w_1$, $A.w_2$, $A.w_3$, $A.w_4$}

W_B={{order}, {3D_softwares},{3D picture},{ 2D picture}}=($B.w_1$, $B.w_2$, $B.w_3$, $B.w_4$).

According to Expressions 4.1 and 4.2,

$$\text{Match}_W(A.w_1, B.w_1) = \text{Match}_W(A.w_2, B.w_2) = \text{Match}_W(A.w_3, B.w_3) = \text{Match}_W(A.w_4, B.w_4) = 1$$

Step 2.3 Calculate the similarity of the number parameters of I/O.

Let D_A= {{0.40,0.30,0.30},{0.20,030,0.10,0.10}}={$A.d_1$, $A.d_2$ }, and D_B = {{0.40,0.10,0.10},{0.55,0.55,0.15,0.15}={$B.d_1$, $B.d_2$}

Then

- According to Expressions 4.9 and 4.10, $\text{Match}_D(A.d_1, B.d_1) = 0.8752$
- According to Expressions 4.9 and 4.11, $\text{Match}_D(A.d_2, B.d_2) = 0.4444$

Step 2.4 Calculate the similarity of the entity class parameters of the I/O.

Set A. ParameterList = Ac_1 and *B. ParameterList* = Bc_1. According to Expressions 4.12 and 4.13

$Ac \cap Bc$ = {conducting wire, air gap, rotor diameter, bias, magnetic flux intensity}.

Ac / Bc = {magnetic material, static bearing capacity}.

Bc/Ac = {static load capacity}.

$a\,(Ac_1, Bc_1) = a$ (*A. ParameterList, B. ParameterList*) = 0.5000.

Therefore, according to Expressions 4.12 and 4.13, the similarity between Ac_1 and Bc_1 is

$$\text{Match}_C(Ac_1, Bc_1) = \frac{5}{5 + 0.5 \times 2 + 0.5 \times 1} = 0.7692.$$

So the similarity between *A. ParameterList* and *B. ParameterList* is 0.7692.

Step 2.5 Calculate the I/O-matching degree of *A* and *B*.

Let the weight of each I/O parameter be the same, according to Expression 4.16, $\text{Match}_{i/o}(A,B)$ =0.8698.

4.6.3 Step 3: QoS-matching

Step 3.1 Calculate the similarity of each QoS parameter.

From Table 4.1, it is known that:

$A.QoS=\{A.QoS_1, A.QoS_2, A.QoS_3\}$
$=\{C\{140, 185\}, Trust\{0.30,0.10,0.10\}, Ma\{0.25,0.25,0.15,0.15\}\}$.
$B.QoS=\{B.QoS_1, B.QoS_2, B.QoS_3\}$
$=\{C\{165, 215\}, Trust\{0.20,0.10,0.10\}, Ma\ \{0.20,0.30,0.10,0.10\}\}$.

Apparently, $A.QoS_1$ and $B.QoS_1$ are *numerical* interval parameter, $A.QoS_2$ and $B.QoS_2$ are *triangular fuzzy number,* and $A.QoS_3$ and $B.QoS_3$ are *trapezoidal fuzzy number.* Therefore,

- according to Expression 4.6, $\text{Match}_D(A.QoS_1, B.QoS_1)$ $\|[140,185]\| / \|[165,215]\| = 0.9000$,
- according to Expressions 4.9 and 4.10, $\text{Match}_D(A.QoS_2, B.QoS_2) = 0.7866$, and
- according to Expressions 4.9 and 4.11, $\text{Match}_D(A.QoS_3, B.QoS_3) = 0.8607$.

Step 3.2 Calculate the QoS-matching degree of *A* and *B*.

Set $\omega_{QoS_1} = 0.2$, $\omega_{QoS_2} = 0.5$ and $\omega_{QoS_3} = 0.3$, and according to Expression 4.17,

$\text{Match}_{QoS}(A,B)$ = 0.2*0.9000+0.5*0.7866+0.3*0.8607=0.8315.

4.6.4 Step 4: Integrated-matching

According to the above proposed matching method and algorithms in Section 4.5, the system first calculates the *basic-matching* values of *A* and *B*, then calculates the *I/O-matching* value and *QoS-matching* value, and lastly calculates the entire matching value, i.e., *integrated-matching*. The corresponding matching values

Table 4.2 Corresponding matching result between A and B

	Matching Value	Weights	Threshold	Compared with Threshold	Whether Qualified?
Basic-matching	0.9608	0.20	0.85	>	Yes
I/O-matching	0.8698	0.45	0.75	>	Yes
QoS-matching	0.8315	0.35	0.80	>	Yes
Integrated-matching	0.8746	/	0.80	>	Yes

are shown in Table 4.2. From the result shown in Table 4.2, it can be concluded that the advertising service *B* is qualified and can be selected as a candidate resource service.

4.7 Performance Results and Discussion

In order to validate a proposed method, a set of experiments are conducted on the experimental MGrid prototype platform, "magnetic bearing resources sharing and service system under manufacturing grid environment (MBRSSP-MGrid)" (Tao *et al.*, 2008a), which was developed by us. In the MBRSSP-MGrid, all users can publish their idle resources (including equipment resources, software resources, human resources, application resources, technique resources, service resources, etc.) through the *resource service publication center* of MBRSSP-MGrid. Users can also search the resources or services (e.g., *remote parameterized design service of magnetic bearing*) they require via the *resource and service optimal-allocation* center of MBRSSP-MGrid. The experiments and evaluations focused on the accuracy and efficiency of our proposed method.

4.7.1 Accuracy

The described *resource services match and search method* (hereinafter referred to as RSMS in the following experimentations) is compared with *UDDI keyword matching* (hereinafter referred to as UDDI in the following experimentations). In our experiments, a set of 50 resource services were selected for a submitted task, of which 20 services were relevant to the task. *Precision* and *recall,* which are the standard measures that have been used in information retrieval for measuring the accuracy of a search method or search engine, are selected as the criteria to test the accuracy of RSMS. Let N_{Rel} be the set of relevant resource services, N^{Ret} be the set of returned resource services, and $N_{\text{Rel}}^{\text{Ret}}$ be the set of returned relevant resource services. In this work, recall, R_{ecall}, and precision , P_{recision}, are defined as follows (Li *et al.*, 2008a):

$$R_{ecall} = \frac{N^{\text{Ret}}}{N_{\text{Rel}}}, \; P_{\text{recision}} = \frac{N_{\text{Rel}}^{\text{Ret}}}{N^{\text{Ret}}}$$

During the implementations, two conditions are considered: (I) the 50 candidate resource services have formal describing information, without fuzzy describing information, entity class describing information, etc., and at least one of the parameters of each candidate resource service is assigned an *exact* match with the submitted query of task; and (II) without any of the above constraints. Each condition has 10 tests, and each test is executed 10 times, the result of each test is the average of 10 times executions, as shown in Figure 4.8.

It can be concluded from Figure 4.8 that RSMS performed better in the tests both under condition (I) and condition (II). In condition (I), the recall of UDDI can reach 100 percent because of the two constraints. But its precision is lower than RSMS because UDDI only considers keyword matching, and the matching of other kinds of describing information for resource service is ignored. For example, if a candidate resource has five describing parameters, only one is keyword, the other four are fuzzy numbers, entity class, sentence, etc. UDDI only matches 1 parameter, and RSMS matches all 5 parameters. Apparently, the precision of RSMS is higher than UDDI.

Figure 4.8(b) illustrates that UDDI does not reach a recall of 90 percent and 100 percent because of its limitations in resource service matching. For example, if there is no keyword or exact match between a resource service and task, then the matching degree of this resource service is zero when using UDDI. As a result, the resource service cannot be returned as a candidate for selection.

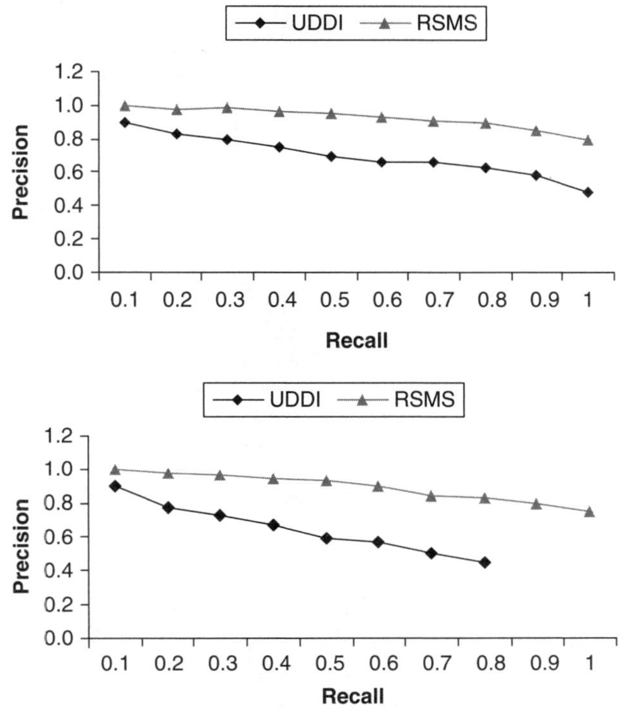

Figure 4.8 Comparisons of Performance between RSMS and UDDI (a) Results under condition (I), (b) Results under condition (II).

4.7.2 Efficiency

In order to measure the *efficiency* of the proposed method, the time between the submission of a batch of tasks and the return of the matched results is measured, which is referred to as *resource service matching response time* (RSMS-*time)* in this work. A batch of 10, 20, and 30 tasks are submitted, and the required resource services for each task vary from 2 to 5. During the matching process, each resource service in RSIC (resource service information center) can only be assigned to one task. A match for a task is successful only if all required resource services of the task are found and matched. The whole matching process stops only when all submitted tasks are matched to related resource services.

The experiments are repeated 20 times, and the *RSMS-time* is the average over the 20 executions. The results shown in Figure 4.9 indicate that the proposed method can handle resource

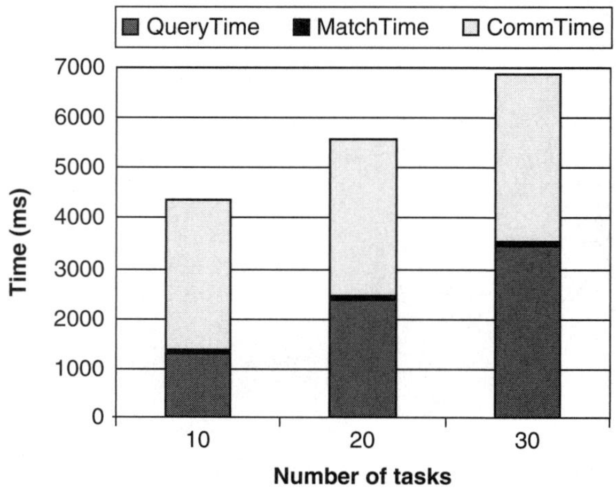

Figure 4.9 Response time for resource services matching.

service matching for tens of tasks in several seconds. In order to test the practical efficiency of the proposed matching algorithms, the *RSMS-time* is broken down into three parts: (a) the time spent on communication, denoted as *CommTime,* (b) the time used for matching, denoted as *MatchTime,* and (c) the time spent on querying resource services information from RSIC or the corresponding repository, denoted as *QueryTime.*

From Figure 4.9 one can see that most time is spent on communications. It is almost the same for different number of tasks because all task submissions make the same service call with different parameters. The second largest part of the time is spent on querying resource service repository. It increases linearly with number of tasks. This is because in our implementation, the system makes one query for each task's request. The smallest part of time, tens milliseconds, is spent on matching and finding qualified resource services for tasks. This result indicates that the proposed matching method is time efficient.

4.8 Summary

Resource services match and search (RSMS) is the key in implementing a real-time MGrid. Current works concerning service match and search primarily concentrate on abstract service discovery mechanisms and methods. The research to practical similarity matching algorithms (SMAs) between basic describing information of services is insufficient. Without the supports of SMAs for basic describing information of services, the upper service discovery cannot be realized. In this chapter, a resource services match and search mechanism is proposed. It is suitable for an MGrid environment where resource services are described not only using common describing information such as keyword, but also are described using sentence information, fuzzy number information, entity class information, etc. The primary works and contributions of this chapter are as follows:

(1) The describing information of resource services are classified into four categories: (a) word concept information, (b) sentence information, (c) number information, including number interval and fuzzy number, and (d) entity class (or data structure) information. The similarly matching algorithms (SMAs) for the four kinds of basic describing information are described, including (a) word mat*ching algorithms (WMAs)*, (b) *sentence matching algorithms (SeMAs)*, (c) *number matching algorithms (NMAs)*, and (d) *entity class matching algorithms (ECMAs)*.

(2) Under the supports of the proposed SMAs, the process of resource services match and search are divided into four steps, they are (a) *basic-matching*, matching the basic information of resource services, such as service name and service description; (b) *I/O-matching*, matching the inputs and outputs information of resource services; (c) *QoS-matching*, matching the QoS information of resource services; and (d) *integrated-matching*, combining the above three matching results and generating an integrated matching result. The algorithms of each step are presented.

(3) A case study is presented to illustrate the application of the SMAs and resource service match and search method described in this chapter. The performance measurements from our prototype implementation indicate that the proposed match and search method for MGrid resource services are accurate and efficient.

This chapter only considered the primary four kinds of basic describing information for resource services, there are some other kinds of describing information that were not considered. Describing information of tasks and resource services involved in matching are assumed symmetrical, i.e., having the same number of parameters and same type of information. In the future, more research into the classification of describing information for resource services in MGird is recommended, as well as the corresponding similarity matching algorithms and their application in MGrid resource services match and search. Further investigation into conflict and failure detection and recovery is also recommended.

5

Resource Service QoS Modeling and Evaluation

5.1 Introduction

Grid resource management controls how to provide the available capacity of resources and services in a grid system for other entities including users, applications, services, middleware, etc. In a service-oriented manufacturing grid (MGrid) architecture (Tao *et al.*, 2008a), resources can no longer be narrowly understood as physical entities, rather they refer to any manufacturing capabilities which can be shared and used in a networked manufacturing environment, including computational resources, equipment resources, human resources, public service resources, material resources, user information resources, application system resources, technical resources, and so forth (Tao *et al.*, 2010a).

MGrid resource service management allocates the submitted manufacturing tasks or resource service requests to appropriate resources within the shortest amount of time and with the highest benefit to both resource provider and demander. It achieves this by searching the qualified resource service in MGrid according to a user's QoS requirement, and realizing the mapping from QoS requirements to resource capabilities. Therefore, an effective QoS management is especially important to resource management and resource optimal allocation. Regardless of the context, QoS management should address the following needs (Al-Ali *et al.*, 2004a; Ding and Tao *et al.*, 2008a):

a. Specifying QoS managements,
b. Mapping QoS requirements to resource capabilities,
c. Negotiating QoS with resource owners,
d. Reserving and allocating resources,
e. Monitoring parameters associated with QoS, and
f. Adapting to varying resource quality characteristics.

In order to realize MGrid resource service management, the evaluation indices system of resources and the classification and modeling of QoS under an MGrid environment are investigated in this chapter. The evaluation model of MGrid resource service QoS is also proposed. The main purpose of this chapter is as follows:

1. To establish the evaluation indices system of resources and resource service, including (a) special evaluation indices (SEIs), (b) individual evaluation indices (IEIs), and (c) general QoS (quality of service) evaluation indices (QoS).
2. To investigate the evaluation framework band method based on Agent for SEIs and IEIs.
3. To investigate the classification and modeling of MGrid resource service QoS from three different perspectives: QoS whole-lifecycle management, MGrid architecture view, and QoS attribute parameters, which are very different from traditional QoS modeling methods and research perspectives.
4. To research the MGrid resource service QoS evaluation models associated with their algorithms.
5. To research a QoS-based MGrid resource service management framework to illustrate how the proposed works can be applied to MGrid resource service management.

5.2 Related Works

"Provide extraordinary quality of service" is one of the three judging criteria of a grid system. QoS management has been studied in various contexts, particularly for computer networks and multimedia applications, network QoS, and end-system QoS, etc. The specific research contents involve architecture of QoS management, QoS controlling policies, QoS negotiation, QoS adaptation, QoS modeling and feedback mechanisms, QoS fault tolerance, etc. There already exist two extensively accepted grid QoS architectures in today's practice: GARA (Foster and Kesselman 1999; Foster *et al.*, 2000a) and G-QoSM (Al-Ali *et al.*, 2003a). Meanwhile, in the field of network QoS, several mature guaranteed technologies have emerged such as IntServ (Braden *et al.*, 1994), DiffServ (Blake *et al.*, 1998), Serv over Differv (Bernet *et al.*, 2000), and DiffServ Over MPLS (Wu and Davari *et al.*, 2005).

When it comes to the measurement and modeling of grid QoS, Al-Ali *et al.*, (2004a) classified the grid QoS into five types: Accounting QoS, Service QoS, Provisional QoS, Service Reliability, and Service Security. Irvine and Levin (1999; 2000) proposed a performance measurement framework for a distributed and heterogeneous network, and they measured some QoS parameters such as priority, version, deadline for completion, and security, which mainly belong to network performance QoS. Zeng *et al.*, (2004a) studied the modeling of some web service QoS parameters such as execution price, execution duration, reputation, successful execution rate, and availability. Moreover, the corresponding brief evaluation algorithms were proposed by them.

However, the above QoS studies primarily focus on the management of computing resources (e.g., CPU, memory, network bandwidth) in a computing grid (CG) environment. Compared with CG, MGrid has characteristics such as more flexible interaction, long life-cycle, multi-party cooperation, resources multiplicity, knowledge, functional complexity, and stronger reliability request (Tao *et al.*, 2008b). As a result, the QoS management in MGrid is even more complex than that in CG. Therefore, this chapter will investigate the classification and modeling of MGrid resource service QoS from three different perspectives: QoS whole-lifecycle management, MGrid architecture view, and QoS attribute parameters, which are very different from traditional QoS modeling methods and research points. The topics in the study also include the QoS evaluation with an emphasis on the trust-QoS evaluation, which is very different from the traditional QoS evaluation perspective. The traditional QoS evaluation only emphasizes the performance QoS but ignores the trust-QoS. A trust-QoS evaluation model is proposed in this chapter. All the works in this chapter aim to promote the research and development of MGrid.

5.3 Evaluation Indices System of MGrid Resource Service

The evaluation and selection of resources or resource service in MGrid involves many indices from all aspects. Besides time, credit, cost, and service, the monitoring and establishing capacity of the manufacturing resources in accordance with the individual requirements from the users, as well as the cooperation capacity and reliability degree, are also key indices we should pay attention to. Hence, the evaluation indices of each kind of resource are divided into three parts: (1) special evaluation indices (SEIs), (2) individual evaluation indices (IEIs), and (3) general quality of service (QoS) evaluation indices (Ding and Tao *et al.*, 2008a).

SEIs are designed for different kinds of resources (equipment resources, human resources, application system resources, material resources, etc.), and they only fit the very kind of resources they are designing for. Some special evaluation indices like the breaking down rate, the requirement of the running environment, action precision, processing size and shape, orientation precision, and position precision of the equipment resources; the version, the capacity of secondary development, opening degree of source code, compatibility, dependability, etc., in the application system resources; the working experience, technique grade, knowledge, age, and the title of a technical post in human resources; the period of validity, the requirements of the employ environment, recycle, and life-span in the material resources.

IEIs are out of the formal library of evaluation indices, but they are established according to the special requirements from the users. So they are very practical and dynamical, and a special role for them will be introduced in Section 5.1.

QoS evaluation indices primarily are concerned with the *time, credit, cost*, and *service* of the manufacturing resources or resource service. This kind of indices can be included in almost all the resources in the resource service evaluation process. The QoS modeling and evaluation method will be investigated in detail in Sections 5.4 and 5.5.

According to the above theories and ideas, the evaluation indices system of equipment resources (Figure 5.1), the evaluation indices

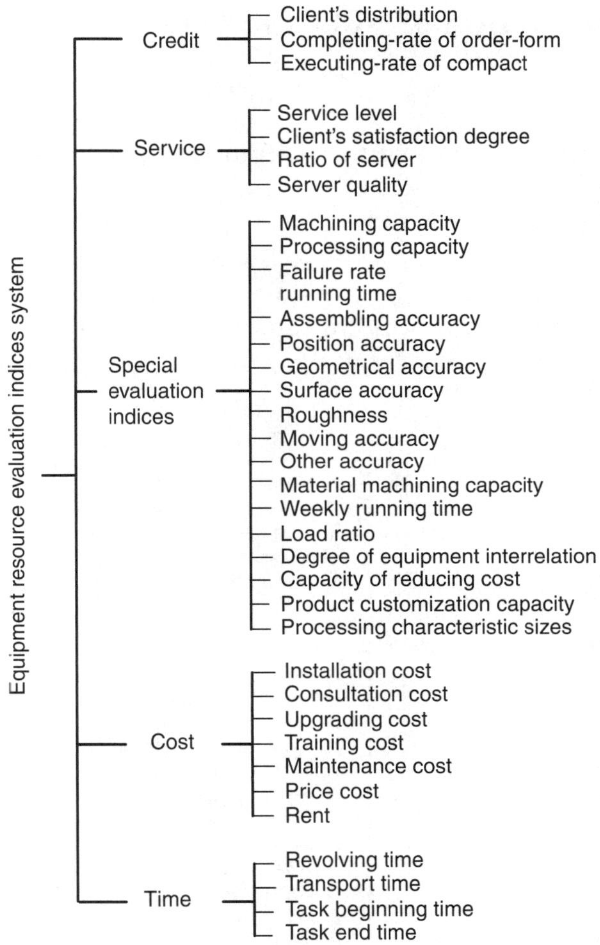

Figure 5.1 Equipment resource evaluation indices system.

system of human resources (Figure 5.2), the evaluation indices system of material resources (Figure 5.3), and the evaluation indices system of application system resources (Figure 5.4) (Ding and Tao *et al.*, 2008a) are designed and established. All these evaluation indices are stored in the library of the comprehensive evaluation indices. During the evaluation process, the system selects the corresponding evaluation indices with the requirements from the user to make up the evaluation indices system with the help of an evaluation agent.

5.4 Evaluation of SEIs and IEIs

5.4.1 Evaluation Framework of SEIs and IEIs

In the traditional evaluating method for SEIs and IEIs, an evaluation experts group is used. The experts usually have different knowledge, different experience, and common responsibility. They work together to evaluate the SEIs and IEIs. Through analyzing, one finds that the evaluation system has the characteristics of synchrony distribution capacity, cooperative capacity, time sequence, activity, self-learning, intellectual computing capacity,

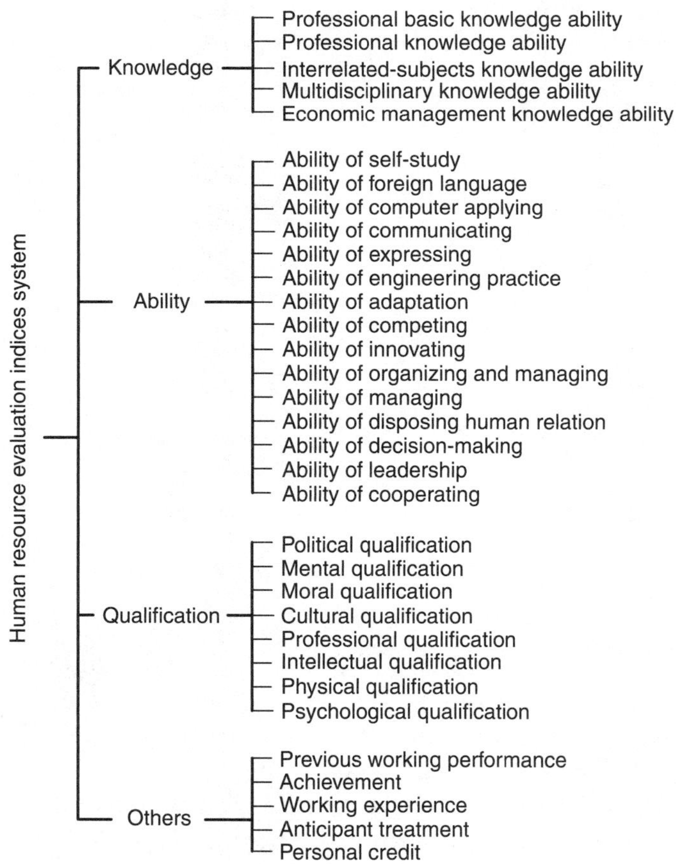

Figure 5.2 Human resource evaluation indices system.

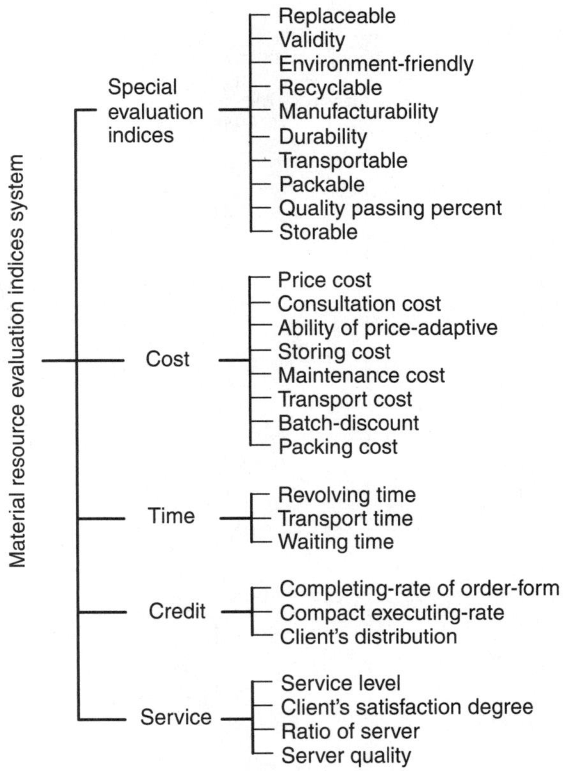

Figure 5.3 Material resource evaluation indices system.

etc., which are all the characteristics of Agent or Agent System. Therefore, in this section, each evaluation expert can be mapped into an agent, and it evaluates the resource according to its nature under the fuzzy evaluation principles which are established by the corresponding expert. The members of the evaluation expert group can share common information and knowledge they acquire through the MGrid platform, in order to create a cooperative working environment of group evaluation. This enables the experts with different knowledge, different experience, and common responsibility to work together to resolve the non-structure problems of grid resources in the evaluation process. The evaluation framework based on Agent is shown in Figure 5.5 (Ding and Tao *et al.*, 2008a).

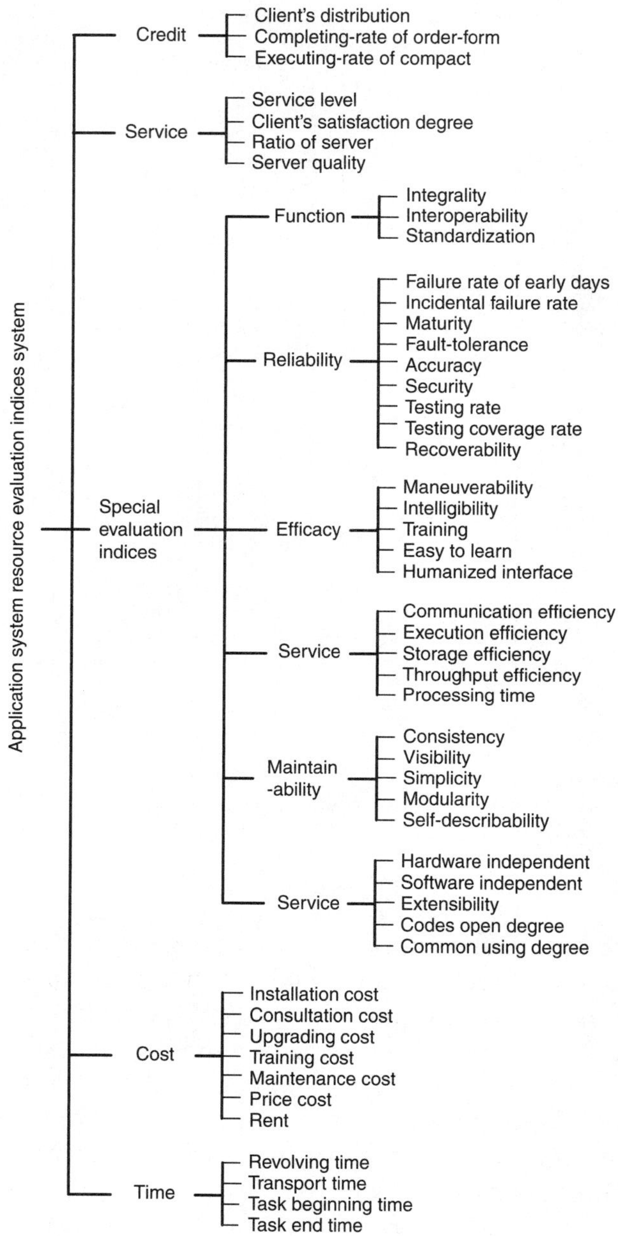

Figure 5.4 Application system resource evaluation indices system.

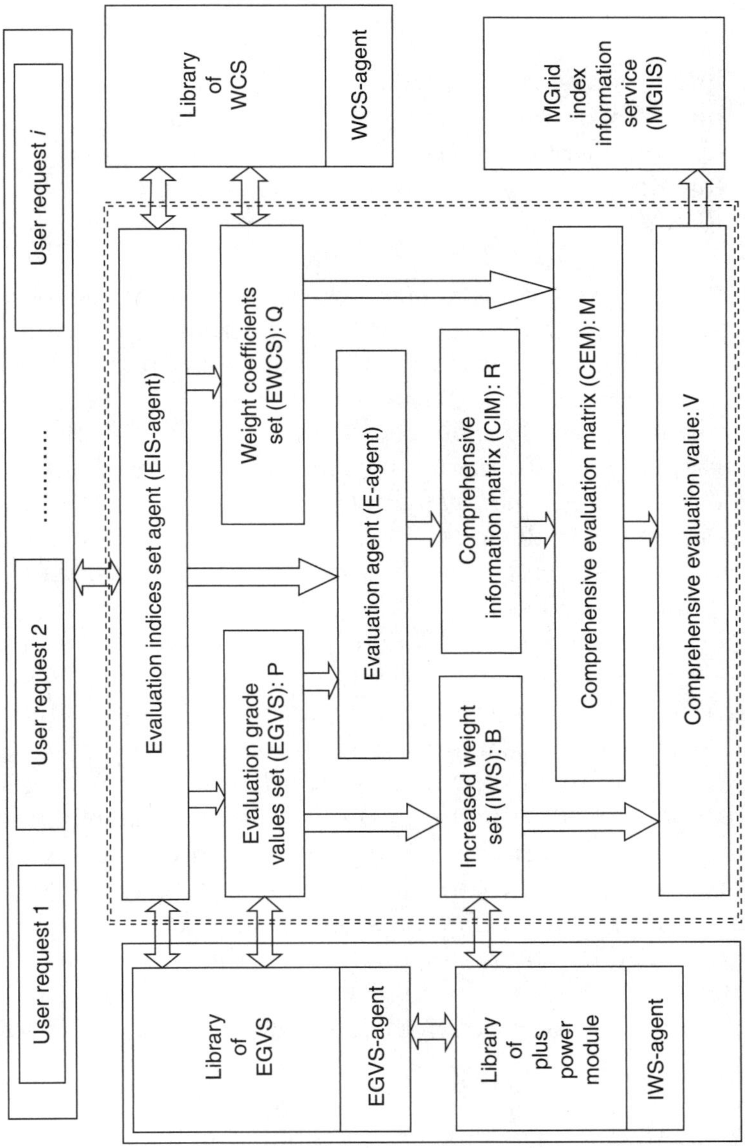

Figure 5.5 SEIs and IEIs evaluation framework based on agent.

5.4.2 Structure Model of Agent

The basic structure model of each agent in Figure 5.5 is shown in Figure 5.6(b). Each agent primarily includes the following five components:

1. Environment sense model,
2. Information processing model,
3. Communication model,
4. Intelligent reason and study model, and
5. Repository.

The *intelligent reason and study model* is the core of each agent, which is also the main difference among the agents in Figure 5.5. It uses the knowledge in the repository to analyze and reason through the rough results of the external environment information output by means of the information processing model after the original processing. It self-learns the required information when need (e.g., in Section 5.4.2, the evaluation system stores and transfers the IEISs by itself). In the following section, the evaluation agent is used as an example to highlight the working principle of each agent (Ding and Tao *et al.* 2008a).

The key function of the evaluation agent is to evaluate each evaluation indices of the resource and output the comprehensive information matrix R. For example, if the evaluation values collectivity P = {excellent, good, acceptable, poor, very poor}, then the main task of the Evaluation Agent is to select a value from P for each evaluation index of the evaluating resource, and then calculate the result according to a certain rule in the repository, and output the comprehensive information matrix R of the evaluating resource. The structure model of the Evaluation Agent is shown in Figure 5.6(a), which primarily has two components, i.e., the Chief Evaluation Agent (CEAgent) and the Evaluation Expert Agent (EEAgent). A detailed working process of the Evaluation Agent is depicted as follows.

1. After the CEAgent accepts the input from the external environment (the input of other agents or direct input), it distributes the evaluating task to a special group of EEAgents, which use the agent technology to evaluate the resource according to its nature under the fuzzy

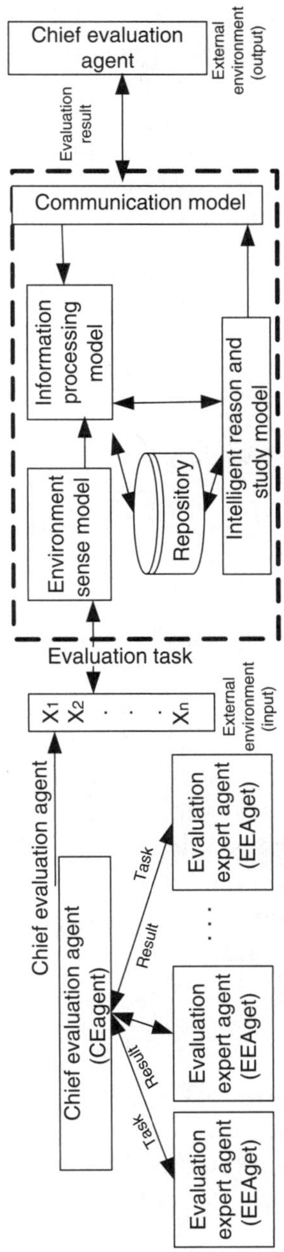

Figure 5.6 The model of evaluation agent.

evaluation principles established by the corresponding experts, according to the property and degree of complexity.
2. The particular evaluating process of each EEAgent is as follows: Let the language variable (the input from external environment) before the rule description be X, and the domain of it be G_X. Let the language variable that depicts the result be Z, and the domain of it be G_Z.

Then the proposed EEAgent decision-making knowledge can be formalized as:

$$\text{IF X} = \text{X}_i\text{; THEN Z} = \text{Z}_i\text{ ; } i = 1,2,3,\ldots t$$

The X_i and Z_i respectively denote the variable of X and Z of the ith rule, and the t expresses the number of the total rules. For example, if there are the five following decision-making knowledge, according to the evaluation index u_i (finish time):

- If the finishing time is earlier, then the evaluation value of the evaluation index is "excellent;"
- If the finishing time is a little bit earlier, then the evaluation value of the evaluation index is "good;"
- If the finishing time is general, then the evaluation value of the evaluation index is "acceptable;"
- If the finishing time is a little bit late, then the evaluation value of the evaluation index is "poor;" and
- If the finishing time is late, then the evaluation value of the evaluation index is "very poor".

3. Each EEAgent returns the evaluation result to the CEAgent.
4. The CEAgent records the submitted results and then computes them according to a certain rule in the repository and exports the comprehensive information matrix. For example, in Section 5.4.3.3, we set r_{iP_m} as follows:

$$r_{iP_m} = \frac{\text{the number of Evaluation expert Agent that selects } P_m \text{ for } U_i}{\text{the total number of Evaluation Expert Agent}}$$

Therefore, the problem of storing and using the agent decision-making knowleddepicted in nature language in the computer is addressed. The basic model and working principles of the other agents are the same as the Evaluation Agent.

5.4.3 Evaluation Process of SEIs and IEIs

5.4.3.1 Setting of Manufacturing Resources Evaluation Indexes Set U

Different kinds of manufacturing resources have different evaluation indices, and different users have different requirements for the same resources providers, which results in the dissimilar and dynamic nature of the evaluation indices set (EIS). So in the evaluating process, we apply the agent technology to establish the evaluation indices set according to the actual problem. The evaluation indices setting model is shown in Figure 5.7 (Ding and Tao *et al.*, 2008a).

First, the user submits the resource requests to MGrid platform through user interface, and the GRAM (Grid Resource Allocation Management) and MDS (Miscomputing Discovery Service) standardize it and then export the external environment input (here it is X) to the EIS-Agent (Evaluation Indices Set Agent). Second, if there are no special requests for the user, the EIS-Agent will make intelligent deductions based on the type of the evaluating resource and user requirements, and select the corresponding indices from the library of evaluation indices for evaluating. If there are some special requests for the user, besides making intelligent deductions based on the type of the evaluating resource and user requirements, and

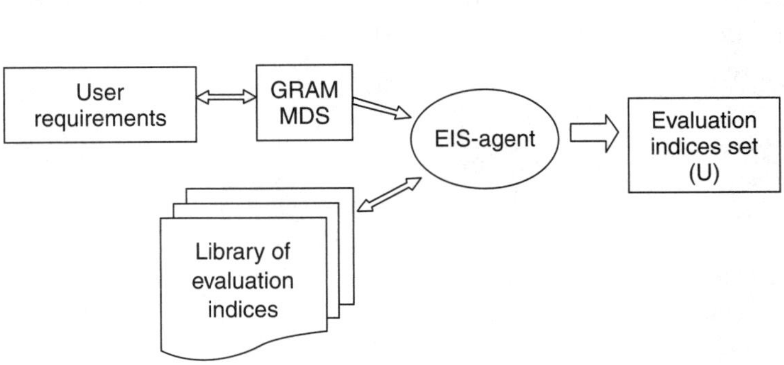

Figure 5.7 Setting of evaluation indices Set^U.

selecting the corresponding indices from the library of evaluation indices, the EIS-Agent will transform the special user request into individual evaluation indices. It then combines the two parts and exports the required evaluation indices set U for later evaluating. At the same time, the EIS-Agent stores the individual evaluation indices in a corresponding evaluation indices library for future use.

Let the evaluation indices be U, and $U = \{U_1\ U_2 \ldots U_i\}$ (i is the number of the evaluation indices, and $i = 1,2,3,\ldots$).

5.4.3.2 *Setting the Evaluation Grade Values Set P*

In the evaluating process, each manufacturing resource evaluation index has a set of evaluation values, such as a five-grade fuzzy evaluation values set: { excellent, good, acceptable, poor, very poor }, and a seven-grade fuzzy evaluation values set: { excellent, very good, good, acceptable, poor, very poor, extremely poor}. Because of the differences of the evaluating criterions and contents of different kinds of manufacturing resources, it is impossible to directly give out the uniform and specific evaluation grade standard in the evaluating process. So we have adopted the EGVS-Agent (Evaluation Grade Values Set Agent) to select the appropriate evaluation indices from the evaluation indices library that has been established in Figure 5.5.

Let the evaluation grade values set be P, and $P = \{P_1\ P_2 \ldots P_n\}$ (n is the number of the evaluation grades).

5.4.3.3 *Establishment of the Comprehensive Information Matrix R*

In the evaluating process, the E-Agent (evaluation agent) is used to evaluate each evaluation index U_i among the evaluation indices set, so one gets the comprehensive information matrix R of the evaluating manufacturing resource:

$$R = \begin{bmatrix} r_{1p_1} & r_{1p_2} & \cdots & r_{1p_{n1}} \\ r_{2p_1} & r_{2p_1} & \cdots & r_{2p_n} \\ \vdots & \vdots & \ddots & \vdots \\ r_{i_ip_1} & r_{i_ip_1} & \cdots & r_{ip_n} \end{bmatrix}$$

where r_{iP_m} denotes the degree of U_i subject to the evaluation grade P_m in the evaluation indices.

5.3.3.4 *Establishment of the Weight Coefficients Set Q*

Weight is decided according to the facts based on the right method, which include statistic-analyzing, expert evaluating, and so on. In the evaluating process, the system uses the WCS-Agent (Weight Coefficients Set Agent) to select the special power weight collectivity according to the specific evaluation indices.

Let the evaluation weight coefficients set be Q and $Q = \{Q_1\ Q_2 \ldots Q_i\}$ ($Q_j > 0$ and $\sum_{j=1}^{i} Q_j = 1$).

5.4.3.5 *Establishment of the Comprehensive Evaluation Matrix M*

$$M = Q \times R$$

5.4.3.6 *Establishment of the Increased Weight Set Matrix B*

The increased weight matrix B of evaluation grade values is the value that takes no account of the fuzzy boundary condition, it reflects the degree of influence that each evaluation act has on the final result. For example, if we define P = {excellent, well, good, general, a little bad, bad, very bad}, then the B can be {90, 80, 70, 60, 50, 30, 20}. In the evaluating process, the system uses the IWS-Agent (Increased Weight Set Agent) to select the special **B**.

We marked the increased weight set matrix as B, and $B = \{B_1\ B_2 \ldots B_i\}$.

5.4.3.7 *The Result of Comprehensive Evaluation Quantitative Value V Based on Matrix M and Matrix B*

$$V = M \times B^T / \sum_{j=1}^{i} m_j$$

We know the comprehensive evaluation result of the evaluating manufacture resources from value V. The bigger V is, the better the corresponding evaluating manufacturing resource suitability is for the users' requirements, and the resource will have priority to be offered to the users. By contrast, the evaluating manufacturing resource with a lower comprehensive quantitative evaluation value will be delayed or not offered to users. As a result, the decision making time for selecting the appropriate resource for specific clients can be effectively reduced, and the quality of service (QoS) can be improved in the MGrid system.

5.5 Classification and Modeling of MGrid QoS

5.5.1 QoS Modeling from the Whole-lifecycle Management of MGrid Resource Service QoS

The whole-lifecycle management (WLCM) of MGrid resource service QoS can be divided into three phases: (a) creating phase, (b) executing phase, and (c) terminating phase (Tao *et al.* 2009c). Each phase has different QoS contents as shown in Figure 5.8.

1. *Creating phase* is the first phase of the WLCM of MGrid resource service QoS, including QoS specification, QoS mapping, QoS requirement negotiation, resource reservation, QoS accounting, etc.
2. *Executing phase* is the core of the WLCM of MGrid resource service QoS. It is responsible for QoS-based resource allocation, QoS evaluation, QoS negotiation

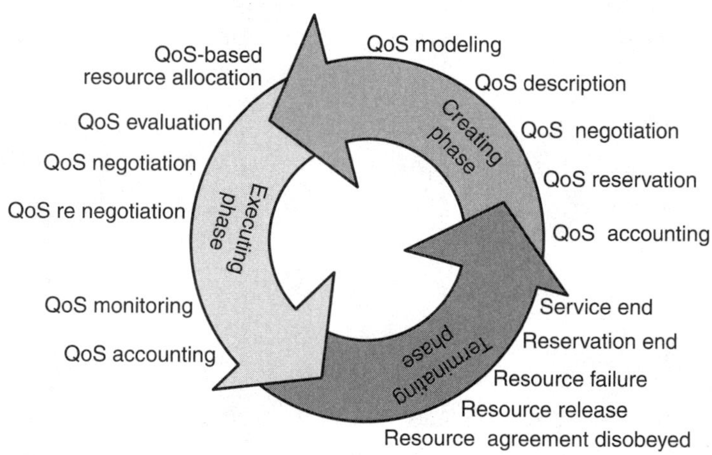

Figure 5.8 MGrid resource service QoS whole-lifecycle management model.

and renegotiation, QoS monitoring, QoS accounting, QoS adaptation, etc.

3. *Terminating phase* is responsible for ending a service when the reservation ends, there is a resource failure or resource release, there is a contravention to the resource agreement, etc.

5.5.2 QoS Modeling from MGrid Architecture Views

In order to enable the proposed MGrid architecture (Tao *et al.* 2009c) to be comprehensively and deeply understood, the architecture of MGrid can be classified into four views:

1. MGrid resource view,
2. MGrid service view (or information view),
3. MGrid executing process view, and
4. MGrid function view.

The MGrid function view impenetrates the other three views to some extent. E.g., in the MGrid resource view, there exist functions such as resource description, resource publication, resource encapsulation, resource monitoring, etc. In the MGrid executing process view, there exist functions such as establishment of workflow, process monitoring, state monitoring, etc. Therefore, we can research MGrid QoS from the perspective of MGrid architecture views, and the QoS in MGrid can also be classified into several corresponding views: (a) MGrid resource QoS view, (b) MGrid service QoS view, and (c) MGrid executing process QoS view. The multi-view QoS model based on MGrid architecture is shown in Figure 5.9.

In an MGrid system, each function is realized with the support and interaction of MGrid resource QoS view, MGrid service QoS view, and MGrid executing process QoS view. MGrid resource QoS view is the foundation. On the one hand, it provides resource information for MGrid service QoS view, and is responsible for information updating and maintaining. On the other hand, it supports resource invoking for MGrid executing process QoS view. MGrid service QoS view is the core, and it is the bridge between MGrid service QoS view and MGrid executing process QoS view. MGrid service QoS view provides the MGrid executing process QoS view with information supports

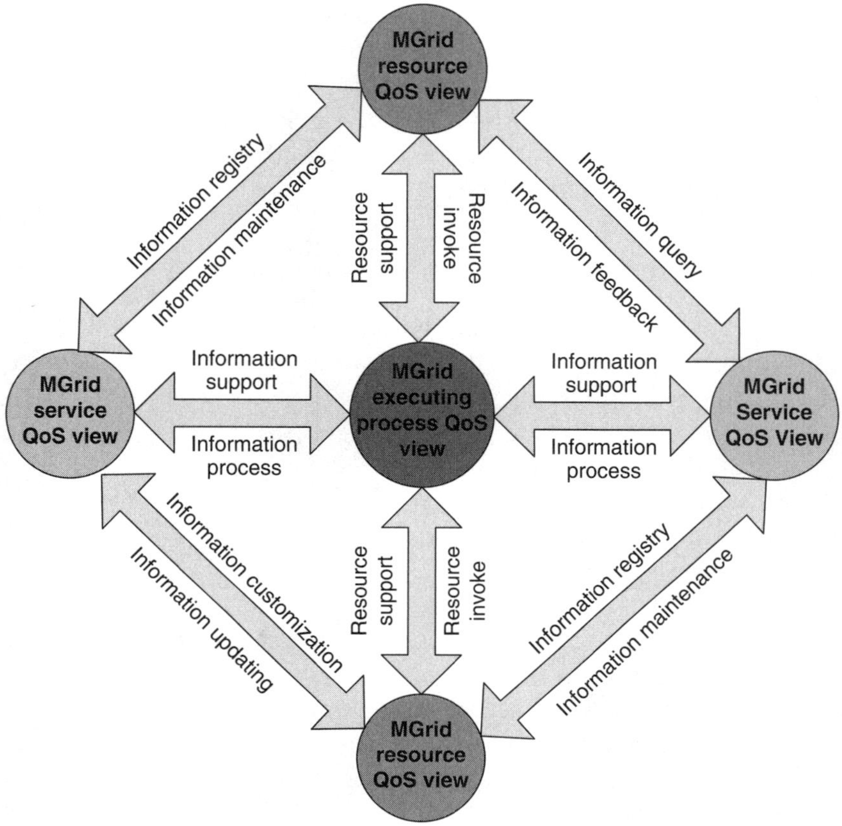

Figure 5.9 Multi-view QoS model based on MGrid architecture.

and information processing. MGrid executing process QoS view is the realizing layer in an MGrid. It realizes each function, such as resource publication, resource discovery, resource registry, and resource matching, under the supports of MGrid resource QoS view and MGrid service QoS view.

5.5.3 MGrid QoS Attribute Parameters Modeling

From the perspective of the degree of influence to MGrid system performance, the QoS attribute parameters in MGrid can be classified into *performance-QoS* and *description-QoS*. The former, *performance-QoS*, primarily includes the parameter and precision with regard to time, such as the final service time limit. The latter, *description-QoS*, is mainly employed to evaluate the validity and reliability of resource service information or manufacturing ability. *Performance-QoS* consist of *time-based performance-QoS, network performance-QoS,* and *service performance-QoS. Description-QoS* primarily includes *security QoS, accounting QoS,* and *trust-QoS.* Each QoS attribute parameter in association with its corresponding description is shown in Table 5.1.

In addition to the above QoS parameters, the resource service execution cost (i.e., *cost*), is an important criteria during MGrid resource service scheduling, and is the fee that a user or resource service requester has to pay for invoking the resource service.

5.6 Evaluation of MGrid QoS Attribute Parameter

There are many grid systems, and researchers (Zeng *et al.* 2004a) have already studied the performance-QoS and presented the corresponding evaluation methods and formulas for them. Therefore, we have omitted the detailed evaluation of some performance-QoS in this study. In order to research QoS-based MGrid management and scheduling, and take into account the importance of measurability, we have only selected some key QoS parameters from Table 5.1, and researched their corresponding evaluation algorithms.

Six QoS parameters are considered in this chapter, and they are *time, cost, reliability, maintainability, trust,* and *function similarity* (Tao *et al.* 2009c).

5.6.1 Time

Time (resource service execution duration): The service execution duration, T, measures the expected delay in seconds or minutes between the moment when a service request is sent, and the moment when the results are received. $T = (T_1) + (T_2) + (T_3)$. T_1 is the service executing time, and measures the expected delay in seconds or minutes between the moment when a task has begun being executed and the moment when the task is finished. Resource service providers advertise their executing time or provide methods to inquire about it. T_2 is the time spent by system middleware, such as *QoSExtraction* (a service in the QoS-based resource service management system as described in Section 5), and is measured by the system. T_3 is the transporting time of service or information,

Table 5.1 MGrid QoS attributes parameters (Zhou and Tao *et al.*, 2007).

QoS Class	QoS Subclass	QoS Parameters	Descriptions
Performance QoS	Time Performance QoS	Service executing time (T_1)	The service executing time measures the expected delay in seconds or minutes between the moment when a task is begin executed and the moment when the task is finished
		Middleware spending time (T_2)	Time spent by system middleware
		Network transporting time (T_3)	Transporting time of service or information, date
		Service responding time (T)	The service responding time measures the expected delay in seconds or minutes between the moment when a service request is sent and the moment when the results are received, and $T = T_1 + (T_2) + (T_3)$
	Service Performance QoS	Robustness	The ability of handling unexpected condition, such as invalid or incorrect inputs
		Transplantation	Transplantation refers to the relevance degree between service and practical physical resource, the higher of the relevance degree the worse of the transplantation
		Resource requirements QoS	The requirements of service to resource
		Success rate	The successful rate of a service is the probability that a request is correctly responded within the maximum expected time.
		Reliability	The executing ability within a specific time and condition, e.g., the accessible rate during time t, average no-failure rate, average failure recovering rate, etc.
		Maintainability	Maintainability is the ability that the resource service can deal with the unexpected conditions, such as decrease of network bandwidth, network disconnection, and so forth.

(Continued)

Table 5.1 (cont.) MGrid QoS attributes parameters (Zhou and Tao *et al.*, 2007).

QoS Class	QoS Subclass	QoS Parameters	Descriptions
Performance QoS	Service Performance QoS	Function similarity.	Function similarity is the feasibility that the functions of the requested resource service can be replaced by other resource service belongs to the same resource service set.
	Network Performance QoS	Net accessible rate	The net connection rate with in a time
		Frequency of processor	Frequency of processor
		Occupancy of processor	Occupancy of processor
		Memory capability/ Bandwidth	Memory capability , I/O accessing speed
		Transmission delay	The transmission delay measures the expected delay in seconds or minutes between the moment when a transmit request is sent and the moment when the transmit are finished
		Loss tolerance	The rate of lost data packet occupies in the whole sent data packet
		Throughput	The maximum bytes the net can transfer in one second
Description QoS		Security QoS	QoS parameters refer to the self security level and the accessing control policies of the service provided by resource service provide
		Accounting QoS	QoS parameters about the service price and its management polices, which can be measured by time, content, and current, etc.
		Trust QoS	The trust relationship between resource service transaction entities (such as RSP, RSD, agent, middleware), including veracity, reliability and so on about resource service information and manufacturing ability.
Other		Cost	Resource service execution cost is the fee that a user or resource service requester has to pay for invoking the resource service.

and is estimated based on past executions of the resource service. $T_3 = (\sum_{i=1}^{n} t_i)/n$, where t_i is a past observation of the transporting time, and n is the number of execution times for the resource service that were observed in the past.

5.6.2 Cost

The execution price (i.e., *cost*) of a resource service is the fee that a user or resource service requester has to pay for invoking the resource service. MGrid resource service provider either advertises the execution price for it when it is published or registered, or provides the means for potential requesters to inquire about it.

5.6.3 Reliability

Reliability is the executing ability within a specific time and condition, e.g., the accessible rate during time *t*, average no-failure rate, average failure recovering rate, etc. The detailed meaning and measuring method of reliability changes accord to the practical requirements and conditions. E.g., if the user emphasizes the accessing reliability (successful accessing rate), then the reliability can be calculated using the formula $Rel = S_{access}(\theta)/\theta$, where $S_{access}(\theta)$ is the total amount of times in which the requested resource service can be successfully accessed or invoked during the time period θ. Similarly, if the user emphasizes the successful execution rate of the requested resource service, then the reliability can be calculated with $\text{Rel} = S_{execute}(\theta)/E_n(\theta)$, where $S_{execute}(\theta)$ is the number of times that the resource service has successfully executed the submitted task during the time period θ, and $E_n(\theta)$ is the total number of times the resource service has been invoked during the time period θ.

5.6.4 Maintainability

Maintainability enables the resource service to deal with unexpected conditions, such as a decrease of network bandwidth, network disconnection, and so forth. Maintainability can be calculated using the following formula:

$$Mn = M_s(\theta) \;/\; F(\theta)$$

Where $M_s(\theta)$ is the number of times that the resource service has successfully processed the unexpected conditions during the time period θ, and $F(\theta)$ is the total number of the unexpected conditions during a time period θ.

5.6.5 Trust-QoS

Trust-QoS is the belief or faith the trusting entity (e.g., user or resource service demander) holds in the trusted entity's ability (e.g., requested resource service or resource service provider) in a given context and time slot, that gives them confidence to carry out a transaction with the trusted entity. The trust-QoS is realized by the concept of *trust-QoS relationship* (Tao *et al.*, 2009d). Each *trust-QoS relationship* denotes a trust value from the trusting entity to the trusted entity for a particular service content and time slot. The *trust-QoS relationship* is a composite of (1) transaction *entities A*, including both trusted entity and trusting entity; (2) manufacturing resource service *contexts C*; (3) *trust type* Γ, including *intra-domain trust-QoS* and *inter-domain trust-QoS*; (4) *trust-QoS request type R*, including strong-trust request, weak-trust request, and no-trust quest; (5) *trust-QoS value E*; (6) *trust-QoS benefit B*; and (7) *time slot T*. Therefore *trust-QoS relationship* can be expressed as $Trust\text{-}QoS = (A,C,\Gamma,R,E,B,T)$. The detailed evaluation model and algorithms of trust-QoS will be described in Chapter 6.

5.6.6 Function Similarity

Function similarity is the possibility that the function of a requested resource service can be replaced by another resource service belonging to the same resource service set (i.e., RSS). Assume resource services RS_a^j and RS_b^j belong to the same resource service set RSS^j (i.e., $RS_a^j, RS_b^j \in RSS^j$). If RS_a^j and RS_b^j are subject to,

$$\bigcup_{i=1}^{k_b} function(RS_b^j) \supseteq \bigcup_{i=1}^{k_a} function(RS_a^j), (k_a, k_b = 1,2,3,\cdots, and\ k_b \geq k_a)$$

then we can say that RS_b^j is a candidate of RS_a^j, or the functions of RS_a^j can be replaced by RS_b^j. $\bigcup_{i=1}^{k_b} function(RS_b^j)$ is the total number of the public functions of RS_a^j, and $\bigcup_{i=1}^{k_a} function(RS_a^j)$ is the number of RS_b^j. Function similarity of resource service RS_i^j ($RS_i^j \in RSS^j$) is calculated using the following formula:

$$FS = C_{RS_i^j} / k_j$$

where $C_{RS_i^j}$ is the number of the candidate of RS_i^j in RSS^j, and k_j is the total number of all the resource services in RSS^j.

5.7 Application Case: QoS-based MGrid Resource Service Management

In order to illustrate how the previous works can be applied to MGrid resource service management, a QoS-based resource service management (QoS-RSM) framework is put forward as shown in Figure 5.10. A brief work flow of QoS-RSM is described as follows (Tao *et al.* 2009c):

1. The user submits a task or resource service request via human-machine interface to the job management system in MGrid;
2. The *QoS parser* specifies and parses the submitted requirement information, and generates the

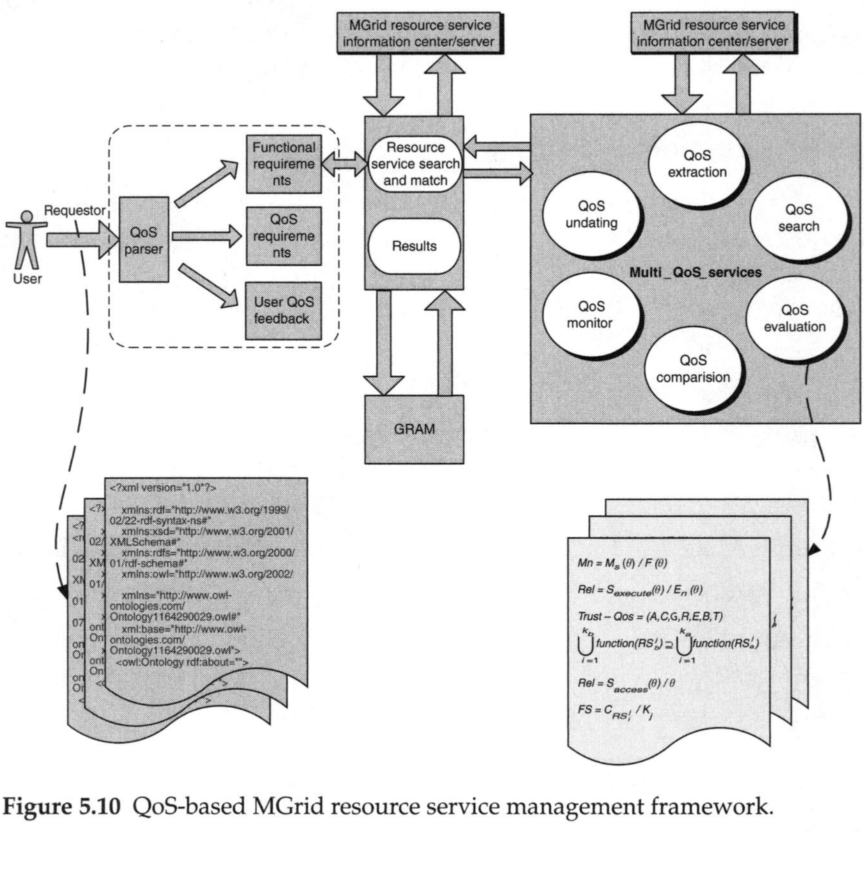

Figure 5.10 QoS-based MGrid resource service management framework.

corresponding QoS requirement information and functional requirement information;
3. Under the support of MGrid resource service *Multi_QoS_Services,* the MGrid resource service search and match agent searches the qualified resource service that can execute the task according to the user's functional requirements and QoS requirements from the resource service information center or server, and generates the set of candidate resource services.
4. GRAM is responsible for the resource service allocation.

The "heart" of the QoS-RSM in Figure 5.3 lies in the *Multi_QoS_Services*. It is an integrated infrastructure that executes resource service QoS searching, resource service QoS extraction, resource service QoS evaluation, resource service QoS monitoring, resource service comparisons, resource service QoS updating, and so forth. *Multi_QoS_Services* consists of six services: *QoS extraction, QoS search, QoS evaluation, QoS comparison, QoS monitor* and *QoS updating*.

1. *QoS extraction* is responsible for extracting QoS information about each candidate resource service required.
2. *QoS search* is responsible for searching the qualified resource service that can execute the task according to the user's QoS requirements.
3. *QoS evaluation* is responsible for evaluating the value of each QoS parameter regarding each candidate resource service.
4. *QoS comparison* is responsible for comparing the required resource service's QoS of each task and the candidate resource service's QoS.
5. *QoS monitor* is responsible for monitoring the resource service states, and deciding what caused the occurrence of a failure by analyzing the state of information of each resource.
6. *QoS updating*. When the QoS values have been stored in a server or related database a long time ago according to a logged timestamp, and they no longer represent the current QoS that the resource service offers, then *QoS updating* is responsible for updating resource service's QoS into a corresponding server or related database.

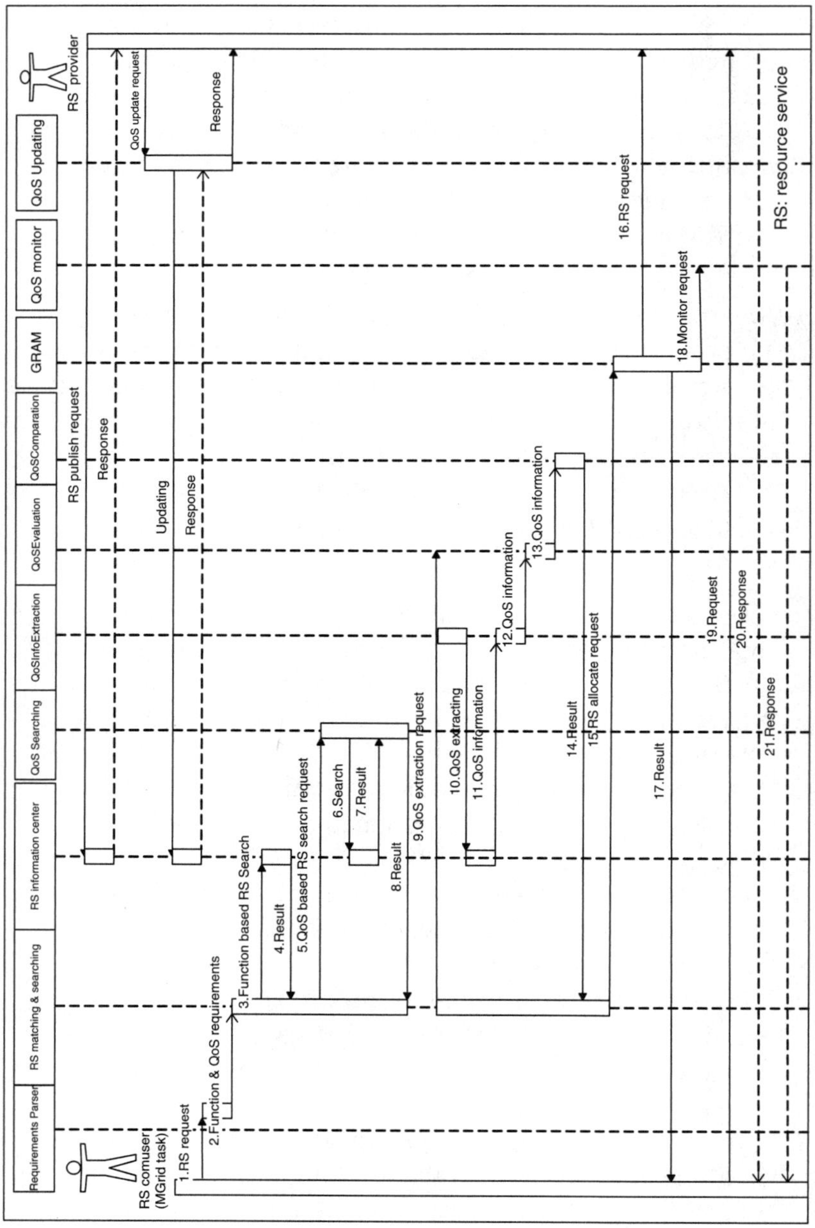

Figure 5.11 Flow of logic within QoS-RSM through a UML sequence diagram.

The work flow of the proposed QoS-RSM associated with its *Multi_ oS_Services* is shown in Figure 5.11.

5.8 Summary

This chapter studied the evaluation indices system of resource and resource service and classified it into three parts, i.e., special evaluation indices (SEIs), individual evaluation indices (IEIs), and general quality of service (QoS) evaluation indices. The evaluation framework band method based on Agent for SEIs and IEIs were studied. The modeling of MGrid QoS from the points of QoS whole-lifecycle management, MGrid architecture views, and QoS attributes parameters were investigated. These are very different from traditional modeling methods and research perspectives. MGrid resource service QoS measurement and evaluation methods were described, particularly for the trust-QoS evaluation model which has rarely been researched in an MGrid field. A QoS-based MGrid resource service management framework was proposed, which illustrated how the proposed works can be applied to MGrid resource service management.

6

Resource Service Trust-QoS Evaluation

6.1 Introduction

MGrid has the following characteristics (Tao *et al.*, 2010a; 2010d): (1) a geographical distribution of resources or entities; (2) resource heterogeneity; (3) an autonomously administering grid domain (or resources domain) with their own resource policies and practices, and where each resource node belongs to an autonomous domain that has respective measurements and methods to manage its resources; (4) a different autonomous domain employing different QoS (Quality of Service) and accessing control strategies, methods, and models to manage and access the resources or resources owned by it; (5) the members (e.g., resources providers and demanders) in MGrid can add/remove into/from it dynamically, which makes it very difficult for an user to confirm whether the identity and ability of the opposing party is reliable or not when they transact with each other; (6) for the Resource Service Demander (RSD), MGrid cares not only about whether the identity of the corresponding Resource Service Provider (RSP) is creditable, but also whether the resources provided by the RSP can provide highly credible QoS service, and whether it can accomplish the submitted manufacturing task successfully and consistantly. Therefore, the resource service management and scheduling in MGrid is even more complicated. When RSD and RSP enact a resource service transaction in MGrid, a series of problems will emerge which should be addressed.

In MGrid, there are two main kinds of users: (a) resource enterprise or resource service provider (RSP), and (b) user enterprise or resource service demander (RSD). The former, RSP publishes its idle resources, products, manufacturing abilities, etc., and provides the manufacturing resource service to meet the user's requirements in MGrid. The later, RSD searches for the optimal manufacturing resource and service required, and selects the corresponding partner to establish the virtual collaboration manufacturing net. The brief process of resource service transaction between RSD and RSP is shown in Figure 6.1.

It can be seen from Figure 6.1 that, when RSD and RSP have a resource service transaction with each other in MGrid system, the following problems will be generated (Tao *et al.*, 2010d; Lin *et al.*, 2004):

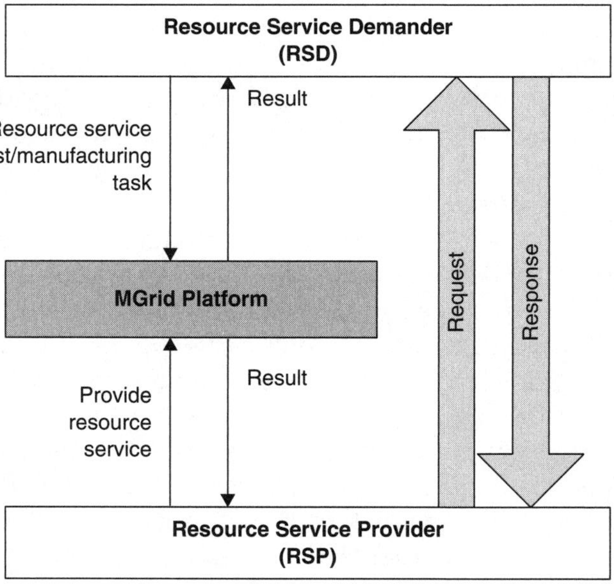

Figure 6.1 Resource service transaction between RSD and RSP.

From the RSD's point of view, before submitting the manufacturing task, there are three problems:

Problem 1 RSD must know whether the RSP (to be visited) who is providing the required resource service is 'trustworthy,' or has the ability to faithfully accomplish the manufacturing task;

Problem 2 If the answer is yes, then RSD must determine if the RSP will trust in it and be willing to accept and execute the submitted manufacturing task or request of the RSD;

Problem 3 RSD must confirm that the RSP will not destroy the submitted manufacturing task, and the latter will not provide baleful results and other uncertainties.

From the RSP's point of view, before the submitted manufacturing task can be executed, there exist two problems:

Problem 4 RSP must check the validity of the manufacturing task submitted by RSD (to be served), and make sure the RSD will not submit a baleful and ruinous task which will destroy the MGrid system;

Problem 5 RSP must be certain that the task was not destroyed and tampered with by a third party before being submitted;

After the submitted task is completed, there exists one other problem:

Problem 6 Both RSD and RSP should check the integrity and validity of the results, and update the trust relationship with the opposing parties mutually agreeing to the practical transaction conditions and results.

Therefore, the baleful actions or results during a resource service transaction resulting from a trust problem concerning RSDs and RSPs will cut down the QoS of MGrid, such as the quality of resources scheduling, the quality of resources management, etc.

In order to address the above problems, we carried out related research on MGrid resource services trust-QoS which will be presented in this chapter. We will also introduce the MGrid resource services trust-QoS concept and relationship mode, propose the evaluation modes of resource services trust-QoS and the quantitative algorithms of trust-QoS evaluation, as well as the real-time and dynamical updating algorithms.

The main objectives of the works presented in this chapter are as follows:

1. Address the trust problem existing in the resource service transaction process in MGrid.
2. Present a MGrid resource service trust-QoS relationship mode that is capable of capturing a comprehensive range of trust relationships that exist in MGrid system.
3. Propose the MGrid resource service trust-QoS evaluation models, including the inter-domain trust-QoS evaluation model and intra-domain trust-QoS evaluation model.
4. Design a trust-time decay function in MGrid resource service trust-QoS evaluation, which makes the trust-QoS evaluation values even more genuine and dynamic.
5. Design a trust-QoS based MGrid resource service scheduling framework and algorithm.

6.2 Related Works

6.2.1 MGrid Resource Scheduling Based on QoS

Some researchers have conducted many related studies on resource scheduling based on QoS. Some typical studies are as follows: Deng *et al.*, (2006a), investigated a grid-based manufacturing resource scheduling system for virtual enterprise, which is composed of an exterior grid and an interior grid in the architecture. They divided the manufacturing resource scheduling into two stages: (a) distributed scheduling, implemented by the exterior grid, and (b) local scheduling, accomplished by the interior grid. But the proposed scheduling system does not pay much more attention to the QoS, especially for trust-QoS. Rasmussen *et al.*, (2007) pointed out that the social integration of the virtual network involves the creation of identities of the participating nodes, the building of trust between them, and the sharing of tacit and explicit knowledge among them, which are the three core dimensions of work in virtual enterprise. They showed that there are three forms of trust in virtual enterprise, including (a) calculus based trust, (b) knowledge based trust, and (c) identification based trust, which should be addressed. But the details of the solutions to these three forms of trust were not presented. Liu *et al.*, (2005a) and Yuan *et al.*, (2005a) researched MGrid resource scheduling based on QoS, mainly focusing on the TQCS (Time, Quality, Cost, Service) factor, and also classified the resource scheduling process into QoS-based resource search, QoS-based resource scheduling, dynamic negotiation, etc. Deng *et al.*, (2006b) proposed a resource characteristic-based scheduling strategy and applied it to the development of a customized artificial joint. He and Sun (2003) studied the self-adapting min-min heuristic grid resource scheduling strategy driven by performance-QoS without considering the trust-QoS.

The above research into MGrid resource scheduling was mainly based on resources' *performance-QoS*, such as time, quality, cost, and service. The quantitative computing methods to these performance QoS indices were not researched, and the existing scheduling methods ignore the important factor of resource service trust-QoS in MGrid. In fact, the user's QoS requirements in the grid system or distributed system can be classified into two categories, *performance-QoS* and *trust-QoS* (Sabata *et al.*, 1997). The former, performance-QoS mainly includes the parameters and precisions with regard to time, such as final service time limit, cost, etc. The latter, trust-QoS is mainly employed to evaluate the validity and reliability of resource service information or manufacturing ability. However, to the best of our knowledge, there is no report about the research done on trust and trust-QoS in the process of resource scheduling in MGrid.

6.2.2 QoS and Trust-QoS

QoS management has been explored in various contexts, particularly for computer networks and multimedia applications, network QoS, end-system QoS, etc. The detailed research contents involve architecture of QoS management (Menasce *et al.*, 2007; Abeni *et al.*, 2005), controlling policies of QoS (Lima *et al.*, 2004), QoS negotiation (Siddiqui *et al.*, 2006; Wang *et al.*, 2006),

QoS adaptation (Al-Ali *et al.*, 2004a), QoS modeling and feedback mechanism (Wu *et al.*, 2006a), QoS fault tolerance (Dasgupta *et al.*, 2006), etc.

Al-Ali *et al.*, (2004a) proposed a G-QoSM, and they classified the grid QoS into five types: accounting QoS, service QoS, provisional QoS, service reliability, and service security; they researched QoS adaptation in a service-oriented grid. Raje *et al.*, (2002) investigated the catalog of QoS parameters, listed eighteen QoS parameters (e.g., throughput, capacity, end-to-end delay, reliability, result, etc.) and proposed a description method to these QoS parameters. They described each QoS parameter from fourteen features of QoS parameters, including name, intent, description, influencing factors, measure, known usages, aliases, related parameters, consequences, levels, technologies, applications, exceptions, and example scenario. However, detailed QoS evaluation methods were not introduced. Zeng *et al.*, (2004a) studied the modeling of some web service QoS parameters such as execution price, execution duration, reputation, successful execution rate, and availability, and they proposed corresponding brief evaluation algorithms. Deora *et al.*, (2006) proposed an expectation-based QoS assessment model. They collected the expectations as well as ratings from the users of a service, then calculated the QoS using only the rates that had similar expectations.

The above QoS studies mainly focus on the performance QoS management (e.g., QoS modeling, QoS attributes parameters evaluation, etc.) for computing resources (e.g., CPU, memory, network bandwidth, etc.) in the computational grid environment. Practically speaking, the QoS in service-oriented environments, particularly for MGrid, can be decomposed into four classes, namely, *time-based* QoS, *network* QoS, *service* QoS, and *description* QoS (Zhou and Tao *et al.*, 2007). The brief classification of MGrid resource service QoS is shown in Figure 6.2 (Tao *et al.*, 2009d; 2010a). The frontal three classes, *time-based* QoS, *network* QoS, and *service* QoS, belong to the *performance* QoS (or *functional* QoS), which play a decisive influence on the serve ability of MGrid resource service. The *descriptions* QoS mainly belong to the MGrid resource service QoS attribute parameters, and have no decisive influence to the serve ability of MGrid resource service. The *descriptions* QoS only provide some necessary QoS information to users, and the users have no way to decide the specific value of the *descriptions* QoS. The trust-QoS belongs to *descriptions* QoS.

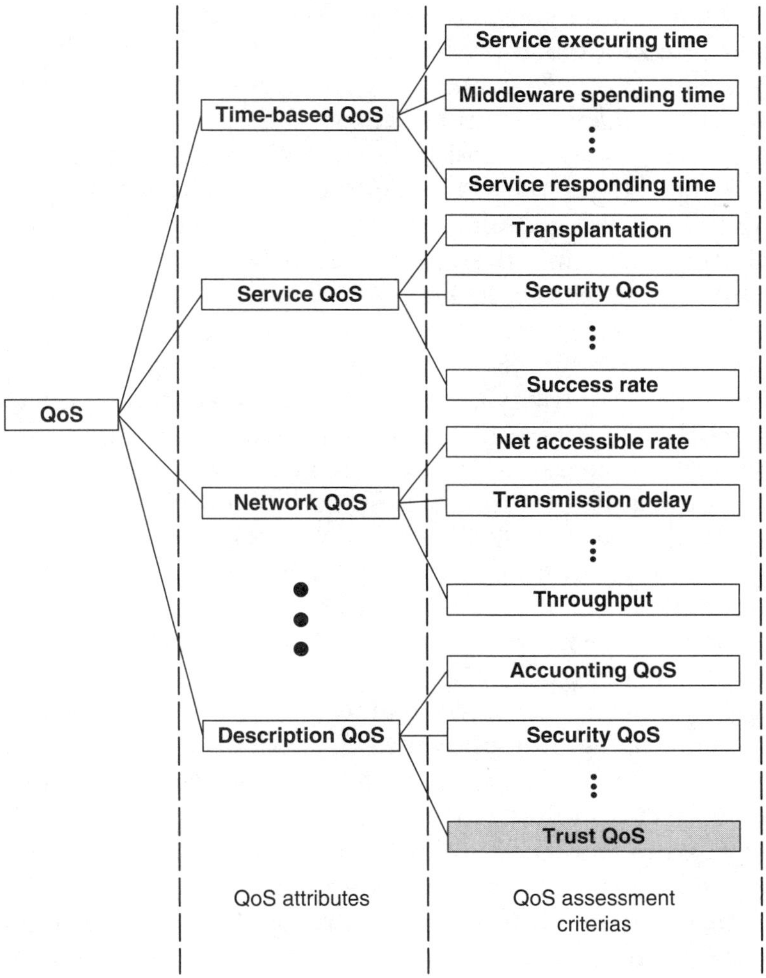

Figure 6.2 Brief classification of MGrid QoS.

Most of the current works concerning the trust problem in the grid system or distributed system have been concentrated in the area of security. The related works are as follows: Li *et al.*, (2003a) considered trust only from the aspect of improving the Grid Security Infrastructure (GSI) to achieve additional authentication means between grid users and grid services; Shen *et al.*, (2006a) analyzed the requirements of trusted computing in grid, and considered the security and safety problems in relation to both software and hardware. They also constructed a way to promote the trusted computing environment for grid through integrating the trusted computing platform. Lin *et al.,* (2004a) used the concept of subjective logic in the context of grid computing using trust relationships to enhance the grid security.

However, few works and little attention have been paid to the area of the integration of trust into grid resources management, such as resource scheduling, resource optimal-selecting, resources monitoring, etc., and the research into MGrid resource service scheduling is especially lacking. Azzedin and Maheswaran (2002) first introduced trust into grid resources management, and pointed out that if an RMS (Resources Management System) maps a resources request strictly according to trust, there could be a severe load imbalance in the grid system. They researched three scheduling heuristics integrating trust that can be used in the grid system, and presented the performance simulation of the three scheduling algorithms. Song *et al.,* (2005a) researched the introduction of trust into grid job scheduling, and proposed a new Space-Time Genetic Algorithm (STGA) for trusted job scheduling, which was very fast and easy to implement. Zhao *et al.,* (2005a) proposed a trust-based scheduling in P2P grids.

But the above research was mainly concerned with the computing resource in the computing grid. Compared with the computing grid, MGrid has characteristics such as more flexible interaction, long life-cycle, multi-party cooperation, resources multiplicity, knowledge, functional complexity, stronger reliability request, etc. (Tao *et al.,* 2008b), making the QoS management in MGrid even more complex. In addition, current research works into QoS are mainly concentrating on the aspect of performance, but the aspect of non-performance QoS (e.g., trust-QoS) is ignored. MGrid aims to solve the distributed and heterogeneous manufacturing resource sharing and operation, with an emphasis on the manufacturing ability, so it has even higher requirements regarding resource service ability and reliability.

Therefore, the research into trust-QoS is very important to MGrid, and it will play a very significant role in realizing a dynamic and executive MGrid system. So in this chapter, we will present the research on modeling of the resource service trust-QoS relationship, trust-QoS evaluation, quantitative algorithms for trust-QoS value evaluation, and its real-time and dynamic updating algorithms as well. All of this research will enhance the success rate and validity of resource scheduling, and reduce the decision-making time needed to select the optimal resource service for users in MGrid.

6.3 Resource Management and Trust Relationship Management in MGrid

In order to enhance the efficiency of the management and service of MGrid resource, improve its flexibility, along with the expansion and fault-tolerance of the MGrid system, in Chapter 3 the MGrid manufacturing resources have been classified into nine categories according to their nature, the demands placed on them, and the ways in which they are engaged. The nine categories are: (1) Human Resources (HR), (2) Equipment Resources (ER), (3) Material Resources (MR), (4) Application System Resources (ASR), (5) Technical Resources (TR), (6) Public Service Resources (PSR), (7) Computational Resources (CR), (8) User Information Resources (UIR), and (9) Other Resources (OR). Then we encapsulate each specific into a *resource service class* model according to related resource categories. According to the manufacturing resource classification model and corresponding resources encapsulation class model in MGrid as shown in Chapter 3, the resources are classified into nine related self-governing domains in a Virtual Organization (VO) in MGrid. Each domain is named with a *Domain Node* (DN) (e.g., HRD: Human Resources Domain, ERD: Equipment Resources Domain, MRD: Material Resources Domain, ASRD: Application System Resources Domain, etc.) in a VO, and the node stands for an entity (such as resources, e.g., machine, computer, clusters, etc; clients, e.g., RSP, RSD, etc.) as *Resource Node* (RN). So a scalable two-level resource management architecture of MGrid manufacturing resource is established, and the relationship and its architecture of DNs and RNs in a VO are shown in Figure 6.3.

DN mainly consists of three parts:

1. *Set of other DNs' information.* This mainly describes the other DNs information, such as performance-QoS, registry information, direct trust relationship, and recommended trust relationship among different DNs. It also has a detailed algorithm for requesting and responding to the resource service from other DNs.

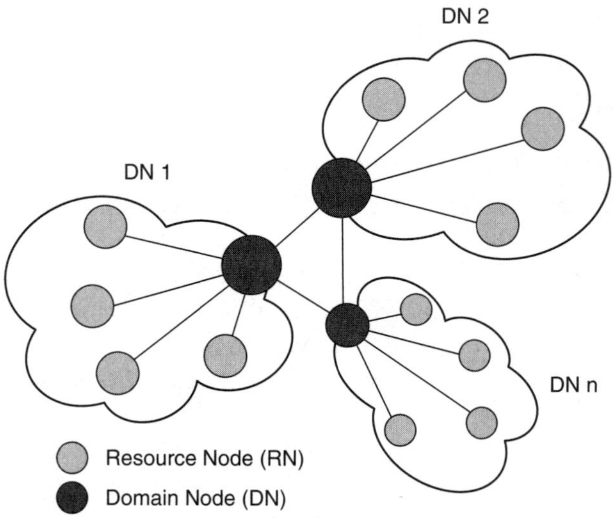

Figure 6.3 The relationship and its architecture of DNs and RNs in a VO (Tao *et al.*, 2006a).

2. *Resource classifier.* This is used to make classifications to the required or registered resources, in order to allocate the information onto the according domain node.
3. *Sets of its RNs information.* This primarily deposits the basic information of each RD in the specific DN, such as QoS information, registry information, direct trust relationship, and recommended trust relationship in the DN itself.

Because of the complexity and diversity of MGrid resources, different management strategies, security strategies, and trust evaluation strategies are adopted in different resource management domains in MGrid. According to the proposed two-layer resources management organization structure in MGrid, the trust relationship management of MGrid resource service can also be divided into two layers correspondingly, including *intra-domain trust relationship* management and *inter-domain relationship management* of resource services. Furthermore, MGrid resource service trust-QoS evaluation is also classified into two layers, they are *intra-domain trust-QoS evaluation model* (see Section 6.5.2) and *intra-domain trust-QoS evaluation model* (see Section 6.5.3).

6.4 MGrid Resource Service Trust-QoS Relationship Model

Trust-QoS is the belief or faith that an entity has in another entity's ability in a given context and time slot, that gives him/her confidence to carry out a transaction with the trusted entity. The trust-QoS is realized by the concept of *trust-QoS relationship*. Each *trust-QoS relationship* denotes a trust value from the trusting entity to the trusted entity, for a particular service content and time slot (Tao *et al.*, 2009d). The *trust-QoS relationship* is a composite of the following seven parts as shown in Figure 6.4:

1. Transaction *entities*, including both trusted entity and trusting entity,
2. manufacturing resource service *contexts*,
3. *trust type*, including *intra-domain trus-QoS* and *inter-domain trus-QoS*,
4. *trust-QoS request type*, including strong-trust request, weak-trust request, and no-trust quest,
5. *trust-QoS value*,
6. *trust-QoS benefit*, and
7. *time slot*.

Therefore, the *trust-QoS relationship* model is expressed as a seven-tuple (Tao *et al.*, 2009d): $(A, C, \Gamma, R, E, B, T)$.

- A is the set of participating entities in MGrid, including *trusting entity*, *trusted entity*, and *recommend entity*. The *trusting entity* is an entity that has faith or belief in the trusted entity. The *trusted entity* is an entity on which faith or belief has been placed by a *trusting entity*. The *recommend entity* is an entity which recommends its faith or belief in the trusted entity to a trusting entity. The entities of MGrid are all components involved in the operation of the MGrid system, which can be related to each other via certain trust relationship, including RSP (Resource Service Provider) and RSD (Resource Service Demander), RD (Resources Domain), RMS (Resources Management System), etc. In addition, an entity must be intelligent.

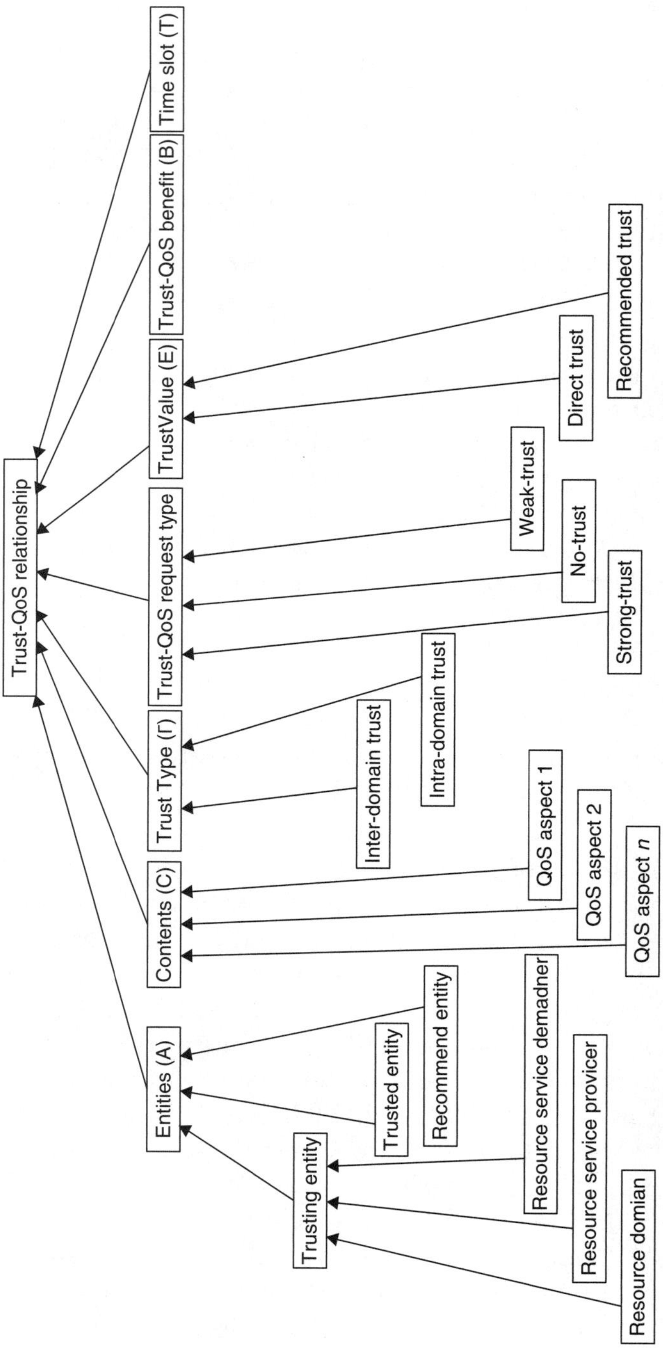

Figure 6.4 MGrid resource service trust-QoS relationship model.

- C is the set of specific service *context* of resource service transaction generated between two entities (trusting entity and trusted entity) in the MGrid system, e.g., data computation, data storage, parameterized design, 'machining a shaft,' etc. *Context* defines the nature of the entity and the entity's functions, service, and service functions. Each *context* has a number of *QoS aspects*. For example, entity A trusts entity B in the context of 'machining a shaft'. There are several *QoS aspects* of 'machining a shaft' such as, 'on-time delivery,' 'customer support,' 'price,' 'precision,' etc.
- Γ is defined as the set of trust type between two entities in MGrid and $\Gamma = \{intra - domain - trust, inter - domain - trust\}$. $intra - domain - trust$ denotes the trust type between two entities belonging to the same resource domain (RD), and $inter - domain - trust$ denotes the trust type between two entities belonging to two different resource domains (RD). Detailed descriptions about them will be presented in Sections 6.5.2 and 6.5.3 respectively.
- R is the set of trust level the trusting entities request from the trusted entities, and $R = \{strong - trust, weak - trust, no - trust\}$. *Strong-trust QoS request* means the requested resource service's trust-QoS level must be higher than the user's requesting trust-QoS level; otherwise the request will be abandoned. The evaluation benefit value to strong-trust QoS request is either the maximal or zero. *Weak-trust QoS request* means the requested resource service's trust-QoS degree level must satisfy the user's requesting trust-QoS requirements furthest. If the requested resource service's trust-QoS degree level cannot reach the user trust-QoS requirements, then a negotiation is needed to ask the user to loosen its QoS request level, and the corresponding evaluation trust-QoS benefit value would be reduced at the same time. *No-trust QoS request* means that the user has no trust-QoS request for the requested resource service, and any qualified resource services which satisfy its other QoS requests can serve it. The evaluation trust-QoS benefit to it is always the maximal.

- E is the integrated trust evaluation value of two entities with a given C and trust type of Γ at time slot t. It is the composite direct trust value (Definition 6.3) and recommended trust (Definition 6.4). The detailed evaluation models and calculating algorithms of E are different according to different trust types Γ, and the two kinds of evaluation model of E is described in detail in Sections 6.5.2 and 6.5.3, including intra-domain trust-QoS evaluation model and inter-domain trust-QoS evaluation model.
- B represents the set of trust-QoS *benefit functions* generated by different QoS under the different trust types and trust relationship, and it is an important index for resource scheduling in MGrid.
- T denotes the time slot, and is a time or a timeframe for which the trust value holds.

6.5 MGrid Resource Service Trust-QoS Evaluation Model

6.5.1 Related Conceptions and Definitions

The set of entities in MGrid (A): Entities in MGrid are all the component parts involved in the operation of an MGrid system, and can be related to each other via certain trust relationships, such as RSD (Resource Service Provider) and RSD (Resource Service Demander), RD (Resources Domain), RMS (Resources Management System), and so forth. The set of entities is denoted as $A = \{a_1, a_2, ..., a_n\}$ $(n = 1,2,3,\cdots)$. The symbols employed later (e.g., x, y, Ax, Ay, By, A, B, etc) all belong to A. An entity must be intelligent.

The set of MGrid resource service context (C): It represents the detailed context of manufacturing resource service transaction in an MGrid system, such as data computation, data storage and transmission, parameterized design, etc.

The set of trust degree (T_D): The trust value in MGrid system is difficult to be expressed with a precise numerical value, so we divided the trust degrees into seven classes and adopted a set of fuzzy logic values $T_D = \{a, b, c, d, e, f, g\}$, to stand for each trust degree value $t_d (t_d \in T_D)$. The trust degrees together with their meanings are shown in Table 6.1.

Table 6.1 Trust degrees and their meanings

Trust Degree	Corresponding Numerical Value	Meaning or Descriptions
a	6	The highest trustworthy
b	5	Very trustworthy
c	4	Trustworthy
d	3	Middle
e	2	Untrustworthy
f	1	Very untrustworthy
g	0	The highest untrustworthy

Definition 6.1, Requirements of resource service QoS : Let $Q = \{q_1, q_2, ..., q_n\}$ $(n = 1, 2, 3, ...)$ denote the vectorial set of QoS attributes required by any entity (RSD or RSP) to the other entity according to the manufacturing task submitted, and each $q_i (q_i \in Q)$ stands for the i th QoS attribute required, such as manufacturing cost, time, machining precision, etc.

Definition 6.2, Resource service trust-QoS: Let $\left((Q,E),(q_1,w_1,e_1),(q_2,w_2,e_2),\cdots,(q_n,w_n,e_n)\right)$ denote the trust-QoS, where e_i denotes the comprehensive trust evaluation value of q_i in the trust evaluation process; w_i represents the weight that e_i (or q_i) takes in E (or Q); E denotes the whole trust degree evaluation of Q, and $E = \sum_{i=1}^{n} e_i w_i$.

Definition 6.3, Direct trust and direct trust-QoS evaluation value d_{td}^* : Direct trust is the belief that trusting entity x holds in trusted entity y's trustworthiness with regard to a given certain $c(c \in C)$ and q $(q \in Q)$, within trust type $\Gamma(\Gamma \in \{intra-domian-trust, inter-domain-trust\})$ at current transacting time t. It mainly depends on the direct resource service transaction between entities x and y in the past. This means that if there was no resource service transaction in the past, then there will be no direct trust. Let d_{td}^* denote the direct trust-QoS evaluation value, and then d_{td}^* can be formulated as:

$$d_{td}^* = d_{td}(x, y, c, q, d_{td}, t) \otimes f^*(\Delta t) \tag{6.1}$$

Where:

- x is the trusting entity and y is the trusted entity, and $x, y \in A$.
- c is the specific resource server context, and $c \in C$.
- q is a specific QoS requirement attribute of the resource service required by entity x (or provided by entity y), and $q \in Q$.
- d_{td} is the direct trust of x with y in the latest transaction at the time, t_{x-y}, and $d_{td} \in T_D$. The calculating algorithms of d_{td} in detail are given in Section 6.7.1.
- t is the current evaluating time slot or the transacting time of current resource service transaction between x and y, and t is later than t_{x-y}.
- $d_{td}()$ is the direst trust evaluation function.
- $f^*(\Delta t)$ shows the trust upgrade or downgrade according to the *trust-time decay function* described in detail in Definition 6.5, and $\Delta t = t - t_{x-y}$.
- Symbol "$\otimes$" denotes the operation 'is downgraded or upgraded by the order'. For example, if $d_{td} = c$, and $f^*(\Delta t) = -2$ (downgraded by two), then $d_{td}^* = c \otimes -2 = e$. Conversely, if $d_{td} = c$, and $f^*(\Delta t) = 2$ (upgraded by two), then $d_{td}^* = c \otimes 2 = a$.

Definition 6.4, Recommended trust and recommended trust-QoS evaluation value r_{td}^* : Recommended trust is the indirect belief that trusting entity x holds in trusted entity y's trustworthiness with regard to a given c ($c \in C$) and q($q \in Q$), within trust type Γ at a current transacting time slot t. It depends on the recommended trust of the third entities (or recommender entities). Let r_{td}^* denote the integrated recommended trust-QoS evaluation value, and is decided by each recommender's (z_j's) recommendation $r_{td_j}^*$, r_{td}^*. It can be formulated as:

$$r_{td}^* = \sum_{j=1}^{k} r_{td_j}^* = \sum_{i=1}^{k} r_{td}(z_j, y, c, q, r_{td_j}, t) \otimes f^*(\Delta t) \qquad (6.2)$$

Where:

- y is the recommended (or trusted) entity, and z_j is the j th recommender of y, where $x, y, z_j \in A$; and k is the number of the total recommenders.
- c is the specific resource server context, and $c \in C$.

- q is a specific QoS requirement attribute of resource service required by entity x (or provided by entity y), and $q \in Q$.
- r_{td_j} is the jth recommender's (z_j's) recommendation in the latest interaction at time t_{x-y}, and $r_{td_j} \in T_D$. The calculating algorithms of r_{td_j} in detail are given in Section 6.7.2.
- t is the current evaluating time or the transacting time of current resource service transaction between x and y, and t is later than t_{x-y}.
- $f^*(\Delta t)$ is the trust upgrade or downgrade according to the *trust-time decay function* described in detail in Definition 6.5, and $\Delta t = t - t_{x-y}$.
- $\sum_{j=1}^{k} r^*_{r_{td_j}}$ (n = 1, 2, 3, ...) denotes the combining algorithm of each recommender z_j's recommendation r_{td_j}, and the calculating algorithms of $\sum_{j=1}^{k} r_{td_j}$ in detail are given in Section 6.7.3.

Definition 6.5, Trust-time decay function $f(\Delta t)$ and trust decay grade $f^*(\Delta t)$: In an MGrid system, the trust degree level of any entity $x(x \in A)$ holding in another entity $y(y \in A)$ usually declines as the time t increases. For example, in the last transaction at time t_{x-y} (e.g., t_{x-y} ='20060901'), x's trust evaluation with y is $a(\in T_D)$; but in current interaction at current time t (t is later than t_{x-y}, e.g., t ='20061001'), x's trust evaluation with y declines to c ($c \in T_D$), which downgrades by two grades because x and y have no contact with each other during the time between t_{x-y} and t. Therefore, in the process of trust evaluation, an adjustment function should be employed when the system evaluates the trust-QoS, including both direst trust and recommended trust. We introduced the adjustment function as *trust-time decay function* and marked it with $f(\Delta t)$. $f(\Delta t)$ can be formulated as:

$$f(\Delta t) = e^{-\Delta t/m}, (f(\Delta t) \in [0,1], \Delta t = t - t_{x-y}, m = N_{date}/1.95) \quad (6.3)$$

Where, N_{date} is the number of the days required when trust degree decays from grade a (the highest trustworthy grade) to grade g (the highest untrustworthy grade).The trust-time decay function curves according to different N_{date} (e.g., N_{date} = 30, 60, 90, 180, 360, 540, respectively), and is shown in Figure 6.5. The trust degree levels are divided into seven grades (see Table 6.1). Therefore, for each entity y, the trust degree of entity x holding in another entity y could decay from level a to level f, and so there are seven decay values with regard to the value of $f(\Delta t)$. Let $f^*(\Delta t)$ denote the *trust decay grade* values, and $f^*(\Delta t)$ is the mapping value according to $f(\Delta t)$, where $f^*(\Delta t) \in \{0, -1, -2, -3, -4, -5, -6\}$. The mapping of the values from $f(\Delta t)$ to $f^*(\Delta t)$ is shown in Table 6.2.

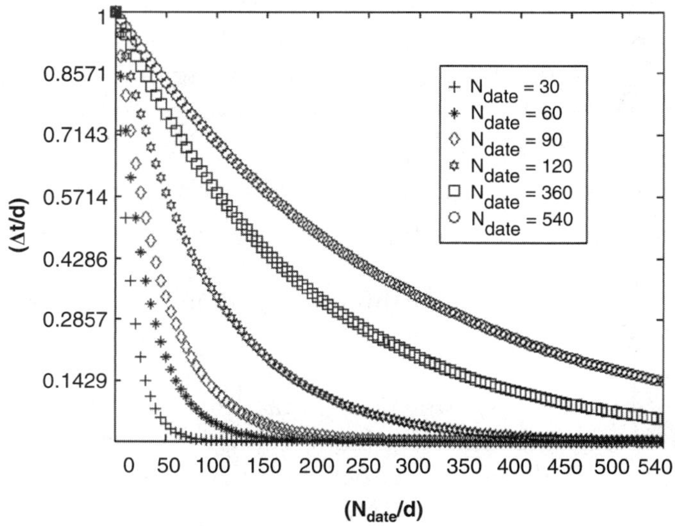

Figure 6.5 Trust-time decay function curves according to different N_{date} (N_{date}= 30, 60, 90, 180, 360, 540 respectively).

Table 6.2 The mapping of the values from $f(\Delta t)$ to $f^*(\Delta t)$

Time Decay Values	Trust Decay Grade Values $f^*(\Delta t)$
$0 \le f(\Delta t) \le 1/7$	–6
$1/7 < f(\Delta t) \le 2/7$	–5
$2/7 < f(\Delta t) \le 3/7$	–4
$3/7 < f(\Delta t) \le 4/7$	–3
$4/7 < f(\Delta t) \le 5/7$	–2
$5/7 < f(\Delta t) \le 6/7$	–1
$6/7 < f(\Delta t) \le 1$	0

6.5.2 Intra-domain Trust-QoS Evaluation Model of Resource Service

Provided that *Ax* and *Ay* stand for two entities in the same resources domain *A*, when *Ax* and *Ay* have a resource service transaction with regard to a given service content (c ($c \in C$)) and $q(q \in Q)$ at current transacting time t, then the trust relationship type between A_x and A_y are defined as an *intra-domain-trust*. So the trust type Γ in a trust-QoS model between the two entities is $intra-domian-trust$, that is $\Gamma = intra-domian-trust$.

Let $E = T_{x \to y}$ denote the intra-domain trust evaluation value of A_x with A_y, so $E = T_{x \to y}$ in the trust-QoS relationship model. $T_{x \to y}$ consist of two parts: (a) d_{td}^*, the direct trust-QoS degree evaluation value between A_x and A_y with regard to a certain $c(c \in C)$) and $q(q \in Q$ at time t, and $d_{td}^* = d_{td}(x, y, c, q, d_{td}, t) \otimes f^*(\Delta t)$; and (b) r_{td}^*, the comprehensive combined recommended trust-QoS evaluation value of each recommender's individual recommended trust with A_y, and $r_{td}^* = \sum_{j=1}^{k} rt(z_j, y, c, q, r_{td_j}, t) \otimes f^*(\Delta t)$. d_{td}^* and r_{td}^* have been described in detail in Definitions 6.3 and 6.4. Let α, β be the weights of d_{td}^* and r_{td}^* respectively. So the *intra-domain trust-QoS* evaluation value of A_x with A_y, $T_{x \to y}$, can be formulated as:

$$T_{x \to y} = \alpha \times d_{td}^* + \beta \times r_{td}^* \tag{6.4}$$

Therefore, the detailed *intra-domain trust-QoS evaluation model* can be formulated as:

$$\begin{aligned} T_{x \to y} &= \alpha \times \left(d_{td}(x, y, c, q, d_{td}, t) \otimes f^*(\Delta t) \right) \\ &\quad + \beta \times \left(\sum_{j=1}^{k} r_{td}(z_j, y, c, q, r_{td_j}, t) \otimes f^*(\Delta t) \right) \end{aligned} \tag{6.5}$$

6.5.3 Inter-domain Trust-QoS Evaluation Model of Resources Service

Provided that *Ax* and *By* stand for two entities in two different resource domains *A* and *B* respectively, when *Ax* and *By* do a resource service transaction with regard to a certain $c(c \in C)$) and

$q(q \in Q)$ at current transacting time t, then the trust relationship type between Ax and By are defined as an *inter-domain-trust*. So the trust type Γ in a trust-QoS relationship model between the two entities is *inter – domian–trust*, that is Γ = *inter – domian–trust*.

Let T^*_{x-y} denote the inter-domain trust evaluation value of Ax about By, so $E = T^*_{x\to y}$ in the trust-QoS relationship model this time. $T^*_{x\to y}$ also consists of two parts: (a) $T_{A\to B}$, the integrative trust evaluation of resource domains A with B, and $T_{A\to B} = \alpha^* \times d_{td}{}^* + \beta^* \times r_{td}{}^*$, where $d_{td}{}^* = d_{td}(A,B,c,q,d_{td},t) \otimes f^*(\Delta t)$ denotes the direct trust evaluation value of A about B with regard to a certain $c(c \in C)$) and $q(q\in Q)$atcurrenttranslatingtime$t$; $r_{td}{}^* = \sum_{j=1}^{k} r_{td}(Z_j, B, c, q, r_{td_j}, t) \times f^*(\Delta t)$ denotes the comprehensive combined recommended trust evaluation value of each recommender's (Z_j's) individual recommended trust with B, and let α^*, β^* be the weights of $d_{td}{}^*$ and $r_{td}{}^*$ respectively; and (b) W_y^B, the trust weight that entity By occupies in resources domain B. So the inter-domain trust evaluation value of Ax about By, $T^*_{x\to y}$, can be formulated as:

$$T^*_{x\to y} = W_y^B \times T_{A\to B} = W_y^B \times \left(\alpha^* \times d_{td}{}^* + \beta^* \times r_{td}{}^*\right) \tag{6.6}$$

Therefore, the detailed *inter-domain trust evaluation model* can be represented as:

$$T^*_{x\to y} = W_y^B \times \begin{pmatrix} \alpha^* \times \left(d_{td}(A,B,c,q,d_{td},t) \otimes f^*(\Delta t)\right) + \\ \beta^* \times \left(\sum_{j=1}^{k} r_{td}(Z_j,B,c,q,r_{td_j},t) \otimes f^*(\Delta t)\right) \end{pmatrix} \tag{6.7}$$

6.6 Data Structure Design

MGrid resource service contexts (C) and QoS attributes aspect (Q): As described in the trust-QoS relationship model, C is the set of specific service *context* of resource service transaction generated between two entities in an MGrid system. E.g., data computation, data storage, parameterized design, "machining a shaft", etc. *Context* defines the nature of the entity and the entity functions, service, and service functions, and each *Context* has the following features (Tao *et al.*, 2009d):

- Name: indicates the name of the context.
- Type: describes the type of the context.
- Functional specification: provides a brief functional information description of the context.
- Influencing factions: depicts the factors to execute this context (mainly designed for trust-QoS based scheduling).
- Relate QoS aspects: indicates other related context.
- Trust-QoS level: indicates the possible trust-QoS levels offered by an entity.
- Application area: describes the application area in which the context has been used.
- Anticipate consequences: describes the effects if this context is executed.
- Measure: describes the unit in which to measure the context if possible.

Each *context* has a number of *QoS attributes aspects*, and *QoS attributes aspects* of a context are the QoS assessment criteria that can be used for quality measure.

Example 1. Entity A trusts entity *B* in the context of 'machining a shaft'. There are several *QoS aspects* of 'machining a shaft' such as 'on-time delivery,' 'customer support,' 'price,' 'precision,' etc.

Example 2. Entity A trusts entity *B* in the context of 'manage the order data'. There are several *QoS aspects* of 'manage the order data' such as 'real-time,' 'safety,' 'price,' 'integrality,' etc.

Direct trust degree data structure (D_{DT}): As stated before, direct trust evaluation is based upon the experience and results of previous transactions in an MGrid system. Therefore, it was designed so that each entity maintains a table of direct trust degree data structure, named (D_{DT}). Provided that the 'outcome of trust evaluation grade' of the direct trust degree of entity x with entity y, d_{td}, with regard to a given $c(c \in C)$, $q(q \in Q)$), within trust type Γ at time t, will be one alpha from the ordered set $T_D = \{a, b, c, d, e, f, g\}$, representing 'The highest trustworthy, 'Very trustworthy', 'Trustworthy', 'Middle', 'Untrustworthy', 'Very untrustworthy', 'The highest untrustworthy', respectively. So, each evaluated entity's direct trust degree (with regard to a certain resource service context c at time t) is an associated 7-tuple $s = (s_a, s_b, s_c, s_d, s_e, s_f, s_g)$, where S_i is an accumulator for trust degree evaluation when $t_d = i$. Let $S = \{(s_a, s_b, s_c, s_d, s_e, s_f, s_g)\}$ and D_{DT} can be defined as:

$$D_{\mathrm{DT}} = (A, C, Q, S, T) \tag{6.8}$$

Where A is the set of entities including both trusting entity and trusted entity; T stands for the set of transaction times. E.g., the direst trust degree data structure of entity x about entities y_1, y_2, y_3 is as follow:

$$\begin{aligned} D_{\mathrm{DT}} = \{ \\ \big((x,\ y_1),\ c_1,\ q_1,\ (\{1\},\ \{\},\ \{\},\ \{\},\ \{\},\ \{\},\ \{\}),\ 20060101\big), \\ \big((x,\ y_2),\ c_1,\ q_1,\ (\{\},\ \{\},\ \{1\},\ \{\},\ \{\},\ \{\},\ \{\}),\ 20060101\big), \\ \big((x,\ y_3),\ c_1,\ q_1,\ (\{\},\ \{\},\ \{\},\ \{1\},\ \{\},\ \{\},\ \{\}),\ 20060101\big)\} \end{aligned}$$

This result demonstrates that the direct trust degree evaluation of entity x with entities y_1, y_2, y_3, with regard to a given $c(c \in C), q(q \in Q)$, within trust type Γ at time '20060101' , are a, c, d respectively.

Recommended trust degree data structure (D_{RT}): As stated before, recommended trust is the indirect belief that any entity x holds in another entity y of its trustworthiness, with regard to a given$c(c \in C), q(q \in Q)$, within Γ at time t. It depends on the recommended trust of the third entities. Similarly, with direct trust evaluation there are also seven grades of the outcome of recommended trust evaluation, and each entity maintains a table of recommended trust degree data structure named D_{RT}. Therefore, because the recommended trust depends on the third entities, on the one hand, a same recommender entity may lie or give out contradictory recommendations about the same recommended entity to different entities. On the other hand, there may be a gap between the trust entity's own evaluation and the recommender's recommendation about the same evaluated entity. For example, if entity z recommends to x that entity y's trusted degree is a (the highest trustworthy degree), while x's evaluation based on its own experience about y's trusted degree is merely c (trustworthy), then future recommendations from z need to be adjusted accordingly. That means x's evaluation with y downgrades z's recommendation by two (the recommended trust degree adjustment value is –2). We denote this recommended trust adjustment as *Semantic Distance* (sd), and mark its value with $s_{d_{t_d}}(t_d \in T_D)$ and $s_{d_a} = -2$ in this example.

Hence, the final recommended trust that entity x actually accepts from z should be $c(c = a \otimes -2)$ in the process of this trust evaluation. Therefore, in order to obtain a reasonable recommended trust degree from the set of recommendations, each recommended trust degree must be adjusted properly. Let G denote the set of possible adjustment values, and $G = \{-6, -5, -4, -6, -2, -1, 0, 1, 2, 3, 4, 6\}$. Therefore, each recommender entity's recommended trust degree of evaluated entity with regard to a specific c, q, Γ at time t, is an associated 7-tuple $S_D = \{(s_{d_a}, s_{d_b}, s_{d_c}, s_{d_d}, s_{d_e}, s_{d_f}, s_{d_g})\}$. Each represents the adjustment of the recommender's recommend trust degree of i when $t_d = i$ and the domain for values in s_d is the set G. Let $S_D = \{(s_{d_a}, s_{d_b}, s_{d_c}, s_{d_d}, s_{d_e}, s_{d_f}, s_{d_g})\}$ and D_{RT} be defined as:

$$D_{RT} = (A, C, Q, S_D, T) \tag{6.9}$$

Where A is the set of entities, including trust entity and trusted entity; T stands for the set of transaction time. E.g., the recommended trust data structure of three recommenders (z_1, z_2, z_3) with recommended entity y is as follows:

$$\begin{aligned}
D_{RT} = \{ \\
&\big((y,\ z_1),\ c_1,\ q_1,\ (\{-1\},\ \{\},\ \{\},\ \{\},\ \{\},\ \{\},\ \{\}),\ 20060101\big), \\
&\big((y,\ z_2),\ c_1,\ q_1,\ (\{\},\ \{\},\ \{-2\},\ \{\},\ \{\},\ \{\},\ \{\}),\ 20060101\big), \\
&\big((y,\ z_3),\ c_1,\ q_1,\ (\{\},\ \{\},\ \{\},\ \{1\},\ \{\},\ \{\},\ \{\}),\ 20060101\big)\}
\end{aligned}$$

The result demonstrates that the recommended trust degree evaluation of the three recommenders (z_1, z_2, z_3) about entities y's are a, c, d, respectively, with given c_1, q_1 and Γ at time '20060101'. And the corresponding adjustment values of the three recommended trust degrees (a, c, d) are $sd_a = -1$, $sd_c = -2$ and $sd_d = 1$, respectively. Therefore, the final recommend trust degrees that entity x actually accept from z_1, z_2, z_3 should be $b(a \otimes s_{d_a} = b), e(c \otimes s_{d_c} = e)$ and $c(d \otimes s_{d_d} = c)$, respectively.

6.7 Trust-QoS Evaluating and Updating Algorithms

The brief evaluation workflow of intra-domain-trust evaluation and inter-domain-trust evaluation is shown in Figures 6.6 and 6.7, respectively (Tao *et al.*, 2009d). From Figures 6.6 and 6.7 it can be concluded that the direct trust-QoS evaluation algorithms, the recommended trust-QoS evaluation algorithms, and the comprehensive combining recommended trust-QoS evaluating algorithms, are at the core to realizing the evaluation of resource service trust-QoS value. Therefore, in the remainder of this section we will describe the above algorithms in detail, respectively.

6.7.1 Direct Trust-QoS Evaluating Algorithm

To evaluate the direct trust degree (d_{td}^*) of entity x with y, with given $c(c \in C)$, $q(q \in Q)$, $\Gamma(\Gamma = intra - domian - trust)$ at the current transacting time t, there are three steps to follow.

1. Firstly, obtain the corresponding direct trust data records $((x, y), c, q, s, t)$ in D_{DT}, and calculate the previous direct trust degree d_{td} $(d_{td} = t_d \in T_D)$ as being the last transaction time t_{x-y}. The value of d_{td} is the subscript of S_{td} where S_{td} is the largest member in s (Abdul-Rahman *et al.*, 2000), where

$$\exists d_{td} \in T_D, \forall s_{td} \in s, s_{td} = \max(s) \text{ then, } d_{td} = t_d \tag{6.10}$$

2. Secondly, calculate the values of $f(\Delta t)$ and $f^*(\Delta t)$. Obtain the time t_{x-y} according to S_{td} which is the largest member in S from D_{DT}, and calculate the value of $f(\Delta t) = f(t - t_{x-y})$ using formula (6.7), and compute the trust-time decay grade $f^*(\Delta t)$ according to Figure 6.5 and Table 6.2.
3. Lastly, calculate the final direct trust degree d_{td}^*.

$$d_{td}^* = d_{td} \otimes f(\Delta t) \tag{6.11}$$

6.7.2 Recommended Trust-QoS Evaluating Algorithm

To evaluate the recommended trust degree r_{td}^* of entity z with y with given $c(c \in C)$, $q(q \in Q)$, $\Gamma(\Gamma = intra - domian - trust)$, etc., at time t, there are three steps to follow.

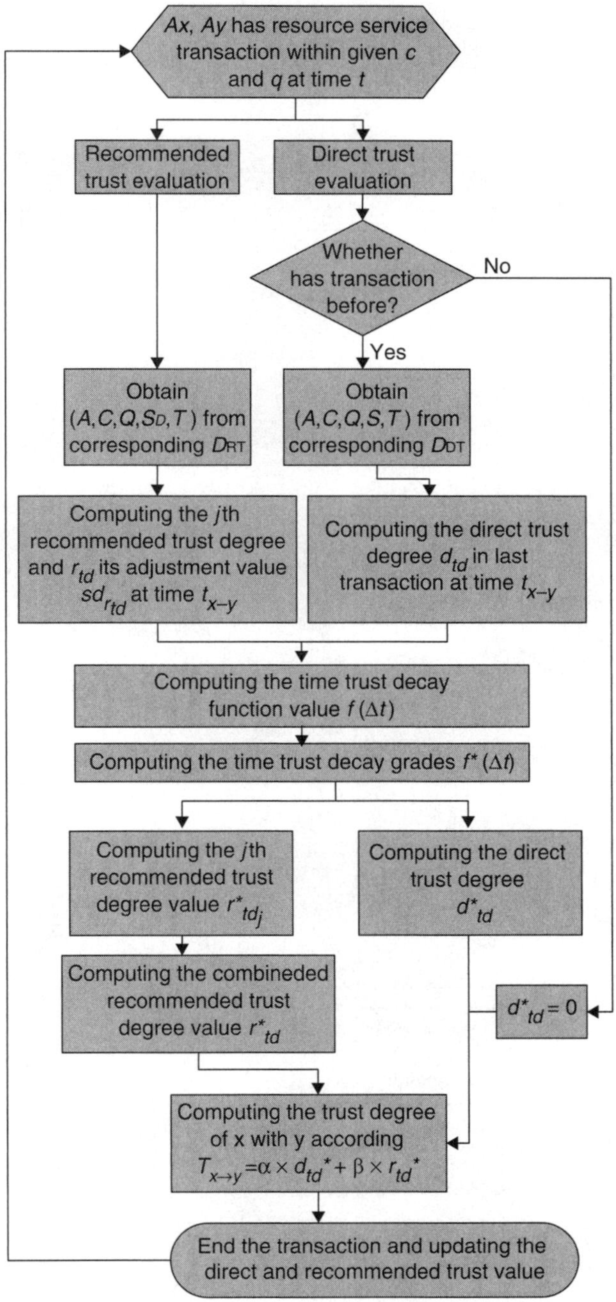

Figure 6.6 Brief workflow of intra-domain-trust evaluation.

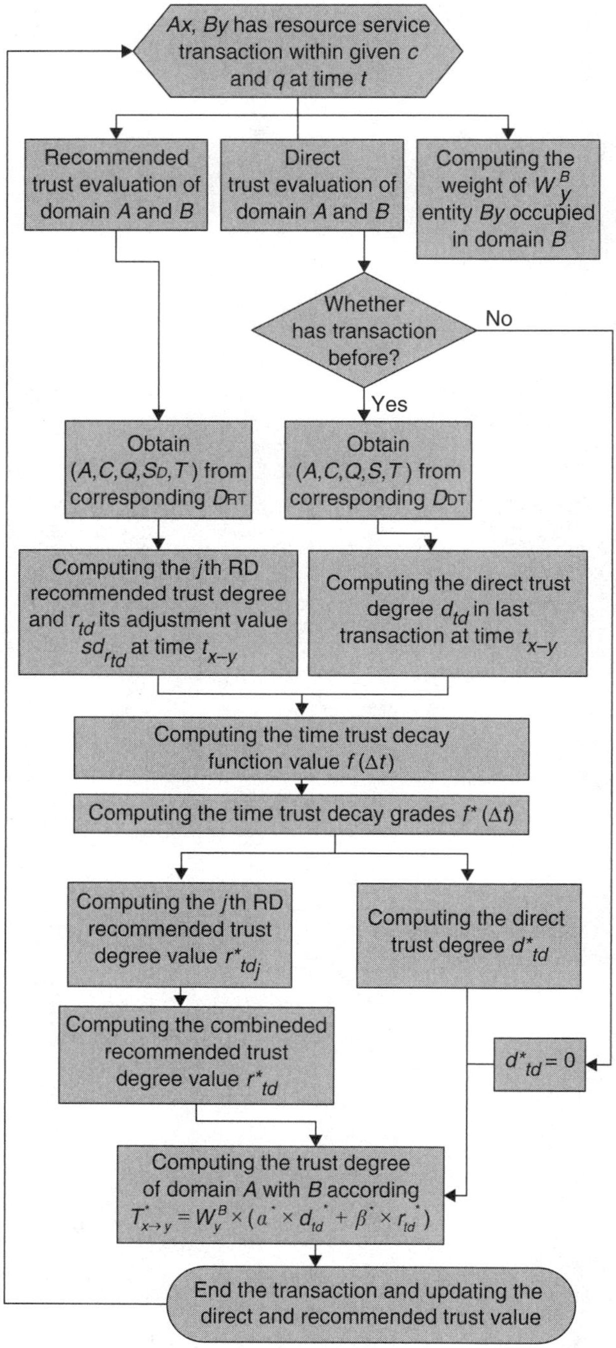

Figure 6.7 Brief workflow of inter-domain-trust evaluation.

1. Firstly, obtain the recommended trust data records $((y,z),c,q,s_d,t)$ in D_{RT} together with the recommended trust degree $r_{td}(r_{td} = t_d \in T_D)$, and calculate the recommended trust degree of r_{td}'s adjustment value sd_{t_d}.
2. Then compute the values of $f(\Delta t)$ and $f^*(\Delta t)$. Obtain the time $t_{x\text{-}y}$ and calculate the value of $f(\Delta t) = f(t-t_{x\text{-}y})$ using Formula 6.3 and the trust downgrade $f^*(\Delta t)$ according to Table 6.2.
3. Lastly, calculate the final recommended trust degree r_{td}^*.

$$r_{td}^* = r_{td} \otimes s_{d_{t_d}} \otimes f^*(\Delta t) \tag{6.12}$$

6.7.3 Comprehensive Combining Recommended Trust-QoS Evaluating Algorithm

Sometimes entity x may accept more than one recommender for a particular trusted entity y. To evaluate the final comprehensive recommended trust degree r_{td}^* of y, the system should combine each recommended trust degree and generate a comprehensive recommendatory trust degree.

Suppose $z_1, z_2, \ldots z_k$ ($k = 1, 2, 3 \ldots$) are the recommenders of y and $r_{td_1}^*, r_{td_2}^*, r_{td_3}^*, \ldots, r_{td_k}^*$ (i.e., $r_{td_i}^*$ is the recommended trust value for recommending entity Z_i) denote the corresponding individual recommended trust degree of each recommender about y according to Expressions 6.3 and 6.4. Let $w_1, w_2, w_3, \ldots, w_k$ be the corresponding individual weights of each recommender according to $r_{td_1}^*, r_{td_2}^*, r_{td_3}^*, \ldots, r_{td_k}^*$ (i.e., w_i is the weight for $r_{td_i}^*$). For each recommended trust degree $r_{td_i}^* (r_{td_i}^* = t_d \in T_D)$, sum the weights of the recommenders who recommend t_d ($t_d \in T_D$) and let sum_a denote the sum result (e.g., Sum_a denotes the sum result of the weights of all the recommenders who recommend a). Let W_{t_d} be the set whose members are the weights associated with recommenders who recommended t_d and C_{t_d} be the final combined recommended trust degree of y, and so we have

$$\forall r_{td_i}^* = t_d \in T_D, \forall w_i \in W_{t_d}, (Sum_{t_d} = \sum_{i=1}^{W_{td}} w_i) \Rightarrow (C_{t_d} = t_d) \tag{6.13}$$

Therefore, the final combined recommended trust degree of $y, C_{r_{td}}$ is the subscript value or index of Sum_{t_d} with the highest value (Abdul-Rahman *et al.*, 2000).

For example, assume z_1, z_2, z_3 z_4 to be recommenders of y, and the respective individual recommended trust degree of them are $r_{td_1}^* = b, r_{td_2}^* = c, r_{td_3}^* = d, r_{td_4}^* = b,$ and the weights for each recommender to be $w_1 = 2$, $w_2 = 9$, $w_3 = 5$, $w_4 = 6$. According to Expression 6.13, $Sum_b = 8$, $Sum_c = 9$, $Sum_d = 5$, and obviously, $Sum_c > Sum_b > Sum_d >$ and so $C_{t_d} = c,$ which means the final combined recommended trust degree of x with y is 'c'.

6.8 Real-time and Dynamical Updating Algorithm of Trust-QoS Degree

The trust degree between two entities is changed according to the date and conditions such as resource service trade, etc. So the trust degree between two entities should be updated real-time and dynamically after a resource service interaction, including both direct trust degree and recommended trust degree.

After a resource service interaction with entity y, the trust degree for entity x in D_{DT} and D_{RT} should be updated real-time and dynamically according to the interaction result.

For example, suppose the direct trust degree data records of x with y, with given $c_1(c_1 \in C)$, $q_1(q_1 \in Q)$, Γ at time t_1 to be:

$$D_{\mathrm{DT}} = \{((x,y),\ c_1,\ q_1,\ (0,\ 0,\ 1,\ 0,\ 0,\ 0,\ 0,),\ t_1)\},$$

It shows that the direct trust degree of x with y is c.

Suppose the recommended trust data records of recommender entities z_1, z_2, z_3 with y at time t_1 to be:

$$\begin{aligned}
&D_{\mathrm{RT}} = \{\\
&((y,z_1),\ c_1,\ q_1,\ (\{-1\},\ \{\},\ \{\},\ \{\},\ \{\},\ \{\},\ \{\}),\ t_1),\\
&((y,z_2),\ c_1,\ q_1,\ (\{\},\ \{\},\ \{-2\},\ \{\},\ \{\},\ \{\},\ \{\}),\ t_1),\\
&((y,z_3),\ c_1,\ q_1,\ (\{\},\ \{\},\ \{\},\ \{1\},\ \{\},\ \{\},\ \{\}),\ t_1)\}
\end{aligned}$$

It demonstrates that the recommended trust degree of z_1, z_2, z_3 with y is $r_{td_1} = b$, $r_{td_2} = e$, and $r_{td_3} = c$ respectively.

When entities x and y trade with each other at time t_2(t_2 is later than t_1), the system calculates the direct trust degree of x with y as $d_{td}^* = b$, and the combined recommended trust degree is $r_{td}^* = c$ according to Expressions 6.10–6.13.

The system results demonstrate that the direct trust degree d_{td}^* of entity x with y changes from the previous 'c' to current 'b', which is upgraded by one ($d_{td}^* = d_{td}^* \otimes 1$). So after the interaction at timet_2, the direct trust of x with y can be updated as follows:

$$D_{\mathrm{DT}} = \{(x,y), c_1, q_1, (\{\}, \{1\}, \{\}, \{\}, \{\}, \{\}, \{\}, t_2)\}$$

At the same time, the system results demonstrate that the combined recommended trust degree of x with y is $r_{td}^* = c$ at time t_2, so the new recommended trust degree adjustment values of z_1, z_2, z_3 with y are -1, 2, 0. Therefore, the recommended trust degree of z_1, z_2, z_3 with y can be updated as follows:

$$\begin{aligned}
&D_{\mathrm{RT}} = \{\\
&\big((y,z_1),\ c_1,\ q_1,\ (\{\},\ \{-1\},\ \{\},\ \{\},\ \{\},\ \{\},\ \{\}),\ t_2\big),\\
&\big((y,z_2),\ c_1,\ q_1,\ (\{\},\ \{\},\ \{\},\ \{\},\ \{2\},\ \{\},\ \{\}),\ t_2\big),\\
&\big((y,z_3),\ c_1,\ q_1,\ (\{\},\ \{\},\ \{0\},\ \{\},\ \{\},\ \{\},\ \{\}),\ t_2\big)\}
\end{aligned}$$

After this process, we realize the real-time updating of direct trust degree and recommended trust degree dynamically.

6.9 Trust-QoS Evaluation Case Study

It is assumed that an entity (RSD-Resources Services Demander) named x, looks for magnetic hard disk experimental equipment (a kind of special experimental equipment needed in the development of a magnetic hard disk) through a MGrid system platform. When it submits its requirement to the MGrid system platform, the system finds out that there are five entities' QoS (five RSPs-Resource Services Providers) in accordance with x's QoS requirements, namely a_1, a_2, a_3, a_4, a_5. In order to select the optimal entity to provide the required resource service for x , reduce the decision-making time for x to select the best resource provider to fulfill its task, and enhance the QoS of MGrid system, the system evaluates the trust-QoS degree of each candidate entity according to the evaluation models and algorithms proposed above, and selects the optimal one to accomplish x's task. The corresponding D_{TD} and D_{RT} about x and the five entities (a_1, a_2, a_3, a_4, a_5) in the database of MBRSSP-MGrid are as follows:

$D_{DT} = \{\}$.

$D_{RT} = \{$

$((a_1,b_1),\ c_1,\ q_1,\ (\{-1\},\ \{\},\ \{\},\ \{\},\ \{\},\ \{\},\ \{\}),\ 20060810),$

$((a_1,b_2),\ c_1,\ q_1,\ (\{\},\ \{\},\ \{-1\},\ \{\},\ \{\},\ \{\},\ \{\}),\ 20060910),$

$((a_1,b_3),\ c_1,\ q_1,\ (\{0\},\ \{\},\ \{\},\ \{\},\ \{\},\ \{\},\ \{\}),\ 20060830),$

$((a_1,b_4),\ c_1,\ q_1,\ (\{-1\},\ \{\},\ \{\},\ \{\},\ \{\},\ \{\},\ \{\}),\ 20060920),$

$((a_1,b_5),\ c_1,\ q_1,\ (\{\},\ \{0\},\ \{\},\ \{\},\ \{\},\ \{\},\ \{\}),\ 20060715),$

$((a_2,b_6),\ c_1,\ q_1,\ (\{0\},\ \{\},\ \{\},\ \{\},\ \{\},\ \{\},\ \{\}),\ 20060909),$

$((a_2,b_7),\ c_1,\ q_1,\ (\{\},\ \{\},\ \{\},\ \{2\},\ \{\},\ \{\},\ \{\}),\ 20060815),$

$((a_2,b_8),\ c_1,\ q_1,\ (\{\},\ \{\},\ \{\},\ \{0\},\ \{\},\ \{\},\ \{\}),\ 20060819),$

$((a_2,b_9),\ c_1,\ q_1,\ (\{\},\ \{\},\ \{1\},\ \{\},\ \{\},\ \{\},\ \{\}),\ 20060916),$

$((a_3,b_{10}),\ c_1,\ q_1,\ (\{\},\ \{\},\ \{\},\ \{-1\},\ \{\},\ \{\},\ \{\}),\ 20060816),$

$((a_3,b_{11}),\ c_1,\ q_1,\ (\{\},\ \{0\},\ \{\},\ \{\},\ \{\},\ \{\},\ \{\}),\ 20060710),$

$((a_3,b_{12}),\ c_1,\ q_1,\ (\{\},\ \{\},\ \{\},\ \{-1\},\ \{\},\ \{\},\ \{\}),\ 20060806),$

$((a_3,b_{13}),\ c_1,\ q_1,\ (\{0\},\ \{\},\ \{\},\ \{\},\ \{\},\ \{\},\ \{\}),\ 20060923),$

$((a_3,b_{14}),\ c_1,\ q_1,\ (\{\},\ \{\},\ \{\},\ \{-1\},\ \{\},\ \{\},\ \{\}),\ 20060804),$

$((a_3,b_{15}),\ c_1,\ q_1,\ (\{\},\ \{0\},\ \{\},\ \{\},\ \{\},\ \{\},\ \{\}),\ 20060901),$

$((a_4,b_{16}),\ c_1,\ q_1,\ (\{\},\ \{\},\ \{\},\ \{2\},\ \{\},\ \{\},\ \{\}),\ 20060817),$

$((a_4,b_{17}),\ c_1,\ q_1,\ (\{\},\ \{\},\ \{\},\ \{0\},\ \{\},\ \{\},\ \{\}),\ 20060903),$

$((a_4,b_{18}),\ c_1,\ q_1,\ (\{0\},\ \{\},\ \{\},\ \{\},\ \{\},\ \{\},\ \{\}),\ 20060830),$

$((a_4,b_{19}),\ c_1,\ q_1,\ (\{-1\},\ \{\},\ \{\},\ \{\},\ \{\},\ \{\},\ \{\}),\ 20060916),$

$((a_4, b_{20}),\ c_1,\ q_1,\ (\{0\},\ \{\},\ \{\},\ \{\},\ \{\},\ \{\},\ \{\}),\ 20060824),$

$((a_4, b_{21}),\ c_1,\ q_1,\ (\{\},\ \{0\},\ \{\},\ \{\},\ \{\},\ \{\},\ \{\}),\ 20060710),$

$((a_4, b_{22}),\ c_1,\ q_1,\ (\{\},\ \{\},\ \{\},\ \{0\},\ \{\},\ \{\},\ \{\}),\ 20060816),$

$((a_4, b_{23}),\ c_1,\ q_1,\ (\{0\},\ \{\},\ \{\},\ \{\},\ \{\},\ \{\},\ \{\}),\ 20060904),$

$((a_5, b_{24}),\ c_1,\ q_1,\ (\{\},\ \{0\},\ \{\},\ \{\},\ \{\},\ \{\},\ \{\}),\ 20060716),$

$((a_5, b_{25}),\ c_1,\ q_1,\ (\{0\},\ \{\},\ \{\},\ \{\},\ \{\},\ \{\},\ \{\}),\ 20060823),$

$((a_5, b_{26}),\ c_1,\ q_1,\ (\{\},\ \{-1\},\ \{\},\ \{\},\ \{\},\ \{\},\ \{\}),\ 20060819),$

$((a_5, b_{27}),\ c_1,\ q_1,\ (\{\},\ \{\},\ \{\},\ \{0\},\ \{\},\ \{\},\ \{\}),\ 20060809),$

$((a_5, b_{28}),\ c_1,\ q_1,\ (\{\},\ \{\},\ \{0\},\ \{\},\ \{\},\ \{\},\ \{\}),\ 20060826),$

$((a_5, b_{29}),\ c_1,\ q_1,\ (\{\},\ \{0\},\ \{\},\ \{\},\ \{\},\ \{\},\ \{\}),\ 20060709),$

}

From the system results, we can know that x has no resource service trade with a_1, a_2, a_3, a_4, a_5 so the direct trust relationships among them are empty (i.e., D_{DT} is empty), So the system only needs to consider the recommended trust evaluation value when estimating the comprehensive trust values between x and a_1, a_2, a_3, a_4, a_5.

Each recommended trust degree concerning a_1, a_2, a_3, a_4, a_5 is shown in Table 6.3. Set $N_{date} = 180$ and the current system date is "20061001" in the evaluating process.

Table 6.3 Recommendations of each candidate entities to entity x

RSP	Recommender	Recommended Trust Degree	s_{td}	$\lvert f(\Delta t) \rvert$	Adjusted Recommendation	Weights
a_1	b_1	a	−1	−1	c	8
	b_2	c	−1	0	d	6
	b_3	a	0	−1	b	9
	b_4	a	−1	0	b	2
	b_5	b	0	−2	d	4

(Continued)

Table 6.3 (cont.) Recommendations of each candidate entities to entity x

RSP	Recommender	Recommended Trust Degree	s_{td}	$\|f(\Delta t)\|$	Adjusted Recommendation	Weights
a_2	b_6	a	0	0	a	9
	b_7	d	2	–1	c	0
	b_8	d	0	–1	e	5
	b_9	c	1	0	b	7
a_3	b_{10}	d	–1	–1	f	6
	b_{11}	b	0	–2	d	8
	b_{12}	d	–1	–1	f	3
	b_{13}	a	0	0	a	0
	b_{14}	d	–1	–1	f	1
	b_{15}	b	0	–1	c	8
a_4	b_{16}	d	2	–1	c	5
	b_{17}	d	0	–1	e	1
	b_{18}	a	0	–1	b	2
	b_{19}	a	–1	0	b	0
	b_{20}	a	0	–1	b	7
	b_{21}	b	0	–2	d	5
	b_{22}	d	0	–1	e	6
	b_{23}	a	0	–1	b	9
a_5	b_{24}	b	0	–2	d	3
	b_{25}	a	0	–1	b	4
	b_{26}	b	–1	–1	d	6
	b_{27}	d	0	–1	e	8
	b_{28}	c	0	–1	d	7
	b_{29}	b	0	–2	d	9

According to the recommendations shown in Table 6.3, the system calculates each final trust degree of a_1, a_2, a_3, a_4, a_5. by using Expressions 6.12 and 6.13, and the results are shown in Table 6.4.

The final recommended trust value shown in Table 6.4 demonstrates that a_2 has the highest trust degree among a_1, a_2, a_3, a_4, a_5 followed by a_1 and a_4, then a_5, and a_3 the last. So it will be better for entity x to choose a_2 to accomplish its task.

Table 6.4 The final comprehensive evaluated trust value of a1, a2, a3, a4, a5

RSP	Sum_a	Sum_b	Sum_c	Sum_d	Sum_e	Sum_f	Sum_g	**Final Comprehensive Evaluated Trust Value**
a_1	0	**11**	8	10	0	0	0	b
a_2	**9**	7	0	0	5	0	0	a
a_3	0	0	8	8	0	**10**	0	f
a_4	0	**18**	5	5	7	0	0	b
a_5	0	4	0	**25**	8	0	0	d

6.10 Application Case: Trust-QoS Based MGrid Resource Service Scheduling

6.10.1 Requirements of Trust-QoS Based MGrid Resource Service Scheduling

Trust-QoS based MGrid resource scheduling must meet the following needs (Tao *et al.*, 2009a):

1. Specifying and classifying the QoS requirements submitted by user;
2. Mapping out user QoS requirements to resource service capabilities;
3. Searching qualified resource services according to user's QoS requirements and generating associated candidate resource service set;
4. Negotiating QoS requirements with user when requirements cannot be exactly met;

5. Measuring and evaluating each candidate resource service's QoS attribute parameter;
6. Optimal-selection evaluating each candidate resource service according to user's requirements based on the QoS evaluation results, and selecting the optimal resource service from the candidate resource service set to execute the task; and
7. Monitoring the resource state, and enabling it to deal with the emerged problems and failures.

6.10.2 Trust-QoS Based MGrid Resource Service Scheduling Framework

In order to address the above problems in MGrid resource service scheduling, a trust-QoS based MGrid resource service scheduling framework is proposed as shown in Figure 6.8. The working flow of it is as follows (Tao *et al.*, 2009a):

1. A user submits its task or resource service request via human-machine interface of the job management system in an MGrid.
2. The QoS requirement parser describes and specifies the QoS information submitted and generates the corresponding QoS information.
3. MGrid resource service QoS classifier classifies the QoS requests into three categories: strong trust-QoS request, weak trust-QoS request, and no trust-QoS request, and sends them to the corresponding request queue.
4. MGrid resource service search agent searches the qualified resource service that is able to execute the task according to user QoS requirements from the information center or server, and generates the set of candidate resource service;
5. MGrid QoS evaluation middleware evaluates each candidate resource service's QoS attribute parameter, and submits the results to MGrid resource service optimal-selection middleware.
6. MGrid trust-QoS benefit evaluation middleware evaluates the corresponding resource service's anticipated trust-QoS benefit if it was executed by an assigned resource.

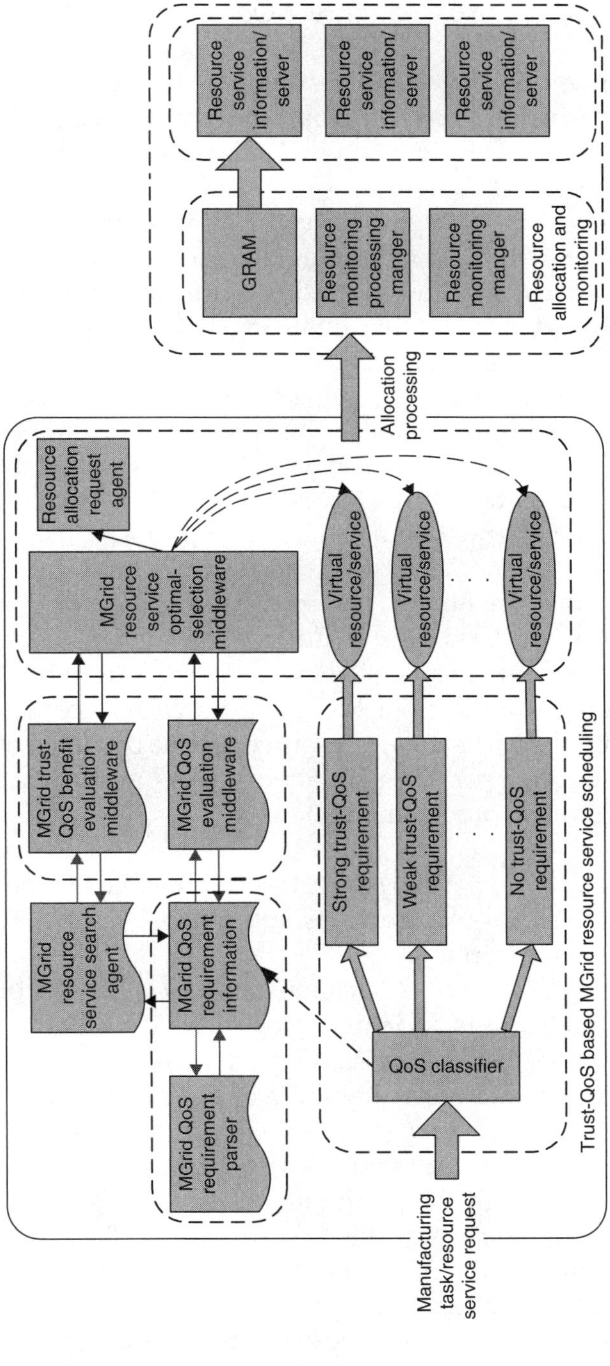

Figure 6.8 Trust-QoS based MGrid resource service scheduling framework.

7. MGrid optimal-selection middleware selects the optimal resource service to execute the submitted task according to the QoS and trust-QoS benefit evaluation results.
8. Resource allocation request agent transforms the manufacturing task into a special language for resource service allocation and requests GRAM to do the allocation.
9. Resources monitoring manager is responsible for monitoring the resource states and deciding the occurrence of a failure by analyzing state information of each resource, and communicating with the resources monitor processing manager.
10. Resources monitor processing manager is responsible for displaying the state of the resource and the type of failure that occurred, and requesting whether the job executed by the resources where failure occurred should be migrated and rescheduled or not.

The main components of the proposed scheduling framework consists of a MGrid QoS requirement parser, a MGrid QoS classifier, a MGrid resource service search agent, a resource service QoS evaluation middleware, a MGrid resource service trust-QoS benefit evaluation middleware, a resource service optimal-selection evaluating middleware, a resource allocation request agent, and a resources monitoring processing manager.

1. *MGrid QoS requirement parser:* It is responsible for specifying the QoS requirements submitted by user into standard description.
2. *MGrid QoS classifier:* It is responsible for classifying the user QoS requests into three categories: (a) *strong trust-QoS request*, (b) *weak trust-QoS request*, and (c) *no trust-QoS request*, and assigning them to the corresponding request queue. *Strong trust-QoS request* means the requested resource service's trust-QoS level must equal or be higher than the request's trust-QoS level; otherwise, the request will be abandoned. The evaluation benefit value to *strong trust-QoS request* is either the maximal or zero. *Weak trust-QoS request* means the requested resource service's trust-QoS degree level

must satisfy the request's trust-QoS requirements furthest. If the requested resource service's trust-QoS degree level cannot reach the request's trust-QoS, then a negotiation is needed to ask the user to lose its QoS request level. As a result, the corresponding evaluation trust-QoS benefit value would drop. *No trust-QoS request* means that the user has no trust-QoS request to the requested resource service, and any qualified resource services that satisfy its other QoS requests can serve it. The evaluation trust-QoS benefit to it is always the maximal.

3. *MGrid resource service agent:* It provides resource service searching and matching mechanisms and algorithms such as words-matching algorithm, numeric-matching algorithm, and ontology-conception matching algorithms. It is responsible for discovering the resource services that meet the RSD's (user) requirements from related resources encapsulation model, and generating an associated candidate resource service set.
4. *Resource service QoS evaluation middleware:* It provides resource service evaluation indices service, including performance-QoS, trust-QoS evaluation indices, general evaluation indices, and special evaluation indices. It is responsible for capturing the trust relationship between users and resource services providers in an MGrid, and providing QoS quantitative evaluation and real-time and dynamic updating services, such as resource service performance-QoS evaluation service, and trust-QoS evaluation service.
5. *MGrid resource service trust-QoS benefit evaluation middleware*: It evaluates the anticipative benefit the system can obtain when a request is completed. The evaluation result is the basic support for the trust-QoS based MGrid resource service scheduling. The detailed evaluating functions and algorithms are described in later sections.
6. *Resource service optimal-selection evaluating middleware*: It provides comprehensive optimal-selection evaluation algorithm service, and resource service composition service. It is responsible for making a comprehensive evaluation to each candidate resource

service's manufacturing ability, QoS, and reliability; and selecting the optimal resource service according to the evaluation result in order to execute the submitted manufacturing task. It aims to reduce the decision-making time of selecting the optimal resource service for users and improve the efficiency and qualities of an MGrid system. The specific evaluation models and algorithms have been studied in the authors' previous work.

7. *Resource allocation request agent*: It is responsible for transforming the manufacturing task into a special language for resources allocation and requesting GRAM to do the allocation.
8. *Resources monitoring manager*: It is responsible for monitoring the resource states and deciding the occurrence of a failure by analyzing the state of information of each resource, and communicating with the *resource monitor processing manager.*
9. *Resource monitor processing manager*: It is responsible for displaying the state of resources and the type of failures that occurred to the user, and requesting whether the job executed by the resources where failure occurred should be migrated and rescheduled or not.

6.10.3 Trust-QoS Based MGrid Resource Service Scheduling Algorithms

In an MGrid system, we consider that it is a benefit to the system when a service (or a request) is completed. The benefit value is demonstrated by the user's evaluation (or degree of satisfaction) to the resource service (or the resource service provider). We define this evaluation as *trust-QoS benefit*. The value of *trust-QoS benefit* is decided by two factors: (a) the request benefit weight, which refers to the importance (or weight) of the user request. Different requests should be assigned different weight while evaluating its corresponding trust-QoS benefit. E.g., the benefit weight of *operation request* (e.g., request a resource to execute a task such as computing) must be higher than *review request* (e.g., just review the resource's information without any further request); and (b) the user's judgment to the resource service.

Provide that an RSD's requesting trust-QoS degree level is D_{TD}, and the corresponding RSP's trust-QoS degree level is P_{TD}, then the

RSD's judgment to the RSP can be evaluated by adopting an evaluation function $\varphi(P_{TD})$, where:

(a) when the RSD's request belongs to *strong trust-QoS request*:

$$\varphi(P_{TD}) = \begin{cases} 0 & (P_{TD} < D_{TD}) \\ 1 & (P_{TD} \geq D_{TD}) \end{cases} \tag{6.14}$$

(b) when the RSD's request belongs to *weak trust-QoS request*:

$$\varphi(P_{TD}) = \begin{cases} 0 & (P_{TD} \leq D_{TD\_\min}) \\ \dfrac{(D_{DT} - P_{DT})}{D_{TD} - D_{TD\_\min})} & (D_{TD\_\min} < P_{TD} < D_{TD}) \\ 1 & (P_{TD} \geq D_{TD}) \end{cases} \tag{6.15}$$

(c) when the RSD's request belongs to *no trust-QoS request*:

$$\varphi(P_{TD}) = 1 \tag{6.16}$$

The corresponding trust-QoS benefit evaluation function curves are shown in Figure 6.9.

(a) strong trust-QoS (b) weak trust-QoS (c) no trust-QoS

Definition 6.6, Trust-QoS benefit: If the RSD's judgment to requested resource service d_i (or RSD_i) is $\varphi(P_{TD_i})$, and the request benefits weight of the RSD's request is B_{w_i}, the trust-QoS benefit of d_i is formulated as (Tao *et al.*, 2009a):

$$B_{i,j} = B_{w_i} * \varphi(P_{TD_j}) \tag{6.17}$$

Where

- $B_{i,j}$ denotes the trust-QoS benefit of d_i;
- B_{w_i} denotes the benefits weight of the RSD's request;
- $\varphi(P_{TD_i})$ denotes the corresponding anticipative evaluation of RSD to RSP's service d_i, and P_{TD_i} denotes the providing trust-QoS degree level of service d_i.

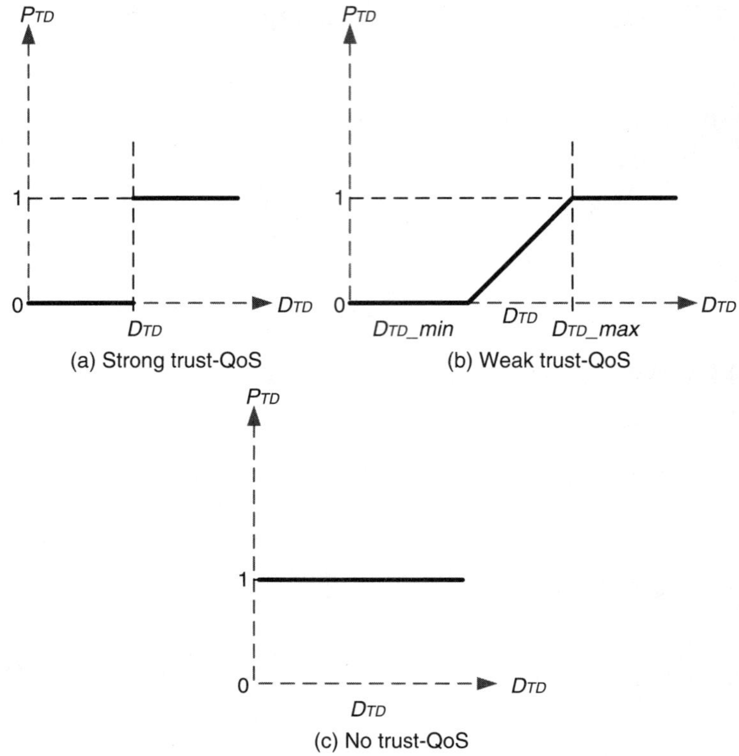

Figure 6.9 Trust-QoS benefit evaluation function curves.

Provide that there are n resources service requests and m qualified resource services that can execute the user's task during time T, while P $=\{p_1, p_2, p_3,\ldots, p_m\}$ denotes the set of the qualified resource services and $\mathbf{D} = \{d_1, d_2, d_3,\ldots, d_n\}$ denotes the set of resource service requests. Assume that the anticipative trust-QoS benefit the system obtained was $B_{i,j}$ if d_j is assigned to execute p_j. Therefore, the trust-QoS based MGrid resource service scheduling has to find a resource service allocation map, *map*(i, j), between resource service p_j and request d_j, and make the *total trust-QoS benefit* the biggest value. In other words, to make sure the following formulation has the maximal value:

$$B_{total} = \sum_{i=1}^{n} B_{i,j} = \sum_{i=1}^{n} B_{w_i} * \varphi(P_{TD_i}) \tag{6.18}$$

Provide that **D** was the resource service request set matrix, and **P** was the resource service set matrix. Matrix $B_{i,j}$ denotes the anticipative trust-QoS benefit information of each request executed by a different resource service *d*. E.g., the *i*th row and *j*th column date, $B_{i,j}$, denotes the anticipative trust-QoS benefit evaluating value generated when resource service p_j is assigned to execute task d_j. And the pseudocode of the proposed trust-QoS based MGrid resource service scheduling algorithm is briefly described as follows:

Trust-QoS based MGrid resource service scheduling algorithms
Inputs: $D, P, B_{i,j}$
Outputs: Resource service scheduling mapping *map*(*i*, *j*)

1. classify the user request set into three subsets, D_1, D_2, D_3 according to the rule of strong trust-QoS request, weak trust-QoS request, and no trust-QoS request;
2. set $k = 1$;
3. If D_k is not empty
4. make Matrix *B* empty
5. For each d_i in D_k
6. set $B_{i,\max}$ is the maximal of the *i*th row in Matrix $B_{i,j}$
7. For each p_j in *P*
8. if p_j satisfies the QoS requirement of d_i
9. calculate the corresponding trust-QoS benefit B_{ij} according to formulation (17);
10. calculate the value of $B^*_{ij} = B_{i,j}/B_{i'\max}$ and save the result in matrix *B*
11. End if
12. End for
13. If no service in p_j satisfy the request of d_i, delete d_i fromD_k
14. End if
15. End for
16. select the maximal value of B^*_{ij}, and record the corresponding subscript combination map (*i*, *j*)
17. assign p_j to execute d_i, and delete p_j from *P*
18. If D_k is empty
19. $k = k+1$
20. End if
21. Until ($k > 3$)

6.11 Summary

The current research on resource scheduling strategy is mainly based on resource performance-QoS (Quality of Service), but the factor of trust-QoS is ignored, which will result in unreasonable and unpractical scheduling results. In this chapter, in order to enhance the validity and success rate of resource scheduling in Manufacturing Grid (MGrid) system, provide high credible resource service abilities and results to user, the concept of resource service trust-QoS is presented, and the trust-QoS was introduced into MGrid resource service scheduling together with the important roles it plays. The trust problems existing in the resource service transaction between Resource Service Demanders (RSD) and Resource Service Providers (RSP) are put forward. The trust-QoS relationship model which is capable of capturing a comprehensive range of trust relationships exist in MGrid system is put forward. Then a two-layer resource service trust-QoS evaluation model are put forward, including intra-domain trust-QoS evaluation model and inter-domain trust-QoS evaluation model. The quantitative evaluating algorithms of trust-QoS degree value are proposed and described in detail, as well as the real-time and dynamic updating algorithms of trust-QoS degree value. Finally, a trust-QoS based MGrid resource service scheduling framework and associated realizing algorithms are proposed to illustrate the application.

7

Resource Service Optimal-selection and Composition Framework

7.1 Introduction

Due to the complexity of the manufacturing process, a submitted manufacturing task (or resource service request) in a distributed manufacturing system, especially in a manufacturing grid (MGrid) system, can be classified into two categories:

1. Single resource service request task (SRSRTask), which can be completed by invoking only one resource service, and
2. Multiple resource service request task (MRSRTask), which has to be implemented by invoking several resource services in a certain sequence (Tao *et al.*, 2008c).

SRSRTask specifies the system has to search out all the resource services that qualify for its function requirements and select the optimal one to execute it, namely, *resource service optimal-selection (RSOS)*; While on MRSRTask, the system has to invoke several resource services in a certain sequence and construct a composite resource service (CRS) to execute it (Tao *et al.*, 2008c). In other words, a submitted task usually can be divided into several subtasks that cannot be decomposed anymore and that can be executed by a single resource service as shown in Figure 7.1(a), namely, *resource service composition*. The above two problems defined as resource service optimal-selection and composition (RSOSC) problem are covered in this Chapter.

Let T denote an MGrid task and $T = \{ST^1, ST^2, \ldots, ST^j, \ldots ST^N\}$, where N is the total number of the decomposed subtasks (if T is a MRSRTask, then $N = 1$) and ST^j denotes the jth ($j = 1,2,3,\ldots,N$) subtask of T. It is assumed that there are M_j candidate resource services available for ST^j, and the corresponding candidate resource service set (CRSS) is $CSSS^j = \{RS_1^j, RS_2^j, \cdots, RS_i^j, \cdots, RS_{M_j}^j\}$, where RS_i^j is the ith candidate resource service for ST^j. Therefore, if the system wants to execute T, it needs to select one candidate resource service from each CRSS (i.e., $RSSS^j$), respectively, and generate a new composite resource service (CRS) to execute T, and $\mathrm{CRS} = \{RS_{i_1}^1, RS_{i_2}^2, \cdots, RS_{i_j}^j, \cdots, RS_{i_1}^{N^\bullet}\}$ ($RS_{i_j}^j \in RSSS^j$). There are $\prod_{j=1}^{N} M_j$ possible composite paths for T in theory (Tao *et al.*, 2008c).

Therefore, the problem of RSOSC can be further expressed as: (a) searching out all the candidate resource services and generating the corresponding resource service set $RSSS^j$ for each subtask ST^j, and selecting the optimal one to execute T if T is a SRSRTask; and (b) generating all the potential resource service compositon executing paths (RSCEP) and selecting the optimal path to execute T if T is a MRSRTask. The simple flowchart from task decomposition to RSCEP generation is shown in Figure 7.1.

It can be concluded from Figure 7.1 that, if one wants to realize the RSOSC, the following problems have to be addressed (Tao *et al.*, 2009e; Tao *et al.*, 2011a):

1. Realize the MGrid task decomposition and decompose each MGrid task into several subtasks which cannot be decomposed anymore and can be executed by invoking only a single resource service, namely, *MGrid task decomposition*.
2. Search out all the corresponding candidate resource services from the server or corresponding resource service center for each decomposed subtask ST^j according to the functional requirements, and generate the

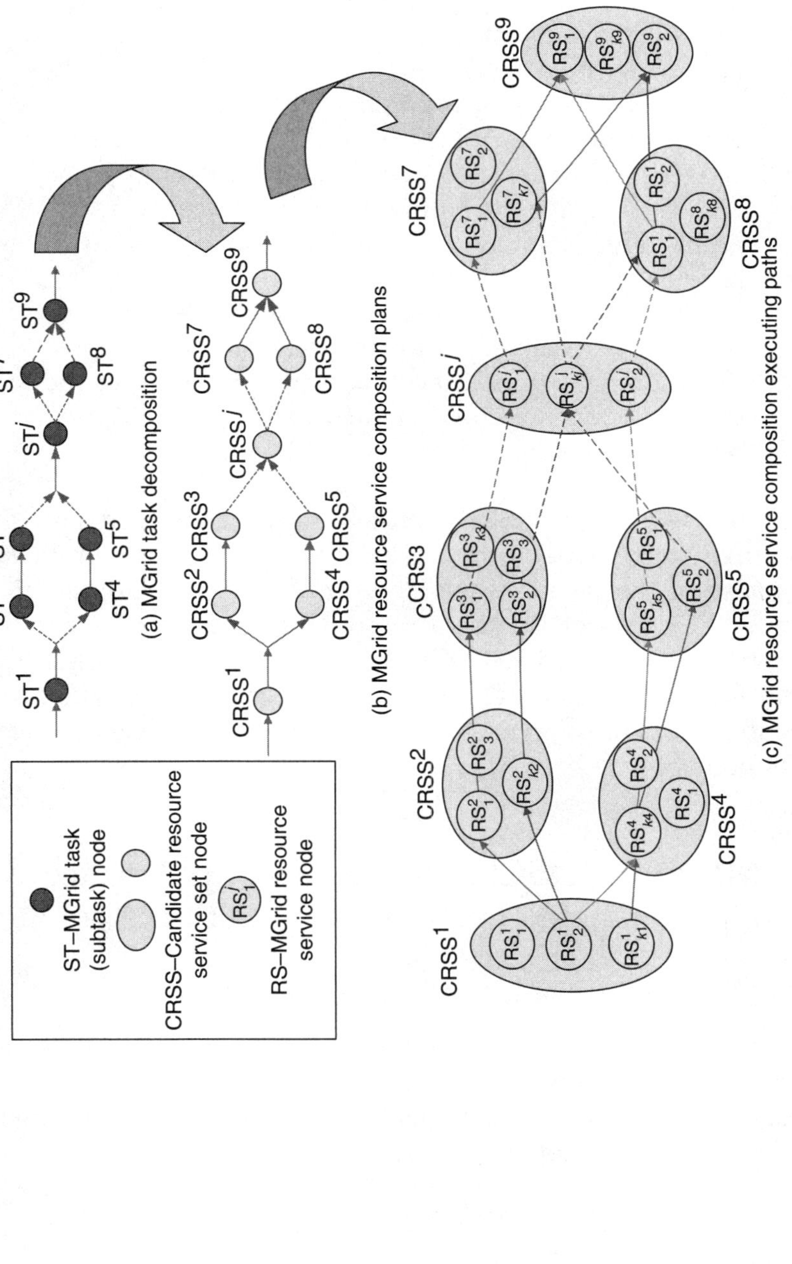

Figure 7.1 Simple flowchart from task decomposing to RSCEP generation (Tao *et al.*, 2008c).

corresponding candidate resource service set (i.e., $CRSS^j$), namely, *resource service match and search (RSMS)*.
3. Retrieve the QoS information of each candidate resource service, and synthetically evaluate the overall quality of the candidate resource service, so as to provide the data and information support for resource service optimal selection and composition, namely, *resource service QoS synthetically processing*.
4. Synthetically evaluate the overall quality of each candidate resource service in each $CRSS^j$ for subtask ST^j, and rank them according to the overall quality evaluation value, namely, *resource service optimal selection and raking*.
5. Select one candidate resource service from each CRSS ($CRSS^j$), respectively, and generate all the composite resource service executing paths, then select the optimal one to execute T, namely, *resource service composition and optimal-selection*.

In other words, the RSOSC problem can be divided into the above five steps. Motivated by addressing the above problems and realizing the resource service optimal-selection and composition, a MGrid resource service optimal-selection and composition framework (MGrid-RSOSCF) is proposed in this Chapter. The MGrid-RSOSCF consists of five layers, and each layer provides corresponding functions and services to address one of the problems mentioned above.

7.2 MGrid-RSOSCF: MGrid Resource Service Optimal-selection and Composition

7.2.1 Architecture of MGrid-RSOSCF

The following Figure 7.2 shows the architecture of the proposed MGrid-RSOSCF. It consists of five layers. Each layer provides corresponding services and functions for addressing one subproblem mentioned in the introduction. The five layers are:

1. Task layer (*T-Layer*), which is responsible for *MGrid task decomposition*,

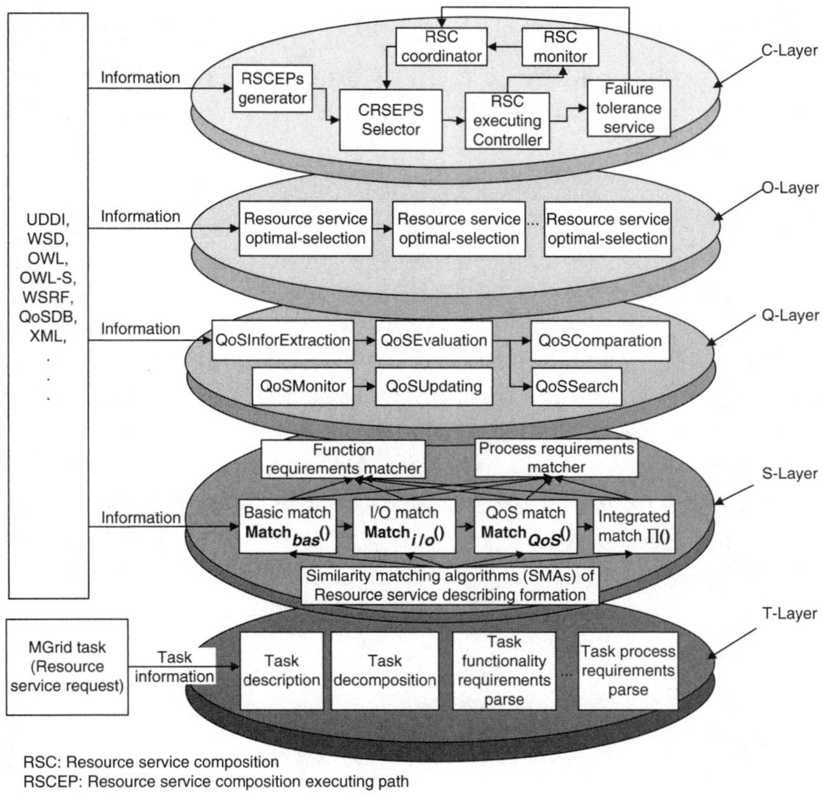

Figure 7.2 Architecture of MGrid-RSOSCF (Tao *et al.*, 2009e).

2. Resource service match and search layer (*S-Layer*), which is responsible for *MGrid resource service match and search*,
3. Resource service QoS synthetically processing layer (*Q-Layer*), which is responsible for *MGrid resource service QoS synthetically processing*,
4. Resource service optimal-selection layer (*O-Layer*), which is responsible for *MGrid resource service optimal-selection*, and
5. Resource service composition layer (*C-Layer*), which is responsible for *MGrid resource service composition and optimal-selection*.

Brief functions of each layer are presented in the following sections.

7.2.2 Life-cycle of MGrid Resource Service Composition

The MGrid resource service composition process can be classified into three phases from the service composition life-cycle as shown in Figure 7.3, and they are:

1. the Design phase (*Des-phase*),
2. the Deploy phase (*Dep-phase*), and
3. the Execution and Monitor phase (*E&M-phase*).

Des-phase. In the design phase, a user's requirement or task is analyzed, including the operation of the task description, task decomposition, task function requirement extraction, and the task process requirement extraction. The input of *Des-phase* is a complex task or user requirement, and its output is an abstract composite service.

Dep-phase. The Dep-phase is responsible for matching and searching for the qualified element for candidate service according

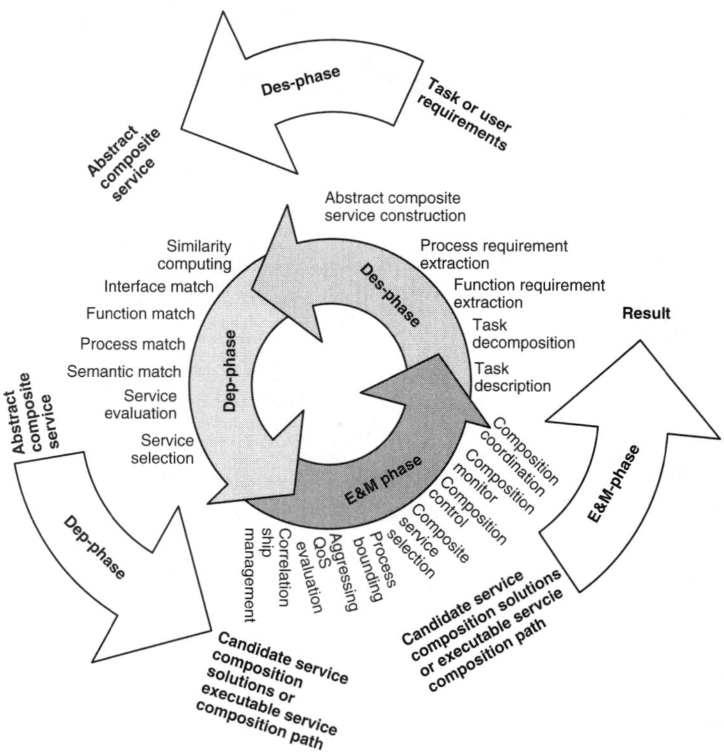

Figure 7.3 Life-cycle of MGrid resource service composition.

to the function and flow requirements generated in the Des-phase, the evaluation of the candidate service quality, and then the selection of the optimal service, and generating the corresponding service composition solutions. The operations involved in Dep-phase are similarly computing, interface matching, function matching, flow matching, semantic matching, service evaluation and ranking, and service selection. The input of Dep-phase is the abstract composite service generated in Des-phase, and its outputs are the candidate service composition solutions or the executable service composition path (ESCP).

E&M-phase. In the E&M phase, the comprehensive quality of all candidate service composition solutions or ESCP generated in the Dep-phase are evaluated, and their aggressing QoS are evaluated while considering the correlations among the element services. The optimal service composition solution or ESCP is selected and the related concrete services are bounded and invoked by this service composition engine. The operations involved in the E&M phase, including optimal-selection of service composition solutions or ESCP, are aggressing QoS calculation and evaluation, correlation management, service composition coordinating, service composition monitoring, and service composition control, etc.

7.3 T-Layer: Task Layer

The lowest level of the architecture of MGrid-RSOSCF is the *T-Layer*, which is responsible for accepting manufacturing tasks or resource service requests submitted by the user and realizing its digital description, and then decomposing them into corresponding subtasks if applicable.

7.3.1 Basic Models of MGrid Task

An MGrid task is the objects that are executed in order to solve a problem in MGrid. The objects are executed by invoking some resource service in the MGrid system. An MGrid task is anything that needs resource service (Nabrzyski *et al.*, 2003). It can be a simple request (e.g., bandwidth, a simple calculation, a query, etc.), an application request (e.g., a structure design, a simulation, etc.), or an application aggregate (e.g., large-scale computing, remote immersion, etc.). It also can be a product or a part for design or manufacturing, and so forth.

Due to the complexity of manufacturing, a manufacturing task often cannot be completed by invoking a single resource service. It must be executed by a composite resource service that is composed by a series of resource services. So an MGrid task can be decomposed into several subtasks (as shown in Figure 7.1[a]), and each subtask cannot be decomposed anymore and can be executed by a single resource service. Let T be an MGrid task, and $T = \{ST^1, ST^2, \ldots, ST^N\}$, where N is the total number of the decomposed subtasks and ST^j denotes the j-th ($j = 1, 2, 3, \ldots, N$) subtask.

Analyzing the composition process of a whole task or the connection between the decomposed subtasks, one can abstract the following four basic MGrid task composition process models (Tao *et al.*, 2010a):

1. ***Sequence model***. As shown in Figure 7.4[a], in a *sequence model*, one subtask's output is the other subtask's input, and each subtask has one and only one other subtask's output as its input (e.g., the outputs of subtask ST^1 is the inputs of ST^2, and ST^2 only have one inputs in Figure 7.4[a]), the corresponding model of *sequence model* can be expressed as $T = \{ST^1, ST^2, ST^3, \ldots, ST^n\}$.

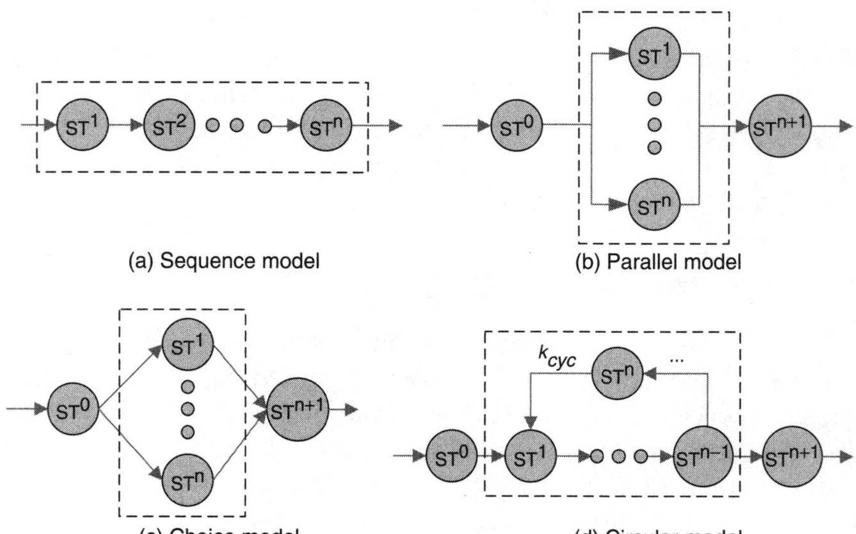

Figure 7.4 Basic MGrid task composition process models.

2. ***Parallel model.*** In *parallel model* (highlighted by dashed pane in Figure 7.4[b]), there is only one inputs (or only one subtask's output outside the *parallel model*, e.g., the outputs of subtask ST^0 in Figure 7.4 [b]) for the module. All subtasks (i.e., ST^1, ST^2, ST^3, …, ST^n, suppose there are n subtasks in parallel model) within the parallel module share the same inputs. The aggregate or the union of all subtasks' (i.e., $ST^1, ST^2, ST^3, \ldots, ST^n$) outputs within the parallel model is the inputs of the subtask (e.g., ST^{n+1}) which is the director successor of the choice model. The corresponding model of *parallel model* can be expressed as $T = \{ST^0, (ST^2 \lozenge\lozenge ST^3 \lozenge\lozenge \ldots \lozenge\lozenge ST^n), ST^{n+1}\}$. The symbol "$\lozenge\lozenge$" denotes the operator of permutation and combination of all subtasks within parallel model.
3. ***Choice model.*** In *choice model* (highlighted by dashed pane in Figure 7.4[c]), there is only one outputs or only one subtask's output outside the choice module, e.g., the outputs of subtask ST^0 in Figure 7.4(c). All the subtasks (i.e., ST^1, ST^2, ST^3, …, ST^n, it is supposed there are n subs tasks for choice) within this choice module share the same inputs. And any subtask's outputs in the choice module can be selected independently as the inputs of the other subtask (e.g., subtask ST^{n+1}) which is the director successor of the choice model. The corresponding model of *choice model* can be expressed as $T = \{ST^0, (ST^2 \Theta ST^3 \Theta \ldots \Theta ST^n), ST^{n+1}\}$. The symbol "$\Theta$" denotes the operator of choice.
4. ***Choice model.*** In *choice model* (highlighted by dashed pane in Figure 7.4[d]), there is only one outputs (or only one subtask's output outside the *circular module*, e.g., the outputs of subtask ST^0 in Figure 4(d)) for the module. Only the beginning subtask (i.e., subtask ST^0 in Figure 7.4[d]) takes exclusive inputs outside the *circular model* as its inputs. The outputs after k_{cyc} times circular action among subtasks within the *circular model* will be taken as the inputs of the subtask (e.g., subtask ST^{n+1}) which is the director successor of the *circular model*. The corresponding model of the *circular model* can be expressed as $T = \{ST^0, \xi^{k_{cyc}} (ST^1, ST^2, ST^3, \ldots, ST^n), ST^{n+1}\}$. The symbol "$\xi^{k_{cyc}}$" denote the operator of k_{cyc} times circulation.

Almost all of the MGrid task can be constructed by the above four basic task composition process models. For example, the task model of Figure 7.5(a) can be expressed as T = {(ST^1, (ST^2 ◊◊ ST^4), ST^5) ◊◊ (ST^3, (ST^4 ◊◊ ST^6), ST^7)}; and the task model of Figure 7.5(b) can be expressed as T = {RS^1, (ST^3 ◊◊ (ST^2, (ST^4 ◊◊ ST^5))), ST^6), ST^7}.

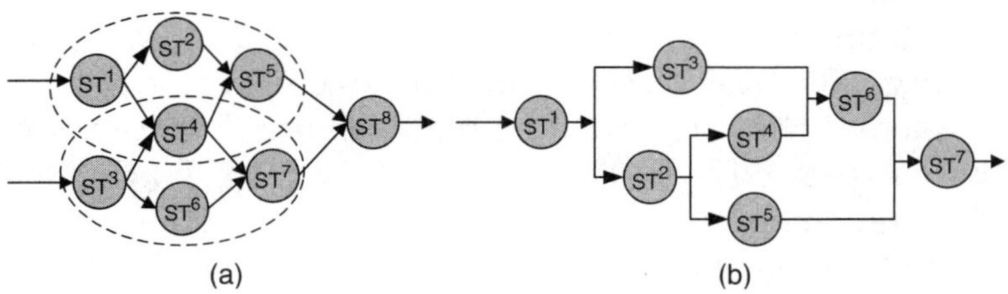

Figure 7.5 Two examples of mixed MGrid task.

Table 7.1 The main functions and services involved in T-Layer

Functions and Services	Description
taskDescription	Describing the accepted task requirements into standard describing file according to corresponding task describing language and standards.
taskDecomposition	Decomposing task into several subtasks that can be executed by invoking one resource service using function based task decomposing method, flow based task decomposing method, and so on.
functionRequirementParse	Analyzing the task describing information, and extracting the related functionality requirements so as to provide date and information for the operation of resource service match and search in S-Layer.
processRequirementParse	sssssAnalyzing the task describing information, and extracting the related process requirements for resource service composition operation in C-Layer.

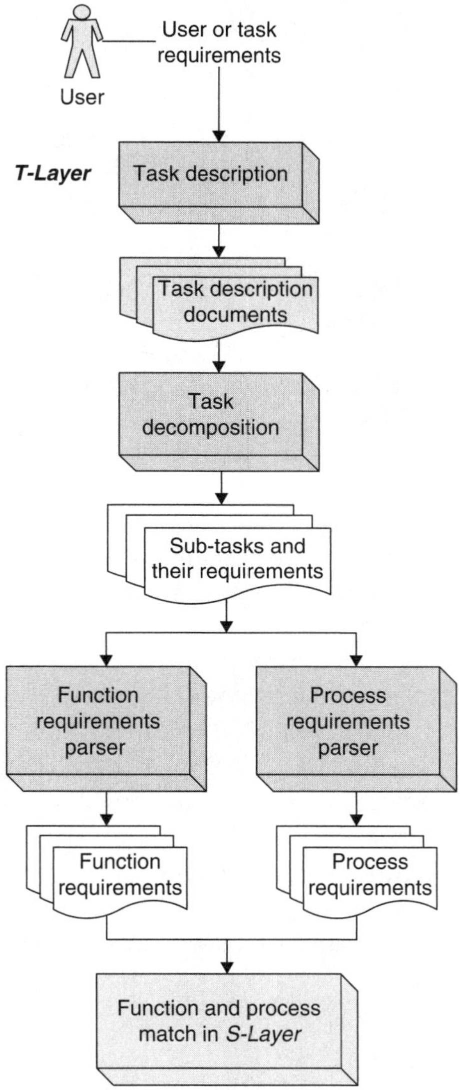

Figure 7.6 Main functions in *T-Layer*.

7.3.2 Main Functions and Services in T-Layer

The main services and components provided by *T-Layer* are *task-Description, functionRequirementParse, processRequirementParse*, and *taskDecomposition*. The corresponding description of these services and components are shown in Table 7.1, and the workflow of the T-Layer is shown in Figure 7.6.

7.4 S-Layer: Resource Service Match and Search Layer

Above the T-Layer sits the *S-Layer*, which takes charge of searching the qualified resource services from a server or resource service center, that meet the task's (or subtask's) functionality requirements.

The MSAgent is designed for searching, and matches the qualified service according to the task requirements generated in the T-layer, including function requirements, process requirements, capability requirements, and so on. It generates CRSS for each task or subtask. Based on the service provided by S-layer, it consists of the following two core components.

1. *Basic algorithms depository (AlgDep).* It provides various similarity matching algorithms (MAs) of resource service as described in Chapter 4, and various searching algorithms.
2. *Resource service match engine (MatchEngine).* It consists of the RS-Matcher which provides basic match ($Match_{bas}()$), I/O match ($Match_{i/o}()$), QoS match ($Match_{QoS}()$), and integrated match ($\Pi()$) as described in Chapter 4, the function requirement matcher (FRMatcher), process requirement matcher (PRMatcher), and ability requirement match agent (ARMatcher).

The detailed match and search mechanisms and algorithms can be seen in Chapter 4.

7.5 Q-Layer: Resource Service QoS Synthetically Processing Layer

Q-Layer is the next layer, which manages the extraction of each candidate resource service's corresponding QoS information from the related database (e.g., QoS database designed by developer) or resource service description (such as OWL-S registered in UDDI). Also, it provides QoS evaluating services and QoS comparing services, and provides data and information to the *O-Layer* and *C-Layer*.

7.5.1 Main Functions and Services in Q-Layer

The resource service matching and searching mainly depends on functional factors of resource service. However, resource service optimal-selection and composition mainly depends on nonfunctional factors (QoS) of resource service. Therefore, the processing of QoS of resource service is necessary for the realization of resource service optimal-selection and composition.

The Q-Layer is designed to provide comprehensive process services for the specific operation to QoS of the resource service. The detailed services involved in Q-Layer are QoS extraction, QoS search, QoS updating, QoS evaluation, QoS monitor, and QoS comparison as shown in Table 7.2. The *QoSEvaluation* and *QoSSearch* were investigated in Chapter 5 and Chapter 6, and the *QoSMonitor* and *QoSUpdating* is easy to achieve. Therefore, the realization of *QoSInforExtraction* and *QoSComparison* will be the focus of this chapter.

7.5.2 QoS Extraction

In MGrid, the QoS characteristics for the resource service are stored in two ways: (a) stored in a database, i.e., QoSDB (QoS database), which contains each resource service key and the value of the QoS characteristics that belong to it, and (b) encapsulated in the OWL-S description of the resource service. The OWL-S file can be appropriately modified so that it provides QoS information for a resource service. We adopted the idea for the service of QoS extraction from Diamadopolou, *et al.*, (2008), in order to retrieve and deliver the QoS information from various recording and storage sources. The designed *QoSExtraction* service retrieves the QoS information via the following three approaches (Tao *et al.*, 2011a):

1. *QoS extraction directly from* OWL-S. When the QoS values found in OWL-S are kept, the *QoSExtraction* access the modified OWL-S using its URI that was returned. The function *GetSourceCode* is invoked for each resource service within the parameters of the address of the modified OWL-S, and then copies from this address the modified OWL-S of the service to a StreamReader which is returned. Then the

Table 7.2 The main functions and services involved in Q-Layer (Tao *et al.*, 2011a)

Function and Services	Description
QoSInforExtraction	Extracting the QoS information for each candidate resource service from related service describing file (e.g., OWL-S, WSDL, and WSRF) or related database (e.g., QoS database designed by developer).
QoSEvaluation	Evaluating the QoS parameter value of each candidate resource service, such as time and trust-QoS.
QoSSearch	Searching the qualified resource services that can execute a task according to its QoS requirements.
QoSComparison	Comparing the required resource service's QoS with the candidate resource service's QoS, and getting rid of some low quality candidate resource service from its RSS, so as to reduce the complexity of resource service optimal-selection and composition in upper layers.
QoSMonitor	Monitoring the QoS states of the selected resource service.
QoSUpdating	Updating the resource service's QoS information in its corresponding describing file or database when it changed.

StreamReader is read and the QoS found is stored to an array, as shown in Figure 7.7(a).

2. *QoS extraction directly from QoSDB.* When the QoS values have not been stored in UDDI registry, the *QoSExtraction* access the QoSDB where the QoS information that belongs to each resource service is stored. The function *FindQoS* is invoked that takes as input a ServiceKey and based on this ServiceKey, the QoS values that belong to it are returned. Then the QoS values are stored to an array, as shown in Figure 7.7(b).

3. *Updated QoS extraction directly from QoSDB*. When the QoS values have been stored in UDDI registry a long time ago according to a logged timestamp, and they no longer represent the current QoS that the resource service offers, the *QoSExtraction* access the QoSDB where the current QoS information that belongs to each resource service is stored. The function *FindQoS*

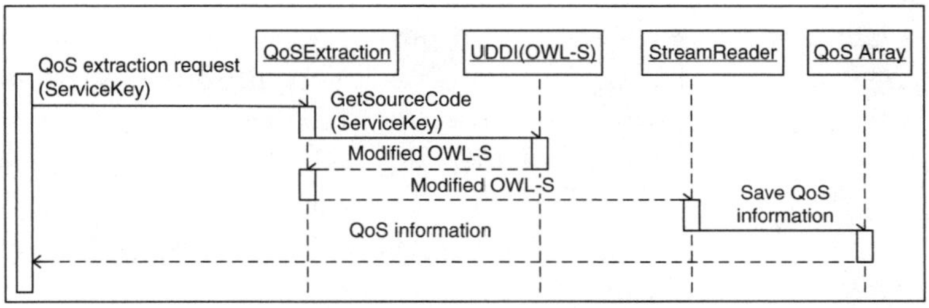

(a) QoS information extraction directly from WSDL

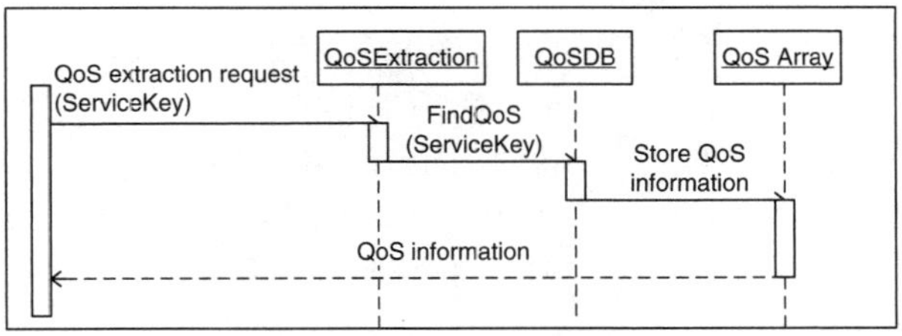

(b) QoS information extraction directly from QoSDB

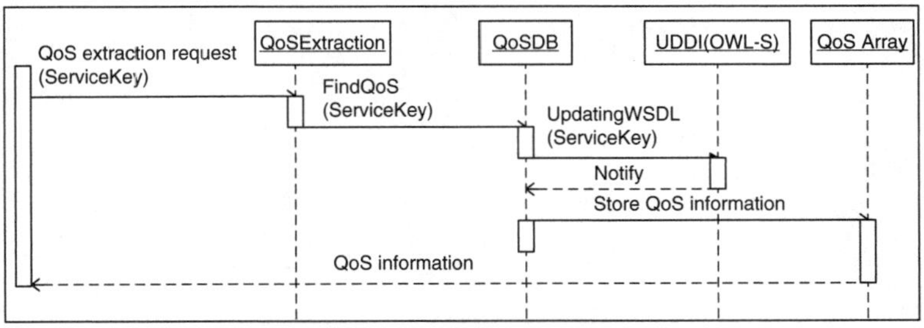

(c) Updated QoS information extraction directly from QoSDB

Figure 7.7 MGrid resource service QoS information extraction approaches. (a) QoS information are stored directly in OWL-S, and QoS extraction directly from OWL-S; (b) QoS information are not stored in OWL-S and QoS extraction directly from QoSDB; (c) QoS information are stored in OWL-S ago, but current updated QoS information are stored in QoSDB, so updated QoS extraction directly from QoSDB and updating the OWL-S to UDDI.

is invoked that takes as input a ServiceKey, and based on this ServiceKey, the QoS values that belong to it are returned. Then the QoS values are stored to an array, as shown in Figure 7.7(c). At the same time, the system is updating the OWL-S description with the current QoS from the QoSDB to UDDI.

7.5.3 QoS Comparison

The *QoSComparison* service compares the QoS of the candidate resource service and the subtask's requirements, so as to get rid of the candidate resource with unqualified QoS characteristics. Let the QoS set for subtask ST^j requirements be $Q^{ST^j} = \{q_1^{ST^j}, q_2^{ST^j}, \cdots, q_{i^*}^{ST^j}, \cdots, q_K^{ST^j}\}$ (in this study, $K = 6$, and q_1, q_2, q_3, q_4, q_5, q_6 denote *Time(T)*, *Cost(C)*, *Reliability(Rel)*, *Maintainability(Ma)*, *Trust-QoS(Trust)*, and *Functional Similarity(FS)*, respectively. In practice, it is extensible according to the requirements, and the corresponding weight set is $W^{ST^j} = \{w_1^{ST^j}, w_2^{ST^j}, \cdots, w_{i^*}^{ST^j}, \cdots, w_K^{ST^j}\}$, where $0 \le w_{i^*}^{ST^j} \le 1$, $\sum_{i^*=1}^{k} w_{i^*}^{ST^j} = 1$. Accordingly, the candidate resource service RS_i^j's QoS capabilities are $Q^{RS_i^j} = \{q_1^{RS_i^j}, q_2^{RS_i^j}, \cdots, q_{i^\bullet}^{RS_i^j}, \cdots, q_K^{RS_i^j}\}$.

Because $q_{i^*}^{ST^j}$ and $q_{i^\bullet}^{RS_i^j}$ have the same unit, one can compare $q_{i^\bullet}^{RS_i^j}$ with $q_{i^*}^{ST^j}$ for each i^* from 1 to K. Let the QoS comparison between the candidate resource RS_i^j and subtask ST^j be defined as $Compare_{RvT}$, where

$$Compare_{R\vee T} = \sum_{i=1}^{k} \left(\frac{w_{i^\bullet}^{ST^j}}{q_{i^\bullet}^{ST^j}} * q_{i^\bullet}^{RS^j}\right) \tag{7.1}$$

If RS_i^j subject to

$$\begin{cases} Compare_{R\vee T} \ge \zeta_{R\vee T} \\ T(\mathrm{RS}_i^j) \le T_0 \\ \mathrm{C}(\mathrm{RS}_i^j) \le \mathrm{C}_0 \\ Rel(\mathrm{RS}_i^j) \ge Rel_0 \\ Ma(\mathrm{RS}_i^j) \ge M_0 \\ Trust(\mathrm{RS}_i^j) \ge Trust_0 \\ FS(\mathrm{RS}_i^j) \ge FS_0 \end{cases} \tag{7.2}$$

then RS_i^j is qualified to ST^j's QoS requirements. In Expression 7.2, $\zeta_{R \vee T}$ is the threshold value initialized by system or user. Rel_0, M_0, $Trust_0$, and FS_0 are the lowest requirement of ST^j QoS requirements of *Reliability, Maintainability Trust-QoS,* and *Functional Similarity* respectively, and T_0 and C_0 are the highest value of *Time* and *Cost* of ST^j requirements, respectively. $T(RS_i^j)$, $C(RS_i^j)$, $Rel(RS_i^j)$, $Ma(RS_i^j)$, $Trust(RS_i^j)$ and $FS(RS_i^j)$ are the QoS value of *Time, Cost, Reliability, Maintainability, Trust-QoS,* and *Functional Similarity*, respectively, of RS_i^j extracted by *QoSExtraction* service.

If RS_i^j cannot satisfy the Expression 7.2, then RS_i^j will be deleted from ST^j's candidate resource service set, i.e., $CRSS^j$.

7.6 O-Layer: Resource Service Optimal-selection Layer

The *O-Layer* is the next layer, which is responsible for the integration of ranking the candidate resource service for each subtask, and selecting the best resource service to serve SRSRTask according to the QoS information provided by *Q-Layer*. In addition, *O-Layer* provides *C-Layer* with data and information support. MADM (multiple attribute decision making), AHP (analytic hierarchy process), and IFS-based methods are usually used in *O-Layer* to comprehensively evaluate and rank the candidate resource service.

7.6.1 Simple Resource Service Optimal-selection Method

After getting all the QoS information of each candidate resource service in the corresponding candidate resource service set, the next step is to integrate the evaluation of the overall quality of each candidate resource service, and rank them according to their overall quality.

Let the QoS set for ST^j requirements be $Q^{ST^j} = \{q_1^{ST^j}, q_2^{ST^j}, q_{i^*}^{ST^j}, \cdots, q_K^{ST^j}\}$, and the corresponding weight set be $W^{ST^j} = \{w_1^{ST^j}, w_2^{ST^j}, \cdots, w_{i^*}^{ST^j}, \cdots, w_K^{ST^j}\}$, where $0 \le w_{i^*}^{ST^j} \le 1$ and $\sum_{i^*=1}^{k} w_{i^*}^{ST^j} = 1$. It is assumed that after QoS comparison, there are M candidate resource services that qualified for ST^j, so the set of corresponding candidate resource service set can be expressed as $CRSS^j = \{RS_1^j, RS_2^j, \cdots, RS_i^j, \cdots, RS_M^j\}$. The candidate resource service RS_i^j's QoS capabilities information is $Q^{RS_i^j} = \{q_1^{RS_i^j}, q_2^{RS_i^j}, \cdots, q_{i^*}^{RS_i^j}, \cdots, q_k^{RS_i^j}\}$.

So the QoS information matrix, $\mathbf{Q}^{CRSS^j}$ of $CRSS^j$ can be built by merging the QoS vectors of all these candidate resource service, and

$$\mathbf{Q}^{CRSS^j} = \begin{bmatrix} q_1^{RS_1^j} & q_1^{RS_2^j} & \cdots & q_1^{RS_i^j} & \cdots & q_1^{RS_M^j} \\ q_2^{RS_1^j} & q_2^{RS_2^j} & \cdots & q_2^{RS_i^j} & \cdots & q_2^{RS_M^j} \\ \vdots & \vdots & \ddots & \vdots & \ddots & \vdots \\ q_{i^\bullet}^{RS_1^j} & q_{i^\bullet}^{RS_2^j} & \cdots & q_{i^\bullet}^{RS_i^j} & \cdots & q_{i^\bullet}^{RS_M^j} \\ \vdots & \vdots & \ddots & \vdots & \ddots & \vdots \\ q_K^{RS_1^j} & q_K^{RS_2^j} & \cdots & q_K^{RS_i^j} & \cdots & q_K^{RS_M^j} \end{bmatrix} \tag{7.3}$$

in which, the $i^\bullet$ th row corresponds to the $i^\bullet$th candidate resource service RS_i^j's QoS capability and the ith column corresponds to the ith QoS parameter of each candidate resource service.

During the resource service optimal-selection process, the MADM method is adopted to comprehensively evaluate and rank the candidate resource service. There primarily are two phases in applying MADM.

Scaling phase. Some of the QoS criteria are positive, meaning that the bigger the QoS criterion the higher the quality or better for the user, e.g., *reliability, maintainability, trust-QoS,* and *Functional Similarity*; and some other QoS criteria are negative, meaning that the smaller the QoS criterion the better for the user, e.g., *time* and *cost*. Therefore, a scaling function is employed to adjust different QoS criterions. The scaling function is as follows (Zeng *et al*, 2004a):

$$v_{i^\bullet}^{RS_i^j} = \begin{cases} \dfrac{q_{i^*}^{\max} - q_{i^*}^{RS_i^j}}{q_{i^*}^{\max} - q_{i^*}^{\min}} & (q_j^{\max} - q_j^{\min} \neq 0, i = 1) \quad (a) \\ \dfrac{q_{i^*}^{RS_i^j} - q_{i^*}^{\min}}{q_{i^*}^{\max} - q_{i^*}^{\min}} & (q_j^{\max} - q_j^{\min} \neq 0, i \neq 1) \quad (b) \\ 1 & (q_j^{\max} - q_j^{\min} = 0) \end{cases} \tag{7.4}$$

In the above equations, $q_{i^\bullet}^{\max}$ is the maximal value of a QoS criterion in matrix Q^{CRSS^j} while $q_{i^\bullet}^{\min}$ is the corresponding minimal value. For positive QoS criteria, values are scaled according to Expression 7.4(b). While, for negative QoS criteria, values are scaled according to Expression 7.4(a). After the adjustment, one can get the following scaled matrix V^{CRSS^j}:

$$\mathbf{V}^{CRSS^j} = \begin{bmatrix} v_1^{RS_1^j} & v_1^{RS_2^j} & \cdots & v_1^{RS_i^j} & \cdots & v_1^{RS_M^j} \\ v_2^{RS_1^j} & v_2^{RS_2^j} & \cdots & v_2^{RS_i^j} & \cdots & v_2^{RS_M^j} \\ \vdots & \vdots & \ddots & \vdots & \ddots & \vdots \\ v_{i^\bullet}^{RS_1^j} & v_{i^\bullet}^{RS_2^j} & \cdots & v_{i^\bullet}^{RS_i^j} & \cdots & v_K^{RS_M^j} \\ \vdots & \vdots & \ddots & \vdots & \ddots & \vdots \\ v_K^{RS_1^j} & v_K^{RS_2^j} & \cdots & v_K^{RS_i^j} & \cdots & v_K^{RS_M^j} \end{bmatrix} \tag{7.5}$$

Weighting phase. During the process of resource service rank and optimal-selection, different QoS criterion will play different roles according to various requirements and users. So each QoS criterion should be assigned a different corresponding weight. Let the weight coefficients set be $\mathbf{W}^{ST^j} = \{w_1^{ST^j}, w_2^{ST^j}, \cdots, w_{i^*}^{ST^j}, \cdots, w_K^{ST^j}\}$, where $0 \le w_i^{ST^j} \le 1$ and $\sum_{i=1}^{k} w_i^{ST^j} = 1$. Therefore, the overall QoS score for each candidate resource service in $CRSS^j$ can be computed using the following formula:

$$\left\{ \sum_{i^*=1}^{K} (v_{i^\bullet}^{RS_1^j} * w_{i^*}^{ST^j})/K, \cdots, \sum_{i^*=1}^{K} (v_{i^\bullet}^{RS_i^j} * w_{i^*}^{ST^j})/K, \cdots, \sum_{i^*=1}^{K} (v_{i^\bullet}^{RS_M^j} * w_{i^*}^{ST^j})/K \right\} \tag{7.6}$$

Therefore, regarding subtask ST^j, the MGrid system will select the optimal candidate resource service from $CRSS^j$ which has the maximal compressive QoS score to execute. If there are several resource services with the same maximal score, then one of them is selected randomly. Otherwise, if there is no resource service that satisfied the requirements of the given task, the MGrid system will return an unsuccessful message to the user and advise the user to resubmit or loosen its requirements. The system will rank the candidate resource services in $CRSS^j$ according to their comprehensive QoS evaluation score.

7.6.2 The Other Resource Service Optimal-selection Methods

In the above method, the user's feeling is not taken into account in the resource service optimal-selection process in an MGrid system. In addition, the weight of a resource service's (or task's) QoS criterion is hard to set using the above method. Because different users have different views on the importance of a resource service's (or task's) QoS criterion, and although they may make the same evaluation (e.g., *very important*) to the importance to a QoS criterion, the definitions of these evaluation terms (e.g., *very important*) may differ. Hence, it is important to aggregate different users' evaluations on the importance of each QoS criterion.

Therefore, in Chapter 8, a more reasonable and practical method that takes into account the above problems for resource service optimal-selection will be described.

7.7 C-Layer: Resource Service Composition Layer

C-Layer is responsible for resource service composition. It selects one resource service from the candidate resource service set of each subtask and assembles them in a certain order so as to construct a new composite resource service, or resource service composition executing path (RSCEP). Furthermore, it has to select the optimal RSCEP to execute the task under multiple objectives and multiple constraints.

7.7.1 RSC-Engine: Resource Service Composition Engine

The resource service composition engine (RSC-Engine) is the core of the *C-Layer*. It is responsible for (Tao *et al.*, 2010a):

a. Generating the resource service composition executing paths, i.e., RSCEPs;
b. Providing RSCEP optimal selecting algorithms based on aggressing QoS and intelligent computing, and selecting the optimal CRESP from all candidate

RSCEPs under multiple objectives and multiple constraints;

c. Launching and controlling the execution of service composition;

d. Monitoring the running state and performance information of selected resource services;

e. Coordinating and adjusting the composition solutions according to the monitoring results.

The RSC-Engine is designed according to the above requirements, as illustrated in Figure 7.8. It primarily consists of the following five components:

1. resource service composition executing paths generator (*RSCEP-Generator*),

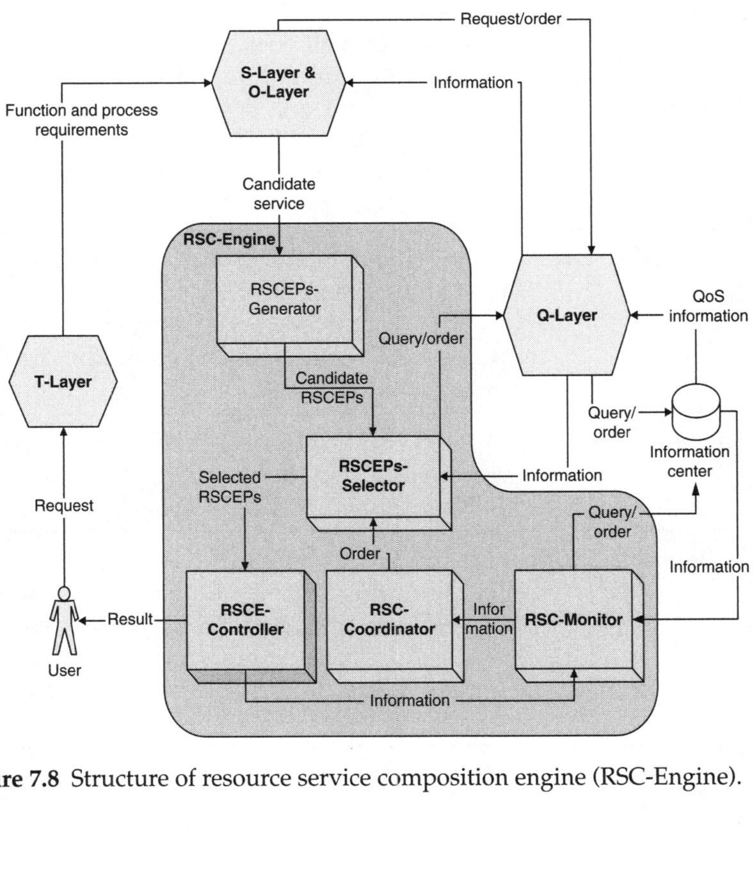

Figure 7.8 Structure of resource service composition engine (RSC-Engine).

2. resource service composition executing paths selector (*RSCEP-Selector*),
3. resource service executing controller (*RSCE-Controller*),
4. resource service monitor (*RSC-Monitor*), and
5. resource service coordinator (R*SC-Coordinator*).

Each component is designed to address one of the problems mentioned above, and the function is described in Table 7.3.

Table 7.3 The main services and components embedded in C-Layer

Name	Description
RSCEP-Generator	Generating the resource service composition executing paths.
RSCEP-Selector	Providing various algorithms for selecting the optimal one from all possible RSCEPs under multi-objective and multi-constraints, such as GA (genetic algorithm), PSO (partial swarm optimization) (Tao *et al.*, 2008a), and QMAEA (Quantum multi-agent evolutionary algorithms) (Tao *et al.*, 2010d).
RSCE-Controller	Receiving the RSCEP information and sending invoking message to the selected resource services, and launching and controlling the implementation of RSCEP, and collecting the executing results.
RSC-Monitor	Monitoring the state information of the resource services involved in the selected RSCEP and the available resource services in MGrid.
RSC-Coordinator	Adjusting the selected RSCEP according to the feedback information from *RSC-Monitor*. Under some conditions, e.g., better resource services are available, some resource services in RSCEP are removed, and executing task's requirements changed.

7.7.2 RSCEP-Generator

After transferring the MCRS (Mixed composite resource service) into SCRS (Sequence composite resource service), the next problem to be addressed is to select one candidate resource service from each CRSS, and generate the corresponding resource service composition paths. There are five steps to generating the MGrid resource service composition executing path (RSCEP):

Step 1: (Transformation). Transfer MCRS into SCRS according to the transforming rules in the authors' previous work (Tao *et al.*, 2008c).

Step 2: (Generation of candidate resource service set for a task). Generate the candidate resource service set $RSS=\{RSS^1, RSS^2,\cdots, RSS^j,\cdots, RSS^N\}$ for task $T = \{ST^1, ST^2, \ldots, ST^N\}$ according to the match and search algorithms provided in *S-Layer*, where $RSS^j=\{RS_1^j, RS_2^j,\cdots, RS_3^j,\cdots, RS_{M_j}^j\}$ is the candidate resource service for executing subtask ST^j.

Step 3: (Delete the poor quality resource service from related RSS). Comparing the QoS requirement of ST^j and the QoS performance of the candidate resource services using *QoSComparison* as described in Section 7.5.3, and delete the poor quality candidate resource services from the candidate list.

Step 4: (Generating the original RSCEP) Generate all possible RSCEP for T, and let the number of possible RSCEP be K_{path} and the set of RSCEPs be $\mathbf{P_{ath}}=\{P_1, P_2,\cdots, P_i,\cdots, P_{K_{path}}\}$, where P_i is the ith candidate RSCEP and $P_i=\{\mathrm{RS}_{i_1}^1, \mathrm{RS}_{i_2}^2,\cdots, \mathrm{RS}_{i_j}^j,\cdots, \mathrm{RS}_{i_1}^{N^*}\}$.

Step 5: (Resource service connection checking). For each RSCEP P_i, it is assumed that the inputs of resource service RS_i^{j+1} is the outputs of resource service RS_i^j (RS_i^{j+1} is the direct successor of RS_i^j). So the connection between RS_i^j and RS_i^{j+1} must be taken into account during the process of RSCEPs optimal-selection. This is because although the quality of RS_i^j and RS_i^{j+1} are very good respectively, if they cannot effectively be connected, the RSCEP cannot be selected to execute the task. E.g., if the data type of the requested inputs parameter of RS_i^{j+1} is "string", but the data type of the outputs of RS_i^j

is "double float", then RS_i^j and RS_i^{j+1} cannot work effectively, therefore the RSCEP which contains RS_i^j and RS_i^{j+1} is not an effectual path for MGrid task T and should be deleted from $\mathbf{P}_{ath}$. Set ζ_{con} $(0<\zeta_{con})$ is the connecting threshold value initialed by the system or user. The following algorithm is designed to check the connection of each candidate path in $\mathbf{P}_{ath}$:

Inputs: $\mathbf{P}_{ath}$, ζ_{con}.
Outputs: new RSCEP set ($\mathbf{P}_{ath}^*$) after checking the connection of each candidate RSCEP

1 **For** each P_i in $\mathbf{P}_{ath}$
2 **For each** $\mathrm{RS}_{i_j}^j$ in P_i,
3 Executing function $\phi(\mathrm{RS}_{i_j}^j, \mathrm{RS}_{i_{j+1}}^{j+1})$ // $\phi()$ is the connection checking function designed by us
4 **If** $(\phi(\mathrm{RS}_{i_j}^j, \mathrm{RS}_{i_{j+1}}^{j+1}) < \zeta_{con})$,
5 Delete P_i from $\mathbf{P}_{ath}$ //the connection of P_i is unqualified to user's requirements
6 **End if**
7 **End For**
8 **End for**

After following the above steps, it is supposed the number of the remainder candidate RSCEP is M^* $(M^* \le K_{path})$, and the RSCEP set is $\mathbf{P}^*_{ath} = \{P_1, P_2, \cdots, P_j, \cdots, P_{M^*}\}$.

7.7.3 RSCEP-Selector

The main aim of RSCEP-selector is to select the optimal RSCEP from $\mathbf{P}_{ath}^*$ to execute the task. It is an NP complete problem. The optimal-selection of an RSCEP can be realized using the MCDM technique on the QoS matrix of RSCEP if the scale of the problem is not too big. There are three steps to realize the optimal-selection of RSCEP using MSDM.

Set the QoS criteria of each element resource service $\mathrm{RS}_{i_j}^j$ in P_i is $Q^{\mathrm{RS}_{i_j}^j} = (q_1^{\mathrm{RS}_{i_j}^j}, q_2^{\mathrm{RS}_{i_j}^j}, \cdots, q_{i^\bullet}^{\mathrm{RS}_{i_j}^j}, \cdots, q_k^{\mathrm{RS}_{i_j}^j})$, so the whole QoS criteria of P_i can be calculated using the following formula (Tao *et al.*, 2008a):

$$\begin{cases} T(P_i) = \sum_{j=1}^{N^\bullet} T(RS_{i_j}^j) = \sum_{j=1}^{N^\bullet} q_1^{RS_{i_j}^j} \\ C(P_i) = \sum_{j=1}^{N^\bullet} C(RS_{i_j}^j) = \sum_{j=1}^{N^\bullet} q_2^{RS_{i_j}^j} \\ Rel(P_i) = \prod_{j=1}^{N^\bullet} Rel(RS_{i_j}^j) = \prod_{j=1}^{N^\bullet} q_3^{RS_{i_j}^j} \\ Ma(P_i) = \prod_{j=1}^{N^\bullet} Ma(RS_{i_j}^j) = \prod_{j=1}^{N^\bullet} q_4^{RS_{i_j}^j} \\ Trust(P_i) = \sum_{j=1}^{N^\bullet} Trust(RS_{i_j}^j) \Big/ N^\bullet = \sum_{j=1}^{N^\bullet} q_5^{RS_{i_j}^j} \Big/ N^\bullet \\ FS(P_i) = \sum_{j=1}^{N^\bullet} FS(RS_{i_j}^j) \Big/ N^\bullet = \sum_{j=1}^{N^\bullet} q_6^{RS_{i_j}^j} \Big/ N^\bullet \end{cases} \tag{7.7}$$

Therefore, the comprehensive QoS information matrix, $\mathbf{QP}_{k*M^*}$, of $\mathbf{P}^*_{\mathbf{ath}}$ can be established as follows:

$$\mathbf{QP}_{k*M^*} = \begin{Bmatrix} q_1^{p_1} & q_1^{p_2} & \cdots & q_1^{p_j} & \cdots & q_1^{p_{M^\bullet}} \\ q_2^{p_1} & q_2^{p_2} & \cdots & q_2^{p_j} & \cdots & q_2^{p_{M^\bullet}} \\ \vdots & \vdots & \ddots & \vdots & \ddots & \vdots \\ q_i^{p_1} & q_i^{p_2} & \cdots & q_i^{p_j} & \cdots & q_i^{p_{M^\bullet}} \\ \vdots & \vdots & \ddots & \vdots & \ddots & \vdots \\ q_k^{p_1} & q_k^{p_2} & \cdots & q_k^{p_j} & \cdots & q_k^{p_{M^\bullet}} \end{Bmatrix} \tag{7.8}$$

The *j*th column of $\mathbf{QP}_{k*M^*}$ is the corresponding QoS criteria value of candidate RSCEP P_i.

Set the QoS set for MGrid task T requirements so that $\mathbf{QT}_{N*k} = \{q_1^T, q_2^T, \cdots, q_{i^*}^T, \cdots, q_K^T\}$, and the corresponding weight set is $W^{ST^j} = \{w_1^T, w_2^T, \cdots, w_{i^*}^T, \cdots, w_K^T\}$, where $0 \le w_{i^*}^T \le 1$ and $\sum_{i^*=1}^{k} w_{i^*}^T = 1$. Therefore, the overall QoS score, V^{P_j}, for each RSCEP P_j in $\mathbf{P}^*_{\mathbf{ath}}$ can be computed as:

$$V^{P_j} = \sum_{i=1}^{k} \left(w_i^{t_1} * \frac{q_i^t - q_i^{p_j}}{q_i^t - q_i^{ave}} \right) + \sum_{i^*=1}^{k-i} \left(w_{i^*}^{t_1} * \frac{q_{i^*}^{p_j}}{q_{i^*}^t} \right) \tag{7.9}$$

where i = 3, 4, 5, 6 and i^* = 12, $q_i^{p_j}$ belong to negative QoS criteria (i = 3, 4, 5, 6), and $q_{i^*}^{p_j}$ belong to positive QoS criteria. $q_i^{ave} = (\sum_{j=1}^{M^*} q_i^{p_j}) / M^*$ is the average of all the elements in ith row in $\mathbf{QP}_{k*M^*}$

Therefore, for task T, the system will select the RSCEP which has the maximal compressive QoS score to execute. If there are several RSCEPs with the same maximal score, then one of them is selected randomly. If there is no RSCEP that satisfying the requirements of the given task, the MGrid system will return the user an unsuccessful message and advise the user to resubmit or loosen its requirements.

However, when the scale of the RSCEP selection problems are sizable and complicated, the above method simply cannot be used to find the optimal solution, and some intelligent optimization should be designed and used to resolve them. Intelligent optimization will be investigated in Chapter 8.

7.7.4 RSCE-Controller: Resource Service Composition Executing Controller

RSCE-Controller is primarily responsible for launching and controlling MGrid resource service, and collecting the executing results of the resource service. It realizes the executing of resource service composition solutions, or RSCEPs, via controlling the resource service. Figure 7.9 is an example of when the *RSCE-Controller* is used to control the execution of a composite service with two element resource services (i.e., *Resource_Service_1* and *Resource_Service_2*). In the example, the *Resource_Service_2* is executed after *Resource_Service_1*.

7.7.5 RSC-Monitor: Resource Service Composition Monitor

RSC-Monitor is primarily responsible for interacting with the resource services involved in the composition or the selected RSCEP, and monitoring their state and performance information. It also receives the status information of the available resource services in the MGrid resource service center or depository. When any abnormalities or failures with the selected resource service is detected by the *RSC-Monitor* (e.g., quit from MGrid system or unavailable suddenly, QoS or performance become bad), it will inform the *RSC-Coordinator* immediately so that the system can make related responses or adjustments to the executing composition solutions or RSCEP.

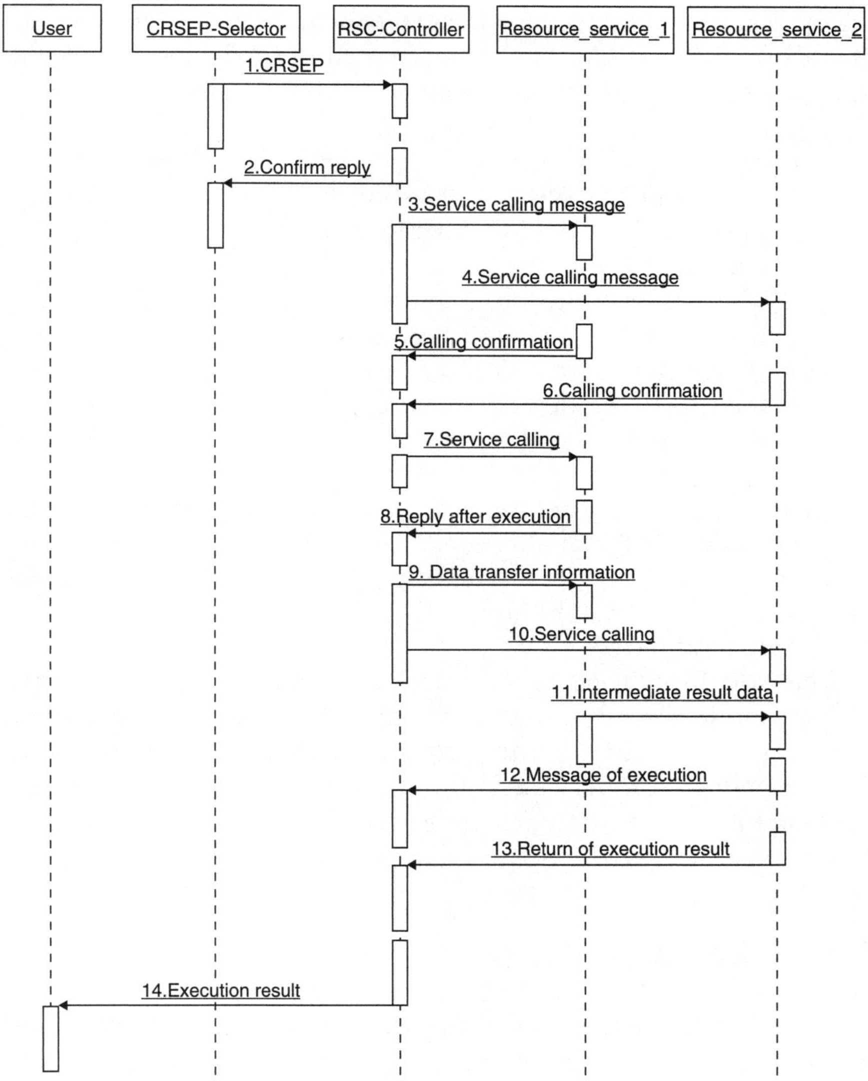

Figure 7.9 Executing process of a RSCEP by RSC-Controller (Ni 2006).

7.7.6 RSC-Coordinator: Resource Service Composition Coordinator

Resource service composition solutions, or RSCEPs, is launched and executed under the control of the *RSCE-Controller*. However, whether the resource service composition solutions or RSCEPs can be successfully executed, and if the expected results can be generated and obtained, is affected and decided by the environment of entire system. For example, before or during the implementation, the following three conditions would have occurred:

1. The selected resource service nodes in the composition quit from the system, resulting in the selected composition solutions or RSCEPs not being successfully executed.
2. The performance parameters or QoS of the selected resources changed, and they were unable to meet the requirements for executing the task.
3. Some new resource service of a higher quality was added into the system.

Each of the above conditions would have a heavy influence on the successful or high quality execution of the selected composition solutions or RSCEPs. Therefore, before or during the implementation of composition solutions or RSCEPs, some dynamic adjustments are required.

The *RSC-Coordinator* is primarily designed to make decisions, and dynamically and in real time adjust the selected composition solutions or RSCEP according to the feedback information from *RSC-Coordinator* under some conditions (e.g., better resource services are available, some resource service in RSCEP are removed, and the executing task's requirements changed) if necessary.

7.8 Summary

In order to address the resource service optimal-selection and composition problem in the manufacturing grid (MGrid) system, and provide high quality service to users, an MGrid resource service optimal-selection and composition framework (MGrid-RSOSCF) is investigated in this chapter. The process of resource service optimal-selection and composition is divided into five steps in MGrid-RSOSCF: (1) decomposing the submitted manufacturing task into several subtasks (i.e., single resource service requested task) if the submitted task is a multiple resource service requested task; (2) searching out the qualified resource service for each decomposed subtask and generating the corresponding candidate resource service set; (3) retrieving, evaluating, and comparing the quality of service (QoS) for each candidate resource service, and providing data for service optimal-selection and composition if the submitted task is a single resource service requested task; (4) synthetically evaluating the overall quality of each candidate resource service, ranking them, and selecting the optimal one for the task if the submitted manufacturing task is an multiple resource service requested task; and (5) selecting one candidate resource service from each candidate resource service set and constructing a new composite resource service according to the submitted task requirements, and collecting all the possible resource service composite execution paths (RSCEP) and selecting the optimal paths to execute the task. The proposed MGrid-RSOSCF consists of five layers and each layer provides the corresponding necessary services and algorithms to address each problem mentioned above. The five layers are: (1) *T-Layer* is responsible for MGrid task decomposition, (2) *S-Layer* is responsible for resource service match and search, (3) *Q –Layer* is responsible for QoS processing, (4) *O-Layer* is responsible for evaluating and ranking the candidate resource service, and (5) *C-Layer* is responsible for resource service composition and optimal-selection. The case study and algorithms performance comparison demonstrate that the proposed methods are sound regarding their success rate and executing efficiency.

In the following chapters, the resource service optimal-selection based on IFS (intuitionistic fuzzy set) and non-functionality QoS (Chapter 8), correlation relationship management in resource service composition (Chapter 9), resource service composition optimal-selection based on improved PSO (Chapter 10), flexibility management in resource service composition (Chapter 11), and resource service composition network (Chapter 12) will be investigated.

8

Resource Service Optimal-selection Based on Intuitionistic Fuzzy Set and Non-functionality QoS

8.1 Introduction

As covered in previous chapters, in an MGrid system there are primarily two kinds of users: (a) resource enterprise or resource service provider (RSP), and (b) user enterprise or resource service demander (RSD), as shown in Figure 8.1. The former, RSP, publishes its idle resource, product, and manufacturing ability, and also provides a manufacturing resource service to meet the user's requirements. The latter, RSD, searches the optimal manufacturing resource and service required, and selects the corresponding partner to establish a collaboration manufacturing net.

One of the key technologies to realize resource service exchange in an MGrid is resource service selection (RSS). There are two steps to realize RSS:

1. Searching the qualified resource services according to the functionality requirements of a task submitted by a user, and generating a candidate resource service

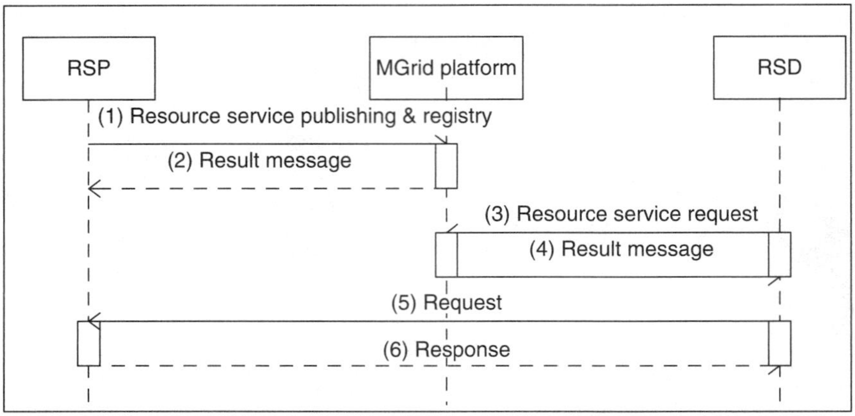

Figure 8.1 Resource services transaction between RSD and RSP in MGrid (Tao *et al.* 2009a).

set (CRSS), namely, *resource service match and search (RSMS)*, as introduced in Chapter 4.

2. Selecting the optimal resource service from CRSS to execute the task, namely, *resource service optimal-selection (RSOS)*.

A number of studies for addressing RSMS have been carried out in a distributed system such as a computing grid system. The key technologies overcoming RSMS are searching engine and matching algorithms. The searching engine makes use of matching algorithms to retrieve related resource services according to the user's requirements. Some researchers investigate RSMS from the perspectives of behavior signature (Shen and Sun, 2005), service community (Perryea and Chuang, 2006), context of service (Balken *et al.*, 2007; Doulkeridis *et al.*, 2007; Lee C and Helal, 2003; Raverdy and Issarny, 2005), user experience (Kokash *et al.*, 2007), role-based interaction model (Fernández *et al.*, 2007), met-interaction (Vitvar *et al.*, 2007), market-orientation (Yu *et al.*, 2006), UML (Zisman and Spanoudakis, 2006), expertise (Haase *et al.*, 2008), and goal-based web service discovery with sophisticated semantic matchmaking (Stollberg *et al.*, 2007), etc.

The above studies put the emphasis on resource service discovery from the aspect of the functionality requirements of a task. What is being ignored is the method that addresses how to select the optimal resource service from CRSS generated by RSMS, i.e., the problem of RSOS. It is expected that there exist hundreds of outsourced services with different QoS properties that offer the same business function, and, as a result, the user faces the problem of choosing the optimal one among numerous candidate resource services. For example, if a user wants to search a 'parametric design service' for a product in an MGrid system, the RSMS may find out a large number of (parametric design) services providing similar functional characteristics, such as the service developed by SolidWorks, Unigraphics, Pro/Engineer, and CATIA. Multiple services with similar functional characteristics point out the problem of RSOS. Users not only expect the selected resource services to meet functional aspects but they also demand a good quality of services such as reliability, security, and trust (Lin *et al.*, 2008a). It is therefore important to address the problem of RSOS in an MGrid system.

Existing research efforts on RSOS have been undertaken in the direction of QoS-aware resource service optimal-selection as shown in Table 8.1.

Table 8.1 Brief summary of web service selection

References	Methods
Menasce (2004)	QoS of component web services in terms of cost and execution time are studied. It employs probability techniques to measure the cost and execution time of component web services by considering different execution scenarios such as parallel, sequential, fastest-predecessor-triggered and so on. It helps in selecting appropriate component web services for web service composition. However, it does not consider any consensus nor does it take into account QoS attributes with *fuzzy* definitions.
Jaeger *et al.* (2004, 2005)	A mechanism for composite web services with pattern-based QoS aggregation is proposed. The QoS aggregation is used to verify that a set of services satisfy the QoS requirement for the required composite web services. In this approach the aggregation of QoS for service composition is defined by using a number of pre-defined composition patterns which include QoS ratings. The concept of the composition pattern is inspired by van der Aalst's Workflow Pattern (van der Aalst, 2003).

(Continued)

Table 8.1 (cont.) Brief summary of web service selection

References	Methods
Liu *et al.*(2004b)	An open, fair and dynamic QoS computation model for web service selection is proposed. The model is tested using a QoS registry in a hypothetical phone service provisioning market place application. The aim of this model is to investigate the relationship between QoS value and business criteria, and to study the effectiveness of price and the service sensitivity factors in QoS computation.
Zeng *et al.* (2003a)	The problem of selecting web service is addressed by maximizing user satisfaction expressed as utility functions over QoS attributes.
Sirine *et al.* (2004)	A goal-oriented and interactive composition approach is described that uses matchmaking algorithms to help users filter and select services while building their composition service.
Lin *et al.* (2005b)	They treated the selection of QoS-driven web service with dynamic composition as a fuzzy constraint satisfaction problem and applied an optimal search approach with adjustments to service composition.
Lin *et al.* (2006a)	A QoS consensus moderation approach (QCMA) is proposed in order to perform QoS ratings based on consumers' consensus so as to alleviate the differences on QoS characteristics.
Huang *et al.* (2005b)	Fuzzy group decision-making methods and semantic web technologies are employed to assist consumers in discovering appropriate services by taking into account their expectations and preferences.
Dai and Wang (20067)	The problem of optimally allocating services on the grid to maximize grid service reliability using a genetic algorithm is studied.

However, the methods in Table 8.1 are more suitable to be employed when QoS attributes are functional properties, and are not a good fit for evaluating QoS attributes which are non-functional properties such as reliability, security, and trust. Compared with functional QoS properties, non-functional QoS attributes are not easy to assess due to their complexity and the involvement of ill-structured information. Furthermore, these approaches fall short of addressing the following issues (Tao *et al.*, 2010f; 2010a):

Problem 1: **A user's feeling and transaction experiences are not considered during QoS evaluation of resource service.** Although QoS in web service discovery and selection attracts a lot of attention, most current research emphasizes the objectivity and functionality of QoS information, which is offered by corresponding service providers. In such cases, the QoS information cannot respond to the feeling of users who actually use services. If users' thought or evaluation can be embedded into web service selection, the search quality could be improved. This kind of information is named non-functionality QoS information in this chapter.

Problem 2: **The issue on how to combine time fact into QoS evaluation is not considered.** The QoS information considered in the above research is treated as static. They are provided by the owner of the resource service and never change with time. In fact, QoS of resource services are dynamically changing with time and transaction results. Take the QoS criterion, trust, for an example. If the trust of a resource service is *'very trustworthy'* in its last transaction at '20080101', and from then on, there was no transaction till the current time, e.g., '20080831', then the trust may decay into *'trustworthy'*. Hence the dynamic of QoS should be considered during QoS evaluation.

Problem 3: **The issues associated with aggregating different service consumers' and providers' evaluation on the importance of each QoS criterion are not considered.** Different users have different views on the importance of a QoS criterion, and although they may make the same evaluation (e.g., *very important*) to the importance to a QoS criterion, the definitions of these evaluation terms (e.g., *very important*) may differ. Hence, it is important to aggregate different users' evaluations on the importance of each QoS criterion.

Motivated by addressing the above issues, and realizing resource service selection, this chapter puts the emphasis on RSOS from the aspect of non-functionality QoS based on intuitionistic fuzzy set (IFS). In this chapter, a resource service selection framework is proposed based on the work in Chapter 4. The user's feeling is taken into account in regards to RSOS in the MGrid system. Non-functionality QoS evaluation of resource service is based on the user's feeling and transaction experiences using IFS. Furthermore, the dynamic of non-functionality QoS is considered, and a time decay function is introduced into the evaluation of non-functionality. A new method for RSOS based on IFS and non-functionality QoS is proposed, and the procedures are presented in detail. A practice case is used to illustrate the developed method and procedure. The primary works and contribution of this chapter are as follows:

1. The evaluation methods for non-functionality QoS of resource service are proposed. The non-functionality QoS evaluations of resource service are based on user's feeling and transaction experiences using intuitionistic fuzzy set (IFS). Furthermore, the dynamics of non-functionality QoS are considered, and a time decay function is introduced in non-functionality QoS evaluation.
2. A new method for resource service optimal-selection based on IFS and non-functionality QoS is proposed, and the procedures are presented in detail. A case study is presented to illustrate the application of the proposed method. The performance analysis and comparison results indicate that the proposed method is excellent on both effectiveness and efficiency.

8.2 Framework of Resource Service Selection

The responsibility of resource service selection (RSS) is to select the optimal resource service according to the requirements after analyzing the required resource QoS submitted by the user. The resource service selection mechanism and its architecture in MGrid are shown in Figure 8.2. The model is primarily based on two key components: RSMS and RSOS. As mentioned in the introduction, RSMS is responsible for searching the resource service according to the functionality

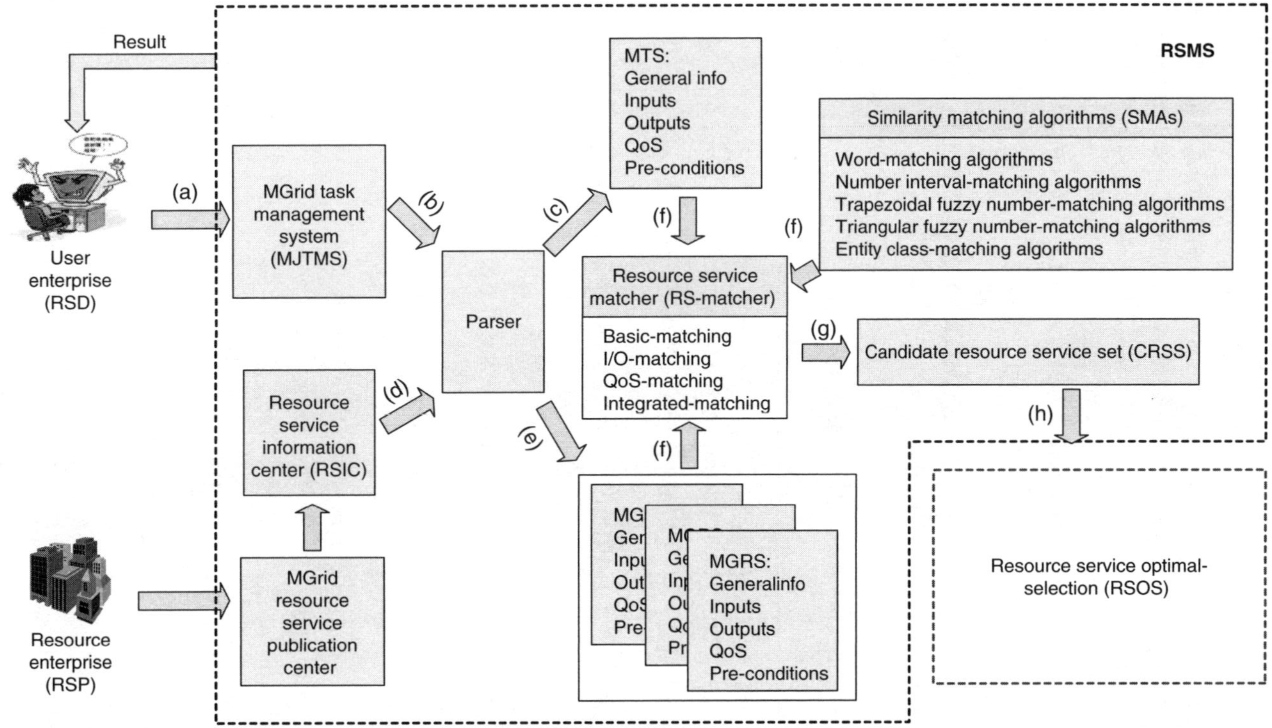

Figure 8.2 Framework for resource service selection in MGrid.

requirements of a task submitted by the user, and generating candidate resource service set (CRSS). However, RSOS is responsible for selecting the optimal resource service from CRSS to execute the task. The workflow has been briefly described in Chapter 4.

In MGrid resource service selection system, RSOS plays a very important role. It decides the quality of the selected resource service and the QoS of the entire MGrid. The key technologies and algorithms for realizing RSMS are presented in Chapter 4. In this chapter, the component of RSOS implementation is emphasized.

8.3 Resource Service Optimal-selection Based on IFS in MGrid

8.3.1 Symbols and Notations

D_A	an arbitrary resource service domain
n	the number of the resource services in D_A, $n = 1,2,3,\cdots$
M	the number of candidate resource services, $M = 1,2,3,\cdots$
RS^m	the mth candidate resource service, and $m = 1,2,3,\cdots,M$
J	the number of QoS criteria, $J = 1,2,3,\cdots$
C	QoS criteria set, $C = (c_1, c_2, c_3, \cdots, c_J)$
c_j	the Jth QoS criterion of a candidate resource service, and $j = 1,2,3\cdots,J$
$E_j^m(t_k)$	the integrated evaluation about C_j of RS^m at periods t_k, and $E_j^m(t_k)$ is an IFS
t_c	the current time of the transaction
t_0	the earlier time of a transaction
t_k	time periods of k, $k = 1,2,3,\cdots,P$
E_j^m	the synthetic evaluation about C_j of RS^m from t_0 to t_c and E_j^m is an IFS
$f(t_k)$	time decay function of $E_j^m(t_k)$ in E_j^m from t_0 to t_c.
N	the number of experts participating in evaluation of the importance of QoS criteria, $N = 1,2,3,\cdots$
U^i	the ith expert, and $i = 1,2,3,\cdots,N$
$U_w^i(c_j)$	Ui's evaluation of the importance of QoS criteria Cj, and $U_w^i(c_j)$ is an IFS

$w^i(c_j)$ the weight of $U_w^i(c_j)$
$S(U_w^i(c_j), U_w^{i^*}(c_j))$ the similarity between $U_w^i(c_j)$ and $U_w^i(c_j)$, $i^* = 1,2,3,\cdots,N$
$A(U_w^i(c_j))$ the average agreement degree of $U_w^i(c_j)$
w_j the weight of QoS criteria C_j
W the weight of QoS criteria, and $W = \{w_1, w_2, \cdots, w_j, \cdots, w_J\}$
V_{RS^m} the synthetic QoS evaluation score of RS^m

8.3.2 Preliminaries on Intuitionistic Fuzzy Set (IFS)

Fuzzy set theory, proposed by Zadeh (1972), has proved to be very effective in handling the vagueness and uncertainty intrinsically existing in the knowledge possessed by people or implied in numerical data (Xiong *et al.*, 2002; Zhang WG, Xiao, 2009). In order to understand the following sections, the basic concept of fuzzy set and IFS are introduced briefly in this section.

Definition 8.1 Let a set Z be fixed, a fuzzy set F is given by (Zadeh 1972) as follows:

$$F = \{< x, \mu_F(x) > | \, x \in Z\} \tag{8.1}$$

where

$$\mu_F : Z \to [0,1], \; x \in Z \to \mu_F(x) \in [0,1] \tag{8.2}$$

and $\mu_F(x)$ denotes the degree of membership of the element x to the set Z.

Definition 8.2 Let a set Z be fixed, an IFS A in Z is given by Atanassov (1986) as an object having the following form:

$$A = \{< x, \mu_A(x), \upsilon_A(x) > | \, x \in Z\} \tag{8.3}$$

where the functions

$$\mu_A : Z \to [0,1], \; x \in Z \to \mu_A(x) \in [0,1] \tag{8.4}$$

and

$$\upsilon_A : Z \to [0,1], \; x \in Z \to \upsilon_A(x) \in [0,1] \tag{8.5}$$

with the condition

$$0 \le \mu_A(x) + \upsilon_A(x) \le 1, \; \forall x \in Z \tag{8.6}$$

$\mu_A(x)$ and $\upsilon_A(x)$ denote the degree of *membership* and the degree of *non-membership* of the element $x \in Z$ to the set Z, respectively. In addition, for each IFS A, if

$$\pi_A(x) = 1 - \mu_A(x) - \upsilon_A(x) \tag{8.7}$$

then $\pi_A(x)$ is called the degree of indeterminacy (i.e., the degree of uncertainty) of x to A, or is called the degree of hesitancy of x to A. Especially if $\pi_A(x) = 0$ for all $x \in Z$, then IFS A is reduced to a fuzzy set.

For every two IFSs A and B, the following relations and operations are valid (Kumar *et al.*, 2000; Wang, 2009):

$$A \cap B = \{< x, \min\left(\mu_A(x), \mu_B(x)\right), \max\left(\upsilon_A(x), \upsilon_B(x)\right) > | \, x \in Z\} \tag{8.8}$$

$$A \cup B = \{< x, \max\left(\mu_A(x), \mu_B(x)\right), \min\left(\upsilon_A(x), \upsilon_B(x)\right) > | \, x \in Z\} \tag{8.9}$$

$$A \oplus B = \{< x, \mu_A(x) + \mu_B(x) - \mu_A(x)\mu_B(x), \upsilon_A(x)\upsilon_B(x) > | \, x \in Z\} \tag{8.10}$$

$$A \otimes B = \{< x, \mu_A(x)\mu_B(x), \upsilon_A(x) + \upsilon_B(x) - \upsilon_A(x)\upsilon_B(x) > | \, x \in Z\} \tag{8.11}$$

Definition 8.3 The similarity between two IFSs A and B is defined as (Hung and Yang, 2008; Xu, 2007):

$$S(A,B) = 1 - \frac{1}{2}\left(| \mu_A(x) - \mu_B(x) | + | \upsilon_A(x) - \upsilon_B(x) |\right) \tag{8.12}$$

Apparently, $0 \le S(A,B) \le 1$.

Definition 8.4 Let $\lambda \in [0,1]$, $A \in IFS$, according to Equation 8.11.

$$A^2 = \{< x, (\mu_A(x))^2, 1 - (1 - \upsilon_A(x))^2 > | \, x \in Z\} \tag{8.13}$$

$$A^3 = \{< x, (\mu_A(x))^3, 1 - (1 - \upsilon_A(x))^3 > | \, x \in Z\} \tag{8.14}$$

In general, the operator of multiplication between λ and A is defined as follows (Kumar *et al.*, 2000):

$$\lambda A = \{< x, \left(1 - \left(1 - \mu_A(x)\right)^{\lambda}\right), \left(\upsilon_A(x)\right)^{\lambda} > | \, x \in Z\} \tag{8.15}$$

Definition 8.5 Let $\boldsymbol{\lambda} = (\lambda_1, \lambda_2, \cdots, \lambda_i, \cdots, \lambda_n)$, $\forall \lambda_i \in [0,1]$ and $\Lambda = (A_1, A_2, \cdots, A_i, \cdots, A_n)$, $\forall A_i \in IFS$. For the usage of the later in this chapter, the operator $'\Theta'$ between $\boldsymbol{\lambda}$ and Λ is defined as follows:

$$\lambda\Theta\Lambda = (\lambda_1 A_1) \oplus (\lambda_2 A_2) \oplus \cdots \oplus (\lambda_i A_i) \oplus \cdots \oplus (\lambda_n A_n) \quad (8.16)$$

Apparently, the result of $\lambda\Theta\Lambda$ is still an IFS.

8.3.3 RSOS Based on IFS

The system searches out the M candidate resource services (CRS) for a job T_{ask} that cannot be decomposed further using the method in Chapter 4, and the CRS set is $RS = \{RS^1, RS^2, \cdots, RS^m, \cdots, RS^M\}$. The non-functionality QoS criteria set of each resource service is defined as $C = (c_1, c_2, c_3, \cdots, c_J)$. The problem of RSOS in MGrid is to select the optimal one from RS to execute T_{ask}. In this work, the user's feeling is integrated into non-functionality QoS evaluation of resource service using IFS. Based on the basic operators of IFS, a new method for addressing the RSOS problem based on IFS and non-functionality, QoS is proposed in the following subsections.

8.3.3.1 Evaluating c_j of RS^m at Time Periods t_k

Resource services in MGrid can be classified into two kinds: *old resource service* (ORS) and *new resource service* (NRS). ORS is the resource service that has been invoked or used at least one time. NRS is the resource service that has been published recently and has never been invoked or used since it was registered in MGrid.

It is assumed that there are n resource services existing in the resource service domain D_A and, $RS^m \in D_A$,

$$\text{where } D_A = \{RS^1, RS^2, \cdots, RS^{m-1}, RS^m, RS^{m+1}, \cdots, RS^n\}$$

Let $\{(E_1^1, E_2^1, E_3^1, \cdots, E_J^1), (E_1^2, E_2^2, E_3^2, \cdots, E_J^2), \cdots, (E_1^n, E_2^n, E_3^n, \cdots, E_J^n)\}$ be the current evaluation of QoS criteria of the n resource services in D_A. If RS^m is a NRS in D_A, the QoS evaluation of RS^m, E_j^m, is set to be the original value, and

$$E_j^m = \frac{1}{n-1}(E_j^1 \oplus E_j^2 \oplus \cdots E_j^{m-1} \oplus E_j^{m+1} \cdots \oplus E_j^n) \quad (8.17)$$

If RS^m is an ORS, the evaluation values of its QoS criteria are calculated based on the experiences of the users who transacted with it. There are two common ways to get initial uncertainty values for an object from an agent (or user/expert) according to Resconi and Kovalerchuk (2008). In the first one, each agent makes a crisp evaluation of the proposition and then frequencies are computed for a set of agents. In the second one, the agent directly assigns an uncertainty value (membership value, probability, etc.). In this chapter, the first one is used. Let N_K be the total number of users in MGrid who had a transaction with RS^m during time periods t_k, the number of the users who made a positive and negative evaluation to C_j of RS^m be N_P and N_N, respectively, and the number of the users who did not make a evaluation be N_E, and $N_P + N_N + N_E = N_k$. Then the integrated evaluation, $E_j^m(t_k)$ of the c_j of RS^m at periods t_k can be represented with an IFS as follows (Tao *et al.*, 2010f):

$$E_j^m(t_k) = \{< x, \mu_{E_j^m(t_k)}(x), \upsilon_{E_j^m(t_k)}(x) > | x \in Z\} \tag{8.18}$$

where

$$\mu_{E_j^m(t_k)}(x) = N_P / N_k \tag{8.19}$$

$$\upsilon_{E_j^m(t_k)}(x) = N_N / N_k \tag{8.20}$$

$$\pi_{E_j^m(t_k)}(x) = 1 - \mu_{EA_j^m(t_k)}(x) - \upsilon_{E_j^m(t_k)}(x) = N_E / N_k \tag{8.21}$$

For example, from August 1–8, 2008, denoted as periods t_k, there are 100 users who have had transactions with RS^1 of which 78 made a positive evaluation, and 11 made a negative evaluation against c_1 of RS^1. Therefore, the integrated evaluation of c_1 of RS^1 at periods t_k can be expressed using an IFS (0.78, 0.11).

8.3.3.2 Calculating the Synthetic Evaluation of C_j of RS^m form t_0 to t_c

In general, the evaluation of a user who has a transaction with RS^m should decay with time. For example, in the last transaction at time t_0 (e.g., '20070101'), user U_i's evaluation to *trust* of RS^m was very *'trustworthy'*, but at current time t_c (e.g., '20080815'), U_i's evaluation should be decreased a certain extent when considering U_i's evaluation while calculating RS^m's trust because U_i and RS^m had no transaction from t_0 to t_c. Therefore, an adjustment function should be employed when the system evaluates QoS criteria. Let $f(t_k)$ be the time decay value of $E_j^m(t_k)$, and $f(t_k)$ is expressed as follows (Tao *et al.*, 2010f):

$$f(t_k) = 1 - e^{-\Delta t_k / \theta},\ f(t_k) \in [0,1], \Delta t_k = t_c - t_k, \theta = N_{date}/1.95 \tag{8.22}$$

Where N_{date} is the time required when the performance of a QoS criterion decays from the best to the worst; e.g., the trust decays from the *'highest trustworthy'* to the *'highest untrustworthy'*. The time decay function curves according to different N_{date} are shown in figure 8.3.

Let $f(t)=(f(t_1),f(t_2),\cdots,f(t_k),\cdots,f(t_p))$ and $E_j^m(t)=(E_j^m(t_1),E_j^m(t_2),\cdots,E_j^m(t_k),\cdots,E_j^m(t_p))$, then the synthetic evaluation, E_j^m, about c_j of RS^m from t_0 to t_c can be formulated as:

$$\begin{aligned} E_j^m &= f(t)\Theta E_j^m(t) \\ &= f(t_1)\cdot E_j^m(t_1)\oplus f(t_2)\cdot E_j^m(t_2)\oplus\cdots\oplus f(t_k)E_j^m(t_k) \\ &\quad \oplus\cdots\oplus f(t_p)E_j^m(t_p) \end{aligned} \tag{8.23}$$

According to Equation 8.15:

$$f(t_k)E_j^m(t_k)=\{<x,\left(1-\left(1-\mu_{E_j^m(t_k)}(x)\right)^{f(t_k)}\right), \left(\upsilon_{E_j^m(t_k)}(x)\right)^{f(t_k)}>|\, x\in Z\} \tag{8.24}$$

According to Equations 8.11, 8.16, and 8.24,

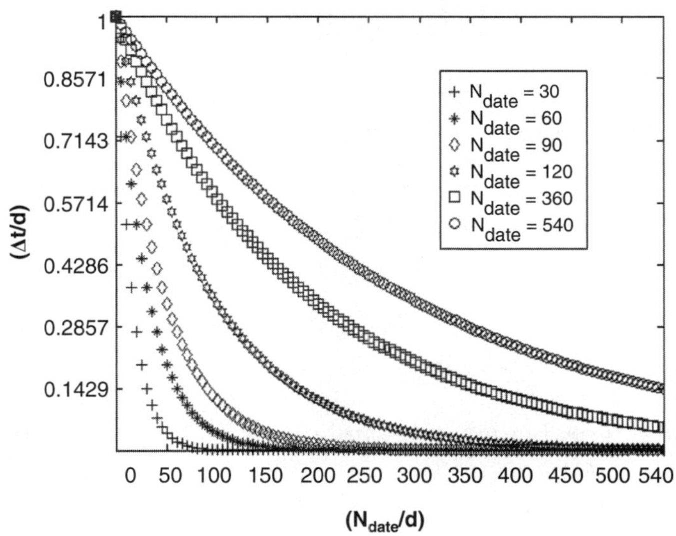

Figure 8.3 Time decay function curves according to different (=30, 60, 90, 180, 360, 540 respectively).

$$\mu_{E_j^m} = 1 - \prod_{k=1}^{p}\left(1 - \mu_{E_j^m(t_k)}\right)^{f(t_k)} \tag{8.25}$$

$$\upsilon_{E_j^m} = \prod_{k=1}^{p}\left(\upsilon_{E_j^m(t_k)}\right)^{f(t_k)} \tag{8.26}$$

it can be easily verified that

$$0 \le 1 - \prod_{k=1}^{p}\left(1 - \mu_{E_j^m(t_k)}\right)^{f(t_k)} + \prod_{k=1}^{p}\left(\upsilon_{E_j^m(t_k)}\right)^{f(t_k)} \le 1 \tag{8.27}$$

$$0 \le 1 - \prod_{k=1}^{p}\left(1 - \mu_{E_j^m(t_k)}\right)^{f(t_k)} \le 1 \tag{8.28}$$

$$0 \le \prod_{k=1}^{p}\left(\upsilon_{E_j^m(t_k)}\right)^{f(t_k)} \le 1 \tag{8.29}$$

Hence, E_j^m is still an IFS for any positive real number $f(t_k)$. Therefore, Equations 8.11, 8.16, 8.23, and 8.24 can be rewritten as follows:

$$E_j^m = \left\{ < x, 1 - \prod_{k=1}^{p}\left(1 - \mu_{E_j^m(t_k)}(x)\right)^{f(t_k)}, \prod_{k=1}^{p}\left(\upsilon_{E_j^m(t_k)}(x)\right)^{f(t_k)} > \mid x \in Z \right\} \tag{8.30}$$

8.3.3.3 *Determining the Weights of QoS Criteria*

The procedure to determine the weights of QoS criteria is organized into 5 steps based on Hu *et al.*, (2005a), Huang *et al.*, (2005b), and (Lin *et al.*, 2008a).

1. Each expert U_i uses the symbolic linguistic terms defined in Table 8.2 to evaluate the importance of each QoS criteria c_j. Let $U_w^i = (U_w^i(c_1), U_w^i(c_2), \cdots, U_w^i(c_j), \cdots, U_w^i(c_J))$ Let be U_i's evaluation of the importance of QoS criterion C of a resource service.
2. This step calculates the similarity between any two experts' (e.g., U^i and U^{i^*}) evaluation (i.e., U_w^i and U^{i^*}) for each specific c_j. Let the similarity between U^i and U^{i^*}, $Sim(U_w^i, U_w^i)$ be

Table 8.2 Linguistic terms for the importance of a QoS criterion (Wang 2009)

Linguistic terms	IFS
Very important	[0.9, 0.1-x]
Important	[0.7, 0.3-x]
Medium	[0.5, 0.5-x]
Unimportant	[0.3, 0.7-x]
Very unimportant	[0.1, 0.9-x]
It's does not matter	[0.0, 0.0-x]

$$Sim(U_w^i, U_w^{i^*}) = \{S(U_w^i(c_1), U_w^{i^*}(c_1)), S(U_w^i(c_2), U_w^{i^*}(c_2)), \cdots, S(U_w^i(c_J), U_w^{i^*}(c_J))\} \tag{8.31}$$

According to Equation 8.12:

$$S(U_w^i(c_j), U_w^{i^*}(c_j)) = 1 - \frac{1}{2}(|\mu_{U_w^i(c_j)} - \mu_{U_w^{i^*}(c_j)}| + |\upsilon_{U_w^i(c_j)} - \upsilon_{U_w^{i^*}(c_j)}|) \tag{8.32}$$

3. This step builds an agreement matrix, $(AMc_j)_{N \times N}$, for showing each similarity between each pair of experts.

$$(AMc_j)_{N\times N} = \begin{bmatrix} 1 & S(U_w^1(c_j), U_w^2(c_j)) & \cdots & \cdots & S(U_w^1(c_j), U_w^N(c_j)) \\ S(U_w^2(c_j), U_w^1(c_j)) & 1 & \cdots & \vdots & SU_w^2(c_j), U_w^N(c_j)) \\ \vdots & \vdots & \ddots \ \vdots & \vdots & \vdots \\ S(U_w^i(c_j), U_w^1(c_j)) & S(U_w^i(c_j), U_w^2(c_j)) & \cdots & \cdots & S(U_w^i(c_j), U_w^N(c_j)) \\ \vdots & \vdots & \vdots \ \vdots & \ddots & \vdots \\ S(U_w^N(c_j), U_w^1(c_j)) & S(U_w^N(c_j), U_w^2(c_j)) & \cdots & \cdots & 1 \end{bmatrix} \tag{8.33}$$

This step calculates the average agreement degree, $A(U_w^i)$, for each single expert U_i

$$A(U_w^i) = \{A(U_w^i(c_1)), A(U_w^i(c_2)), \cdots, A(U_w^i(c_J))\} \tag{8.34}$$

where

$$A(U_w^i(c_j)) = \frac{1}{N}\sum_{i^*=1}^{N} S(U_w^i(c_j), U_w^{i^*}(c_j)) \tag{8.35}$$

4. This step calculates the relative average agreement degree $w^i(c_j)$, i.e., the weight of $U_w^i(c_j)$, for each single expert U^i.

$$w^i(c_j) = \frac{A(U_w^i(c_j))}{\sum_{i^*=1}^{N} A(U_w^{i^*}(c_j))} \tag{8.36}$$

5. This step calculates the final weight of each QoS criterion c_j. Let W be the weight vector of QoS criteria C, and $W = \{w_1, w_2, \cdots, w_j, \cdots, w_J\}$, where

$$\begin{aligned} w_j = {} & (U_w^1(c_j) \otimes w^1(c_j)) \oplus \cdots \oplus (U_w^i(c_j) \otimes w^i(c_j)) \\ & \oplus \cdots \oplus (U_w^N(c_j) \otimes w^N(c_j)) \end{aligned} \tag{8.37}$$

and

$$\mu_{w_j} = 1 - \prod_{i=1}^{N} \left(1 - \mu_{U_w^i(c_j)}\right)^{w^i(c_j)} \tag{8.38}$$

$$\upsilon_{w_j} = \prod_{i=1}^{N} \left(\upsilon_{U_j^i(c_j)}\right)^{w^i(c_j)} \tag{8.39}$$

8.3.3.4 *Evaluating Fuzzy Synthetic of Each CRS*

The QoS performance rating matrix, *QPRM*, for all candidate resource services is shown as follows:

$$QPRM = [E_j^m] = \begin{array}{c} \\ RS^1 \\ RS^2 \\ \vdots \\ RS^m \\ \vdots \\ RS^M \end{array} \begin{array}{c} \begin{array}{cccccc} c_1 & c_2 & \cdots & c_j & \cdots & c_J \end{array} \\ \begin{bmatrix} E_1^1 & E_2^1 & \cdots & E_j^1 & \cdots & E_J^1 \\ E_1^2 & E_2^2 & \cdots & E_j^2 & \cdots & E_J^2 \\ \vdots & \vdots & \ddots & \vdots & \ddots & \vdots \\ E_1^m & E_2^m & \cdots & E_j^m & \cdots & E_J^m \\ \vdots & \vdots & \ddots & \vdots & \ddots & \vdots \\ E_1^M & E_2^M & \cdots & E_j^M & \cdots & E_J^M \end{bmatrix} \end{array} \tag{8.40}$$

Then we get the weighted rating matrix, *QPRMW*, of RS^m as follows:

$$QPRMW = [V_j^m] = [w_j \otimes E_j^m] = \begin{matrix} \\ RS^1 \\ RS^2 \\ \vdots \\ RS^m \\ \vdots \\ RS^M \end{matrix} \begin{matrix} c_1 & c_2 & \cdots & c_j & \cdots & c_J \\ \end{matrix} \begin{bmatrix} V_1^1 & V_2^1 & \cdots & V_j^1 & \cdots & V_J^1 \\ V_1^2 & V_2^2 & \cdots & V_j^2 & \cdots & V_J^2 \\ \vdots & \vdots & \ddots & \vdots & \ddots & \vdots \\ V_1^m & V_2^m & \cdots & V_j^m & \cdots & V_J^m \\ \vdots & \vdots & \ddots & \vdots & \ddots & \vdots \\ V_1^M & V_2^M & \cdots & V_j^M & \cdots & V_J^M \end{bmatrix} \quad (8.41)$$

8.3.3.5 *Calculating the Closeness Coefficient of Each CRS*

According to Wang (2009), define $E^+ = \{E_1^+, E_2^+, \cdots, E_j^+, \cdots E_J^+\}$ and $E^- = \{E_1^-, E_2^-, \cdots, E_j^-, \cdots, E_J^-\}$ as the *positive idea solution* (PIS) and *negative idea solution* (NIS), respectively, where

$$E^+ = (E_1{}^+, E_2{}^+, \cdots, E_J{}^+)$$

$$= \left\{(\max(\mu_{V_1^m}), \min(\upsilon_{V_1^m})), (\max(\mu_{V_2^m}), \min(\upsilon_{V_2^m})), \cdots, (\max(\mu_{V_J^m}), \min(\upsilon_{V_J^m}))\right\} \quad (8.42)$$

$$E^- = (E_1{}^-, E_2{}^-, \cdots, E_J{}^-)$$

$$= \left\{(\max(\mu_{V_1^m}), \min(\upsilon_{V_1^m})), (\max(\mu_{V_2^m}), \min(\upsilon_{V_2^m})), \cdots, (\max(\mu_{V_J^m}), \min(\upsilon_{V_J^m}))\right\} \quad (8.43)$$

According to Equation 8.12, the similarities of RS^m with PIS (i.e., E^+) and NIS (i.e., E^-) are defined as follows:

$$d(E^+, RS^m) = 1 - \frac{1}{2}\sum_{j=1}^{J}\left(|\mu_{E_j^+}(x) - \mu_{V_j^m}(x)| + |\upsilon_{E_j^+}(x) - \upsilon_{V_j^m}(x)|\right) \quad (8.44)$$

$$d(E^-, RS^m) = 1 - \frac{1}{2}\sum_{j=1}^{J}\left(|\mu_{E_j^-}(x) - \mu_{V_j^m}(x)| + |\upsilon_{E_j^-}(x) - \upsilon_{V_j^m}(x)|\right) \quad (8.45)$$

According to Xu and Yager (2007), the final synthetic QoS evaluation scores of RS^m are calculated as follows:

$$V_{RS^m} = \frac{d(E^-, RS^m)}{d(E^+, RS^m) + d(E^-, RS^m)} \tag{8.46}$$

8.3.3.6 Ranking the Order of Candidate Resource Services (CRSs)

Ranking all candidate resource service RS^m according to the closeness coefficients (i.e., synthetic QoS evaluation) V_{RS^m}, the greater the value of V_{RS^m} is, the better the RS^m is.

8.3.4 Data Structure Design

Data structure for recording resource service's QoS evaluation in a RSD (D_{RS}): As stated before, at a time period t_k, the QoS evaluation values of an ORS are calculated based on the experiences of the users who transacted with it. Therefore, we designed each entity of resource service by maintaining a table of data structure, which is responsible for recording the transaction evaluation results, named D_{RS}. Clearly, D_{RS} is decided by the factors of time period, the resource service, the specific QoS criteria, and the evaluation results from users. Hence, D_{RS} is defined as (Tao et al. 2009a):

$$D_{RS} = (RS,\ \ C,\ \ E,\ \ T) \tag{8.47}$$

where

- RS denotes the resource service set (RSS).
- C denotes the QoS criteria set of resource service, $C = (c_1, c_2, c_3, \cdots, c_J)$.
- E denotes the QoS criteria evaluation and $E = (\{N_P\}, \{N_N\}, \{N_E\})$, where N_P, N_N and N_E are the number of users who made a positive evaluation, negative evaluation, and did not make an evaluation or had no idea to a specific QoS criterion $c_j \in C$ of a $RS^m \in RS$ at a time period $t_k \in T$, respectively.
- T stands for the set of transaction time periods.

For example, the data structure of RS^1's QoS evaluation is as follows:

$$D_{RS^m} = \begin{Bmatrix} (RS^1, & c_1, & (\{121\},\{31\},\{11\}), & 20060301), \\ (RS^1, & c_2, & (\{129\},\{29\},\{5\}), & 20060301), \\ (RS^1, & c_3, & (\{135\},\{18\},\{10\}), & 20060301) \end{Bmatrix}$$

This illustrates that at time '20060301', 121 users made a positive evaluation to c_1 of RS^1, 31 users made a negative evaluation, and 11 users did not make an evaluation. N_P, N_N and N_E on c_2 and c_3 of RS^1 are 129, 29, 35 and 135, 18, 10, respectively. According to Equations 17–20, at time '20060301', the integrated evaluation to c_1 c_2, and c_3 of RS^1 with IFS are (c_1, 0.74, 0.19), (c_2, 0.79, 0.18) and (c_3, 0.82, 0.11), respectively.

8.4 Case Study

Let us assume that a user (or RSD) RSD^1 looks for a *parameterized radial magnetic design service* on a MGrid system platform. The system searches out five resource services, RS^1, RS^2, RS^3, RS^4, and RS^5, that qualify for its functional requirements using the search and match mechanism in Tao *et al.*, (2008a). In order to select the optimal resource service to serve RSD^1 and reduce the decision-making time for RSD^1 to select the best resource service, the system can evaluate the non-functionality QoS of the five resource services, and select the optimal one from them based on the above proposed method.

The six non-functionality QoS criteria selected in this case study are *Trust, Reliability, Availability, Scalability, Accuracy,* and *Security,* which is defined as $C=\{c_1,c_2,c_3,c_4,c_5,c_6\}=$ {*Trust, Reliability, Availability, Scalability, Accuracy, Security*}.

8.4.1 Step 1: Extracting the Related Data of RS^1, RS^2, RS^3, RS^4, and RS^5

The corresponding D_{RS} about RS^1, RS^2, RS^3, RS^4, and RS^5 in the database are as follows:

$$D_{RS^1}=\begin{Bmatrix}(RS^1, & c_1, & (\{99\},\{16\},\{4\}), & 20080520),\\ (RS^1, & c_2, & (\{87\},\{24\},\{8\}), & 20080520),\\ (RS^1, & c_3, & (\{103\},\{14\},\{2\}), & 20080520),\\ (RS^1, & c_4, & (\{81\},\{32\},\{6\}), & 20080520),\\ (RS^1, & c_5, & (\{87\},\{23\},\{9\}), & 20080520),\\ (RS^1, & c_6, & (\{99\},\{13\},\{7\}), & 20080520),\end{Bmatrix}$$

$$
D_{RS^1} = \begin{Bmatrix}
(RS^1, & c_1, & (\{132\},\{25\},\{8\}), & 20080620), \\
(RS^1, & c_2, & (\{123\},\{36\},\{6\}), & 20080620), \\
(RS^1, & c_3, & (\{129\},\{29\},\{7\}), & 20080620), \\
(RS^1, & c_4, & (\{97\},\{45\},\{23\}), & 20080620), \\
(RS^1, & c_5, & (\{101\},\{41\},\{23\}), & 20080620), \\
(RS^1, & c_6, & (\{121\},\{23\},\{21\}), & 20080620), \\
(RS^1, & c_1, & (\{124\},\{21\},\{9\}), & 20080720), \\
(RS^1, & c_2, & (\{126\},\{23\},\{5\}), & 20080720), \\
(RS^1, & c_3, & (\{135\},\{19\},\{0\}), & 20080720), \\
(RS^1, & c_4, & (\{111\},\{34\},\{9\}), & 20080720), \\
(RS^1, & c_5, & (\{99\},\{41\},\{14\}), & 20080720), \\
(RS^1, & c_6, & (\{124\},\{23\},\{7\}), & 20080720), \\
(RS^1, & c_1, & (\{165\},\{32\},\{154\}), & 20080820), \\
(RS^1, & c_2, & (\{169\},\{31\},\{12\}), & 20080820), \\
(RS^1, & c_3, & (\{179\},\{28\},\{5\}), & 20080820), \\
(RS^1, & c_4, & (\{135\},\{34\},\{43\}), & 20080820), \\
(RS^1, & c_5, & (\{159\},\{32\},\{21\}), & 20080820), \\
(RS^1, & c_6, & (\{167\},\{33\},\{12\}), & 20080820),
\end{Bmatrix}
$$

$$
D_{RS^2} = \begin{Bmatrix}
(RS^2, & c_1, & (\{15\},\{6\},\{2\}), & 20080620), \\
(RS^2, & c_2, & (\{14\},\{7\},\{2\}), & 20080620), \\
(RS^2, & c_3, & (\{17\},\{4\},\{2\}), & 20080620), \\
(RS^2, & c_4, & (\{12\},\{4\},\{7\}), & 20080620), \\
(RS^2, & c_5, & (\{13\},\{7\},\{3\}), & 20080620), \\
(RS^2, & c_6, & (\{14\},\{7\},\{2\}), & 20080620), \\
(RS^2, & c_1, & (\{32\},\{16\},\{8\}), & 20080720), \\
(RS^2, & c_2, & (\{35\},\{13\},\{8\}), & 20080720),
\end{Bmatrix}
$$

$$D_{RS^2} = \left\{\begin{matrix} (RS^2, & c_3, & (\{41\},\{11\},\{4\}), & 20080720), \\ (RS^2, & c_4, & (\{36\},\{5\},\{15\}), & 20080720), \\ (RS^2, & c_5, & (\{36\},\{9\},\{11\}), & 20080720), \\ (RS^2, & c_6, & (\{33\},\{8\},\{15\}), & 20080720), \\ (RS^2, & c_1, & (\{58\},\{16\},\{5\}), & 20080820), \\ (RS^2, & c_2, & (\{61\},\{13\},\{5\}), & 20080820), \\ (RS^2, & c_3, & (\{70\},\{9\},\{0\}), & 20080820), \\ (RS^2, & c_4, & (\{49\},\{25\},\{5\}), & 20080820), \\ (RS^2, & c_5, & (\{59\},\{13\},\{7\}), & 20080820), \\ (RS^2, & c_6, & (\{56\},\{13\},\{10\}), & 20080820), \end{matrix}\right\}$$

$$D_{RS^3} = \left\{\begin{matrix} (RS^3, & c_1, & (\{31\},\{5\},\{3\}), & 20080620), \\ (RS^3, & c_2, & (\{27\},\{7\},\{5\}), & 20080620), \\ (RS^3, & c_3, & (\{35\},\{3\},\{1\}), & 20080620), \\ (RS^3, & c_4, & (\{30\},\{5\},\{4\}), & 20080620), \\ (RS^3, & c_5, & (\{29\},\{4\},\{6\}), & 20080620), \\ (RS^3, & c_6, & (\{32\},\{5\},\{2\}), & 20080620), \\ (RS^3, & c_1, & (\{39\},\{4\},\{0\}), & 20080720), \\ (RS^3, & c_2, & (\{40\},\{3\},\{0\}), & 20080720), \\ (RS^3, & c_3, & (\{41\},\{2\},\{0\}), & 20080720), \\ (RS^3, & c_4, & (\{28\},\{7\},\{8\}), & 20080720), \\ (RS^3, & c_5, & (\{28\},\{6\},\{9\}), & 20080720), \\ (RS^3, & c_6, & (\{25\},\{10\},\{8\}), & 20080720), \\ (RS^3, & c_1, & (\{41\},\{45\},\{1\}), & 20080820), \\ (RS^3, & c_2, & (\{36\},\{49\},\{2\}), & 20080820), \\ (RS^3, & c_3, & (\{62\},\{20\},\{5\}), & 20080820), \end{matrix}\right\}$$

$$D_{RS^3} = \begin{Bmatrix} (RS^3, & c_4, & (\{37\},\{42\},\{8\}), & 20080820), \\ (RS^3, & c_5, & (\{31\},\{45\},\{11\}), & 20080820), \\ (RS^3, & c_6, & (\{38\},\{43\},\{6\}), & 20080820), \end{Bmatrix}$$

$$D_{RS^4} = \begin{Bmatrix} (RS^4, & c_1, & (\{168\},\{57\},\{9\}), & 20080520), \\ (RS^4, & c_2, & (\{183\},\{47\},\{4\}), & 20080520), \\ (RS^4, & c_3, & (\{221\},\{13\},\{0\}), & 20080520), \\ (RS^4, & c_4, & (\{199\},\{23\},\{12\}), & 20080520), \\ (RS^4, & c_5, & (\{156\},\{34\},\{44\}), & 20080520), \\ (RS^4, & c_6, & (\{171\},\{44\},\{19\}), & 20080520), \\ (RS^4, & c_1, & (\{148\},\{43\},\{23\}), & 20080620), \\ (RS^4, & c_2, & (\{147\},\{46\},\{21\}), & 20080620), \\ (RS^4, & c_3, & (\{165\},\{43\},\{6\}), & 20080620), \\ (RS^4, & c_4, & (\{123\},\{43\},\{48\}), & 20080620), \\ (RS^4, & c_5, & (\{131\},\{36\},\{47\}), & 20080620), \\ (RS^4, & c_6, & (\{137\},\{65\},\{12\}), & 20080620), \\ (RS^4, & c_1, & (\{128\},\{33\},\{17\}), & 20080720), \\ (RS^4, & c_2, & (\{130\},\{33\},\{15\}), & 20080720), \\ (RS^4, & c_3, & (\{145\},\{31\},\{2\}), & 20080720), \\ (RS^4, & c_4, & (\{129\},\{27\},\{22\}), & 20080720), \\ (RS^4, & c_5, & (\{137\},\{31\},\{10\}), & 20080720), \\ (RS^4, & c_6, & (\{123\},\{44\},\{11\}), & 20080720), \\ (RS^4, & c_1, & (\{118\},\{13\},\{1\}), & 20080820), \\ (RS^4, & c_2, & (\{111\},\{18\},\{3\}), & 20080820), \\ (RS^4, & c_3, & (\{121\},\{9\},\{2\}), & 20080820), \\ (RS^4, & c_4, & (\{106\},\{23\},\{3\}), & 20080820), \\ (RS^4, & c_5, & (\{114\},\{15\},\{3\}), & 20080820), \\ (RS^4, & c_6, & (\{113\},\{13\},\{6\}), & 20080820), \end{Bmatrix}$$

$$D_{RS^5} = \begin{Bmatrix} (RS^5, & c_1, & (\{379\},\{45\},\{8\}), & 20080520), \\ (RS^5, & c_2, & (\{356\},\{59\},\{17\}), & 20080520), \\ (RS^5, & c_3, & (\{396\},\{23\},\{13\}), & 20080520), \\ (RS^5, & c_4, & (\{345\},\{34\},\{53\}), & 20080520), \\ (RS^5, & c_5, & (\{320\},\{65\},\{47\}), & 20080520), \\ (RS^5, & c_6, & (\{347\},\{57\},\{28\}), & 20080520), \\ (RS^5, & c_1, & (\{333\},\{32\},\{14\}), & 20080620), \\ (RS^5, & c_2, & (\{343\},\{23\},\{13\}), & 20080620), \\ (RS^5, & c_3, & (\{353\},\{21\},\{5\}), & 20080620), \\ (RS^5, & c_4, & (\{313\},\{24\},\{42\}), & 20080620), \\ (RS^5, & c_5, & (\{343\},\{29\},\{7\}), & 20080620), \\ (RS^5, & c_6, & (\{318\},\{45\},\{16\}), & 20080620), \\ (RS^5, & c_1, & (\{299\},\{23\},\{11\}), & 20080720), \\ (RS^5, & c_2, & (\{303\},\{23\},\{7\}), & 20080720), \\ (RS^5, & c_3, & (\{313\},\{19\},\{1\}), & 20080720), \\ (RS^5, & c_4, & (\{279\},\{34\},\{20\}), & 20080720), \\ (RS^5, & c_5, & (\{305\},\{24\},\{4\}), & 20080720), \\ (RS^5, & c_6, & (\{287\},\{32\},\{14\}), & 20080720), \\ (RS^5, & c_1, & (\{211\},\{31\},\{12\}), & 20080820), \\ (RS^5, & c_2, & (\{201\},\{34\},\{19\}), & 20080820), \\ (RS^5, & c_3, & (\{221\},\{21\},\{12\}), & 20080820), \\ (RS^5, & c_4, & (\{187\},\{39\},\{28\}), & 20080820), \\ (RS^5, & c_5, & (\{189\},\{34\},\{31\}), & 20080820), \\ (RS^5, & c_6, & (\{184\},\{36\},\{34\}), & 20080820), \end{Bmatrix}$$

From the data, it is known that RS^2 and RS^3 have had no history transaction before '20080520'.

8.4.2 Step 2: Evaluating the QoS Criteria of Each Candidate Resource Service at Different Time Periods

According to Equations 8.17–8.20, the evaluation results of the QoS criteria of RS^1, RS^2, RS^3, RS^4 and RS^5 are calculated and the results are shown in Table 8.3.

8.4.3 Step 3: Calculating the Synthetic QoS Criteria of Each Candidate Resource Service

Let N_{date}= 210 and the current transacting time be '20080820'. According to Equations 8.22 and 8.30 the synthetic QoS evaluation of RS^1, RS^2, RS^3, RS^4 and RS^5 are calculated and the results are shown in Table 8.4.

8.4.4 Step 4: Determining the Weights of Each QoS Criterion

The original evaluations of users for the importance of QoS criteria are shown in Table 8.5.

According to Equations 8.31–8.36, the relative weights of each QoS criterion of each user are calculated and their results are shown in Table 8.6.

According to Equations 8.37–8.39, the final weights of $C=(c_1,c_2,c_3,\cdots,c_J)$ are as follows:

$$W=\{w_1,w_2,\cdots,w_j,\cdots,w_J\}$$

=(0.8125,0.0000),(0.8296,0.0000),(0.8341,0.0000),(0.5848,0.0000),(0.7922,0.1753),(0.7790,0.1685)

According to Equation 8.41, the $QPRMW$ of RS^1, RS^2, RS^3, RS^4, RS^5 is calculated and the results are shown in Table 8.7.

8.4.5 Step 5: Calculating the Closeness Coefficient of Each CRS

According to Equations 8.42 and 8.43, the *positive idea solution* is

$$E^+=\{E_1^+,E_2^+,\cdots,E_j^+,\cdots E_J^+\}$$

Table 8.3 QoS evaluation of candidate resource services RS^1, RS^2, RS^3, RS^4 and RS^5

Time	CRSs	Trust c_1	Reliability c_2	Availability c_3	Scalability c_4	Accuracy c_5	Security c_6
20080520	RS^1	(0.83,0.13,0.03)	(0.73,0.20,0.07)	(0.87,0.12,0.02)	(0.68,0.27,0.05)	(0.73,0.19,0.08)	(0.83,0.11,0.06)
20080520	RS^4	(0.72,0.24,0.04)	(0.78,0.20,0.02)	(0.94,0.06,0.00)	(0.85,0.10,0.05)	(0.67,0.15,0.19)	(0.73,0.19,0.08)
20080520	RS^5	(0.88,0.10,0.02)	(0.82,0.14,0.04)	(0.92,0.05,0.03)	(0.80,0.08,0.12)	(0.74,0.15,0.11)	(0.80,0.13,0.06)
20080620	RS^1	(0.80,0.15,0.05)	(0.75,0.22,0.04)	(0.78,0.18,0.04)	(0.59,0.27,0.14)	(0.61,0.25,0.14)	(0.73,0.14,0.13)
20080620	RS^2	(0.65,0.26,0.09)	(0.61,0.30,0.09)	(0.74,0.17,0.09)	(0.52,0.17,0.30)	(0.57,0.30,0.13)	(0.61,0.30,0.09)
20080620	RS^3	(0.79,0.13,0.08)	(0.69,0.18,0.13)	(0.90,0.08,0.03)	(0.77,0.13,0.10)	(0.74,0.10,0.15)	(0.82,0.13,0.05)
20080620	RS^4	(0.69,0.20,0.11)	(0.69,0.21,0.10)	(0.77,0.20,0.03)	(0.57,0.20,0.22)	(0.61,0.17,0.22)	(0.64,0.30,0.06)
20080620	RS^5	(0.88,0.08,0.04)	(0.91,0.06,0.03)	(0.93,0.06,0.01)	(0.83,0.06,0.11)	(0.91,0.08,0.02)	(0.84,0.12,0.04)

(Continued)

Table 8.3 (cont.) QoS evaluation of candidate resource services RS^1, RS^2, RS^3, RS^4 and RS^5

20080720	RS^1	(0.81,0.14,0.06)	(0.82,0.15,0.03)	(0.88,0.12,0.00)	(0.72,0.22,0.06)	(0.64,0.27,0.09)	(0.81,0.15,0.05)
20080720	RS^2	(0.57,0.29,0.14)	(0.63,0.23,0.14)	(0.73,0.20,0.07)	(0.64,0.09,0.27)	(0.64,0.16,0.20)	(0.59,0.14,0.27)
20080720	RS^3	(0.91,0.09,0.00)	(0.93,0.07,0.00)	(0.95,0.05,0.00)	(0.65,0.16,0.19)	(0.65,0.14,0.21)	(0.58,0.23,0.19)
20080720	RS^4	(0.72,0.19,0.10)	(0.73,0.19,0.08)	(0.81,0.17,0.01)	(0.72,0.15,0.12)	(0.77,0.17,0.06)	(0.69,0.25,0.06)
20080720	RS^5	(0.90,0.07,0.03)	(0.91,0.07,0.02)	(0.94,0.06,0.00)	(0.84,0.10,0.06)	(0.92,0.07,0.01)	(0.86,0.10,0.04)
20080820	RS^1	(0.78,0.15,0.07)	(0.80,0.15,0.06)	(0.84,0.13,0.02)	(0.64,0.16,0.20)	(0.75,0.15,0.10)	(0.79,0.16,0.06)
20080820	RS^2	(0.73,0.20,0.06)	(0.77,0.16,0.06)	(0.89,0.11,0.00)	(0.62,0.32,0.06)	(0.75,0.16,0.09)	(0.71,0.16,0.13)
20080820	RS^3	(0.47,0.52,0.01)	(0.41,0.56,0.02)	(0.71,0.23,0.06)	(0.43,0.48,0.09)	(0.36,0.52,0.13)	(0.44,0.49,0.07)
20080820	RS^4	(0.89,0.10,0.01)	(0.84,0.14,0.02)	(0.92,0.07,0.02)	(0.80,0.17,0.02)	(0.86,0.11,0.02)	(0.86,0.10,0.05)
20080820	RS^5	(0.83,0.12,0.05)	(0.79,0.13,0.07)	(0.87,0.08,0.05)	(0.74,0.15,0.11)	(0.74,0.13,0.12)	(0.72,0.14,0.13)

Table 8.4 QoS evaluation of candidate resource services RS^1, RS^2, RS^3, RS^4, RS^5

CRSs	Trust c_1	Reliability c_2	Availability c_3	Scalability c_4	Accuracy c_5	Security c_6
RS^1	(0.97,0.02)	(0.96,0.02)	(0.98,0.02)	(0.90,0.04)	(0.92,0.04)	(0.96,0.02)
RS^2	(0.86,0.08)	(0.88,0.07)	(0.95,0.04)	(0.81,0.05)	(0.86,0.05)	(0.84,0.05)
RS^3	(0.92,0.06)	(0.91,0.07)	(0.97,0.02)	(0.81,0.08)	(0.78,0.07)	(0.81,0.10)
RS^4	(0.96,0.02)	(0.96,0.03)	(0.99,0.01)	(0.95,0.02)	(0.96,0.02)	(0.95,0.03)
RS^5	(0.99,0.01)	(0.98,0.01)	(0.99,0.00)	(0.96,0.01)	(0.98,0.01)	(0.97,0.01)

Table 8.5 The evaluation of users for the importance of QoS criteria.

	Trust c_1	Reliability c_2	Availability c_3	Scalability c_4	Accuracy c_5	Security c_6
u_1	(0.90,0.06,0.04)	(0.90,0.10,0.00)	(0.90,0.10,0.00)	(0.30,0.50,0.20)	(0.70,0.20,0.10)	(0.70,0.20,0.10)
u_2	(0.70,0.20,0.10)	(0.70,0.20,0.10)	(0.90,0.10,0.00)	(0.70,0.20,0.10)	(0.90,0.10,0.00)	(0.90,0.10,0.00)
u_3	(0.90,0.08,0.02)	(0.90,0.10,0.00)	(0.70,0.20,0.10)	(0.90,0.10,0.00)	(0.70,0.20,0.10)	(0.70,0.20,0.10)
u_4	(0.50,0.20,0.30)	(0.70,0.10,0.20)	(0.90,0.10,0.00)	(0.10,0.50,0.40)	(0.90,0.10,0.00)	(0.90,0.10,0.00)
u_5	(0.70,0.20,0.10)	(0.70,0.20,0.10)	(0.70,0.10,0.20)	(0.00,0.00,1.00)	(0.90,0.10,0.00)	(0.30,0.50,0.20)
u_6	(0.50,0.20,0.30)	(0.70,0.30,0.00)	(0.90,0.00,0.10)	(0.90,0.10,0.00)	(0.50,0.50,0.00)	(0.50,0.50,0.00)
u_7	(0.70,0.20,0.10)	(0.90,0.00,0.10)	(0.50,0.20,0.30)	(0.70,0.20,0.10)	(0.90,0.10,0.00)	(0.90,0.10,0.00)
u_8	(0.90,0.10,0.00)	(0.90,0.10,0.00)	(0.90,0.10,0.00)	(0.50,0.40,0.10)	(0.50,0.30,0.20)	(0.70,0.10,0.20)
u_9	(0.90,0.00,0.10)	(0.70,0.10,0.20)	(0.70,0.20,0.10)	(0.00,0.00,1.00)	(0.90,0.10,0.00)	(0.70,0.20,0.10)
u_{10}	(0.90,0.10,0.00)	(0.90,0.10,0.00)	(0.90,0.10,0.00)	(0.50,0.50,0.00)	(0.30,0.50,0.20)	(0.90,0.10,0.00)

Table 8.6 The weight of each user to each QoS criterion

	Trust c_1	**Reliability c_2**	**Availability c_3**	**Scalability c_4**	**Accuracy c_5**	**Security c_6**
U_1	0.1024	0.1026	0.1048	0.1137	0.1057	0.1076
U_2	0.1046	0.1000	0.1048	0.1256	0.1071	0.1047
U_3	0.1027	0.1026	0.0995	0.1043	0.1057	0.1076
U_4	0.0881	0.0987	0.1048	0.0948	0.1071	0.1047
U_5	0.1046	0.1000	0.0981	0.0450	0.1071	0.0698
U_6	0.0881	0.0948	0.0995	0.1043	0.0893	0.0858
U_7	0.1046	0.0974	0.0796	0.1256	0.1071	0.1047
U_8	0.1024	0.1026	0.1048	0.1256	0.0923	0.1032
U_9	0.0999	0.0987	0.0995	0.0450	0.1071	0.1076
U_{10}	0.1024	0.1026	0.1048	0.1161	0.0714	0.1047

Table 8.7 QoS evaluation of candidate resource services RS^1, RS^2, RS^3, RS^4, RS^5.

CRSs	Trust c_1	Reliability c_2	Availability c_3	Scalability c_4	Accuracy c_5	Security c_6
RS^1	(0.78,0.02)	(0.80,0.02)	(0.82,0.02)	(0.52,0.04)	(0.73,0.20)	(0.75,0.18)
RS^2	(0.70,0.08)	(0.73,0.07)	(0.79,0.04)	(0.47,0.05)	(0.68,0.22)	(0.66,0.21)
RS^3	(0.75,0.06)	(0.76,0.07)	(0.81,0.02)	(0.47,0.08)	(0.62,0.24)	(0.63,0.26)
RS^4	(0.78,0.02)	(0.79,0.03)	(0.82,0.01)	(0.55,0.02)	(0.76,0.19)	(0.74,0.19)
RS^5	(0.80,0.01)	(0.82,0.01)	(0.83,0.00)	(0.56,0.01)	(0.78,0.18)	(0.75,0.18)

$$= \{(\max(\mu_{V_1^m}), \min(\upsilon_{V_1^m})), (\max(\mu_{V_2^m}), \min(\upsilon_{V_2^m})), \cdots, (\max(\mu_{V_6^m}), \min(\upsilon_{V_6^m}))\}$$

={(0.80,0.01), (0.82,0.01),(0.83,0.00),(0.56,0.01),(0.78,0.18), (0.75,0.08)},

and *negative idea solution* is

$$E^- = \{E_1^-, E_2^-, \cdots, E_j^-, \cdots, E_J^-\}$$

$$= \{(\min(\mu_{V_1^m}), \max(\upsilon_{V_1^m})), (\min(\mu_{V_2^m}), \max(\upsilon_{V_2^m})), \cdots, (\min(\mu_{V_6^m}), \max(\upsilon_{V_6^m}))\}$$

= {(0.70,0.08), (0.73,0.07),(0.79,0.04),(0.47,0.08),(0.62,0.24),(0.63, 0.26)}.

According to Equations 8.44–8.46, the similarities of RS^m with E^+ and E^-, and the closeness coefficient are calculated and the results are shown in Table 8.8.

8.4.6 Step 6: Ranking the Order of CRSs

Apparently, according to the result of V_{RS^m} in Table 8.8,

$$RS^5 \succ RS^4 \succ RS^1 \succ RS^2 \succ RS^3$$

Thus, the user (i.e., RSD^1) should better select RS^5 to execute the task, rather than RS^4, RS^1, RS^2, and RS^3 in sequence.

Table 8.8 The final evaluation result of resource services RS^1, RS^2, RS^3, RS^4, RS^5

m	$d(E^+, RS^m)$	$d(E^-, RS^m)$	V_{RS^m}
m=1	0.8800	0.6340	0.5813
m=2	0.6059	0.9081	0.4002
m=3	0.5838	0.9303	0.3856
m=4	0.9220	0.5921	0.6089
m=5	1.0000	0.5140	0.6605

8.5 Performance Analysis and Discussion

8.5.1 Scalability and Efficiency

In order to test the scalability (i.e., the relationship between the number of data and processing time) of the proposed method, a set of experiments were conducted. In our experiments, the number of candidate resource services for each task, i.e., $N_{\text{Candidate}}$ or M defined in Section 8.3.1, varies from 500–5000 with an increment of 500, and the number of QoS (i.e., J defined in Section 8.3.1) for each resource service changes from 4–20 with an increment of 4. The experiments are implemented in MATLAB 7.4 on an AMD2.2GHz with 2.0G RAM under Microsoft Windows XP. The algorithms were coded with MATLAB 7.4 and saved in a MATLAB file. It was then executed directly in MATLAB 7.4. The result of each test is an average of ten executions. The summary of the results is shown in Figure 8.4.

It can be concluded from Figure 8.5, that with the increase of the number of QoS criteria, the entire processing time of the proposed method under the same amount of candidate resource services for each task increases to a small extent. When the number of candidate resource services for each task is 500, and number of QoS criteria is 20, i.e., $N_{\text{Candidate}} = 500$ and $J = 20$, the entire processing time for the proposed method is about 6 seconds. When $N_{\text{Candidate}} \leq 2000$ and $J \leq 12$, the entire processing time for the proposed method is within 15 seconds. This solving scale is suited for the requirements of most RSOS problems. Furthermore, most of the RSOS problems in MGrid are addressed by high performance computer, which shortens the processing time under the same data scale sharply. Hence, the efficiency and scalability of the proposed approach is apparent.

8.5.2 Effectiveness

In order to validate the performance of the proposed method, in addition to the above case study, a set of experiments were conducted. *Recall* and *Precision*, which are the standard measures that have been used in information retrieval for measuring the accuracy of a search method or search engine, were borrowed and selected as the criteria to test the accuracy of the proposed method. But the specific meanings of *Recall* and *Precision* are different with those in

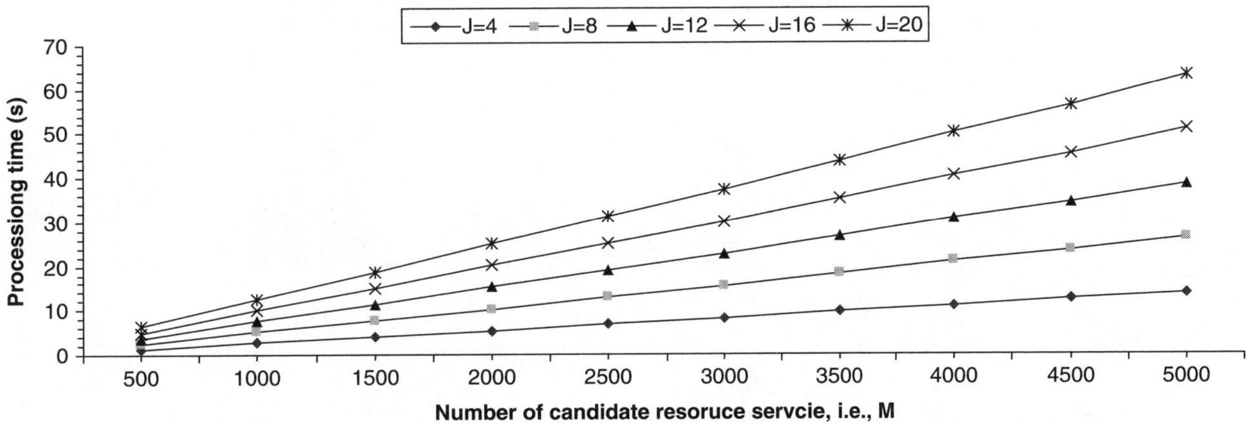

Figure 8.4 Scalability (i.e., relationships between the number of data and processing time) of the proposed method.

Li *et al.*, (2008a). Let $N_{\text{Candidate}}$ be the number of all candidate resource resources for a task; $N_{\text{Qualified}}$ be the number of the resource services that qualify for the requirements of the task among all candidate resource resources; $N_{\text{Front}}(N_{\text{Front}} = 1,2,3,\cdots$ and $N_{\text{Front}} \leq N_{\text{Qualified}})$ be the number of the resource services located in front in the recommend list generated by using the proposed method, and; $N_{\text{Qualified}}^{\text{Front}}$ $(N_{\text{Qualified}}^{\text{Front}} = 1,2,3\cdots$ and $N_{\text{Qualified}}^{\text{Front}} \leq N_{\text{Font}})$ be the number of the qualified resource services among N_{Front}. In this chapter, recall, R_{ecall}, and Precision, $P_{recision}$, are defined as follows:

$$R_{ecall} = \frac{N_{\text{Qualified}}^{\text{Front}}}{N_{\text{Qualified}}}, \quad P_{recision} = \frac{N_{\text{Qualified}}^{\text{Front}}}{N_{\text{Front}}}$$

Let *QualifiedRate* be the ratio of $N_{\text{Qualified}}$ and $N_{\text{Candidate}}$ and $QualifiedRate = \dfrac{N_{\text{Qualified}}}{N_{\text{Candidate}}}$.

In our experiments, the number of candidate resource services is set to 500 for each task, i.e., $N_{\text{Candidate}} = 500$. The *QualifiedRate* varies from 10–70% with an increment of 20% and R_{ecall} changes from 0.1–1 with an increment of 0.1. The parameters of each candidate resource service are generated by computer in regards to specific *QualifiedRate*. The precision of the proposed method under different *QualifiedRate* and R_{ecall} is shown in Figure 8.5, and the result of each test is the average of ten executions.

It can be concluded from Figure 8.5 that when recall is under 0.4, the precision of the proposed method is close to 100%. It means that, when $N_{\text{Candidate}} = 500$, $N_{\text{Qualified}} = 50$, and $R_{ecall} = 0.4$, the first 20 resource services located in front in the recommend list generated by using the proposed method are qualified for the user's requirements. This resolving scale is suited for the requirements of most RSOS problems in the MGrid system. Hence, the effectiveness of the proposed approach is apparent.

8.5.3 Comparison with the Method of Bedi (Bedi *et al.*, 2007)

We compared our proposed method with that of Bedi (Bedi *et al.*, 2007). At first we tried to compare the performance of the two methods but failed. That's because the original data source is impossible to be used by the two methods. The main differences are as follow.

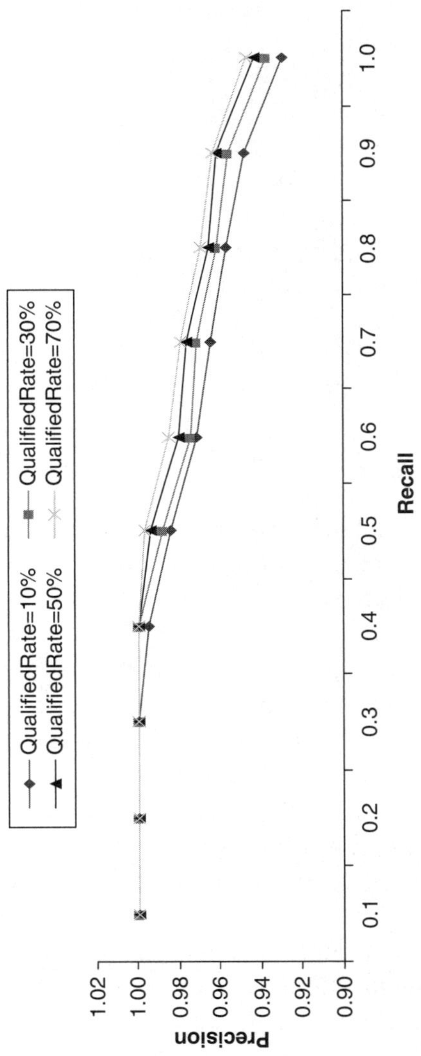

Figure 8.5 Performance of the proposed method.

Both methods use IFS to represent the evaluation of criteria to a product, but their ways of computing and the data they use are totally different. In our proposed method, the evaluation of criteria to a product (e.g., a resource service) is based on the evaluation or experiences of inexhaustible users who transacted with it, as described in Section 8.3.3.1. However, in Bedi *et al.*, (2007), the computing of the evaluation of criteria (i.e., trust) to a product is primarily based on a "preference list" and "uncertain list" using temporal ontology, which are maintained by the recommender agent as shown in Section 3.3 (Bedi *et al.*, 2007). Apparently, the data source is different and is impossible to be shared by the two methods.

Even though we can compare their performances by using different data sources, there is no meaning in fact. That is because we cannot conclude that our method is better than Bedi's, or Bedi's is better than ours, based on the comparison results generated by different data sources. Hence, we compared our proposed method with Bedi's from the procedural aspect. Based on the comparison of the realized processes of the two methods, we think the advantages of our proposed method are as follows:

- More users can be involved in the evaluation of a product, and their evaluation and transaction experiences with a resource service can be considered during the evaluation. Compared with Bedi's method, it is easier to get the original evaluation to a criterion with IFS. We can compare Equation 8.19 in Section 8.3.3.1 to Equation 8.1 in Bedi *et al.*, (2007). It is easy to maintain and record the evaluation data of users in our proposed method. In the author's proposed method, it only needs to record the number of users who made *negative* and *positive* evaluations to a product. As the number of users incrementally increases, the data size to be recorded remains the same. But in Bedi *et al.*, (2007), each recommender agent needs to maintain a "preference list" and an "uncertain list", and the corresponding ontology must be developed and used. As the number of recommenders (or recommender agents) incrementally increases, more data or a bigger data size needs to be maintained.
- The evaluation of criteria to a product is more dynamic because of the introduction of time-decay function

during the evaluation process. In general, the evaluation of a user who has had a transaction with a product or service should decay with time. For example, in the last transaction at time t_0(e.g., '20070101'), the evaluation of user U_i to the *trust* of a product is very '*trustworthy*', but at current time t_c(e.g., '20080815'), U_i's evaluation decreases to a certain extent when considering U_i's evaluation when calculating the product's current trust degree, because U_i and RS^m has no transaction from t_0 to t_c. It is the same in real life. For example, James told his good friend Jack that the food at the "Cherry Restaurant" was very good and inexpensive. Jack also believed this after he had just finished eating there. But let us assume that after that day, James moved to another city for two years and never visited the "Cherry Restaurant" again. Two years later, when James meets Jack, and tells Jack the food at the "Cherry Restaurant" is very good and inexpensive, maybe Jack does not believe it this time. This is because he does not know what has happened to the "Cherry Restaurant" during the two years he was away.

8.6 Summary

In the manufacturing grid (MGrid) system according to the functional requirements of a task, there exist a lot of resource services which have similar functional characteristics. Multiple resource services with similar functional characteristics raise the concern over resource service optimal-selection (RSOS). It is important to select the optimal resource services according to their non-functionality characteristics or quality of service (QoS). However, non-functionality QoS attributes are not easy to measure due to their complexity and the involvement of ill-structured information. In this chapter, the user's feeling is taken into account in RSOS in an MGrid system. The non-functionality QoS evaluation of resource services is based on the users' feeling and transaction experiences using intuitionistic fuzzy set (IFS). Furthermore, the dynamics of non-functionality QoS is considered, and a time-decay function is introduced into the non-functionality QoS evaluation.

A new method is proposed for resource service optimal-selection based on IFS and non-functionality QoS, and the procedures are presented in detail. A practice case is used to illustrate the proposed method and procedure, and the performance and advantage of the proposed method are discussed.

9

Correlation Relationship Management in Resource Services Composition

9.1 Introduction

During resource services composition in the manufacturing grid (MGrid), the flow of resource services composition is usually first established according to the logical functions of resource services, and then the system selects the optimal one to execute. There always exist a lot of resource services with similar functions but different non-functionality properties in the process of resource services composition. In order to enhance the quality of resource service composition, non-functionality properties, i.e., quality of service (QoS), are usually considered. The QoS of resource services composition can be taken as a criterion to decide which one is to be selected. In the practical operation, the QoS of each flow of resource services composition is computed, then the flow of resource services composition having the highest QoS can be obtained based on the users' demands. In previous studies, when computing the QoS of resource services composition, most works treat resource services involved in composition as being mutually independent.

However, most existing works about QoS-based resource service composition treat the resource services involved in composition as independent from each other, and their correlations are usually ignored. In fact, most resource services have some correlations with each other, and the correlations can badly affect the quality of resource service composition. In this chapter, the three kinds of correlation relationship in resource services composition are investigated. The impact of each kind of correlation relationship on the whole QoS of resource services composition is investigated. A case study is conducted that indicates the application and effectiveness of the work.

9.2 Related Works

Current evaluation of QoS is mainly based on the QoS criteria, and the QoS of the whole resource services composition is arrived at through building the effective QoS evaluation model. Zeng *et al.*, (2004a) proposed five generic QoS criteria for elementary resource services: execution price, execution duration, reputation, successful execution rate, and availability. A five dimensional QoS evaluation model based on these five attributes is presented. Liu *et al.*, (2004b) thought that the QoS model should also include domain specific criteria except for generic criteria, so they proposed an extensible QoS model including both generic and domain specific criteria.

Considering the correlations existing among composite resource services, a few studies have been focused on the research of correlation-aware resource services composition. Dai *et al.*, (2006a) proposed that there are three kinds of relations between two composite resource services (i.e., combinable relation, business entity relation, and statistical relation), and established a resource service correlation-aware QoS model by taking relation degree as a measure of the QoS of resource services composition. Based on the domain ontology, Ni *et al.*, (2007a) defined concept sets similarity, which is used to obtain the correlation and correlation degree among composite resource services. In order to obtain a workflow model of resource services composition, Cardoso and Sheth (2003) proposed the computing method with regard to semantic similarity between output and input concept properties. Tao *et al.*, (2010e) and Ye *et al.*, (2008a) studied the business entity relation among composite resource services, and established a QoS description model supporting the composite resource services correlation respectively.

However, the above research on composite resource services correlations and QoS models are far from comprehensive. These works only consider one kind of combinable correlations. Namely only when the output of the previous resource service is the input of latter resource service can these two resource services be composed together. Other kinds of combinable correlations are ignored. Furthermore, the discrepancies among business entity correlation also have different influences on the QoS of resource services composition. But the previous works only considered one kind of business entity correlation which has a benefit to QoS of resource services composition. In addition, there is still very little research on statistical correlation. Dai *et al.*, (2006a) measure the statistical correlation between two composite resource services through a statistical correlation angle, and have not considered the fuzzy issue of statistics. In view of these deficiencies, a detailed analysis of three kinds of correlations relationships in resource services composition is presented in this chapter, including the establishment of a new QoS computation model of resource services composition.

9.3 Motivation

In order to show the concept of resource service composition, an example about "online purchasing automobiles parts (*OPAP*)" is presented. The whole flow of online procurement is shown in Figure 9.1 (Guo and Tao *et al.*, 2010), which contains five elementary resource services, i.e., *OMS*, *PS*, *SSS*, *EPS*, *DAS*, and *EPS*, and each elementary resource service has a series of candidate resource services set (*CSS*). The flow of resource services composition is as follows:

1. Through "order management service" (*OMS*), the automobile agent online submits the information of the required types, quantity, delivery date, etc.;
2. The automobile manufacturers deal with the information of the order by "PDM services" (*PS*) to generate a corresponding list of product data, including procurement list and production list;
3. According to the procurement list, the appropriate suppliers are selected by "suppliers select services"

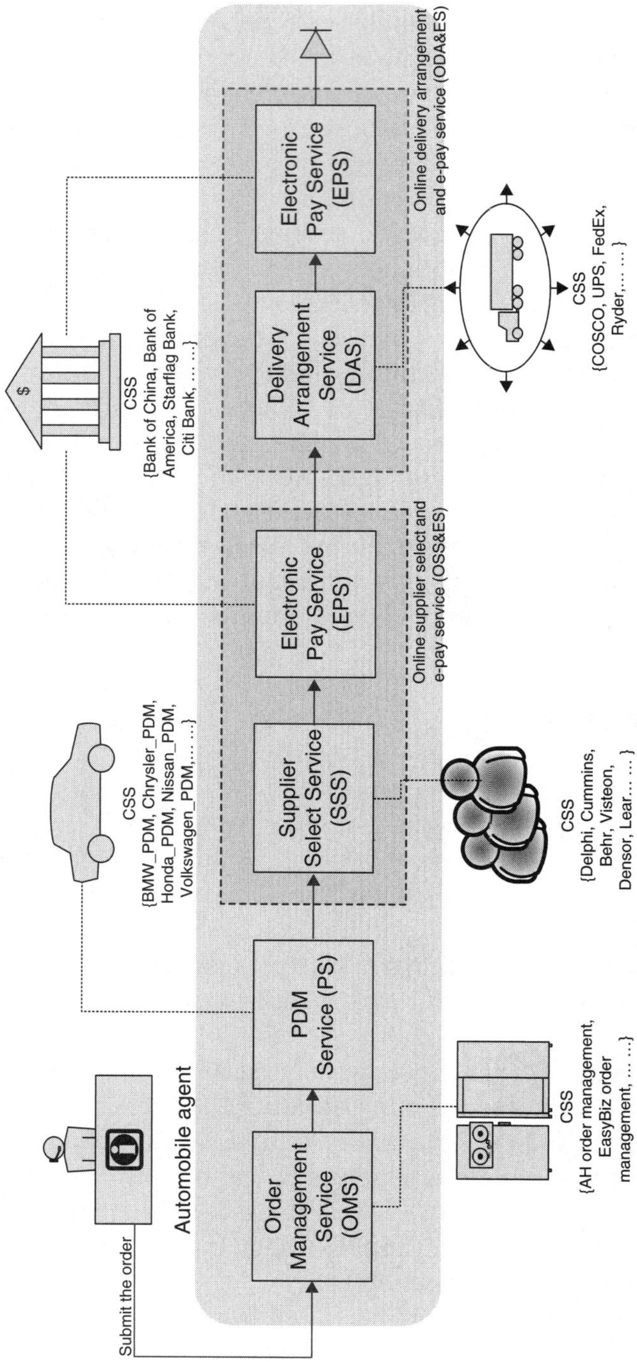

Figure 9.1 Flow of online purchasing automobiles parts.

(*SSS*), and advance payment is paid for them by "electronic pay services" (*EPS*);
4. Finally, the delivery plan of procurement parts is arranged by "Delivery arrangement services" (*DAS*), and after advance payment is paid for the deliverers by "Electronic pay services" (*EPS*), the deliverers deliver the parts to the processing assembly workshop assigned by manufacturers.

In this process, the automobile manufacturers wish to obtain the parts within the shortest time and at the lowest price, which is necessary in order for each resource service provider to provide a higher QoS of resource services composition, so that the QoS of the entire resource services composition is the highest. But in the actual resource services composition, these resource service providers are not completely independent. There always exist correlations among them, which influences the entire quality of the resource services composition.

Take the "execution price" as an example. Without considering correlation, if the automobile manufacturer selects *Cummins Corporation* and *USP Corporation* as the supplier and deliverer, 100,000 and 50,000 is needed to pay for two Corporations, respectively, and the total cost of composition is 150,000. If the two companies have a certain cooperation correlation, only 90,000 and 45,000 are needed to pay for them, respectively, and the total cost of these kinds of parts is 135,000. Apparently, in this example, the quality of the resource services composition is improved by reducing the "execution price" when considering the correlation of composite resource services.

In order to find what influences to QoS of resource services composition the correlations generate, the following three issues are emphasized in this chapter:

1. How to abstractly describe the elementary resource services considering the correlation in the composite flow of OPAP.
2. The type and the representation methods of correlations among composite resource services.
3. How to establish the QoS models of resource services composition considering the correlation.

9.4 Correlation Relationship among Resource Services

It is assumed that there are m elementary resource services in a resource services composition, and each elementary resource service has n_m candidate resource services. For any two resource services in the candidate resource service set, according to the influence of correlation among resource services to QoS, the correlation between two elementary resource services is defined as $Correlation(S_i, S_j) = <CoC, BuC, StC>$, where CoC, BuC, and StC denote combinable correlation , business entity correlation, and statistical cooperate correlation, respectively (Dai *et al.*, 2006a). The type and representation methods of correlations among composite resource services are discussed in detail in the following section, as well as the influence to QoS in the flow of "*OPAP*".

9.4.1 Combinable Correlation

9.4.1.1 *Definition and Model*

In resource services composition, let S_j be the successor (Zeng *et al.*, 2004a) of S_i, and *Combinable correlation* is defined as whether there exists data logical correlation between S_i and S_j, i.e., whether there exists a certain similarity in concepts and properties between S_i's output Out_i and S_j's input In_j. *Combinable correlation degree* is the degree of their similarity.

Considering the correlations of input and output among composite resource services (see Figure 9.2), the model of Combinable correlation is defined as follows (Guo and Tao *et al.*, 2010):

$$CoC(S_i,S_j) = \{<Out_i,In_j> \big| Out_i = In_j, Out_i \subset In_j, Out_i \supset In_j, \\ Out_i \cap In_j \neq \{Out_i, In_j\} \,\&\, Out_i \cap In_j \neq \phi\} \qquad (9.1)$$

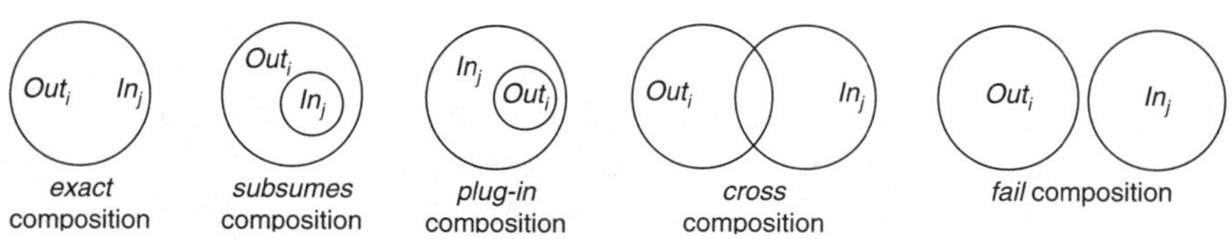

Figure 9.2 Correlation of input and output between resource services.

Where (i) $Out_i = In_j$, (ii) $Out_i \supset In_j$, (iii) $Out_i \subset In_j$ and (iv) $Out_i \cap In_j \neq \{Out_i, In_j\}$ & $Out_i \cap In_j \neq \phi$ denote the combinable correlation between S_i and S_j and are (i) *exact* composition, (ii) *subsumes* composition, (iii) *plug-in* composition, and (iv) *cross* composition, respectively. Otherwise, $Out_i \cap In_j \neq \phi$ means S_i and S_j is *fail* composition.

9.4.1.2 Combinable Correlation Degree

Whether two resource services are Combinable is decided by the similarity between their concepts, properties, and values in the function description of resource services. These descriptions can be used to measure the combinable correlation degree influencing QoS of resource services composition.

For *exact* composition and *subsumes* composition, the output of the preceding resource service is sure to satisfy the input of the latter resource service, so their correlation degree is 1; for *fail* composition, the correlation degree of two composite resource services is 0.

However, for *plug-in* composition and *cross* composition, the combinable correlation degree is uncertain. e.g., in "*OPAP*" in Figure 9.1, if candidate resource services *BMW, PDM,* and *Lear Corporation* are *plug-in* composition or *cross* composition, although they are combinable, their reputation will be reduced. The end user may not select *Lear Corporation* to provide the parts, which reduces the availability of resource service *SSS*. Hence, it is necessary to calculate Combinable correlation degree to find what influence will be generated to QoS.

(1) Calculating semantic similarity of concept sets of input and output by comparing the level relation of concepts.

Let $C_i = \{c_{i1}, c_{i2}, \cdots c_{in}\}$ and $C_i = \{c_{j1}, c_{j2}, \cdots c_{jm}\}$ be concepts sets of output O_i and input I_i respectively, where $c_{ik} \in O_{ik}$, $k = 1, 2, \cdots, n$, and $c_{jl} \in I_{jl}$, $l = 1, 2, \cdots m$. Semantic tree about automobile is shown in Figure 9.3, and the directed edge expresses the correlation between father concept and son concept.

All directed edges in domain ontology are regarded as a vector space, and the number of the directed edges is the dimension of the space. It is assumed that if concept A is the direct son concept of concept B, then the concept distance is 1. Let the number of the directed edges be N, and the vector space can be expressed as $V = (1,1,1, \cdots, 1)_{N\times1}$. The distance of two concepts is defined as

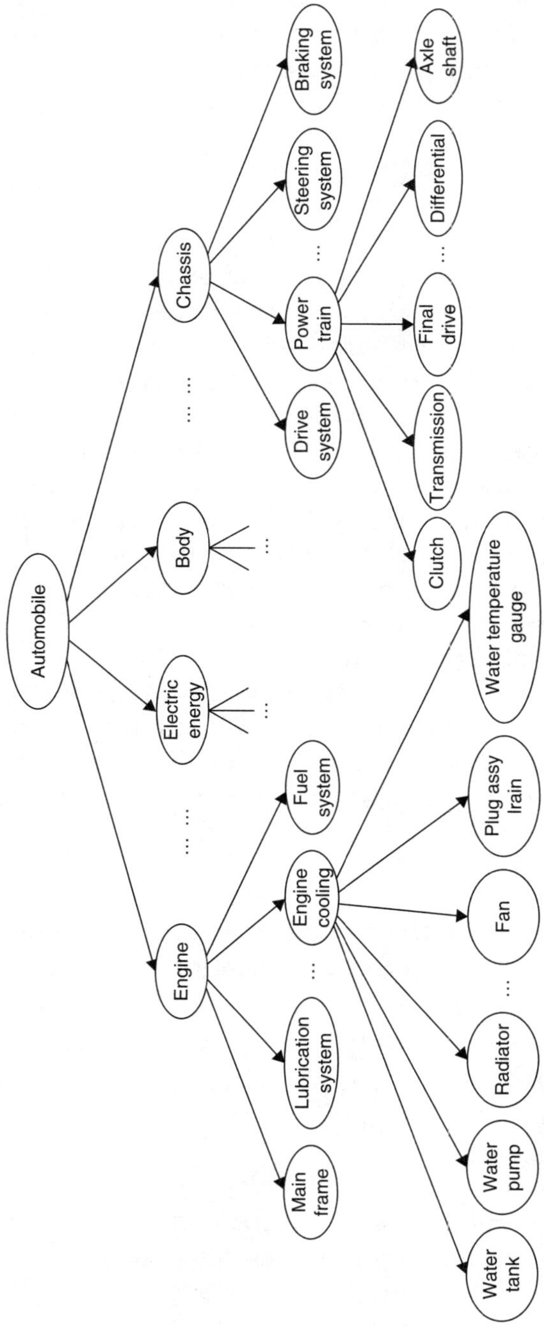

Figure 9.3 Local semantic tree about automobile.

$d = \left| (\underbrace{1,1,1,\cdots,1}_{N_1},0,\cdots 0) \right|_{N\times 1}$, where N_1 is the number 1 as well as the directed edges of the shortest distance of two concepts, then the semantic similarity of two concepts is $Sim_1(c_{ik}, c_{jl}) = 1 - d/|V|$. Therefore, the semantic similarity of concepts sets between C_i and C_j is defined as follows (Guo and Tao *et al.*, 2010):

$$Sim_1(C_i, C_j) = \begin{cases} 1, & C_i \equiv C_j \ and \ C_i \supset C_j \\ n / \sum_{k=1}^{n} \dfrac{1}{\max\limits_k Sim_1(c_{ik}, c_{jl})}, & C_i \subset C_j \\ u / \sum_{k=1}^{n} \dfrac{\lambda}{\max\limits_k Sim_1(c_{ik}, c_{jl})}, & C_i \cap C_j \neq \{C_i, C_j\} \ and \ C_i \cap C_j \neq \phi \\ 0, & C_i \cap C_j = \phi \end{cases} \tag{9.2}$$

where $\lambda = \begin{cases} 1, & \max\limits_k Sim_1(c_{ik}, c_{jl}) \geq \xi \\ 0, & \max\limits_k Sim_1(c_{ik}, c_{jl}) < \xi \end{cases}$ and $\xi \in [0,1]$. When the semantic similarity of two concepts is larger than or equal to ξ, retaining their values of semantic similarity; when the semantic similarity of two concepts is less than ξ, not retaining their values of semantic similarity. u is the number when λ is 1.

(2) Calculating semantic similarity of properties sets of input and output by comparing the properties of concepts.

Let $PO_i = \{p_{c_{i1}}, p_{c_{i2}}, \cdots, p_{c_{in}}\}$ and $PI_j = \{p_{c_{j1}}, p_{c_{j2}}, \cdots, p_{c_{jm}}\}$ be properties sets of output O_i and input I_j respectively, where $p_{c_{ik}}$ is the properties set of c_{ik} and $p_{c_{jl}}$ is the properties set of c_{jl}, $k = 1,2,\cdots,n$ and $r = 1,2,\cdots,m$. $N_{\mathrm{Pro}}(*)$ is the number of properties of concepts, and the semantic similarity of properties sets of concepts is $Sim_2(p_{c_{ik}}, p_{c_{jl}}) = \dfrac{N_{\mathrm{Pro}}(p_{c_{ik}}) \cap N_{\mathrm{Pro}}(p_{c_{jl}})}{N_{\mathrm{Pro}}(p_{c_{ik}}) \cup N_{\mathrm{Pro}}(p_{c_{jl}})}$ (Cardoso and

Sheth, 2003). The semantic similarity of properties sets of input and output is defined as follow:

$$Sim_2(PO_i, PI_j) = \begin{cases} 1, & PO_i \equiv PI_j \text{ and } PO_i \supset PI_j \\ n / \sum_{k=1}^{n} \dfrac{1}{\max_k Sim_2(p_{c_{ik}}, p_{c_{jl}})}, & PO_i \subset PI_j \\ v / \sum_{k=1}^{n} \dfrac{\delta}{\max_k Sim_2(p_{c_{ik}}, p_{c_{jl}})}, & PO_i \cap PI_j \neq \{PO_i, PI_j\} \text{ and } \\ 0, & PO_i \cap PI_j \neq \phi \; PO_i \cap PI_j = \phi \end{cases} \tag{9.3}$$

Where $\delta = \begin{cases} 1, & \max_k Sim_2(p_{c_{ik}}, p_{c_{jl}}) \geq \mu \\ 0, & \max_k Sim_2(p_{c_{ik}}, p_{c_{jl}}) < \mu \end{cases}$ and $\mu \in [0,1]$. When the semantic similarity of two properties sets is larger than or equal to μ, retaining their values of semantic similarity; when the semantic similarity of two properties sets is less than μ, not retaining their values of semantic similarity. ν is the number when δ is 1.

(3) Calculating semantic similarity of properties value sets.

In the flow of *"OPAP"* in Figure 9.1, the properties value of input and output, such as the delivery date, cost, and so on, can be expressed by interval numbers. Let $dO_i = \{d_{i_1}, d_{i_2}, \cdots, d_{i_{n'}}\}$ and $dI_j = \{d_{j_1}, d_{j_2}, \cdots, d_{j_{m'}}\}$ be properties value sets of output O_i and input I_j, where d_{i_k} and d_{j_l} is interval of any properties value of output O_i and input I_j, $k = 1, 2, \cdots, n'$, $l = 1, 2, \cdots, m'$. Semantic similarity of properties value sets of concepts is $Sim_3(d_{i_k}, d_{j_l}) = \dfrac{|Ad_{i_k} \cap Bd_{j_l}|}{|Ad_{i_k} \cup Bd_{j_l}|}$. Semantic similarity of properties value sets of input and output is as follows (Guo and Tao *et al.*, 2010):

$$Sim_3(dO_i, dI_j) = \begin{cases} 1, & dO_i \equiv dI_j \text{ and } dO_i \supset dI_j \\ n' / \sum_{k=1}^{n'} \dfrac{1}{\max_k Sim_3(d_{i_k}, d_{j_l})}, & dO_i \subset dI_j \\ \tau / \sum_{k=1}^{n'} \dfrac{\varepsilon}{\max_k Sim_3(d_{i_k}, d_{j_l})}, & dO_i \cap dI_j \neq \{dO_i, dI_j\} \text{ and } dO_i \cap dI_j \neq \phi \\ 0, & dO_i \cap dI_j = \phi \end{cases} \tag{9.4}$$

where $\varepsilon = \begin{cases} 1, & \max_k Sim_3(d_{i_k}, d_{j_l}) \geq \rho \\ 0, & \max_k Sim_3(d_{i_k}, d_{j_l}) < \rho \end{cases}$ and $\rho \in [0,1]$. When the semantic similarity of two properties value sets is larger than or equal to

ρ, retaining their values of semantic similarity; when the semantic similarity of two properties value sets is less than ρ, not retaining their values of semantic similarity. τ is the number when ε is 1.

Hence, the Combinable correlation degree of S_i and S_j is defined as follows:

$$CD_{CoC}(S_i,S_j)=\omega_1 Sim_1(C_i,C_j)+\omega_2 Sim_2(PO_i,PI_j)+\omega_3 Sim_3(dO_i,dI_j) \tag{9.5}$$

where $\omega_i \in [0,1]$ is the weight, and $\omega_1+\omega_2+\omega_3=1$.

9.4.2 Business Entity Correlation

Business entity correlation (*BuC*) refers to the correlation among providers of the composite resource services (*PCS*) involved in the composition. The set of *BuC* is defined as $BuC(S_i,S_j)=\{\wedge_1,\wedge_2,\vee_1,\vee_2\}$ (Guo and Tao *et al.*, 2010), where

- $\wedge_1$ denotes the *preferential cooperation correlation (PCC)*, where PCC means two or more entities in the business alliance influence QoS of entire resource services composition through providing preferential purchasing prices. E.g., it is assumed that Cummins Corporation and COSCO Corporation have a certain cooperation correlation. If they are selected as the supplier and deliverer, they will provide preferential purchasing prices to the manufacturer, which will make the property "cost" of QoS reduce and "reputation" increase.

- $\wedge_2$ denotes the *joint production correlation (JCC)*, where JCC means two or more entities in the business alliance join together by technology unions to produce standardization parts to reduce production cost. E.g., assume that BMW Corporation, Nissan Corporation and Volkswagen Corporation decide to produce standardization parts in Cummins Corporation to reduce production cost and make parts that can be mutually used. Through joint production, not only can they reduce manufacturing cost, improve their satisfaction, and reduce the response time of resource service *SSS*, but they also can improve the availability of parts owing to standardization.

- $\vee_1$ denotes the *benign competition correlation (BCC)*, where BCC brings positive effect to the QoS of the entire resource services composition. E.g., Behr Corporation and Delphi Corporation can also provide the same kind of parts. They attract the users by reducing the price to benefit the users, which will make the property "cost" reduce and "reputation" increase.

- $\vee_2$ denotes the *vicious competition correlation (VCC)*. VCC brings negative effects to the QoS of the entire resource services composition. E.g., Behr Corporation and Delphi Corporation can also provide the same kind of parts. In order to seize the market, they use the off-normal means to minimize the production cost, which reduces the quality of parts, which in turn reduces the reliability and reputation of the resource service.

9.4.3 Statistical Cooperate Correlation

The statistical cooperate correlation (StC) refers to the correlation that two or more resource services are often banded to execute a task.

In the flow of resource services composition, if two or more resource services are often banded, the user will often select them to execute the task together. Then the response time will be shortened. Moreover, the more times these resource services cooperate together, the higher the availability, reliability and reputation are, so that there is a greater probability that they will be selected once more to cooperate together. Let η be the number of times resource services S_i and S_j cooperate together to execute the task successfully in a certain time of historic record, then the probability they are designated to cooperate together again is $P_{\mathrm{StC}} = \eta / E$, where E is the total times that the entire flow of resource services composition executes.

Because *"often"* in *"often banded"* is not an exact value, we use fuzzy number (Li *et al.*, 2009a) to describe the number of *"often"*. If the statistical relationship of resource services cooperation is expressed as never, sometimes, often, and sure, then *"often"* can be mapped to a triangular fuzzy number $[\alpha, \chi, \beta]$, where α, χ, β denotes lower bound, nucleus, and upper bound. The model that resource services S_i and S_j often cooperate is defined as $often_{StC} = [\alpha_{ij}, \chi_{ij}, \beta_{ij}]$, namely after a certain time of historic record, if the times of S_i and S_j cooperating satisfies $[\alpha_{ij}, \beta_{ij}]$, S_i and S_j are called *"often banded"*. The model of statistical cooperate correlation among composite resource services is defined as follows:

$$StC(S_i, S_j) = \{\xi | \xi \geq often_{StC}\} \tag{9.6}$$

According to the above discussion, the influence of three correlations to the QoS of a composition, e.g., *"OPAP"* in Table 9.1 is shown as follows:

Table 9.1 The influence of three correlations to QoS of resource services composition

Correlation \ QoS	Time	Cost	Reliability	Availability	Reputation
Combinable correlation	No	No	No	Yes	Yes
Business entity correlation	Yes	Yes	Yes	Yes	Yes
Statistical correlation	Yes	No	Yes	Yes	Yes

9.5 QoS Computation Model of Correlation-aware Resource Services Composition

In this section, the QoS computation model of correlation-aware resource services composition will be established.

First, it is assumed that there are m elementary resource services, each of which has n candidate resource services. Obviously, the above three correlations are binary relation. Let

$$v_{ij}(S_{ab},S_{ij})=\begin{cases}1, & if\ S_{ab}(\text{CoR})S_{ij}\\ 0, & if\ S_{ab}(\overline{\text{CoR}})S_{ij}\end{cases},$$
$$t_{ij}(S_{ab},S_{ij})=\begin{cases}1, & if\ S_{ab}(\text{BuR})S_{ij}\\ 0, & if\ S_{ab}(\overline{\text{BuR}})S_{ij}\end{cases}, \tag{9.7}$$
$$r_{ij}(S_{ab},S_{ij})=\begin{cases}1, & if\ S_{ab}(\text{StR})S_{ij}\\ 0, & if\ S_{ab}(\overline{\text{StR}})S_{ij}\end{cases}.$$

where $a,i=1,2,\cdots m,b=1,2,\cdots,n_a,j=1,2,\cdots,n_i$ and v_{ij}, t_{ij} r_{ij} denote one of three kinds of correlations between the bth candidate resource service of the ath elementary resource service and the jth candidate resource service of the $ith(i\neq a)$ elementary resource service respectively. If there exist correlations between them, the value is 1. Otherwise, the value is 0. The correlation incidence matrices between the bth candidate resource service of the ath elementary resource service and the other resource service are as follows (Guo and Tao *et al.*, 2010):

$$V_{ab}=\begin{pmatrix}v_{11} & v_{12}\cdots & v_{1n}\\ \cdots\cdots\cdots\cdots & & \\ v_{a-1,1} & v_{a-1,2}\cdots & v_{a-1,n}\\ v_{a+1,1} & v_{a+1,2}\cdots & v_{a+1,n}\\ \cdots\cdots\cdots\cdots & & \\ v_{m1} & v_{m2}\cdots & v_{mn}\end{pmatrix},T_{ab}=\begin{pmatrix}t_{11} & t_{12}\cdots & t_{1n}\\ \cdots\cdots\cdots\cdots & & \\ t_{a-1,1} & t_{a-1,2}\cdots & t_{a-1,n}\\ t_{a+1,1} & t_{a+1,2}\cdots & t_{a+1,n}\\ \cdots\cdots\cdots\cdots & & \\ t_{m1} & t_{m2}\cdots & t_{mn}\end{pmatrix},R_{eb}=\begin{pmatrix}r_{11} & r_{12}\cdots & r_{1n}\\ \cdots\cdots\cdots\cdots & & \\ r_{a-1,1} & r_{a-1,2}\cdots & r_{a-1,n}\\ r_{a+1,1} & r_{a+1,2}\cdots & r_{a+1,n}\\ \cdots\cdots\cdots\cdots & & \\ r_{m1} & r_{m2}\cdots & r_{mn}\end{pmatrix} \tag{9.8}$$

If the number of candidate resource service of each elementary resource service is different, one only needs to make the number of the columns of all correlation incidence matrices be equivalent to the biggest number of candidate resource services and let the insufficient part be 0.

Let the QoS model of resource services be $(q^1,q^2,\cdots,q^{K_p})$, where q^k denotes the kth QoS criteria and K_p is the number of K_p QoS criteria. In this chapter, $K_p = 5$, which denotes time, cost, reliability, availability and reputation respectively, when k=1, 2, 3, 4, 5.

It is assumed that correlation only exists in two resource services. The default value of five QoS criteria of *b*th candidate resource service of the *a*th elementary resource service is $(q_{ab}^1, q_{ab}^2, \cdots, q_{ab}^5)$. Thus the QoS of three kinds of correlation between it and the $ith(i \neq a)$ candidate resource service of the *j*th elementary resource service is as follows (Guo and Tao *et al.*, 2010):

$$\begin{aligned}&[(q_{ij}^1 + \Delta q_{ij}^{1h}) + (q_{ab}^1 + \Delta q_{ab}^{1h}), (q_{ij}^2 + \Delta q_{ij}^{2h}) + (q_{ab}^2 + \Delta q_{ab}^{2h}),\\ &(q_{ij}^3 + \Delta q_{ab}^{3h}) \cdot (q_{ab}^3 + \Delta q_{ab}^{3h}), (q_{ij}^4 + \Delta q_{ab}^{4h}) \cdot (q_{ab}^4 + \Delta q_{ab}^{4h}),\\ &(q_{ij}^5 + \Delta q_{ij}^{5h}) + q_{ab}^5 + \Delta q_{ab}^{5h}]\end{aligned} \tag{9.9}$$

where h = 1, 2, 3 denotes the *h*th correlation, Δq_{ab}^{kh}, Δq_{ij}^{kh} denote respective increment (or decrement) of QoS when the *b*th candidate resource service of the *a*th elementary resource service and the *j*th candidate resource service of the $ith(i \neq a)$ elementary resource service have correlation h and execute together. If there is no correlation, the value of Δq_{ab}^{kh}, Δq_{ij}^{kh} is 0.

Based on the above model, when correlation is considered, for a certain execution plan including candidate resource service b, the computation model of each QoS criteria of elementary resource service a is as follows:

$$\begin{aligned}&CoC: q_a^{k1} = CD_{CoC_{ij}}(q_{ab}^k + \Delta q_{ab}^{k1} v_{ij}), \quad BuC: q_a^{k2} = q_{ab}^k + \Delta q_{ab}^{k2} t_{ij},\\ &StC: q_a^{k3} = q_{ab}^k + \Delta q_{ab}^{k3} P_{StC_{ij}} r_{ij}\end{aligned} \tag{9.10}$$

According to (9.10), the computation model of the *k*th QoS criteria of elementary resource service is as follows:

$$q_a^k = \gamma_1 q_a^{k1} + \gamma_2 q_a^{k2} + \gamma_3 q_a^{k3} \tag{9.11}$$

where $\gamma_i \in [0,1]$ is the weight of these three correlations, and $\sum_{i=1}^{3} \gamma_i = 1$. Here we only consider the controlling relationship among resource services in sequence, other relationship models are similar to it.

Refer to reference (Tao *et al.*, 2010e), we can know the calculating formulations of QoS in resource services composition is as follows:

$$Q_1 = \sum_{a=1}^{m} q_a^1,\ Q_2 = \sum_{a=1}^{m} q_a^2,\ Q_3 = \prod_{a=1}^{m} q_a^3,\ Q_4 = \prod_{a=1}^{m} q_a^4,\ Q_5 = \frac{1}{m}\sum_{a=1}^{m} q_a^5$$

Because resource service has multidimensional QoS, the overall QoS value of resource services composition is as follows:

$$F(Q) = \max\left(\sum_{k=1}^{2} \xi_k \frac{1}{Q_k} + \sum_{k=3}^{5} \xi_k Q_k\right) \tag{9.12}$$

where $\xi_k \in [0,1]$ is the weight corresponding to the need of users, and $\sum_{k=1}^{5} \xi_k = 1$.

9.6 Case Study: Correlation-aware Resource Services Composition

In this section, the validity of the model presented in this chapter is tested by comparing the QoS of correlation-aware with the QoS of no correlation-aware resource services composition in the flow of "*OPAP*". Five QoS properties i.e., *time, cost, reliability, availability and reputation*, are considered in this case study.

It is assumed that there are five elementary resource services in the flow of "*OPAP*", and each elementary resource service has two candidate resource services (see Figure 9.4), i.e., m=5, n=2. As a result, 32 execution plans will be generated. For simplicity, the defaults of QoS of each candidate resource service are mapped to [0, 1] through normalization (see Table 9. 2). Let the weight of the five QoS criteria be 0.1, 0.3, 0.2, 0.2, and 0.2, respectively.

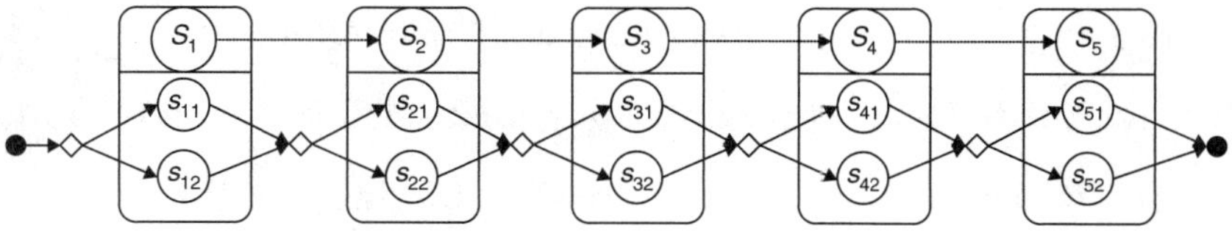

Figure 9.4 Sequence composite resource service process.

Table 9.2 The default of QoS criteria of candidate resource services

Candidate Resource Services \ QoS		Time	Cost	Reliability	Availability	Reputation
S_1	S_{11}	0.6	0.9	0.7	0.1	0.7
	S_{12}	0.4	0.5	0.5	0.8	0.1
S_2	S_{21}	0.3	0.5	0.9	0.1	0.9
	S_{22}	0.4	0.3	0.7	0.8	0.7
S_3	S_{23}	0.8	0.3	0.7	0.8	0.1
	S_{32}	0.3	0.5	0.9	0.1	0.1
S_4	S_{41}	0.3	0.3	0.9	0.1	0.3
	S_{42}	0.6	0.4	0.7	0.8	0.1
S_5	S_{51}	0.3	0.5	0.9	0.1	0.1
	S_{52}	0.5	0.3	0.7	0.8	0.7

First, we calculate the semantic similarity of S_{21} and S_{31}. Local semantic ontology repository is selected from Figure 9.3 in which there are 400 directed edges.

The output sets of candidate resource service S_{21} are O_2 = ({transmission, clutch}, {cost, delivery date}, {[2000$, 3000$]}, and [August 31 2009, September 30 2009])), respectively. The corresponding concept sets, properties sets, and value sets of input and output of S_{21} are C_2= {transmission, clutch}, PO_2 = {<cost, delivery date>, <cost, delivery date>}, dO_2= {[2000$, 3000$], [August 31 2009, September 30 2009], [2000$, 3000$], and [August 31 2009, September 30 2009]}, respectively.

The input sets of candidate resource service S_{31} are I_3= ({power train, steering system, drive system}, {cost, delivery date}, {[2000$, 3500$]}, and [August 31 2009, October 30 2009])), respectively. The corresponding concept sets, properties sets and value sets of input and output of S_{31} are C_3 = {power train, steering system, drive system}, PI_3= {<cost, delivery date>, <cost, delivery date>,<cost, delivery date>}, dI_3 = { <[2000$, 3500$]}, [August 31 2009, October 30 2009]>, <[2000$, 3500$]}, [August 31, October 30]>, <[2000$, 3500$]}, and [August 31 2009, October 30 2009]>}, respectively.

Obviously, $C_2 \subset C_3$, $PO_2 \subset PI_3$, $dO_2 \subset dI_3$. According to Expressions 9.2, 9.3, and 9.4, one has:

$$Sim_1(C_2,C_3) = 2/(\frac{20}{19}+\frac{20}{19}) = 0.95,$$
$$Sim_2(PO_2,PI_3) = 2/(1+1) = 1, \text{and}$$
$$Sim_3(dO_2,dI_3) = 4/(\frac{500}{1500}+\frac{30}{60}+\frac{500}{1500}+\frac{30}{60}) = 0.4$$

Let $\omega_1 = \omega_2 = \omega_3 = \frac{1}{3}$, then $CD_{CoR}(s_{21}, s_{31}) = 0.7833$. Respective increment (or decrement) of QoS of S_{21} and S_{31} are as follows:

$$\Delta q_{21}^{11} = 0, \Delta q_{21}^{21} = 0, \Delta q_{21}^{31} = 0, \Delta q_{21}^{41} = 0.8, \Delta q_{21}^{51} = 0.05$$
$$\Delta q_{31}^{11} = 0, \Delta q_{31}^{21} = 0, \Delta q_{31}^{31} = 0, \Delta q_{31}^{41} = 0.15, \Delta q_{31}^{51} = 0.7$$

Other combinable correlation degrees can be obtained similarly.

It is assumed that candidate resource services S_{31} and S_{51} have preferential cooperation correlation, S_{32} and S_{52}have vicious competition correlation. Respective increment (or decrement) of QoS of S_{31} and S_{51} are as follows:

$$\Delta q_{31}^{12} = -0.6, \Delta q_{31}^{22} = -0.2, \Delta q_{31}^{32} = 0.25, \Delta q_{31}^{42} = 0.15, \Delta q_{31}^{52} = 0.8$$
$$\Delta q_{51}^{12} = -0.2, \Delta q_{51}^{22} = -0.4, \Delta q_{51}^{32} = 0.05, \Delta q_{51}^{42} = 0.8, \Delta q_{51}^{52} = 0.8$$

Respective increment (or decrement) of QoS of S_{32} and S_{52} are as follows:

$$\Delta q_{32}^{12} = 0.2, \Delta q_{32}^{22} = 0.3, \Delta q_{32}^{32} = -0.4, \Delta q_{32}^{42} = -0.02, \Delta q_{32}^{52} = -0.04$$
$$\Delta q_{52}^{12} = 0.1, \Delta q_{52}^{22} = 0.2, \Delta q_{52}^{32} = -0.3, \Delta q_{52}^{42} = -0.4, \Delta q_{52}^{52} = -0.3$$

It is assumed that candidate resource services S_{31} and S_{42}, S_{32} and S_{41} have statistical cooperate correlation, and the probability they are designated to cooperate together are 0.85 and 0.90 respectively. Respective increment (or decrement) of QoS of S_{31} and S_{42} are as follows:

$$\Delta q_{31}^{13} = -0.7, \Delta q_{31}^{23} = 0, \Delta q_{31}^{33} = 0.2, \Delta q_{31}^{43} = 0.15, \Delta q_{31}^{53} = 0.8$$

$$\Delta q_{42}^{13} = -0.5, \Delta q_{42}^{23} = 0, \Delta q_{42}^{33} = 0.2, \Delta q_{42}^{43} = 0.05, \Delta q_{42}^{53} = 0.7$$

Respective increment (or decrement) of QoS of S_{32} and S_{41} are as follows:

$$\Delta q_{32}^{11} = -0.2, \Delta q_{32}^{21} = 0, \Delta q_{32}^{31} = 0.05, \Delta q_{32}^{41} = 0.7, \Delta q_{32}^{51} = 0.8$$

$$\Delta q_{41}^{11} = -0.2, \Delta q_{41}^{21} = 0, \Delta q_{41}^{31} = 0.05, \Delta q_{41}^{41} = 0.8, \Delta q_{41}^{51} = 0.4$$

According to the Model and Objective Function 9.12, the QoS of correlation-aware and no correlation-aware resource services composition of 32 execution plans are shown in Table 9.3, where QoS1 denotes no correlation-aware and QoS2 denotes correlation-aware. The comparing results are shown in Figure 9.5.

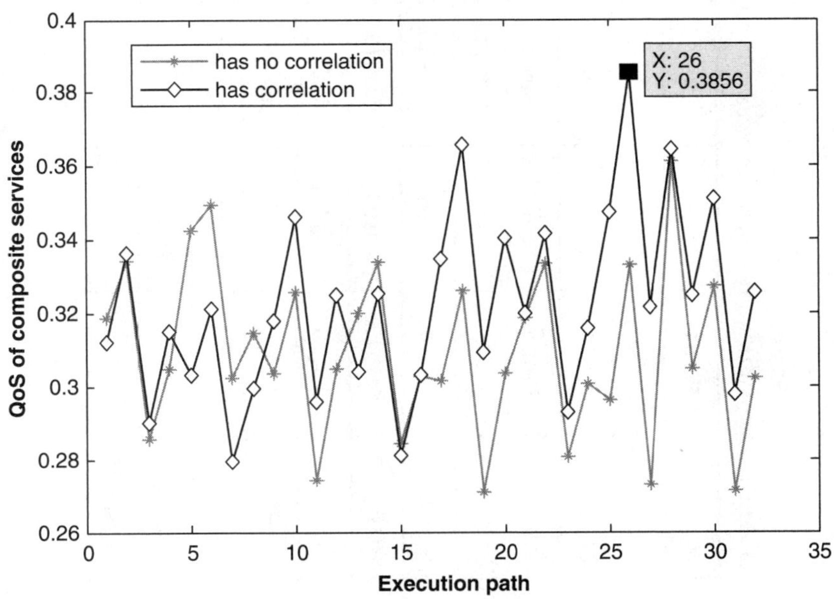

Figure 9.5 Comparison of QoS between correlation-aware and no correlation-aware.

Table 9.3 QoS value of resource services composition

Path	**Path1**	**Path2**	**Path3**	**Path4**	**Path5**	**Path6**	**Path7**	**Path8**
QoS1	0.3189	0.3341	0.2855	0.3050	0.3425	0.3495	0.3022	0.3146
QoS2	0.3121	0.3365	0.2902	0.3150	0.3031	0.3214	0.2795	0.2993
Path	**Path9**	**Path10**	**Path11**	**Path12**	**Path13**	**Path14**	**Path 15**	**Path16**
QoS1	0.3038	0.3256	0.2743	0.3047	0.3201	0.3337	0.2845	0.3029
QoS2	0.3179	0.3463	0.2959	0.3248	0.3039	0.3253	0.2809	0.3034
Path	**Path17**	**Path18**	**Path19**	**Path20**	**Path21**	**Path22**	**Path23**	**Path24**
QoS1	0.3016	0.3261	0.2707	0.3035	0.3186	0.3336	0.2808	0.3007
QoS2	0.3346	0.3656	0.3092	0.3406	0.3201	0.3417	0.2928	0.3159
Path	**Path25**	**Path26**	**Path27**	**Path28**	**Path29**	**Path30**	**Path31**	**Path32**
QoS1	0.2961	0.3332	0.2731	0.3612	0.3048	0.3272	0.2711	0.3025
QoS2	0.3475	0.3856	0.3218	0.3647	0.3250	0.3511	0.2981	0.3256

From Table 9.3 and Figure 9.5, it can be concluded that the quality (i.e., QoS) of correlation-aware resource services composition is higher than that of no correlation-aware resource services composition for the same composite path. When correlations are considered, the maximal QoS value obtained from the 26th execution plan is 0.3856, which is higher than the maximal QoS value 0.3612 obtained from the 28th execution plan whose correlations are not considered. The values of QoS in several execution plans when correlations are considered are less than the values of QoS when correlations are not considered, it is because candidate resource services S_{32} and S_{52} are included in these paths. There is vicious competition correlation between S_{32} and S_{52}, so if they are selected to execute the plan the quality of the resource services composition will be reduced. However, compared to no considering correlation, when the correlations among composite resource services are considered, a higher quality of resource services composition can be achieved.

9.7 Summary

There always exist a lot of web services with similar functions but different non-functionality properties in the process of web services composition. In order to enhance the quality of service composition, QoS are usually considered. However, most existing works about QoS-based service composition treat the services involved in composition as independent from each other, and their correlations are usually ignored. In fact, most services have some correlations with each other, and the correlations can badly affect the entire quality of service composition. In this chapter, the issue of correlation-aware composite resource service in MGrid is considered. Three kinds of correlations relationship (i.e., Combinable correlation, business entity correlation, and statistical cooperate correlation) in services composition are investigated. The impact of each kind of correlation relationship on the entire quality of resource service composition is investigated, and a QoS computation model based on the three correlations is proposed. The case study indicates that the higher quality of services composition can be achieved when considering the correlations in resource services composition.

10

Resource Service Composition Optimal-selection

10.1 Introduction

Liu (2004) researched MGrid resource scheduling based on QoS, primarily including the factors of time, quality, cost, and service (TQCS); and classified the resource scheduling processes into QoS-based resource search, QoS-based resource scheduling, and dynamic negotiation. He and Sun (2003) studied self-adapting min-min heuristic grid resource scheduling strategy driven by performance-QoS. Deng *et al.*, (2006a) investigated a grid-based manufacturing resource scheduling system for virtual enterprises, which is composed of an exterior grid and an interior grid in the architecture. They divided the manufacturing resource scheduling into two stages: (a) distributed scheduling, implemented by the exterior grid, and (b) local scheduling, accomplished by the interior grid. The authors have researched QoS modeling and associated evaluation models (Tao *et al.*, 2009c; 2009d), and trust-QoS based scheduling (Tao *et al.*, 2009a) in the MGrid system.

However, the above research on MGrid scheduling emphasized resource service QoS, particularly in TQCS. The task they considered was only a single resource service request task (SRSRTask), but attention was not paid to the multi-resource service request task (MRSRTask). In other words, *MGrid resource service composition and optimal-selection (MRSCOS),* which is one of the key technologies to realize a dynamic MGrid executing platform, has not been adequately addressed. Therefore, the MGrid scheduling method from the aspect of resource service composition, and the method to realize MRSCOS are investigated in this work.

Due to the complexity of some manufacturing processes, a submitted manufacturing task (or resource service request) in an MGrid can be classified into two types: (a) single resource service request task (SRSRTask), and (b) *multi-resource service request task* (MRSRTask). Generally speaking, SRSRTask is completed by invoking only one resource service while MRSRTask is completed by invoking several resource services in sequence. For an SRSRTask, the system searches all resource service that qualify for its function requirements and selects the optimal one for implementation, namely, *resource service optimal-selection* (RSOS). An MRSRTask usually can be decomposed into several subtasks that cannot be farther decomposed and can be executed by a single resource service. Hence, for an MRSRTask, in addition to searching all qualified resource services according to each subtask, the system has to select one candidate resource service for each subtask and generate a composite resource service (CRS) as shown in Figure 7.1. Then it selects the optimal resource service composite path from all possible paths to execute the task under multi-objectives, e.g., time minimization, cost minimization, and reliability maximization, along with given constraints. The above problem is defined as a *multi-objectives MGrid resource service composition and optimal-selection* (MO-MRSCOS) problem (Tao *et al.*, 2008c). An MO-MRSCOS problem is an NP-complete problem.

Extensive studies have been conducted related to the web service composition problem in service-oriented system and distributed system. Existing research efforts for web service composition have been undertaken in two orthogonal directions: (a) manual composition, and (b) automated composition. Manual composition is an *ad hoc* and time-consuming task, and it is supported by major IT enterprises such as IBM and Microsoft. In this approach, the web service developer selects the outsourced web services that are relevant to their composition intents, and programs the interaction logic of the component services with a low level programming language such as BPEL (Juric *et al.*, 2004). The automated composition has been investigated in both *semantic web* and *workflow communities*. Zeng *et al.*, (2003) presented an automated composition approach based on a rule inference engine. They described the supported functionality of their system flow such as the composition of parallel and alternative control flow or the annotation of services with Quality of Service (QoS) properties. The approach by Berardi *et al.*, (2003) was based on the automatic synthesis of finite state machines. In Berardi *et al.*, (2005) they extended this approach to non-deterministic finite state machines to support real world, external services. Several other approaches (Traverso and Pistore, 2004) treated the problem of automated web services composition as a software/plan synthesis problem or a plan execution problem. To tackle the composition task as a plan synthesis problem, they performed planning using predefined available services as the building blocks of a plan. Kalasapur *et al.*, (2007), employed the service-oriented middleware platform called Pervasive Information Communities Organization (PICO) to model and represent resources as services. Their proposed service composition mechanism models services as directed attributed graphs, maintain a repository of service graphs, and dynamically combine multiple basic services into complex services. In order to allow a user to request a service in an intuitive form (e.g., using a natural language), Fujii and Suda (2005) proposed a semantics-based service composition architecture. Their proposed architecture obtains the semantics of the service requested in an intuitive form, and dynamically composes the requested service based on the semantics of the service. The key components of the proposed architecture, including Component Service Model with Semantics (CoSMoS), Component Runtime Environment (CoRE), and Semantic Graph-Based Service Composition (SeGSeC), are described in detail. In order to address the challenges of scalability, flexibility, and quality-of-service (QoS) management for distributed multimedia service composition, Gu and Nahrstedt (2006) presented a fully decentralized service composition framework, called SpiderNet, which provides statistical multi-constrained QoS assurances and load balancing for service composition and supports directed acyclic graph composition topologies and exchangeable composition orders. A context-based multi-type policy approach for web service composition is investigated in Maamar *et al.*, (2007). The design with object (DWO) approach to web services composition is studied in Liang *et al.*, (2007). Yue *et al.*, (2007) presents an approach for automating geospatial web service composition by employing geospatial semantics in the service-oriented architecture (SOA). Rao *et al.*, (2006a) introduced a method for automatic composition of semantic web services using liner logic (LL) theorem proving. Chi and Lee (2007) proposed a composing platform based on a formal modeling language. An approach to facilitate dynamic and scalable web service composition called Integrated Service Planning and Execution (ISP&E) based on AI planning techniques is put forward by Madhusudan and Uttamsingh (2006). Implementation of architecture for dynamic web service composition and execution based on a domain-independent AI planning framework called Hierarchical Task Network (HTN) planning is also described by them. Sirin *et al.*, (2004a) studied how HTN planning system SHOP2 can be used for web service composition. They first translated OWL-S service descriptions to a SHOP2 domain, and then investigated the composition method using SHOP2.

However, the above research on service composition primarily concentrates on the compositing method and architecture. These studies fall short of addressing the following issues: (a) the correlation among resource services are not considered, (b) the service composing process models are understood to a lesser extent, (c) methods to select an optimal composite path from all possible composite paths are investigated to a lesser extent.

To address the above issues, the correlations among resource services are taken into account during MGrid resource service composition, and a QoS description mode supporting resource service correlation is presented. The basic resource service composite modes (RSCM) for CRS are described, including *sequence model, parallel model, selective model* and *circular model*. The rules for translating a complicated RSCM into a simple sequence RSCM are presented in order to simplify the resolving process and complexity of an MO-MRSCOS problem. Next, the formulation of an MO-MRSCOS problem for minimizing execution time and cost, and maximizing the reliability is introduced.A new MGrid resource service composition and optimal-selection method, based on the principles of particle swarm optimization (PSO), is proposed for solving an MO-MRSCOS problem. Unlike previous works:

1. the proposed PSO algorithms combine the non-dominated sorting technique to achieve the selection of global best position and private best position;

2. the parameters of particle updating formulation in PSO are dynamical generated in order to make a compromise between the global exploration and local exploitation abilities of PSO;
3. *permutation-based* and *objective-based* population trimming operators are applied in PSO to maintain diversity of solutions in population.

10.2 Problem Formulation and Review

10.2.1 Motivating Example

Before introducing the problem, an example is presented to illustrate the problems addressed in this study. The task of a product design and manufacture, denoted by T_{pro}, on an MGrid system is used as an example. Apparently, T_{pro} is an MRSRTask as shown in Figure 10.1, and it has five subtasks such as D&A, MPA, Mfg, DA, and EP. It has to be completed by invoking five resource services such as *design and analysis service* (D&AS), *manufacturing plan and arrangement service* (MPAS), *manufacture service* (MfgS), *delivery arrangement service* (DAS), and *electronic pay service* (EPS). Each resource service has several candidates. In the absence of automated composition of resource services, the user invests considerable resources visiting numerous sites, determining appropriate service providers, entering his/her preferences repeatedly, and integrating or aligning the different type of results coming from different sites. It is preferred that the user enters information once and receives the expected results from the most appropriate services with minimal additional assistance. One possible interaction model is shown in Figure 10.1.

In this scenario, the user sends a single request to a service, CPDMS shown in Figure 10.1, containing the information required about T_{pro}, e.g., the design parameters, functional requirements, price, and delivery address. As it is unlikely to find an *ad hoc* service to meet the user's requirements, the service the user interacts with tries to find several services providing a result for one or more of the needed information. Once these services are discovered, the CPDMS composes the functionalities leading to the needed results and resolves their data types constraints. The CPDMS is defined as a *composite resource service* (CRS).

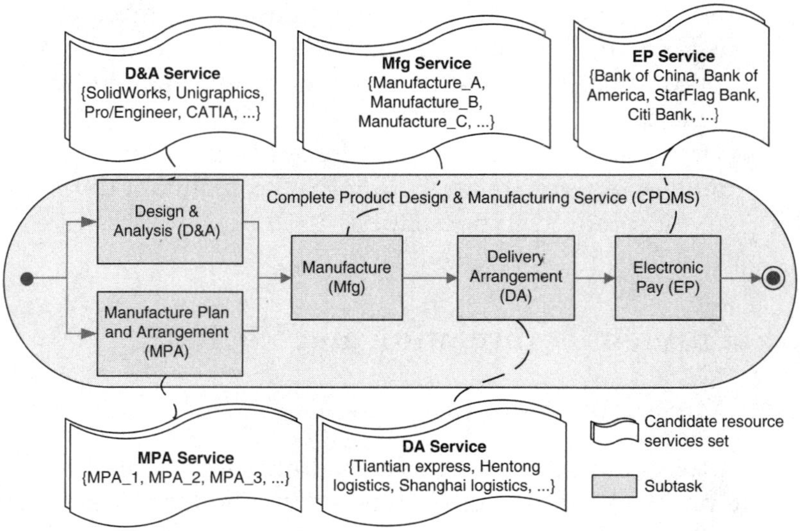

Figure 10.1 State Charts for a Composite Resource Service (CRS). The Complete Product Design and Manufacturing Service (CPDMS) invokes a number of other services, including D&AS, MPAS, MfgS, DAS and EPS (Tao *et al.* 2008c).

In order to realize a CRS such as CPDMS, the following two issues must be considered (Tao *et al.*, 2010e):

1. The correlation among resource services in a CRS.

As a candidate resource service of MfgS (manufacturing service) in Figure 10.1, *Manufacture_B* has a discount policy, if the user selects *Hentong logistics* as the DAS (delivery arrangement service) and *Bank of China* as the EPS (*electronic pay service*), then the service price of *Manufacture_B* will be lower than its bid price (i.e., its default value). In addition, if the system selects *Unigraphics* as the *design and analysis service* (D&AS), its response time (i.e., manufacturing time) will be shorter than its default value. It is demonstrated that the QoS criterion of the selected MfgS in CPDMS will change dynamically depending on the other selected resource service in CPDMS. Therefore, when evaluating the value of aggregation QoS property of a CRS, the correlation between resource services in CRS should be taken into account.

2. Multi-objective MGrid resource service composition optimal-selection (MO-MRSCOS)

Let T_{ask} denote a MGrid task and $T_{ask} = \{ST^1, ST^2, \cdots, ST^j, \cdots, ST^{N_T}\}$, where N_T is the total number of the decomposed subtasks (if T_{ask} is a SRSRTask then $N_{\mathrm{T}}=1$), and ST^{j} denotes the jth ($j=1,2,3,\cdots,N_{\mathrm{T}}$) subtask of T_{ask}. It is assumed that there are M_j candidate resource services available for ST^{j}, and the corresponding candidate resource services set (RSS) is $\mathrm{RSS}^{j}=\{\mathrm{RS}_1^{j},\mathrm{RS}_2^{j},\cdots,\mathrm{RS}_{i_j}^{j},\cdots,\mathrm{RS}_{M_j}^{j}\}$, where $\mathrm{RS}_{i_j}^{j}$ is the ith candidate resource service for subtask ST^{j}. Hence, if the system wants to execute T_{ask}, it needs to select one candidate resource service from each candidate resource services set (i.e., RSS^{j}), respectively, and generate a new CRS to execute task T_{ask}. Apparently, there are $\prod_{j=1}^{N_{\mathrm{T}}} M_j$ possible composite paths for task T_{ask} in theory.

MO-MRSCOS problem is to select one candidate resource service from each corresponding RSS (i.e., RSS^{j}), respectively, and generate all possible composite resource service executing paths (CRSEPs), then select the optimal one from $\prod_{j=1}^{N_{\mathrm{T}}} M_j$ CRSEPs to execute T_{ask} under multi-objective (e.g., time minimization, cost minimization, and reliability maximization) and multi-constraint (e.g., satisfy lowest maintainability, trust-QoS, and function similarity).

This chapter will emphasize addressing the above two issues.

10.2.2 Basic Models of Resource Service Composition and their QoS Evaluation

10.2.2.1 QoS Properties and QoS Description Mode Supporting Service Correlation

Six QoS properties of resource service will be considered in this chapter. They are (1) *Time* (*T*): resource service execution duration, (2) *Cost* (C): the cost of invoking a resource service, (3) *Reliability* (*Rel*), (4) *Maintainability* (*Ma*): the ability for a resource service to deal with unexpected conditions, (5) *Trust-QoS* (*Trust*): the reputation of a resource service evaluated by users, (6) *Function similarity* (*FS*): the possibility that the functions of a requested resource service can be replaced by other resource services which belong to the same resource service set. The detailed definition and evaluation methods of each QoS property can be found in Chapters 5 and 6. The QoS model, Q_{RS}, of resource service *RS* is defined as:

$$Q_{RS}=(q^1,q^2,\cdots,q^i,\cdots,q^{K_q}) \tag{10.1}$$

where K_q is the number of QoS properties and q^i is the *ith* QoS criterion. Apparently, in this chapter $K_q = 6$ and

$$Q_{RS} = (q^1, q^2, q^3, q^4, q^5, q^6) = (T(RS), C(RS), Rel(RS), Ma(RS), Trust(RS), FS(RS)) \tag{10.2}$$

In this study, these QoS properties are classified into two categories: *positive QoS* and *negative QoS*. *Positive QoS* means that the bigger the value of the QoS property the higher the quality of the resource service, such as q^3, q^4, q^5 and q^6; While *negative QoS* means that the smaller the value of the QoS property the higher the quality of the resource service, such as q^1 and q^2.

It is assumed that the set of resource services that affect any QoS properties of a resource service *RS* is $CRSS_{RS}$, and $CRSS_{RS}$ is defined as the *correlation resource service set* **(CRSS)** of resource service *RS*. Considering the influence of correlated resource services on the value of q^i of *RS* in a CRS, in this chapter, each QoS property, q^i, of *RS* is represented as:

$$q^i = \{q(i,0): default; q(i,1): C_{orrelation}(i,1), \cdots, q(i,j): C_{orrelation}(i,j), \cdots, q(i,K_i): C_{orrelation}(i,K_i)\} \tag{10.3}$$

where $K_i (K_i = 0, 1, 2, 3,...)$ is the number of correlations that affect the value of q^i. $q(i, 0)$ is the default value of q^i if it is not affected by other resource services, i.e., $K_i = 0$. Item $C_{orrelation}(i, j)$ is the *j*th boolean correlation statements of q^i, it is used to describe the correlation in regard to q^i between *RS* and other resource services, the value of $C_{orrelation}(i, j)$ is true (i.e., 1) or false (i.e., 0). The representation of $C_{orrelation}(\ )$ is as follow:

$$C_{orrelation}(\ ) ::= selected_{RS} \in CRSS_{RS} \tag{10.4}$$

Item $selected_{RS}$ denotes the selected resource service, and $CRSS_{RS}$ is the corresponding CRSS. There are three operators among: $C_{orrelation}(\ )$ **'not'**, **'and'**, **'or'**.

- **not** $C_{orrelation}(\ )$
- $C_{orrelation}(\ )$ **and** $C_{orrelation}(\ )$
- $C_{orrelation}(\ )$ **or** $C_{orrelation}(\ )$

The priority of the three operations is not $\succ$ and $\succ$ or . Based on the above model, the value of q^i is calculated as follows:

$$q^i = \begin{cases} \max\left(\text{Bool}(i,j) * q(i,j)\right) & (q^i \text{ is a positive QoS}) \\ \min\left(\text{Bool}(i,j) * q(i,j)\right) & (q^i \text{ is a negetive QoS and } \forall \text{Bool}(i,j) = 1) \end{cases} \quad (10.5)$$

where

- $$\text{Bool}(i,j) = \begin{cases} 0 & (\text{if } C_{\text{orrelation}}(i,j) \text{ is false}) \\ 1 & (\text{if } C_{\text{orrelation}}(i,j) \text{ is true}) \end{cases} \quad (10.6)$$

- $$C_{\text{orrelation}}(i,0) \text{ is true if } \sum_{j=1}^{K_i} \text{Bool}(i,j) = 0. \quad (10.7)$$

For example, the QoS description of ***Manufacture_B*** in Figure 10.1 is as follows:

QoS description of *Manufacture_B* in Figure 10.1.
{Execution time = {700ms: *default;* 550ms: *selected* $_{\text{D\&AS}}\in$ {*Unigraphics*}; 450ms: *selected* $_{\text{D\&AS}}\in$ {*Pro/Engineer*}} Cost = {99 USD: *default,* 94 USD: *selected* $_{\text{EPS}}\in$ {*Bank of America*} or *selected* $_{\text{EPS}}\in$ {*Bank of China*}; 87 USD: *selected* $_{\text{DAS}}\in$ {*Hentong logistics* }and *selected* $_{\text{EPS}}\in$ {*Bank of China* }} Reliability={95: *default* } ... }

The above QoS model describes three QoS properties of ***Manufacture_B***: *executiontime*(*i.e.*,q^1),*cost*(*i.e.*,q^2),and*reliability*(*i.e.*,q^3). The value of q^3 is not influenced by other resource services. The value of q^1 has three options: (a) if the selected D&AS is *Unigraphics* then q^1=550ms; (b) if the selected D&AS is *Pro/Engineer,* then q^1=450ms; (c) otherwise, q^1=700ms. The value of q^2 also has three options: (a) if the selected EPS is *Bank of America* or *Bank of China, then* q^2=94USD; (b) if the selected DAS is *Hentong logistics* and the selected EPS is *Bank of China,* then q^2=87USD; (c) otherwise, q^2=99USD.

10.2.2.2 Basic Composite Models of CRS and their QoS Evaluation

In general, there are four basic composite modes for a CRS, they are *sequence mode, parallel model, selective model* and *circular mode,* as shown in the first column in Figures 10.2–10.5, respectively. Almost all CRS can be constructed using these four basic models. Apparently, a CRS primarily has two forms as follows (Tao *et al.*, 2008c):

- *Sequence composite resource service* (SCRS): SCRS is composed by a series of resource services, and the connecting relationship among resource services is series-wound.
- *Mixed composite resource service* (MCRS): In an MCRS, the connecting relationship among resource services are not only simply ordinal, but also parallel, selective, and circular.

Due to the complexity of MCRS itself, the executing paths of an MCRS are even more complex. Hence, it is difficult to solve the MO-MRSCOS problem in MCRS comparatively. Therefore, we will attempt to transfer MCRS into SCRS, and then solve the MO-MRSCOS problem in SCRS. If we can transform the above mentioned four basic composite modes into sequence mode and design relevant algorithms for calculating the corresponding aggregation QoS properties' values, then the MO-MRSCOS problem in MCRS can be transformed into that in SCRS.

Therefore, the following four rules are constructed for transforming MCRS into SCRS and calculating the corresponding aggregation QoS properties' values (Tao *et al.*, 2008c).

Rule 1 (sequence mode): a new resource service set RSS^{seq}= $\{RS_1^{seq}, RS_2^{seq}, \cdots, RS_{M_{seq}}^{seq}\}$ is constructed to provide the functions of the sequence model (highlighted by dashed pane in Figure 10.2).

Addition operator is used to compute the aggregation QoS properties q^i $(i=1, 2, 5, 6)$ of RS^{seq} and multiplication operator is used to compute q^i $(i=3, 4)$, because every component resource service in sequence model should be executed in sequence. It is assumed that there are n nodes, i.e., $\{\text{RSS}^1, \text{RSS}^2, \cdots, \text{RSS}^j, \cdots, \text{RSS}^n\}$, in a *sequence mode* as shown in the first column in Figure 10.2, where $\text{RSS}^j = \{\text{RS}_1^j, \text{RS}_2^j, \cdots, \text{RS}_{i_j}^j, \cdots, \text{RS}_{M_j}^j\}$. The corresponding transforming chart and calculating formulations are shown in Figure 10. 2.

Rule 2 (parallel mode): a new resource service set $\text{RSS}^{\text{com}} = \{RS_1^{\text{com}}, RS_2^{\text{com}}, \cdots, RS_{M_{\text{com}}}^{\text{com}}\}$ is constructed to provide the functions of the *parallel model* (highlighted by dashed pane in Figure 10.3). Maximum operator is used to compute q^i of RS^{com}, addition operator is used to compute q^i $(i=2, 5, 6)$, and multiplication operator is used to compute q^i $(i=3, 4)$, because every component resource service in sequence model should be executed in sequence. It is assumed that there are n nodes, i.e., $\{\text{RSS}^1, \text{RSS}^2, \cdots, \text{RSS}^j, \cdots, \text{RSS}^n\}$, in a *parallel mode* as shown in the first column in Figure 10.3, where $\text{RSS}^j = \{\text{RS}_1^j, \text{RS}_2^j, \cdots, \text{RS}_{i_j}^j, \cdots, \text{RS}_{M_j}^j\}$. The corresponding transforming chart and calculating formulations are shown in Figure 10.3.

Rule 3 (selective mode): a new resource service set $\text{RSS}^{\text{sel}} = \{RS_1^{\text{sel}}, RS_2^{\text{sel}}, \cdots, RS_{M_{\text{sel}}}^{\text{sel}}\}$ is constructed to provide the functions of the *selective model* (highlighted by dashed pane in Figure 10.4). In *selective model* we do not know which elements will be selected, in order to quantificationally calculate the QoS of the *selective model*, the corresponding probability, λ_j $(\sum_{j=1}^{n} \lambda_j = 1)$, that RSS^j been selected is introduced. Addition and multiplication operators are used to compute q^i of RS^{sel}. It is assumed that

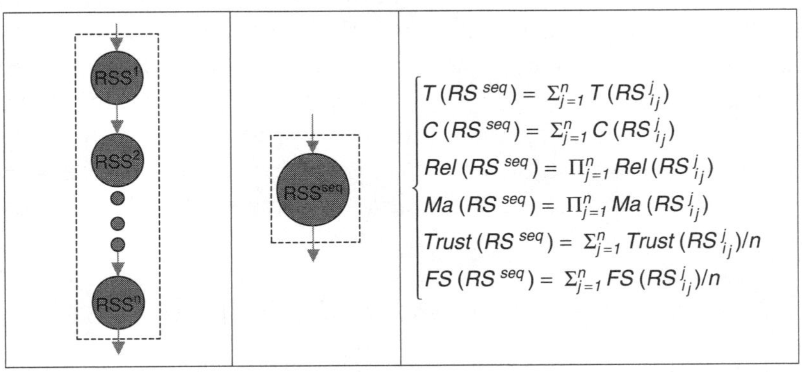

Figure 10.2 Transforming rule of sequence mode.

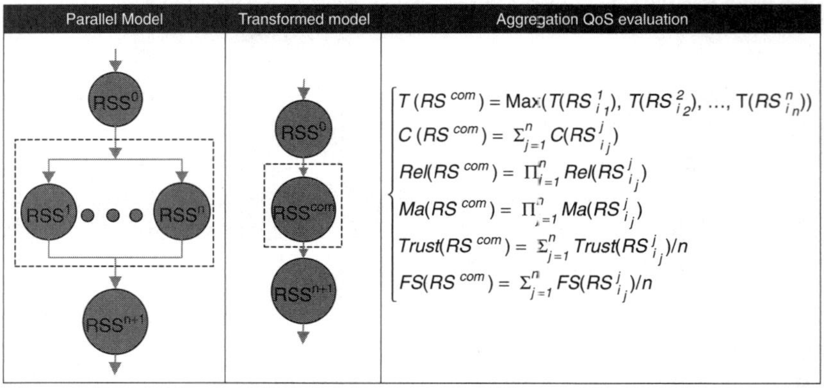

Figure 10.3 Transforming rule of parallel mode.

there are n nodes, i.e., $\{RSS^1, RSS^2, \cdots, RSS^j, \cdots, RSS^n\}$, in a *selective mode* as shown in the first column in Figure 10.4, where $RSS^j = \{RS_1^j, RS_2^j, \cdots, RS_{i_j}^j, \cdots, RS_{M_j}^j\}$. The corresponding transforming chart and calculating formulations are shown in Figure 10.4.

Rule 4 (Circular mode): a new resource service set $RSS^{cyc} = \{RS_1^{cyc}, RS_2^{cyc}, \cdots, RS_{M_{cyc}}^{cyc}\}$ is constructed to provide the functions of the *circular model* (highlighted by dashed pane in Figure 10.5). The primary difference between *sequence model* and the *circular model* is that the elements in *sequence model* are only executed one time, but the elements in *circular model* are executed k_{cyc} times. Apparently, the value of $q^1$and q^2 increase with the executing times, while q^i $(i=3, 4, 5, 6)$ do not change with the execution time. Therefore, addition and multiplication operators are used to compute q^1 and q^2 of RS^{cyc}, multiplication operator is used to compute q^3 and q^4, and addition operator is used to compute q^5 and q^6. It is assumed that there are n nodes, i.e., $\{RSS^1, RSS^2, \cdots, RSS^j, \cdots, RSS^n\}$, in a *circular mode* as shown in the first column in Figure 10.5, where $RSS^j = \{RS_1^j, RS_2^j, \cdots, RS_{i_j}^j, \cdots, RS_{M_j}^j\}$. The corresponding transforming chart and calculating formulations are shown in Figure 10.5.

In the above formulations of aggregation QoS properties, the value of each QoS property of each resource service is calculated by Equation 10.5.

10.2.3 Formulation of MO-MRSCOS Problem

Based on the transforming rules, almost all MCRS can be transformed into SCRS as shown in Figure 10.6. A SCRS can be represented as

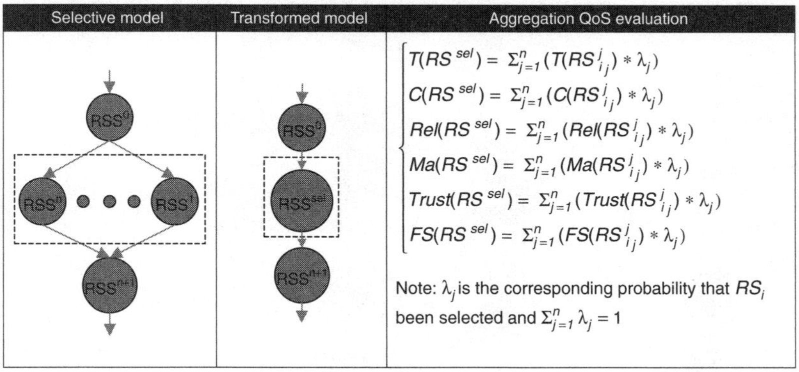

Figure 10.4 Transforming rule of selective mode.

$\mathrm{CRS} = \{\mathrm{RS}^1_{i_1}, \mathrm{RS}^2_{i_2}, \cdots, \mathrm{RS}^j_{i_j}, \cdots, \mathrm{RS}^{\mathrm{N}}_{i_{\mathrm{N}}}\}\ (\mathrm{RS}^j_{i_j} \in \mathrm{RSS}^j)$. The QoS model of $i^{\mathrm{th}} (i = 1, 2, \cdots, \prod_{j=1}^{\mathrm{N}} M_j)$ composite path (hereafter denoted as p_i) of CRS is denoted as $\mathrm{Q}(p_i) = (T(p_i), C(p_i), Rel(p_i), Ma(p_i), Trust(p_i), FS(p_i))$, and $\mathrm{Q}(p_i)$ is calculated as follows (Tao *et al.*, 2008c):

$$\begin{cases} T(P_i) = \sum_{j=1}^{\mathrm{N}} T(\mathrm{RS}^j_{i_j}) \\ C(P_i) = \sum_{j=1}^{\mathrm{N}} C(\mathrm{RS}^j_{i_j}) \\ \mathrm{Re}\,l(P_i) = \prod_{j=1}^{\mathrm{N}} Rel(\mathrm{RS}^j_{i_j}) \\ Ma(P_i) = \prod_{j=1}^{\mathrm{N}} Ma(\mathrm{RS}^j_{i_j}) \\ Trust(P_i) = \sum_{j=1}^{\mathrm{N}} Trust(\mathrm{RS}^j_{i_j}) \Big/ \mathrm{N} \\ FS(P_i) = \sum_{j=1}^{\mathrm{N}} FS(\mathrm{RS}^j_{i_j}) \Big/ \mathrm{N} \end{cases} \tag{10.8}$$

Apparently, there are $\prod_{j=1}^{\mathrm{N}} M_j$ paths, and each path is a possible solution to an MO-MRSCOS problem. An **MO-MRSCOS** problem is to select the optimal one from all possible paths under multi-objective (e.g., time minimization, cost minimization, and reliability maximization) and multi-constraints (e.g., satisfy lowest

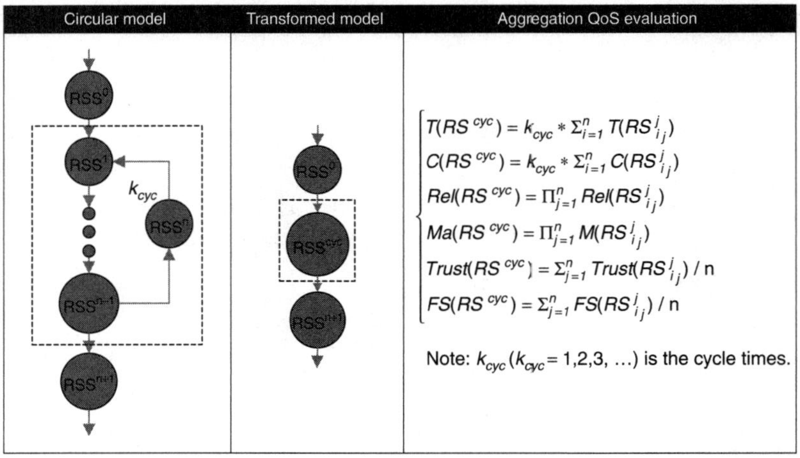

Figure 10.5 Transforming rule of circular mode.

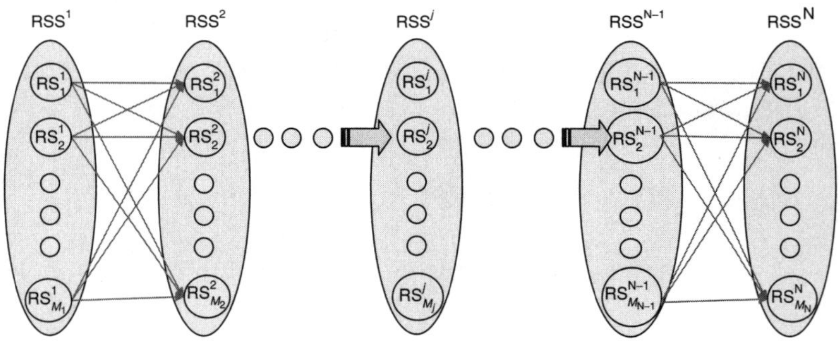

Figure 10.6 Sequence composite resource service (SCRS) process.

maintainability, *trust-QoS*, and *function similarity*). The mathematical formulations of the MO-MRSCOS problem are as follows (Tao *et al.*, 2008c):

$$\textbf{Minimize} \quad T(p_i) \tag{10.9}$$

$$\textbf{Minimize} \quad C(p_i) \tag{10.10}$$

$$\textbf{Maximize} \quad Rel(p_i) \tag{10.11}$$

Subject to

$$Ma(p_{i,j}) \geq Ma_{\circ} \quad \forall j = 1,2,3,\cdots,N \tag{10.12}$$

$$Trust(p_{i,j}) \geq Trust_{\circ} \quad \forall j = 1,2,3,\cdots,N \tag{10.13}$$

$$FS(p_{i,j}) \geq FS_{\circ} \quad \forall j = 1,2,3,\cdots,N \tag{10.14}$$

where $i = 1,2,\cdots,\prod_{j=1}^{N} M_j$; $Ma_{\circ}$, $Trust_{\circ}$ and $FS_{\circ}$ are the lowest requirements of *maintainability, trust-QoS,* and *function similarity,* respectively; $Ma(p_{i,j})$, $Trust(p_{i,j})$ and $FS(p_{i,j})$ are the values of maintainability, trust-QoS, and function similarity of the Jth resource service (i.e. $RS_{i_j}^{j}$) invoked in the ith CRSEP (i.e. $\{RS_{i_1}^{1}, RS_{i_2}^{2}, \cdots, RS_{i_j}^{j}, \cdots, RS_{i_N}^{N}\}$) respectively. Constraints 10.12–10.14 corresponding to the *maintainability, trust-QoS* and *function similarity* should be no less than the user's requirements.

Many studies have been devoted to the subject of multi-objective optimization (MOO), and the proposed methods to address MOO can be classified into three different types (Zhang *et al.*, 2008):

1. A utility function or weighted function is employed to aggregate multiple objectives into a single one, so a MOO problem can be transformed into a mono-objective optimization problem that consists of combining the different objectives into a weighted sum. Methods in this class are based on utility functions or E-Constraint and goal programming.
2. The non-Pareto approach utilizes operators for processing the different objectives in a separated way. It optimizes one objective at a time while imposing constraints on the others. Because it is difficult to determine the order of all the objectives to be optimized, this kind of approach is seldom used.
3. The Pareto-based approach is directly based on the Pareto optimality concept. In this approach, a vector containing all objective values represents the solution fitness, and the concept of Pareto dominance is used

to establish preference between solutions. Its aim is to satisfy two goals: convergence towards the Pareto front while obtaining diversified solutions scattered all over the Pareto front.

In this chapter, the third approach is selected to resolve MO-MRSCOS using PSO.

10.2.4 Pareto Solution of MO-MRSCOS

Definition 10.1, Pareto optimal: Let p_i^* and p_j be two feasible solutions to MO-MRSCOS, and both p_i^* and p_j satisfy Constraints 10.12–10.14. Solution p_i^* is said to dominate p_j if and only if the following statement is true:

$$\left(\begin{pmatrix} T(p_i^*) \le T(p_j) \text{ and} \\ C(p_i^*) \le C(p_j) \text{ and} \\ Rel(p_i^*) \ge Rel(p_j) \end{pmatrix} \text{ and } \begin{pmatrix} T(p_i^*) < T(p_j) \text{or} \\ C(p_i^*) < C(p_j) \text{ or} \\ Rel(p_i^*) > Rel(p_j) \end{pmatrix}\right) \tag{10.15}$$

The above relationship is represented as $p_i^* \succ p_j$, and p_i^* is defined as the *Pareto optimal* solution (or non-dominated solution). If there is no such solution p_j satisfies $p_j \succ p_i^*$. The whole of all Pareto optimal objectives vectors is called the *Pareto front* and the set of all the Pareto optimal decision vectors is called the *Pareto optimal set*.

For an MO-MRSCOS problem, the system may search out a large number of candidate solutions, (a) how to distinguish and classify the *Pareto optimal* solution, i.e., *classification of Pareto front*, and (b) how to compare different *Pareto optimal* solutions and select the best *Pareto optimal* solution, i.e., *comparison of Pareto optimal solution*, are key to using the Pareto-based approach in PSO.

(a) Classification of Pareto front

Let population P be a set of candidate solutions. In order to realize the classification of the Pareto front, the solution P is divided into $K(K=1, 2,...)$ subpopulations, SP_1, SP_2, ..., SP_k according to the dominant relationship among solutions. Any solution in set SP_j is dominated by the solutions in set SP_i if $i<j$, where the solutions in the same set are not dominated to each other. After that, all solutions in the populations are classified into K Pareto fronts, where SP_1 contains all the Pareto-optimal solutions in the current population. According to (Deb *et al.*, 2002), the implementing algorithms of classification of Pareto front are as follows:

***NonDonminateClassify*(P)**	
Step 1)	For each solution $p_i \in P$, set $N_{p_i} = 0$ and $\psi_{p_i} = \phi$
Step 2)	For each solution $p_j \in P\ (j \neq i)$ If $(p_i \succ p_j)$, then let $\psi_{p_i} = \psi_{p_i} \cup \{p_j\}$ Else If $(p_j \succ p_i)$, then let $N_{p_i} = N_{p_i} + 1$
Step 3)	If $N_{p_i} = 0$, then let $SP_1 = SP_1 \cup \{p_i\}$
Step 4)	Set $k = 1$
Step 5)	While $SP_k \neq \phi$, set $\theta = \phi$
Step 6)	For each $p_i \in SP_k$
Step 7)	For each $p_j \in \psi_{p_i}$, $N_{p_j} = N_{p_j} - 1$, If $N_{p_j} = 0$ then $\theta = \theta \cup \{p_j\}$
Step 8)	$k = k + 1$, $SP_k = \theta$
Step 9)	Output SP_k

(b) Comparison of Pareto solutions

After the above classification, a set of candidate solutions P can be represented as $P = \{SP_1, SP_2, ..., SP_K\}$, where any solution in set SP_j is dominated by the solutions in set SP_i ($i<j$ and $i, j = 1, 2, ..., K$). Apparently, the distribution of two different arbitrary solutions, e.g., P_i and P_j, has two conditions: (a) that the two solutions belong to two different Pareto fronts, and (b) that the two solutions belong to the same Pareto front. In this chapter, if two solutions with differing non-domination ranks (i.e., belonging to different Pareto fronts), the system prefers the solution with the lower rank, e.g., $\forall p_i \in SP_i$ and $\forall p_j \in SP_j$, if $i<j$, the system prefers P_i. Otherwise, if both solutions belong to the same Pareto front, then the system prefers the solution with bigger *crowding distance* according to (Deb *et al.*, 2002).

Let $d(P_i)$ be the *crowding distance* of solution P_i in $SP_k(k=1, 2, 3, ..., K$ and $d(P_i)$ is calculated as follows:

CrowdDistanceEvalate **(P)**
Step 1) Perform *NonDonminateClassify*(*P*) and generate $P=\{SP_1, SP_2, \cdots, SP_K\}$
Step 2) Set $k=1$
Step 3) While $SP_k \neq \phi$
Step 4) For each $p_i \in SP_k = \{p_1, p_2, \cdots, p_{N_k}\}$, set $d(p_i)=0$
Step 5) Set $j=1$
Step 6) $Sort \uparrow (SP_k, j)$ // Order the solutions in SP_k according to the jth objective values in ascending order.
Step 7) Set $f_j^{\max} = \max\{f_j(p_i), \forall p_i \in SP_k, \mathrm{k}=2,3,\cdots,N_k-1\}$ and $f_j^{\min} = \min\{f_j(p_i), \forall p_i \in SP_k, \mathrm{k}=2,3,\cdots,N_k-1\}$ // $f_j(p_i)$ be the jth objective value of p_i
Step 8) Set $d(p_1)=d(p_{N_k})=+\infty$
Step 9) For $i=2$ to (N_k-1), $d(p_i)=d(p_i)+(f_j(p_{i+1})-f_j(p_{i-1}))/(f_j^{\max}-f_j^{\min})$
Step 10) If $j=3$, then go to **Step 11)**; Otherwise $j=j+1$ and go to **Step 6)**
Step 11) Output $d(p_i)$
Step 12) $k=k+1$

Based on above idea, a comparison operator, $\succ_C$, between two Pareto optimal solutions is defined as follows: For a set of candidate solutions $P=\{SP_1, SP_2, ..., SP_k\}$, where any solution in set SP_j is dominated by the solutions in set SP_i ($i<j$ and $i, j=1, 2, ..., K$):

$$(1) \quad \forall p_i \in SP_i \text{ and } \forall p_j \in SP_j \Rightarrow p_i \succ_C p_j \tag{10.16}$$

$$(2) \quad \forall p_i \in SP_i \text{ and } \forall p_j \in SP_i, \text{ if } d(p_i) > d(p_j) \Rightarrow p_i \succ_C p_j \tag{10.17}$$

where $p_i \succ_C p_j$ means p_i is better than p_j, and the system prefers p_i. Operator $\succ_C$ will be used for the generation of the best local solution (i.e., *gbest*) and global best solution (i.e., *pbest*) in PSO for MO-MRSCOS in the following section.

10.3 Review of Standard PSO

PSO is a population-based stochastic optimization technique proposed by Kennedy and Eberhart (1995) and Kennedy *et al.*, (2001). This technique has good performance, low computational cost and easy implementation. Hence, PSO has attracted significant attention from researchers around the world since its introduction in 1995, and many successful applications of PSO, ranging from evolving weights form artificial neural networks (Eberhart and Shi, 1998), manufacture and milling (Eberhart and Shi, 1998), reactive power and voltage control (Yoshida *et al.*, 1999), to image registration (Yin 2006), have been reported.

The standard PSO proceeds as follows: Given an optimization function $f(x)$ with N-dimension ($N=1, 2, 3, \ldots$) real-valued decision variables, the PSO searches for the optimal solution by iteratively evolving a swarm of candidate solutions, called particles. The initial swarm is generated randomly. Let each particle be represented by $P_i = (x_{i,1}, x_{i,2}, \cdots, x_{i,N})$, and the quality of particle P_i is evaluated by the fitness function that needs to be optimized. Each particle P_i at - t^{th} $(t = 1,2,3\cdots)$ iteration has two attributes that characterize its status on the N-dimension search space: the j-dimension position ($j=1, 2, \ldots, N$), $x_{i,j}(t)$ and J-dimension velocity, $v_{i,j}(t)$. In each search-iteration, the current position and velocity of each particle are updated by two best values: (a) *pbest* , which is the individual best position (best solution or fitness) that has been achieved so far; and (b) *gbest*, which is another "best" value that is tracked by the particle swarm optimizer, representing the best position among all particles in the population or swarm. The current position and velocity of each particle P_i are updated as follows (Kennedy and Eberhart, 1995):

$$
\begin{aligned}
v_{i,j}(t) = \omega v_{i,j}(t-1) + c_1 r_1 \left(pbest_i(t-1) - x_{i,j}(t-1) \right) \\
+ c_2 r_2 \left((gbest(t-1) - x_{i,j}(t-1) \right)
\end{aligned}
\tag{10.18}
$$

$$
x_{i,j}(t) = x_{i,j}(t-1) + v_{i,j}(t) \tag{10.19}
$$

The velocity, $v_{i,j}(t)$, represents the distance to be traveled by P_i from its current position. Position, $x_{i,j}(t)$, represents P_i's position, $pbest_i(t)$ represents its best previous position, and $gbest(t)$ represents the best position among all particles in the population. Terms r_1 and r_2 are random real numbers drawn from U(0, 1). Parameters c_1 and c_2 are positive constants parameters called acceleration coefficients. Parameter ω is the inertia weight, a user-specified parameter that controls, together with c_1 and c_2, the impact of previous historical values of particle velocities on its current one.

The standard PSO procedure is as follows. Interested readers may refer to Kennedy and Eberhart, (1995) and Kennedy *et al.*, (2001) for details.

1. Initialize a population of particles with random positions and velocities in the search space.
2. Update the velocity and the position of each particle according to Equation 10.18 and 10.19.
3. Map the position of each particle into solution space and evaluate its fitness value according to the desired optimization fitness function. At the same time, update *gbest* and *pbest*.
4. If the stopping criterion is met, terminate search; or else go to 2.

Let

p be the size of the PSO population;

$x_i(t)$ ($x_i(t) = (x_{i,1}(t), x_{i,2}(t), \ldots, x_{i,N}(t))$) be the position of the ith particle of the PSO population at t^{th} ($t = 1,2,3, \ldots$)iteration; this represents a candidate solution for the problem;

$fitness(i)$ be the fitness function of the i^{th}particle;

$v_i(t)$ ($v_i(t) = (v_{i,1}(t), v_{i,2}(t), \ldots, v_{i,N}(t))$) be the traveled distance (or velocity) of the ith particle at t^{th} iteration;

$gbest(t)$ be an index to global-best position;

$pbest_i(t)$ be the position of the local-best position of the ith particle so far during t iterations;

$fitness(pbest_i(t))$ be the local-best fitness for the best position visited by the ith particle.

Step 1. Initialization : For each particle i in the population:

Step 1.1: Initialize $x_i(1)$ randomly.

Step 1.2: Initialize $v_i(1)$ randomly.

Step 1.3: Evaluate $fitness(i)$

Step 1.4: Initialize $gbest(t)$ with the index of the particle with the best fitness among the population

Step 1.5: Initialize $pbest_i(t)$ with a copy of $x_i(1)$.

Step 2. Repeat until a stopping criterion is satisfied.

Step 2.1: Find $gbest(t)$ such that $fitness(gbest_i(t)) \geq fitness(i), \forall i \leq p$

Step 2.2: For each particle i: $pbest_i(t) = x_i(t)$ *iff* $fitness(i) \geq fitness(pbest_i(t)), \forall i \leq p$

Step 2.3: For each particle i: update $v_i(t)$ and $x_i(t)$ according to 10.18 and 10.19

Step 2.4: Evaluate $fitness(i), \forall i \leq p$

Step 3. Output

Figure 10.7 Simple PSO algorithm.

Figure 10.7 shows a simple PSO algorithm. Interested readers may refer to Kennedy and Eberhart, (1995) and Kennedy *et al.*, (2001) for details.

10.4 Improved PSO for MO-MRSCOS Problem

Our creative work for designing the PSO for MO-MRSCOS can be summarized as follows:

- Based on permutation representation, PSO is applied to perform exploration in binary hyperspace.
- A non-dominated sorting technique is applied to get the personal best position (*pbest*) and global best (*gbest*) position in the PSO.
- The parameters of particle updating formulation in PSO are dynamical generated in order to make a compromise between the global exploration and local exploitation abilities of PSO.
- To maintain diversity of solutions in population, *permutation-based* and *objective-based* population trimming operators are applied in PSO.

Next, we will describe PSO for MO-MRSCOS in detail, including representation, particle movement, generation of *pbest* and *gbest*, and population trimming. Then the whole procedure of PSO for MO-MRSCOS problem will be presented.

10.4.1 Representation

One of the key issues in using a successful PSO algorithm is the representation, i.e. finding a suitable mapping between a problem solution and PSO particle. For the considered MO-MRSCOS, a search space of N-dimension is set up for a task which can be decomposed into N subtasks. Therefore, the ith particle can be represented as $(p_i = (x_{i,1}, x_{i,1}, \cdots, x_{i,j} \cdots, x_{i,N})$, where $x_{i,j}$ indicates the index of the allocated resource service for the jth subtask, ST^j, of a task. Each dimension has a discrete set of possible limit values, $\gamma = \{x_{i,j} : 1 \le x_{i,j} \le M_j\}$. M_j is the number of candidate resource services in RSS^j. Figure 10.8 shows a mapping between a possible composition instance and particle position coordinates in PSO. The PSO swarm consists of M particles and the initial swarms generated randomly. Hence, the PSO population is represented as a $M \times N$ two-dimensional array consisting of M particles.

10.4.2 Particle Movement (update)

As described in Section 10.3, swarm intelligence is enriched by tallying the best experiences observed by individual particles and the entire swarm during the evolution. In particular, each particle remembers the best vector with the highest fitness it has visited so far, i.e., *pbest*, and the best vector visited by its neighbors, i.e., *gbest*. At each evolutionary iteration t, particle p_i modifies its velocity $v_{i,j}$ and position $x_{i,j}$ through each dimension j by referring to its personal best experience, $pbest_i(t)$, and the swarm's best experience,

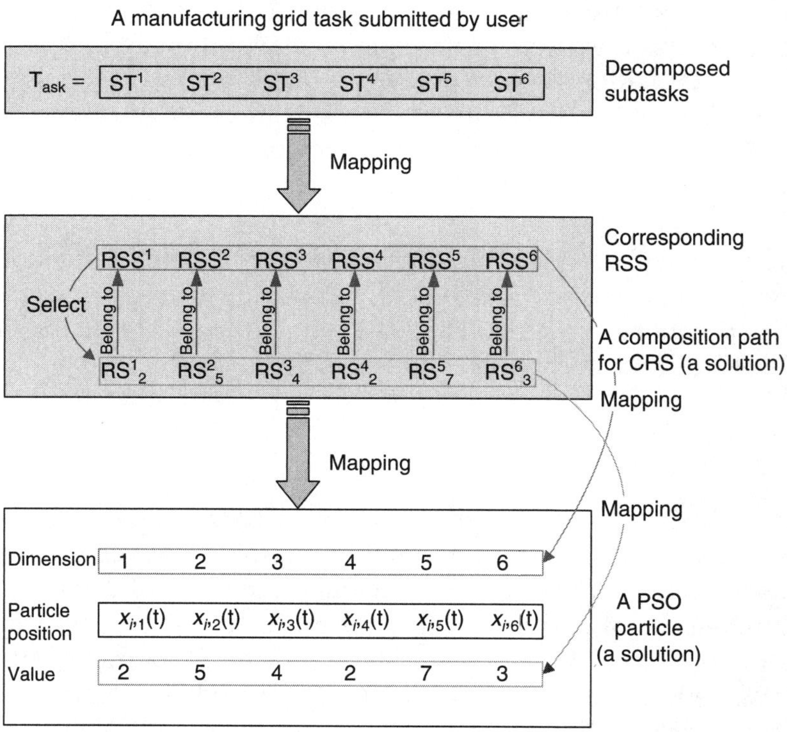

Figure 10.8 Mapping between a possible composition instance and a particle position coordinates in the PSO.

gbest(t), using Equations 10.20 and 10.21 (Kennedy and Eberhart, 1995; Kennedy *et al.*, 2001):

$$v_{i,j}(t) = \omega(t)v_{i,j}(t-1) + c_1(t)r_1(t)\left(pbest_i(t-1) - x_{i,j}(t-1)\right) + c_2(t)r_2(t)\left((gbest(t-1) - x_{i,j}(t-1)\right) \quad (10.20)$$

$$x_{i,j}(t) = x_{i,j}(t-1) + v_{i,j}(t) \quad (10.21)$$

In Equation 10.20, $c_1(t)$ and $c_2(t)$ are positive constant parameters called *acceleration coefficients*, which control the maximum step size a particle can do. Terms $r_1(t)$ and $r_2(t)$ are two random functions drawn from (0,1). *Inertia weight*, $\omega(t)$, is a user-specified parameters that controls, together with $c_1(t)$ and $c_2(t)$, the impact of previous historical values, personal best position, and global best position of particle velocities on its current one. Suitable selection of the $\omega(t)$, $c_1(t)$ and $c_2(t)$ provides a balance between the global and the local searches. The setting of the parameters in Equation 10.20 is discussed in the following section.

10.4.2.1 Selection of Inertia Weight

The role of *inertia weight* is very important in PSO convergence behavior. It is used to control the effect of the previous velocities on the current velocity. It makes a compromise between the global exploration and local exploitation abilities of the swarm. All the population-based search techniques rely on global exploration and local exploitation in order to achieve good performance. A suitable value of the *inertia weight* usually provides a balance between global and local exploration abilities, and consequently a more efficient search to locate the optimum solution can be made. In fact, the PSO search process is a nonlinear and complicated process, and a linearly decreasing or increasing *inertia weight* changes the search from global to local and from local to global linearly.

More exploration is generally needed in the initial stages when the algorithm has very little knowledge about the search space. In contrast, more exploitation is needed during the later stages so that the algorithm is able to exploit the information it has gained so far, and make better global exploration of the search space. Hence, the value of $\omega(t)$ at t^{th} iteration should be allowed to decrease linearly.

In this chapter, $\omega(t)$ is calculated as (Ratnaweera *et al.*, 2004; Tripathi *et al.*, 2007):

$$\omega(t) = (\omega(1) - \omega(N_{exe})) * (N_{exe} - t) / N_{exe} + \omega(N_{exe}) \quad (10.22)$$

where N_{exe} is the maximum number of iterations, t is the current iteration number, $\omega(N_{exe})$ is the *inertia weight* at $N_{exe}{}^{th}$ iteration (i.e., the minimum value) and $\omega(1)$ is that at the first iteration (i.e., the maximum value), and $\omega(t) \in [\omega(1), \omega(N_{exe})]$.

10.4.2.2 *Selection of Acceleration Coefficients*

In the PSO algorithm, each particle keeps track of its coordinates in the problem space which are associated with the best solution (fitness) it has achieved so far. The fitness value is also stored, called *pbest*. Another "best" value tracked by the global version of the PSO is the overall best value and its location obtained so far by any particle in the swarm (i.e. *gbest*). Hence, in the velocity updated Equation 10.18 there are two important *acceleration coefficient* parameters c_1 and c_2, representing the weighting of the stochastic acceleration terms that pull each particle toward *pbest* and *gbest* positions. Parameter c_1 is called *cognitive acceleration coefficient* while c_2 is the *social acceleration coefficient*. At each step in PSO algorithm, c_1 and c_2 control the changing of the velocity of each particle toward its *pbest* and *gbest* locations. Low values allow particles to roam far from target regions before being tugged back, while high values result in abrupt movement toward target regions. Acceleration is weighted by a random term with separate random numbers generated for acceleration toward *pbest* and *gbest* locations. Early experience with PSO (mostly by trial and error) leads the user to set the acceleration constants c_1 and c_2 equal to 2, but it is not a usual rule. In fact, the acceleration to *pbest* location should be higher while the acceleration to *gbest* location should be lower for more exploration at the initial stages. In contrast, the acceleration to *pbest* location should be lower while the acceleration to *gbest* location should be higher at the later stages so that the algorithm is able to exploit the information. Therefore, c_1 is designed as a decreasing function while c_2 is an increasing function (Ratnaweera *et al.*, 2004; Tripathi *et al.*, 2007).

The values of $c_1(t)$ and $c_2(t)$ at t^{th} iteration, are calculated according to Equations 10.23 and 10.24 (Ratnaweera *et al.*, 2004; Tripathi *et al.*, 2007):

$$c_1(t) = \left(c_1(N_{exe}) - c_1(1)\right) * t / N_{exe} + c_1(1) \tag{10.23}$$

$$c_2(t) = \left(c_2(N_{exe}) - c_2(1)\right) * t / N_{exe} + c_2(1) \tag{10.24}$$

where $c_1(N_{exe})$ is the *acceleration coefficients* at $N_{exe}{}^{th}$ iteration (i.e., the minimum value of $c_1(t)$) and $c_1(1)$ is that at the first iteration (i.e., the maximum of $c_2(t)$). They are set by user, and $c_1(t) \in [c_1(N_{exe}), c_1(1)]$. By contrast, $c_2(N_{exe})$ is the *acceleration coefficients* at N_{exe} iteration (i.e., the maximum value of $c_2(t)$) and $c_2(1)$ is that at the first iteration (i.e., the minimum of $c_2(t)$), respectively, and $c_2(t) \in [c_2(N_{exe}), c_2(1)]$.

10.4.2.3 *Biggest Personal Position Value*

Because the PSO particle position, $x_{i,j}$, represents a series of numbers that varies in [1, M_j], all parameters of the J-dimensional particle positions, either initialized or updated during search, must be limited to [1, M_j]. Hence, to avoid infeasible particle positions that can lead to slow PSO search, each parameter of the initialized or updated position that is beyond [1, M_j] can be adjusted according to Equations 10.25 and 10.26:

$$\text{If } x_{i,j}(t) > M_j,\ then\ x_{i,j}(t) = M_j \tag{10.25}$$

$$\text{Else if } x_{i,j}(t) < 1, \text{then } x_{i,j}(t) = 1 \tag{10.26}$$

10.4.2.4 *Biggest Velocity Value*

Oscillating of the particle velocity, $v_{i,j}$, may cause particles to fly outside feasible search space [1, M_j] or $[x_{i,j}^{\min}, x_{i,j}^{\max}]$ when using Equation 10.20. Similarly, the particle velocity based on the current position should also be limited to prevent the updated position from oscillating too heavily. Apparently, the maximum velocity, $v_{i,j}^{\max}$, equals $x_{i,j}^{\max}$ at most. Therefore, the velocity of an updated particle should be limited to $[-v_{i,j}^{\max}, v_{i,j}^{\max}]$ or [$-M_j$, M_j]. In this Chapter, each parameter of the N-dimension particle velocity that is beyond $[-v_{i,j}^{max}, v_{i,j}^{max}]$ is adjusted according to Equations 10.27 and 10.28:

$$\text{If } v_{i,j}(t) > v_{i,j}^{max}, \text{then } v_{i,j}(t) = v_{i,j}^{max} \tag{10.27}$$

$$\text{Else if } v_{i,j} < -v_{i,j}^{max}, \text{then } v_{i,j}(t) = -v_{i,j}^{max} \tag{10.28}$$

10.4.3 Generation of Personal Best (*pbest*)

The term $pbest_i(t)$ is the best performance attained so far by particle p_i in its flight. In the proposed PSO, the current position of $p_i(t)$ is compared with the $pbest_i(t)$ using the comparison operator $\succ_C$ defined in Section 10.2.4, and it replaces the latter if it dominates that solution. The generation of $pbest_i(t)$ for particle p_i at t^{th} iteration is as follows:

Generate_pbest$_i$(t)

$generate\_pbest_i\,(t, p_i(t), pbest_i(t-1))$
{
Step 1) If t=0, $pbest_i(t) = p_i(t)$;
Step 2) If $p_i(t) \succ_C pbest_i(t-1)$, then $pbest_i(t) = p_i(t)$;
Step 3) If $pbest_i(t-1) \succ_C p_i(t)$,
then $pbest_i(t) = pbest_i(t-1)$;
Step 4) Else $pbest_i(t) = p_i(t)$;
Step 5) Return $pbest_i(t)$
}

10.4.4 Generation of Global Best (*gbest*)

The term *gbest* represents the best solution obtained by the swarm. Often the conflicting nature of the multiple objectives involve multiple-objective optimization problems which make the choice of a single optimum solution difficult. To resolve this problem, the concept of non-dominance is employed. And an external archive, Λ, of non-dominated solutions (i.e., Pareto to optimal solutions) produced in the searching process of PSO is maintained, from which a solution is picked up as the *gbest*. In order to enhance the efficiency for selecting the best solutions from the archive, a maximum size, $N_\Lambda (N_\Lambda = 1,2,3,\cdots)$, is specified for the external archive. In each iteration, the member of Λ_{t+1} is produced by a combination of Λ_t and searched solutions P_t. The new non-dominated solutions become a member of the archive when one of the following conditions is met (Lei, 2008; Coello Coello *et al.*, 2004):

1. The new solution dominates some members of the archive.
2. The current archive size is less than N_Λ.
3. The current archive size is equal to N_Λ and the new solution has higher *density value* than at least one member of the archive.

Based on the above idea, the generation algorithms of Λ_{t+1} are as follows:

$ArchiveUpdating(t, \Lambda(t), N_\Lambda, P(t+1))$
Step 1) Let $k=1$
Step 2) Perform $NonDonminateClassify(P(t+1))$ and get $SP_1(t) = \{p_1, \cdots, p_i, \cdots, p_{K_1}\}$ //find all non-dominated solutions in $P(t+1)$, //K_1 is the number of non-dominated solutions
Step 3) Set $\Lambda_{t+1} = \Lambda_t = \{p_1, \cdots, p_j, \cdots, p_{N_a}\}$ // N_a is the current number of members in Λ_{t+1}
Step 4) Set $i=1$
Step 5) For $p_i \in SP_1(t)$, set $N_1 = N_2 = N_3 = 0$
Step 6) Set $j=1$
Step 7) For $p_j \in \Lambda_{t+1}$, If $p_i \succ p_j$, then $\psi = \psi \cup \{p_j\}, N_1 = N_1 + 1$; If $p_j \succ p_i$, then $N_2 = N_2 + 1$; Else $N_3 = N_3 + 1$;
Step 8) If $j = N_a$, then go to **Step 9)**; otherwise $j = j+1$ and go to **Step 7)**

ArchiveUpdating$(t, \mathit{\Lambda}(t), N_\Lambda, P(t+1))$ *(cont.)*
Step 9) If $N_1 = N_a$, then delete all members in Ψ from $\mathit{\Lambda}_{t+1}$ and then $\mathit{\Lambda}_{t+1} = \mathit{\Lambda}_{t+1} \cup \{p_i\}$, $= N_a - N_1 + 1$
Step 10) If $N_3 = 0$ //no members in $\mathit{\Lambda}_{t+1}$ is dominated by p_i If $N_a < N_\Lambda$, then $\mathit{\Lambda}_{t+1} = \mathit{\Lambda}_{t+1} \cup \{p_i\}$ and $N_a = N_a + 1$; // N_Λ is the maximum size of archive If $N_a = N_\Lambda$, then $\mathit{\Lambda}_{t+1} = \mathit{\Lambda}_{t+1} \cup \{p_i\}$ and delete a solution with the minimum *crowding distance* from $\mathit{\Lambda}_{t+1}$
Step 11) If $i = K_1$, then go to **Step 12)**; Otherwise $i = i+1$ and go to **Step 5)**
Step 12) Output $\mathit{\Lambda}_{t+1}$

After $\mathit{\Lambda}_{t+1}$ is generated, then the global best solution, *gbest*(t), is selected from $\mathit{\Lambda}_{t+1}$ using the comparing operator in Section 10.2.4. The generation of *gbest*(t) for particle at t^{th} iteration is as follows:

Generation of *gbest(t)*
generate_gbest$(t, \mathit{\Lambda}_t)$ {
Step 1) Perform *NonDonminateClassfiy*$(\mathit{\Lambda}_t)$ //find all non-dominated solutions in $\mathit{\Lambda}_t$
Step 2) Perform *CrowdDistanceEvaluate*$(\mathit{\Lambda}_t)$ and get the crowding distance of each member in $\mathit{\Lambda}_t$.
Step 3) For each solution $p_i \in \mathit{\Lambda}_t$, set $N_{p_i} = 0$ and $\psi_{p_i} = \phi$
Step 4) Set $i = 1$,
Step 5) Let $j = i + 1$
Step 6) For solution $p_j \in \mathit{\Lambda}_t$, If $(p_i \succ_C p_j)$, then let $\psi_{p_i} = \psi_{p_i} \cup \{p_j\}$ Else If $(p_j \succ_C p_i)$, then let $N_{p_i} = N_{p_i} + 1$
Step 7) If $j = N_a$, then go to **Step 8)**; Otherwise $j = j + 1$ and go to **Step 6)** // N_a is the current number of members in $\mathit{\Lambda}_{t+1}$

Generation of *gbest(t)* (cont.)
Step 8) If $i = N_a - 1$, then go to **Step 9)**; Otherwise $i = i + 1$ and go to **Step 5)**
Step 9) Set $\psi = \phi$
Step 10) For each solution $p_i \in A_t$, if $N_{p_i} = 0$, then let $\psi = \psi \cup \{p_i\}$
Step 11) Select one solution randomly from ψ as the *gbest*(*t*)
Step 12) Return *gbest*(*t*)
}

10.4.5 Population Trimming in PSO

For an MO-MRSCOS problem, PSO algorithms may obtain a population with some same resource service permutations, and the objectives values of different population may be the same. Therefore, two trimming operators, i.e., *permutation-based* and *objective-based* trimming, are employed to remove some redundant solutions so as to enrich the diversity of the population.

(a) *Permutation-based trimming* operators (*PB Trim*()): If there are some populations with the same resource service permutations, the system deletes the redundant solutions and only one resource service permutation is reserved. The deleted permutations will be replaced by some random permutations in the population. After the trimming, all the permutations in the population are different.
(b) *Objective-based trimming* operators (*OB Trim*()): In PSO, after evaluating the quality of the solutions of the population, if there are some solutions with the same objectives values, then the system deletes the redundant solutions and only one is reserved. Those deleted solutions are replaced by some random permutations which have at least one objective value different from all those reserved in the population.

After the above two operations, the diversity of the population is enhanced greatly and effectively.

10.4.6 The Procedures of the Proposed PSO Algorithm for MO-MRSCOS Problem

The procedures of the PSO algorithm for solving an MO-MRSCOS problem are presented in Figure 10.9, where every key element and function have been explained in detail as before.

PSO algorithms for MO-MRSCOS
(1)Initialize ($t=0$)
Step 1) Randomly generate an initial population $P(t)=\{p_1^t, p_2^t, \cdots, p_N^t\}$, where N denotes the size of the population.
Step 2) Initialize the velocity of each particle $v_i(t)$
Step 3) Evaluate each particle in $P(t)$, perform $PBTrim(P(t))$ and $OBTrim(P(t))$.
Step 4) Perform $NonDonminateClassify(P(t))$, select out the non-dominated solutions and store them in A_t
Step 5) Initialize the personal best position of each particle p_i^t. For $i=1$ to N, $pbest_i(t)= p_i^t$.
(2)Repeat ($t=1$ to N_{exe})
Step 6) For $i=1$ to N, $pbest_i(t)= generate\_pbest_i\left(t, p_i^t, pbest_i(t-1)\right)$ // generate the personal best position $pbest_i(t)$ for each particle p_i^t in $P(t)$
Step 7) Perform $generate\_gbest(\mathrm{t}, A(t-1))$ and generate the global best position $gbest(t)$
Step 8) Updating $\omega(t)$, $c_1(t)$, $c_2(t)$ according to Equations 10.26, 10.27, and 10.28, respectively.
Step 9) Computing the velocity of particle p_i^t using Equation 10.24 and use Equations 10.31 and 10.32 to adjust the velocity of p_i^t in order to maintain the particles with the search space in case they go beyond their boundaries.
Step 10) Generating new position of particle p_i^t using Equation 10.25 and use Equations 10.29 and 10.30 to adjust the position of p_i^t in order to maintain the particles with the search space in case they go beyond their boundaries. Generate $P(t+1)=\{p_1^{t+1}, p_2^{t+1}, \cdots, p_N^{t+1}\}$,

PSO algorithms for MO-MRSCOS (cont.)
Step 11) Evaluate each particle in $P(t+1)$, perform trimming operators $PBTrim(P(t+1))$ and $OBTrim(P(t+1))$.
Step 12) Perform $ArchiveUpdating(t, \Lambda(t), P(t+1))$ and produce Λ_t
(3) Output
Step 13) Output Λ_t //return the Pareto optimal font

Figure 10.9 Proposed PSO algorithms for MO-MRSCOS problem.

10.5 Performance Analysis and Discussion

These experiments test the effectiveness of the proposed approach to finding the Pareto resolutions to MO-MRSCOS problems. In these experiments, we will use the following two approaches to generate the final Pareto resolutions to a same MO-MRSCOS problem and compare their results.

1. The first approach is that the correlation among resource services are not taken into account when calculating the aggregation QoS value of a CRS or a solution to an MO-MRSCOS problem. This means that each QoS criterion value of the selected resource service in a solution is its default value. Then the proposed PSO method is used to generate the final Pareto solutions set. This approach is denoted as *NoCorrelation* in this section for the sake of convenience.
2. The second approach considers the correlation among resource services during resource service composition. This means that each QoS criterion value of the selected resource service in a solution is calculated according to Equation 10.5 in Section 10.2.2.1. Similarly, this approach is denoted as *ProposedMethod* in this section.

The *average fitness* of the obtained Pareto solution set is selected as the criterion to test and compare the effectiveness of *NoCorrelation* and *ProposedMethod*. The fitness of a solution p_i, $fitness(p_i)$, in this chapter is defined as follows (Tao *et al.*, 2008c):

$$fitness(p_i) = \left(Z(p_i) + C_{onstraint}(p_i)\right)^{-1} \quad (10.29)$$

where

$$Z(p_i) = \alpha * T(p_i) + \beta * C(p_i) + \varepsilon / Rel(p_i) \quad (10.30)$$

$$C_{onstraint}(p_i) = \begin{pmatrix} \lambda_1 \times \sum_{j=1}^{N} \left(\max\left(0, \frac{Ma_\circ - Ma(p_{i,j})}{Ma_\circ} \right) \right) + \\ \lambda_2 \times \sum_{j=1}^{N} \left(\max\left(0, \frac{Trust_\circ - Trust(p_{i,j})}{Trust_\circ} \right) \right) + \\ \lambda_3 \times \sum_{j=1}^{N} \left(\max\left(0, \frac{FS_\circ - FS(p_{i,j})}{FS_\circ} \right) \right) \end{pmatrix} \quad (10.31)$$

In above expression, $j = 1, 2, 3, \ldots, N$ and $i = 1, 2, \cdots, \prod_{j=1}^{N} M_j$. $Z(p_i)$ is the original minimization objective function of the MO-MRSCOS problem. It combines the three optimization objectives (time minimization, cost minimization, and reliability maximization) into a single function by using three scaling factors, $\alpha, \beta, \varepsilon$. The factors $\alpha, \beta, \varepsilon$ are used as the corresponding weighted parameters to control the relative significance of each objective's functions. They normalize the values of $T(p_i)$, $C(p_i)$ and $Rel(p_i)$ to comparable ranges so that $Z(p_i)$ will not be dominated by a single objective. $C_{onstraint}(p_i)$ is the penalty function that estimates the infeasibility of p_i. It is the quantified amount of mismatch if p_i is infeasible, otherwise $C_{onstraint}(p_i)$ is set to 0. The term $\sum_{j=1}^{N} \left(\max\left(0, \frac{Ma_\circ - Ma(p_{i,j})}{Ma_\circ} \right) \right)$ calculates the total insufficiency amounts of the invoked resource services in p_i against the lowest requirement of *maintainability* if p_i is infeasible. Otherwise it returns zero. Thus, the penalty function causes a proportional amount of penalties on the infeasible solutions whose *maintainability* is under the user's lowest *maintainability* requirement. Similarly, $\sum_{j=1}^{N}\left(\max\left(0,\frac{Trust_{\circ}-Trust(p_{i,j})}{Trust_{\circ}}\right)\right)$ and $\sum_{j=1}^{N}\left(\max\left(0,\frac{FS_{\circ}-FS(p_{i,j})}{FS_{\circ}}\right)\right)$ calculate the total insufficiency amounts of p_i's *trust-QoS*, and *function similarity*, respectively. $\lambda_1,\lambda_2,\lambda_3$ are respectively the weights controlling the relative significance constraints.

The main parameters in proposed PSO algorithms are set as follows: $N_{exe}=30$. $\omega(1)=0.7$ and $\omega(N_{exe})=0.4$ in Equation 10.22 as suggested in van den Bergh (2001). $c_1(1)=2.5, c_1(N_{exe})=0.5$ in Equation 10.23 and $c_1(1)=0.5, c_2(N_{exe})=2.5$ in Equation 10.24 as suggested in van den Bergh (2001). $\alpha=0.001$, $\beta=0.01$ and $\varepsilon=0.25$ in Equation 10.30. $\lambda_1=\lambda_2=\lambda_3=1$ in Equation 10.31.

During the implementations, two conditions are considered:

(I) The percent of correlation resource service is set at 40%, and the number of candidate services of each service varies from 50–450 with an increment of 50.

(II) The number of candidate services of each service is fixed at 250, and the percent of correlation resource service varies changing from 0–100% with an increment of 20%.

Condition (I) has nine tests and condition (II) has six tests, respectively. Each test is executed twenty times. The results of each test is the average of the twenty executions, as shown in Figure 10.10. It can be concluded from Figure 10.10 that *ProposedMethod* shows better performance than *NoCorrelation* in the tests both under condition (I) and condition (II).

In condition (I), along with the increment of the number of candidate resource services, the QoS of CRS obtained by *ProposedMethod* and *NoCorrelation* are both increasing. The reason is that, along with the increment of the number of candidate resource services, there are more candidate composition paths, which result in the improvement of QoS of CRS. But the result produced using *ProposedMethod* is better than that using *NoCorrelation*. This is because *ProposedMethod* considers the correlation among resource service, and the QoS of the selected resource service in a CRS is better than declared (i.e., its default value).

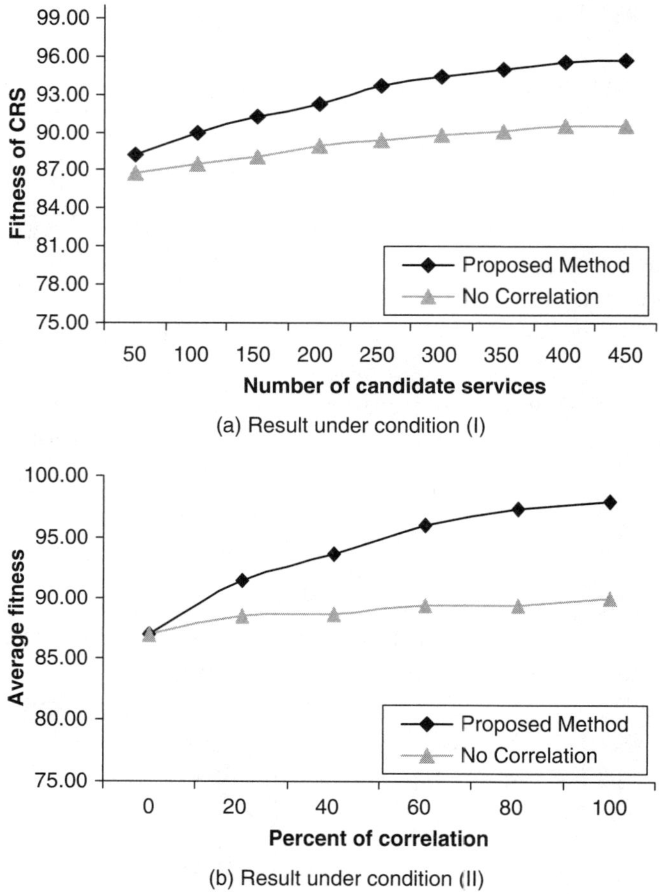

Figure 10.10 Comparison of performance between *NoCorrelation* and *Proposed Method*.

In condition (II), when the number of candidate services of each service is fixed at 250, and the percent of correlation resource service varies from 0–100% with an increment of 20%, the QoS of CRS obtained by *ProposedMethod* is improved. That's because as the increasing of the percentage of correlation resource services, more resource services' quality is affected by other resource services in a CRS. And the actual computing QoS of each selected resource service is better than its default value, and the whole QoS of the CRS is improved. However, because *NoCorrelation* does not consider the correlation among resource service, and the QoS of the selected resource service are the default value, which is not changed along with the increasing of the numbers of correlation resource services. Hence, as the percent of correlation resource services increases, there is almost no change in the QoS of the CRS.

(a) Result under condition (I)
(b) Result under condition (II)

10.6 Summary

For a multi-resource service request task (MRSRTask) in the manufacturing grid (MGrid) system, in addition to the search for all qualified resource services according to each subtask, the system selects one candidate resource service for each subtask. Then the system generates a new composite resource service (CRS) and selects the optimal resource service composite path from all possible paths to execute the task with the given multi-objective (e.g., time minimization, cost minimization, and reliability maximization) and multi-constraints. The above problem is defined as multi-objective MGrid resource service composition and optimal-selection (MO-MRSCOS) problem. In this chapter the formulation is presented for an MO-MRSCOS problem. The correlations among resource services are taken into account during MGrid resource service composition, and a QoS description mode supporting resource service correlation is presented. The basic resource service composite modes (RSCM) for CRS are described, and the principles for translating a complicated RSCM into a simple sequence RSCM are presented in order to simplify the resolution process and complexity of the MO-MRSCOS problem. A new method based on the principles of particle swarm optimization (PSO) is also proposed for solving the MO-MRSCOS problem.

Unlike previous work: (a) the proposed PSO algorithms combine the non-dominated sorting technique to achieve the selection of global best position and private best position; (b) the parameters of particle updating formulation in PSO are dynamical generated in order to make a compromise between the global exploration and local exploitation abilities of PSO; and (c) to maintain diversity of solutions in population, *permutation-based* and *objective-based* population trimming operators are applied in PSO. The experimental results show that the proposed methods are sound from both the efficiency and effectiveness standpoints for solving the MO-MRSCOS problem.

11

Resource Services Composition Flexibility

11.1 Introduction

As shown in Figure 9.1 and described in Chapter 9, when an automobile manufacturer submits a task request for ordering automobiles parts, five elementary resource services are integrated according to a certain flow to execute the process from producing to transporting, and finally these automobiles parts are delivered to the manufacturer. In this flow, any single resource service only has a single-function. In order to obtain the resource services with more powerful functions, these individual resource services need to be integrated to form composite resource services that satisfy the different demands of the users. Namely, resource service composition (RSC) means taking advantage of currently existing resource services to provide a new resource service that does not exist on its own in order to execute a complex task that a single resource service cannot do (Hashemian and Mavaddat, 2006).

However, massive resource services (e.g., various manpower and equipment), and the demand of enterprise users and basis environment (e.g., software/hardware, network resource) are involved during RSC. Therefore, there are increasing resources and resource services in the resource service registry, and an unceasing expansion of the scale of RSC. In view of the complexity of the process of RSC, a series of phenomenon, such as the load imbalance and network interruption, can possibly occur during RSC, as well as a variety of dynamic changes such as the changes of resource services, correlations among resource services (Guo *et al.*, 2010), quality of resource service (QoS), the requirements of enterprise users, etc. All of these make the manufacturing task impossible to be completed efficiently and with high-quality. For example, one invoked resource service may appear abnormal in the execution process of RSC. The reasons may be that the way of transporting needs to be changed due to the weather, or the production of parts cannot be completed due to mechanical or human error, so that other suppliers need to be selected. Hence, in order to make RSC operate in a stable, robust and transparent manner, the introduction of flexibility is an effective solution to the dynamic problem during RSC.

In the whole process of RSC, RSC is needed to make timely responses to various dynamic changes, and the ability to adapt to these changes is namely RSC flexibility. When RSC has flexibility, it will have strong self-organization and dynamic reconfiguration ability. At this time, the users do not need to care about and maintain the underlying detail of RSC. RSC can dynamically coordinate various abnormal changes to complete tasks transparently to realize automation, intelligence, and information of RSC (Guo *et al.*, 2011).

In the meantime, in the actual process of RSC, there always exist many RSC plans satisfying user demand, so the optimal one needs to be selected to execute the user's request, i.e. optimal-selection of RSC. The previous research on optimal-selection of RSC only considered optimal-selection of RSC based on QoS. However, because there exist a lot of dynamic changes during RSC, in order to guarantee the rate of success of RSC, when the optimal RSC plan is selected to execute the task, not only the optimal-selection of RSC based on QoS should be considered, but also optimal-selection of RSC based on flexibility should be taken as a evaluation index. Therefore, it is very necessary to research the measurement methods of RSC flexibility for achieving optimal-selection of RSC.

However, there exists little research on RSC flexibility at present, especially its measurement method. In view of the issue of RSC flexibility, this chapter presents the concept and the classification of RSC flexibility, and the measurement method of RSC flexibility is investigated in order to achieve the optimal-selection of RSC based on flexibility.

11.2 Related Works

RSC is a hot and difficult research point in the resource service-oriented applied research field. Many research works have been done in this field, and many kinds of ways have been put forward to try to solve the issue of RSC. The research that has been done on RSC has a different focus in the industrial and academic communities (Guo *et al.*, 2011).

As described in Chapter 7, industrial circles focused on the description language of RSC and the tools for establishing and running RSC. A number of resource service-oriented flow description languages have been developed, such as WS-BPEL (Web Services Business Process Execution Language), OWL-S (OWL for Services), WSCI (Web Services Choreography Interface), WS-CDL (Web Service Choreography Description Language), etc. All of these play an important role in promoting the development of RSC software tools. Academic circles focused on the constructing methods of RSC, of which the representative methods are based on business flow (Aggarwal *et al.*, 2004; Deng *et al.*, 2004a; Nanda and Karnik, 2003), artificial intelligence and planning (Ponnekanti and Fox, 2002; Schuschel and Weske, 2004), graph theory (Rao *et al.*, 2006a), agent (Maamar *et al.*, 2005; Richards *et al.*, 2003), etc. In addition, there is still a method of RSC based on QoS (Tao *et al.*, 2008c, 2010e; Zeng *et al.*, 2004a). This kind of method is used to select the best execution plan of RSC according to users' different preferences to the performance of resource services after RSC completed construction, i.e., optimal-selection of RSC.

However, the above research only concentrated on the composing methods of RSC. Since the invoking of the resource services has the characteristics of long life-cycle, high complexity, and high reliability, etc., many uncertain dynamic changes may occur during RSC, which make these methods not suitable for addressing RSC. In view of these dynamic changes, this chapter will present research on the RSC flexibility to make RSC able to adapt various dynamic changes to ensure the reliability and rate of success of RSC.

Most of the previous works about the measurement method of flexibility mainly focused on Flexible Manufacturing System (FMS).

Sethi and Sethi (1990) defined 11 different types of flexibility, including machine flexibility, process flexibility, routing flexibility, operation flexibility, etc., which classified the flexibility of manufacturing systems. Zhang *et al.*, (2003a) organized literature on manufacturing flexibility and classified it according to a competence and capability theory. Most of the subsequent research focused on the types of flexibility presented in Sethi and Sethi (1990) and Zhang *et al.*, (2003a).

Jordan and Graves (1995) measured the process of flexibility through the maximum probability of unfulfilled demand for some products and a surplus capacity for other products. Brill and Mandelbaum (1989) defined a measurement method of the flexibility of a machine in relation to a set task that depends on the productivity attainable in each task and the relative importance of each task. In other works, some researchers applied Shannon's entropy (Shannon, 1948) to measure the manufacturing flexibility. Yao (1985) used the entropy to measure routing flexibility, and determine the best routing sequence for parts. Kumar (1987) presented several other entropy methods. Chang *et al.*, (2001, 2004, 2007, 2009) have applied Shannon's entropy to measure the flexibility of single-machine, machine-group, routing, labor, etc. Shuiabi *et al.*, (2005) proposed the entropy as a measure of flexibility for manufacturing operations to indicate a firm's capability to respond to customer demands. There also exist some works, such as Chryssolouris and Lee (1992), Chryssolouris (1996), Alexopoulos *et al.*, (2007a), and Alexopoulos *et al.*, 2010, that thought the flexibility of a manufacturing system is determined by its sensitivity to change, which is quantified by using discounted cash flow (DCF) technologies. In addition, Alexopoulos *et al.*, (2007b) discussed a method of transfer function by the dynamic behavior analogy between a mechanical and a manufacturing system to quantify the flexibility of the manufacturing system. Papakostas *et al.*, (2009) introduced a simulation-based approach pertaining to nonlinear dynamics for assessing the complexity of manufacturing systems for the execution stage, which can identify elements of the system where the complexity index is high.

However, the above methods cannot be directly applied to measuring RSC flexibility due to the difference between the traditional flexibility of the manufacturing system and RSC flexibility. Therefore, in order to cope with the issues of dynamic changes during RSC, this chapter will establish an effective evaluation system to measure RSC flexibility to achieve optimal-selection of RSC based on flexibility.

11.3 The Analysis, Definition and Classification of RSC Flexibility

11.3.1 The Factors Influencing RSC

In order to study RSC flexibility, first the life-cycle of RSC needs to be discussed. This can be classified into the following four phases according to pre-execution, execution, and post-execution (see Figure 11.1) (Guo *et al.*, 2011).

1. Designing phase: In this phase, the user's request will be processed to establish composite resource services. First, users input the task request, then task request parser decomposes the task request to a series of subtasks with flow-dependent correlations (e.g., sequence, parallel, selective, circular). The corresponding sets of candidate resource services can be generated by function match, and the corresponding flow template can be established by flow match, and thus the abstract composite resource service will be constructed.
2. Deploying phase: In this phase, the abstract composite resource service will be mapped into concrete resource services, and finally RSC plans will be generated. The input is the abstract composite resource service in this phase. Several RSC plans are generated according to input/output correlations among composite resource services. The optimal-selection of RSC based on QoS can be realized by extracting, evaluating, and comparing QoS; and the optimal-selection of RSC based on flexibility can be achieved by the measurement of potential RSC flexibility, thus the optimal RSC plan is selected.
3. Execution phase: In this phase, the optimal RSC plan will be executed. The concrete resource services are bound and invoked by the RSC execution engine, and in the meantime the execution process of RSC will be monitored. If

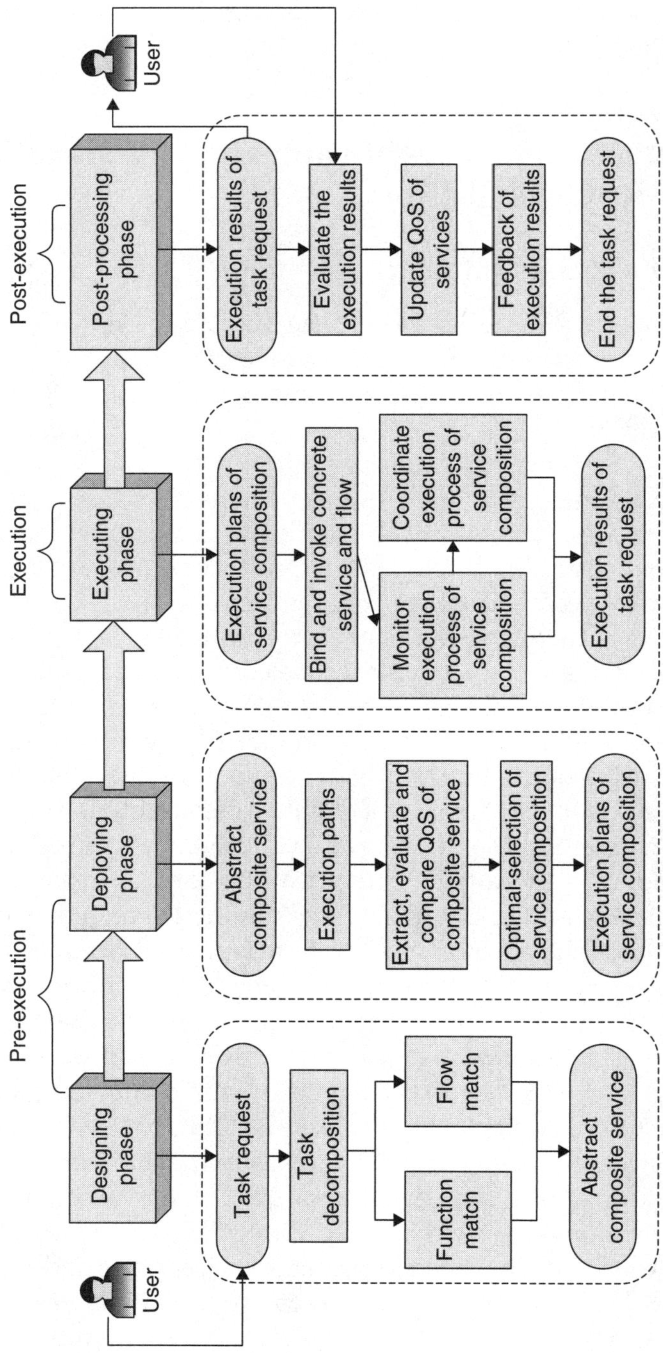

Figure 11.1 The life-cycle of RSC (Guo *et al.*, 2011).

the execution is influenced by the dynamic changes during RSC, RSC can continue to execute by coordination, thus the execution results of RSC are output.

4. Post-processing phase: In this phase, the execution results of RSC will be processed. After the execution results of RSC are submitted to users, users will evaluate the execution results. If the user is not satisfied with the execution results, RSC may be carried on again, or it needs to decide whether feedback to case base that RSC having been executed successfully. Meantime, after the execution of RSC, QoS of some resource services will have changed, so the QoS of these resource services need to be updated.

In the running process of a traditional manufacturing system, many "changing" or "uncertain" factors will happen. If the life-cycle of RSC is also regarded as a system, will there also exist many external or internal uncertain factors influencing the running of RSC? The factors which can influence the process of RSC in the pre-execution, execution, or post-execution of RSC will be analyzed as shown in Table 11.1 (Guo *et al.*, 2011).

In this chapter, task-related, resource service-related, QoS-related, and correlation-related factors will be considered.

11.3.2 The Definition and Classification of RSC Flexibility

As previously analyzed, the process of RSC is dynamic, so the whole RSC should have intelligence and a dynamic reconfiguration ability to adapt to these dynamic changes. In view of this, the definition of RSC flexibility is presented as follows (Guo *et al.*, 2011).

Definition 11.1: The RSC flexibility is defined as the ability that exists in the life-cycle of RSC, that when RSC receives the external force (e.g., network overload) or internal force (e.g., resource service exited), the resulting expected or unexpected dynamic changes can be responded to and the flow of RSC can be autonomously dynamic reconfigured to complete the tasks.

According to the different effects on the life-cycle of RSC, the RSC flexibility can be classified into five types as seen in Table 11.2 (Guo *et al.*, 2011), where their corresponding definitions are presented as well.

Table 11.1 Major dynamic changing factors and examples in the life-cycle of RSC

		Pre-execution of RSC	**Execution of RSC**	**Post-execution of RSC**
task-related factor	the change of tasks request	In view of certain changes of requests, some parameters of tasks are needed to be reset.	If users want to adjust some parameters of tasks, the execution of RSC may be stopped to reset the parameters of tasks.	
	the hang of tasks	The matching resource services cannot be found because QoS of users request is too high, or there are not appropriate flows of RSC to be executed because the tasks that users submit are too special.	Tasks may be hung due to performance of system, etc.	
	the reset of tasks	Before the execution of RSC, if users are not satisfied with the all execution plans, tasks will be reset to obtain better plans.		After the execution of RSC, if users are not satisfied with the results, tasks will be reset to restart RSC to obtain more satisfied results.

	the cancel of tasks	Before the execution of RSC, if users don't need to execute a certain task due to some reasons, they may cancel it.	If the execution time of RSC is too long, which exceeds the anticipation of users, the users may cancel the original task.	
resource service-related factor	the adding of resource services	The adding of new resource service will influence QoS of RSC, so users may redeploy RSC to obtain better RSC plan due to specific performance preferences.	The same as pre-execution of RSC.	
	the exit of resource services	Some resource services may exit due to the fault of machine and human etc., which will influence the design and deployment of RSC.	If there is resource service exiting from RSC, then the RSC cannot continue to be executed.	
	the overloading of resource services		Because all the resource services are sharing, if the resource service that RSC needs to invoke just being invoked by other RSC, then RSC will not continue to be executed.	

(Continued)

Table 11.1 (cont.) Major dynamic changing factors and examples in the life-cycle of RSC

		Pre-execution of RSC	Execution of RSC	Post-execution of RSC
QoS-related factor	the changes of QoS of resource services	If a certain or some properties of QoS of resource services changes, such as time adding, cost increasing, etc., the results of optimal-selection of RSC will be influenced.	The same as pre-execution of RSC.	
Correlation-related factor	the mismatch or changes of composable correlation		Because the complexity of logical relation, it cannot make sure that two composite resource services just can match with each other. Users may change the input/output parameters of resource services at any moment, which will make the original composable correlation break down.	

	the change of business entity correlation	Cooperation correlation among business entities may change into competition correlation, and vice versa, which will influence QoS of RSC.	The same as pre-execution of RSC.	
	the change of statistical cooperate correlation	In a certain period of time, the number of two or more resource services cooperating together is uncertain, so statistical cooperate correlation will change over time.	The same as pre-execution of RSC.	
other external force	transmission of data fail	Network bandwidth insufficiency, transmission line accrue fault, etc.	The same as pre-execution of RSC.	The same as pre-execution of RSC.
	access right limited	Not all resource services can be accessed by all users, some only open to special user.	The same as pre-execution of RSC.	The same as pre-execution of RSC.
	......	......	......	......

Table 11.2 The Types of RSC Flexibility

Type	Definition	Example
Task flexibility	The ability of constructing RSC to execute different tasks.	The requirement of tasks is changed, the parameter of tasks is modified, etc.
Flow flexibility	The ability of different flow of RSC to execute certain task.	The task does not change, but the flow of RSC selected to execute the task can be not used.
Resource service flexibility	The ability of single resource service to execute different tasks and the ease with which resource service can be substituted.	The resource services exits or overloads, which make resource services unavailable.
QoS flexibility	The ability of RSC to maintain the QoS performance level.	The QoS properties of the resource services changed, such as executing time increasing, executing cost enhancing, etc.
Correlation flexibility	The ability of RSC to adapt to the changes of correlations among resource services.	The correlations among resource services changed, which affects the QoS of RSC.

These five types of flexibility will be explained specifically in the next section as well as their measurement methods.

11.4 The Measurement of RSC Flexibility

Considering that there exist a lot of dynamic changes during RSC, if the task is allocated to RSC with more flexibility the task will be more easily executed successfully. The better the RSC flexibility is, the higher the ability of RSC is. Therefore, it is necessary for the optimal-selection of RSC to establish a perfect evaluation system of RSC flexibility to evaluate the potential ability of RSC to adapt to the changes. This evaluation system is also needed in order to find appropriate methods to measure the five types of RSC flexibility as defined in Table 11.2 (Guo *et al.*, 2011).

It has introduced in Section 11.2, that many scholars have used the entropy approach to measure the flexibility of FMS. First, whether the entropy approach can be applied to the measurement of RSC flexibility is analyzed.

The entropy approach was originally proposed by Shannon (1948) in information theory, and now it has been applied to the measurement of FMS successfully. The entropy in information theory can be defined as follows: if there exist a few events $S = \{e_1, e_2, \cdots, e_n\}$ in a system S, and the probability distribution of each event is $P = \{p_1, p_2, \cdots, p_n\}$, then the entropy function of the whole system is (Guo *et al.*, 2011):

$$S(p_1, p_2, \cdots, p_n) = -\sum_{i=1}^{n} p_i \log p_i \tag{11.1}$$

Obviously, the more the uncertainty of variable is, the higher the entropy is. The more chaotic a system is, the higher the entropy is.

It can be seen that Shannon (1984) thought entropy could be taken as a measurement of the uncertainty of a system. The more the uncertainty of the system is, the higher the entropy is. The life-cycle of RSC also can be regarded as a system. Many new flows of RSC may emerge in the process of dynamic reconfiguration of RSC. The more the uncertainty of the system is, the greater the RSC flexibility is. Hence, the RSC flexibility also can be measured by entropy.

Following are several methods (e.g., entropy, etc.) that will be used to analyze and measure the five types of RSC flexibility.

11.4.1 Task Flexibility

Task flexibility refers to the ability of constructing RSC to execute different tasks. The motivation for task flexibility lies in the fact that during the process of resource service composition, the changes of task requirement task hanged, and task canceled, always occur at any moment. When these task-related factors change, the original task will be replaced by the new task. If RSC can be constructed to execute the new different task, it indicates that RSC has the task flexibility.

Task flexibility makes dynamic reconfiguration of RSC be as quick as possible when users change the demand of the task, cancel the task, etc., so that the completing time of the task can be shortened.

Since users can change tasks arbitrarily without worrying about the consequences resulting from task changing, the satisfaction of users can be improved.

In view of the above analysis, the following two attributes of task flexibility should be considered and measured (Guo *et al.*, 2011):

Flexibility 1: According to the total number of tasks that RSC can execute. The more tasks a RSC can execute, the higher probability for the RSC been invoked to execute a new tasks (the new task is changed from its assigned task) when its assigned task is changed, which means the greater flexibility of the RSC is.

Flexibility 2: According to the difference between the number of the task requested by the user and the actual number of the task a RSC completed. The less the difference is, the higher probability that the RSC can execute a new task (the new task is dynamically changed from its assigned task) when its assigned task is dynamically changed, which means the greater flexibility of the RSC is.

The mathematical model of measuring task flexibility is as follows:

In order to measure the RSC task flexibility, take the h same sampling time segments from t_1 to t_h. It is assumed that the flow of RSC set constructed by resource services set is $C = \{C_1, C_2, \cdots, C_l\}$ in sampling time segments, where the number of flows of RSC is l. Meantime, Let the number of execution tasks of each flow of RSC in each time segment be $N_{T_{ki}}$, where k denotes the k *th* flow of RSC, i denotes the i th time segment, which is as follows (Guo *et al.*, 2011):

$$
\begin{array}{l}
\quad\;\; t_1 \quad\; t_2 \quad \cdots \quad t_h \\
C_1 : N_{T_{11}}, N_{T_{12}}, \cdots, N_{T_{1h}} \\
C_2 : N_{T_{21}}, N_{T_{22}}, \cdots, N_{T_{2h}} \\
\cdots \\
C_l : N_{T_{l1}}, N_{T_{l2}}, \cdots, N_{T_{lh}}
\end{array}
$$

For Flexibility 1, the number of tasks executed by C_k is

$$\bar{N}_{T_k} = \frac{1}{h} \sum_{i=1}^{h} N_{T_{ki}} \tag{11.2}$$

Normalize (11.2) as follows:

$$\bar{N}'_{T_k} = \frac{\bar{N}_{T_k}}{\sum_{k=1}^{l} \bar{N}_{T_k}} \tag{11.3}$$

Because the more the number of tasks that C_k can execute, the greater the RSC task flexibility is. Since $0 \le N'_{T_k} \le 1$, use entropy to reflect the average number of tasks that flows of RSC can execute as follows:

$$T_{f_1} = -\sum_{k=1}^{l} e^{-\bar{N}'_{T_k}} \log e^{-\bar{N}'_{T_k}} \tag{11.4}$$

For Flexibility 2, in h sample time segment, let the number of successful execution tasks of each $C_k (k = 1, \cdots, l)$ in each time segment $i (i = 1, \cdots, h)$ be $N'_{T_{ki}}$, which are as follows:

$$\begin{array}{ll} \quad\;\; t_1 \quad\; t_2 \quad \cdots \quad t_h & \quad\;\; t_1 \quad\; t_2 \quad \cdots \quad t_h \\ C_1 : N_{T_{11}}, N_{T_{12}}, \cdots, N_{T_{1h}} & C_1 : N'_{T_{11}}, N'_{T_{12}}, \cdots, N'_{T_{1h}} \\ C_2 : N_{T_{21}}, N_{T_{22}}, \cdots, N_{T_{2h}} & C_2 : N'_{T_{21}}, N'_{T_{22}}, \cdots, N'_{T_{2h}} \\ \cdots & \cdots \\ C_l : N_{T_{l1}}, N_{T_{l2}}, \cdots, N_{T_{lh}} & C_l : N'_{T_{l1}}, N'_{T_{l2}}, \cdots, N'_{T_{lh}} \end{array}$$

For C_k, define vectors $x_{T_k} = \{N_{T_{k1}}, N_{T_{k2}}, \ldots, N_{T_{kh}}\}$ and $x'_{T_k} = \{N'_{T_{k1}}, N'_{T_{k2}}, \ldots, N'_{T_{kh}}\}$, and let $D_{T_k} = \left\| x_{T_k} - x'_{T_k} \right\|$ denote the difference between the number of task request and the actual number of task RSC completed. Normalize D_{T_k} as follows:

$$D'_{T_k} = \frac{\left\| x_{T_k} - x'_{T_k} \right\|}{\sum_{k=1}^{l} \left\| x_{T_k} - x'_{T_k} \right\|} \tag{11.5}$$

Use entropy to reflect the average difference between the number of task request and the actual number of task RSC completed in the sample time is shown as follows:

$$T_{f_2} = -\sum_{k=1}^{l} D'_{T_k} \log D'_{T_k} \tag{11.6}$$

By (11.4) and (11.6), the mathematical model of measuring RSC task flexibility is as follows (Guo *et al.*, 2011):

$$T_f = \alpha_1 T_{f_1} + \alpha_2 T_{f_2} \tag{11.7}$$

where $\alpha_1, \alpha_2 \in [0,1]$ are the weight, and $\alpha_1 + \alpha_2 = 1$.

E.g. 11.4.1 Compute task flexibility of RSC shown in Figure 11.1.

It is assumed that there are five elementary resource services *OMS, PS, SSS, EPS, DAS* in the flow of *"OPAP"*, and the number of candidate resource services that each elementary resource service has is 1, 1, 2, 2, 2, respectively (see Figure 11.2).

These candidate resource services can be composed of 8 flows of RSC, i.e. $l = 8$. Sample one time every two days, and the same sampling time segment is taken as $h = 5$. The number of tasks each flow of RSC can execute in sampling time segment is as follows:

	t_1	t_2	t_3	t_4	t_5		t_1	t_2	t_3	t_4	t_5
C_1:	2,	1,	3,	2,	2	C_5:	1,	3,	4,	0,	2
C_2:	1,	4,	1,	2,	1	C_6:	1,	1,	2,	3,	4
C_3:	2,	2,	1,	3,	1,	C_7:	3,	0,	3,	3,	1
C_4:	3,	2,	4,	1,	2	C_8:	2,	0,	4,	1,	4

For Flexibility 1, the number of tasks the 8 flows of RSC can execute in a certain time segment is 2, 1.8, 1.8, 2.4, 2, 2.2, 2, 2.2, respectively by (11.2). According to (11.3), the normalizing value is 0.1220, 0.1098, 0.1098, 0.1463, 0.1220, 0.1341, 0.1220, 0.1341, respectively. By (11.4), in this time segment, the average number of task flows of RSC can execute is $T_{f_1} = 0.7644$.

It is assumed that in sampling time segment, the number of tasks that each flow of RSC can execute successfully is as follows:

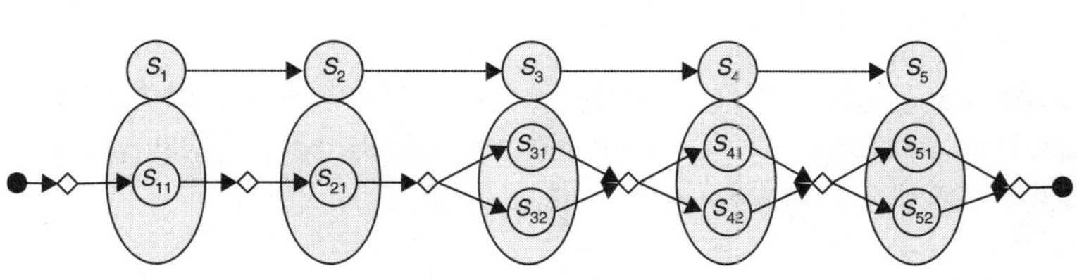

Figure 11.2 Sequence Composite Resource Service Process.

	t_1	t_2	t_3	t_4	t_5		t_1	t_2	t_3	t_4	t_5
C_1 :	2,	1,	2,	1,	1	C_5 :	0,	3,	4,	0,	1
C_2 :	1,	2,	0,	2,	0 ,	C_6 :	1,	1,	2,	3,	3
C_3 :	1,	2,	1,	1,	1	C_7 :	2,	0,	3,	3,	1
C_4 :	3,	0,	4,	0,	1	C_8 :	2,	0,	4,	1,	1

For Flexibility 2, through the above two groups of data, the difference value D_{T_k} of 8 flows of RSC is $\sqrt{3}$, $\sqrt{6}$, $\sqrt{5}$, $\sqrt{6}$, $\sqrt{2}$, 1, 1, 3, respectively, then the value of D'_{T_k} is 0.1133, 0.1603, 0.1463, 0.1603, 0.0925, 0.0654, 0.0654, 0.1963, respectively. Hence, the average difference of number of tasks completed by flows of RSC actually is $T_{f_2} = 2.0114$. Let the weight be $a_1 = 0.7, a_2 = 0.3$, by Flexibility 1 and Flexibility 2, the measurement value of task flexibility is $T_f = 1.1385$.

11.4.2 Resource Service Flexibility

Resource service flexibility refers to the ability of a single resource service to execute different tasks, and the ease with which resource service can be substituted. When the resource services providers submit the resource services to the resource service registry, these resource services may be invoked and composed to complete the users' different tasks. During the RSC, when a certain resource service exits, overloads, etc., or some new resource services with higher QoS added in that makes the users want to replace the resource service, then this resource service needs to be coped with. If the flow of RSC can be reconfigured quickly to deal with this case, it means RSC has resource service flexibility

Several attributes involved in the measurement of RSC flexibility will be analyzed as follows:

When the resource service is unavailable, it is an effective and feasible method to substitute it with other resource services or composite resource services if RSC is required to be reconfigured as soon as possible. There are two schemas for the substitution: single resource service or composite resource services in accordance with certain logical relation. As is shown in Figure 11.3., when resource service RS_3 in original flow is unavailable for some reason, it can be substituted with resource service RS_9 in plan 1, or composite resource services $\{RS_{10}, RS_{11}, RS_{12}, RS_{13}\}$ in plan 2.

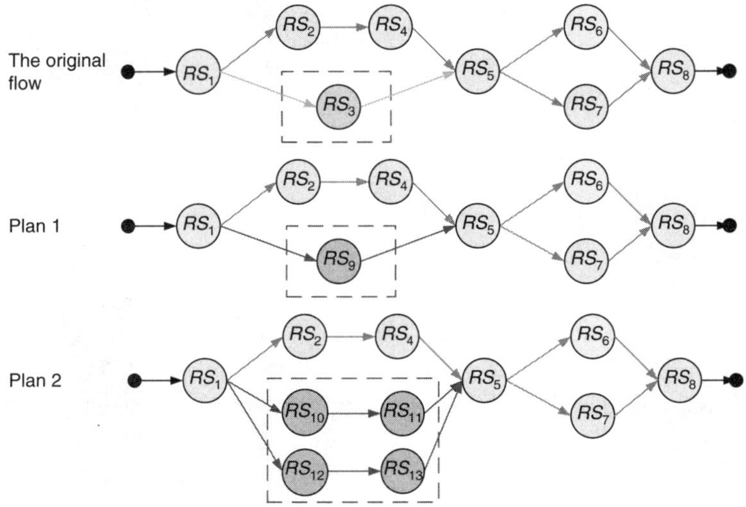

Figure 11.3 Substituting plan of resource service.

For an arbitrary resource service RSi, if it can be invoked to execute more tasks than the other resource services, then when a certain resource (e.g., RSj) in a RSC has some problem and cannot execute the assigned task, then the probability for RSi been selected to replace RSj will be greater than the other resource services. Apparently the probability for the task to be executed successfully will be higher.

In addition, the change of QoS in the flow of RSC also needs to be considered. If QoS of the flow of RSC after substituting the resource service is the same as prior to the QoS of the original flow of RSC due to the addition of resource service, then this kind of replacement is beneficial for users. However, if QoS of the substituting flow of RSC has a certain loss due to resource service exiting, overloading, etc., it is also a point that needs to be considered in order for the measurement of RSC flexibility to maintain the level of QoS of the substituting flow of RSC.

In view of previous analysis, the following three attributes of resource service flexibility should be considered and measured (Guo *et al.*, 2011):

Flexibility 3: The ability of resource services being substituted with other resource service or composite resource services.

Flexibility 4: The ability of resource services executing different tasks.

Flexibility 5: The loss degree of QoS of RSC after resource service (including single resource service or composite resource services) is substituted.

The mathematical model of measuring resource service flexibility is shown below.

For Flexibility 3, the following matrix *Rf* is defined:

$$Rf = \begin{pmatrix} Rf_{11} \; Rf_{12} \cdots Rf_{1n} \\ Rf_{21} \; Rf_{22} \cdots Rf_{2n} \\ \cdots \\ Rf_{n1} \; Rf_{n2} \cdots Rf_{nn} \end{pmatrix} \tag{11.8}$$

where the diagonal elements $Rf_{ij}(i = j)$ is the number that resource service can be substituted with composite resource services, and the non-diagonal elements $Rf_{ij}(i \neq j)$ denotes the ability that resource services substitute with each other, that is the total number that a resource service can be substituted by another resource service. The higher the sum of elements in matrix *Rf* is, the greater the RSC flexibility is.

However, it is not sensitive to use the sum of all the elements of the matrix, because several different matrices can have the same sum of the elements (Iravani *et al.*, 2005). Therefore, the ability to substitute resource service or composite resource services during RSC can be expressed as:

$$S_{f_1} = tan\left(1 - \frac{1}{e^{sum(Rf)}}\right) \tag{11.9}$$

where *sum* (*Rf*) denotes the sum of all elements in matrix *Rf*. When *sum* (*Rf*) increases, the value of S_{f_1} also increases.

For Flexibility 4, it is assumed that there exists resource services set $S = \{S_1, S_2, \ldots, S_n\}$, where *n* is the number of resource services, the tasks set they can deal with in a certain time is $T = \{T_1, T_2, \ldots, T_m\}$, and the elements in matrix $NSf = (NSf_{ij})_{n \times m}$ is as follows:

$$NSf_{ij} = \begin{cases} 1, \textit{the ith resource service can execute the jth task} \\ 0, \textit{otherwise} \end{cases}$$

The same as in Flexibility 3, the ability that a single resource service can execute the task can be reflected as follows:

$$S_{f_2} = tan\left(1 - \frac{1}{e^{sum(NSf)}}\right) \tag{11.10}$$

where *sum* (*NSf*) denotes the sum of all elements in matrix *NSf*. When *sum* (*NSf*) increases, the value of S_{f_2} also increases.

For Flexibility 5, the measurement method will be discussed in the next section, and here ΔQoS_s is used to denote the loss degree of QoS of RSC after resource services (including single resource service or composite resource services) are substituted.

By Flexibility 3, Flexibility 4, and Flexibility 5, the mathematical model of measuring the RSC resource service flexibility is as follows:

$$S_f = \delta_1 S_{f_1} + \delta_2 S_{f_2} + \delta_3 \Delta QoS_s \tag{11.11}$$

where $\delta_1, \delta_2, \delta_3 \in [0,1]$ are the weight, and $\delta_1 + \delta_2 + \delta_3 = 1$.

11.4.3 QoS Flexibility

QoS flexibility refers to the ability of RSC to maintain the QoS performance level. The motivation for QoS flexibility lies in two aspects. First, QoS of resource service is not static and its changes will lead to the change of resource service, For example, if a certain resource service is invoked frequently, the provider of the resource service may increase the price of this resource service to obtain more benefits, which will decrease the QoS of resource service for the user to invoke it. Second, QoS of the flow of RSC may also change. For example, when flow of RSC is unavailable, there is a difference between QoS of the new substituting flow of RSC and that of the original flow of RSC.

In either case, in order to ensure the user's satisfaction with the execution result of the task, RSC is needed to maintain QoS of new flow of RSC at the same or higher level. In this chapter, the difference between QoS of new flow of RSC and the original flow is defined as QoS loss of RSC.

The mathematical model of measuring QoS flexibility is as follows (Guo *et al.*, 2011):

Flexibility 6: The ability for RSC to maintain the QoS performance level after the flow of RSC is substituted.

First, according to the computing method of QoS of RSC in Tao *et al.*, (2008c, 2010e), for each task in task set $T = \{T_1, T_2, \ldots, T_m\}$, QoS of all flows of RSC that can complete the task can be computed as follows:

$$T_1 : QoST_{11}, QoST_{12}, \cdots QoST_{1l_1}$$

$$T_2 : QoST_{21}, QoST_{22}, \cdots QoST_{2l_2}$$

$$\ldots$$

$$T_m : QoST_{m1}, QoST_{m2}, \cdots QoST_{mlm}$$

where $l_1, l_2, \ldots, l_m$ denote the number of different QoS of flows of RSC that can complete task T_j. When task T_j is completed, the biggest QoS loss of flow of RSC is defined as:

$$LossT_j = \frac{\max_{h=1}^{l_j}\{QoST_{jh}\} - \min_{h=1}^{l_j}\{QoST_{jh}\}}{\max_{h=1}^{l_j}\{QoST_{jh}\}} \tag{11.12}$$

where $\max_{h=1}^{l_j}\{QoST_{jh}\}$ and $\min_{h=1}^{l_j}\{QoST_{jh}\}$ denote respectively the maximum and minimum value of QoS of flow of RSC that can complete task T_j. By (11.12), $0 \le LossT_j \le 1$, so the ability that RSC maintains the QoS performance level after the flow of RSC is substituted is defined as follows:

$$\Delta QoS_C = -\sum_{j=1}^{m} LossT_j \log(LossT_j) \tag{11.13}$$

Flexibility 7: The ability that RSC maintains the QoS performance level after resource services is substituted.

It is assumed that the loss of QoS of flow of RSC including resource service S_i after S_i is substituted with resource service or composite resource services when task T_j is executed is as follows:

$$\begin{array}{l} \quad S_1 \qquad\quad S_2 \qquad \cdots \quad S_n \\ S_i : \Delta QoST_{i1}, \Delta QoST_{i2}, \cdots, \Delta QoST_{in}, \\ \quad CS_1 \qquad\quad CS_2 \qquad \cdots \quad CS_q \\ \Delta QoSCST_{i1}, \Delta QoSCST_{i2}, \cdots, \Delta QoSCST_{ip} \end{array}$$

where p is the number that resource service S_i can be substituted. It is assumed that users must select optimal RSC plan to execute, and let the total loss of QoS of flow of RSC including resource service S_i after S_i is substituted with resource service or composite resource services when task T_j is executed is as follows:

$$\Delta QoS(ST)_{ij} = \sum_{k=1}^{n} \Delta QoST_{ik} + \sum_{l=1}^{q} \Delta QoSCST_{il} \tag{11.14}$$

then the loss of QoS of flow of RSC including resource service S_i after S_i is substituted when task set T is executed is as follows:

$$\overline{\Delta QoS(ST)_i} = \sum_{j=1}^{m} \Delta QoS(ST)_{ij} \tag{11.15}$$

Normalize (11.15) as follows:

$$\Delta\overline{QoS(ST)_i}' = \frac{\Delta\overline{QoS(ST)_i}}{\sum_{i=1}^{n} \left|\Delta QoS(ST)_i\right|} \tag{11.16}$$

then the ability that RSC maintains the QoS performance level after resource services is substituted is as follows:

$$\Delta QoS_s = -\sum_{i=1}^{n} e^{\Delta\overline{QoS(ST)_i}'} \log e^{\Delta\overline{QoS(ST)_i}'} \tag{11.17}$$

By Flexibility 6 and Flexibility 7, the mathematical model of measuring RSC QoS flexibility is defined as follows:

$$\Delta QoS_f = \beta_1 \Delta QoS_C + \beta_2 \Delta QoS_s \tag{11.18}$$

where $\beta_1, \beta_2 \in [0,1]$ is the weight and $\beta_1 + \beta_2 = 1$.

E.g. 11.4.2 Compute QoS flexibility of RSC shown in Figure 9.1.

According to the RSC shown in Figure 11.2 in E.g. 11.4.1, there are 8 resource services and 8 flows. It is assumed that in a certain sampling time, there are 4 tasks being executed successfully, i.e. $m = 4$. Here, only time and price of QoS properties are considered, and the default of QoS of 8 candidate resource services in Figure 11.2 is normalized and shown in Table 11.3.

Table 11.3 The default of QoS properties of candidate resource Services

Resource Service	S_1	S_2	S_3		S_4		S_5	
QoS	S_{11}	S_{21}	S_{31}	S_{32}	S_{41}	S_{42}	S_{51}	S_{52}
Time	0.6	0.3	0.8	0.3	0.3	0.6	0.3	0.5
Cost	0.9	0.5	0.3	0.5	0.3	0.4	0.5	0.3

It is assumed that in the sampling time, flows of RSC that can execute the 4 tasks are respectively as follows:

$$T_1 : s_{11} \rightarrow s_{21} \rightarrow s_{31} \rightarrow s_{41} \rightarrow s_{51}, \quad s_{11} \rightarrow s_{21} \rightarrow s_{31} \rightarrow s_{42} \rightarrow s_{51}$$
$$T_2 : s_{11} \rightarrow s_{21} \rightarrow s_{32} \rightarrow s_{41} \rightarrow s_{51}, \quad s_{11} \rightarrow s_{21} \rightarrow s_{32} \rightarrow s_{42} \rightarrow s_{52}$$
$$T_3 : s_{11} \rightarrow s_{21} \rightarrow s_{31} \rightarrow s_{42} \rightarrow s_{52}, \quad s_{11} \rightarrow s_{21} \rightarrow s_{32} \rightarrow s_{42} \rightarrow s_{51}$$
$$T_4 : s_{11} \rightarrow s_{21} \rightarrow s_{31} \rightarrow s_{41} \rightarrow s_{52}, \quad s_{11} \rightarrow s_{21} \rightarrow s_{32} \rightarrow s_{41} \rightarrow s_{52}$$

where "→" denotes the control flow, then according to the computing methods of QoS of RSC in Tao *et al.*, (2010e):

$$Q_{time} = \sum_{i=1}^{5} q^i_{time}, \; Q_{cost} = \sum_{i=1}^{5} q^i_{cost}, F(Q) = \xi_1 Q_{time} + \xi_2 Q_{cost}$$

the weight of time and cost are $\xi_1 = 0.5$ a $\xi_2 = 0.5$, then QoS of flow of RSC that the 4 tasks correspond to is as follows:

$$T_1 : s_{11} \rightarrow s_{21} \rightarrow s_{31} \rightarrow s_{41} \rightarrow s_{51}, (QoS : 2.4)$$
$$s_{11} \rightarrow s_{21} \rightarrow s_{31} \rightarrow s_{42} \rightarrow s_{51} (QoS : 2.7)$$
$$T_2 : s_{11} \rightarrow s_{21} \rightarrow s_{32} \rightarrow s_{41} \rightarrow s_{51}, (QoS : 2.4)$$
$$s_{11} \rightarrow s_{21} \rightarrow s_{32} \rightarrow s_{42} \rightarrow s_{52} (QoS : 2.45)$$
$$T_3 : s_{11} \rightarrow s_{21} \rightarrow s_{31} \rightarrow s_{42} \rightarrow s_{52}, (QoS : 2.6)$$
$$s_{11} \rightarrow s_{21} \rightarrow s_{32} \rightarrow s_{42} \rightarrow s_{51} (QoS : 2.45)$$
$$T_4 : s_{11} \rightarrow s_{21} \rightarrow s_{31} \rightarrow s_{41} \rightarrow s_{52}, (QoS : 2.4)$$
$$s_{11} \rightarrow s_{21} \rightarrow s_{32} \rightarrow s_{41} \rightarrow s_{52} (QoS : 2.35)$$

By (11.12), when tasks $T = \{T_1, T_2, T_3, T_4\}$ are completed, the biggest loss of QoS of flow of RSC is 0.1111, 0.0204, 0.0577, 0.0208, respectively: By (11.13), the ability that RSC maintains the QoS performance level after the flow of RSC is substituted is $\Delta QoS_C = 0.5687$.

It can be seen from Figure 11.2 that the candidate resource services can be substituted with each other is S_{31} and S_{32}, S_{41} and S_{42}, as well as S_{51} and S_{52}. In this example, there are no resource services substituted with composites resource services, so it is assumed that the loss of QoS of flow of RSC substituted with composites resource services is 0. By (11.14) and (11.15), the loss of QoS after resource services S_{11}, S_{21}, S_{31}, S_{32}, S_{41}, S_{42}, S_{51}, S_{52} are substituted when executing the task set $T = \{T_1, T_2, T_3, T_4\}$ is 0,0,–0.2,0,0,–0.35,0,–0.2, respectively. By (11.16) and (11.17), the ability that RSC maintains the QoS performance level after resource service is substituted is $\Delta QoS_s = 0.7011$. Let the weight be $\beta_1 = 0.5$ and $\beta_2 = 0.5$, then the measurement value of QoS flexibility is $\Delta QoS_f = 0.6349$.

E.g. 11.4.3 Compute resource service flexibility of RSC shown in Figure 9.1. Consider RSC in Figure 11.2., by (11.8), it can be obtained that

$$Rf = \begin{pmatrix} 0 & 0 & 0 & 0 & 0 & 0 & 0 & 0 \\ 0 & 0 & 0 & 0 & 0 & 0 & 0 & 0 \\ 0 & 0 & 0 & 1 & 0 & 0 & 0 & 0 \\ 0 & 0 & 1 & 0 & 0 & 0 & 0 & 0 \\ 0 & 0 & 0 & 0 & 0 & 1 & 0 & 0 \\ 0 & 0 & 0 & 0 & 1 & 0 & 0 & 0 \\ 0 & 0 & 0 & 0 & 0 & 0 & 0 & 1 \\ 0 & 0 & 0 & 0 & 0 & 0 & 1 & 0 \end{pmatrix}$$

By (11.9), one has $S_{f_1} = 1.5489$.

According to the task set and flow of RSC defined in E.g. 11.4.2,

$$NSf = \begin{pmatrix} 1 & 1 & 1 & 1 \\ 1 & 1 & 1 & 1 \\ 1 & 0 & 1 & 1 \\ 0 & 1 & 1 & 1 \\ 1 & 1 & 0 & 1 \\ 1 & 1 & 1 & 0 \\ 1 & 1 & 1 & 0 \\ 0 & 1 & 1 & 1 \end{pmatrix}$$

By (11.10), one has $S_{f_2} = 1.5574$.

By (11.11) and ΔQoS_s obtained in E.g. 11.4.2, let $\delta_1 = \delta_2 = 0.2$, $\delta_3 = 0.6$, the measurement value of resource service flexibility is $S_f = 0.8898$.

11.4.4 Correlation Flexibility

In the actual RSC, there generally exist correlations influencing each other among composite resource services, which can affect QoS of RSC. In Chapter 9, the correlation between two elementary resource services is defined as composable correlation (CoC), business entity correlation (BuC), and statistical cooperate correlation (StC).

The existence of the three kinds of correlation not only influences QoS of RSC, but also they are not invariable. When the correlations among resource service change, the QoS of RSC also changes. If the change of a certain correlation is beneficial to QoS of RSC, the user may consider substituting the original executing resource service with resource service having this kind of correlation. But if the correlations of the executing resource service become harmful to QoS of RSC, the user may search for other resource service to replace this resource service to enhance the QoS of RSC. In order to avoid the resulting waste of time, cost, and energy due to the change of correlations among resource services, it is necessary to introduce the correlation flexibility that can make RSC adapt to the change of correlations.

Which aspects of correlation will be analyzed and the mathematical model of measuring correlation flexibility are presented as follows (Guo *et al.*, 2011):

Flexibility 8: The measurement of correlation flexibility from the global view is considered. This is because when composable correlation, business entity correlation, or statistical cooperate correlation change, all the results will be affecting QoS of flow of RSC. The more the number of the correlation among composite resource services is, when the correlations among composite resource services changes, the more the number of the resource service being used to substitute other resource services is, and the smaller the loss degree of QoS of RSC. Therefore, correlation flexibility should be measured by the number of correlations among resource services and the effects (positive or negative) on QoS of RSC.

Let set G denotes the information of resource service with correlation, i.e., all the resource services with a certain correlation are taken as an array to be stored in G. For example, if resource services $\{RS_1, RS_2, RS_3\}$ have business entity correlation and resource services $\{RS_4, RS_5\}$ have statistical cooperate correlation, then $G = \{\{RS_1, RS_2, RS_3\},\{RS_4, RS_5\}\}$.

Let $D_1 = (d_{ij}^1)_{n\times r}$ and $D_2 = (d_{ij}^2)_{n\times r}$, where r denotes the total number of array in G, and

$$d_{ij}^1 = \begin{cases} 1, \text{ if the } ith \text{ service is included in the } jth \text{ array of G} \\ 0, \text{ otherwise} \end{cases}$$

Because some resource services with correlations will enhance QoS, some will reduce QoS, and the correlation that can reduce QoS is harmful to RSC flexibility. Let

$$d_{ij}^2 = \begin{cases} 2, \text{ if the } ith \text{ service is included in the } jth \text{ array of G} \\ \quad \text{which is beneficial to QoS} \\ \frac{1}{2}, \text{ if the } ith \text{ service is included in the } jth \text{ array of G} \\ \quad \text{which is harmful to QoS} \\ 1, \text{ otherwise} \end{cases}$$

Let matrix $D = (d_{ij})_{n\times r}$ be:

$$D = D_1 \circ D_2 \tag{11.19}$$

Then the ability value of RS_i that has correlation with other resource services which has certain impact on QoS can be expressed as:

$$d_i = \sum_{j=1}^{r} d_{ij} \tag{11.20}$$

Normalize (11.20), then

$$d_i' = \frac{d_i}{\sum_{i=1}^{n} d_i} \tag{11.21}$$

Thus, the average information quantity of correlation among resource services is:

$$(C\,\&\,B\,\&\,S)_f = -\sum_{i=1}^{n} e^{-d_i'} \log e^{-d_i'} \tag{1.22}$$

Flexibility 9: Only after the number of correlations among resource service and the effect on QoS of RSC are considered in Flexibility 8, then the matching problem that resource service is substituted with other resource service can be considered., That is, whether the substituted resource service can satisfy input/output correlation with other resource services to compose the flow of RSC. In the meantime, according to the discussion in Section 11.3.1, when composable correlation occurs mismatched, e.g. the output of RS_i cannot satisfy the input of RS_j, another resource service is needed to be matched as the input of RS_j. Therefore, the number of resource services that can satisfy composable correlation should be also considered for the measurement of correlation flexibility. Moreover, the more the number of candidate resource services with composable correlation is, the greater the correlation flexibility is.

As shown in Figure 11.4., there are two kinds of composable correlation for RSC including four resource services $\{RS_1, RS_2, RS_3, RS_4\}$. In composable correlation (a), the output of RS_1 can be taken as the input of RS_3, and the output of RS_2 can be taken as the input of RS_4. In composable correlation (b), not only (a) is satisfied, but also the output of RS_1 can be taken as the input of RS_4. It can be seen that the RSC flexibility with composable correlation (b) is greater

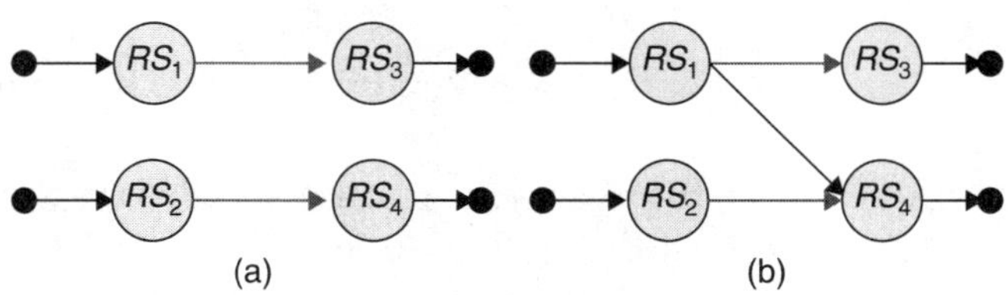

Figure 11.4 Two kinds of composable correlation.

than (a). That is because in a flow of RSC, if the composable correlation between RS_1 and RS_3 is mismatched due to some reason, it is possible for (b) to continue to execute the original flow of RSC through replacing RS_3 with RS_4, neither for (a).

The two matrices $CoCf^{in} = (CoCf_{ij}^{in})_{n\times l}$ and $CoCf^{out} = (CoCf_{ij}^{out})_{n\times l}$ are presented, where

$$CoCf_{ij}^{in} = \begin{cases} 1, \text{the } ith \text{ service is taken as the input of a certain} \\ \quad \text{service in the } jth \text{ flow of RSC} \\ 0, \text{otherwise} \end{cases}$$

$$CoCf_{ij}^{out} = \begin{cases} 1, \text{the } ith \text{ service is taken as the output of a certain} \\ \quad \text{service in the } jth \text{ flow of RSC} \\ 0, \text{otherwise} \end{cases}$$

The higher the sum of elements in two matrices is, the greater the composable correlation flexibility is. Similar to the discussion in Section 11.4.3, the measurement of composable correlation flexibility can be expressed as:

$$CoC_f = \tan\left(1 - \frac{1}{e^{sum(CoCf^{in})}}\right) + \tan\left(1 - \frac{1}{e^{sum(CoCf^{out})}}\right) \tag{11.23}$$

By (11.22) and (11.23), the mathematical model of measuring correlation RSC flexibility is defined as follows:

$$CC_f = \lambda_1 (C \,\&\, B \,\&\, S)_f + \lambda_2 CoC_f \tag{11.24}$$

where $\lambda_1, \lambda_2 \in [0,1]$ are the weight, and $\lambda_1 + \lambda_2 = 1$.

E.g. 11.4.4 Compute correlation flexibility of RSC shown in Figure 9.1.

Considering RSC in E.g. 11.4.1, it can be seen from the control logical correlation of 8 candidate resource services in RSC that the candidate resource services with composable correlation are S_{11} and S_{21}, S_{21} and S_{31}, S_{21} and S_{32}, S_{31} and S_{41}, S_{31} and S_{42}, S_{32} and S_{41}, S_{32} and S_{42}, S_{41} and S_{51}, S_{41} and S_{52}, S_{42} and S_{51}, S_{42} and S_{52} respectively.

It is assumed that S_{11} and S_{31}, S_{32} and S_{51}, S_{41} and S_{52} has business entity correlation respectively, and the correlation between S_{11} and S_{31} is positive to QoS of RSC, the correlation between S_{32} and S_{51}, S_{41} and S_{52} is negative to QoS of RSC.

It is assumed that S_{42} and S_{52}, S_{31} and S_{42} have statistical cooperate correlation respectively.

Therefore, $G=\{\{S_{11}, S_{21}\},\{S_{21}, S_{31}\},\{S_{21}, S_{32}\},\{S_{31}, S_{41}\}, \{S_{31}, S_{42}\},\{S_{32}, S_{41}\}, \{S_{32}, S_{42}\},\{S_{41}, S_{51}\},\{S_{41}, S_{52}\},\{S_{42}, S_{51}\},\{S_{42}, S_{52}\},\{S_{11}, S_{31}\},\{S_{32}, S_{51}\},\{S_{41}, S_{52}\}, \{S_{42}, S_{52}\},\{S_{31}, S_{42}\}\}$, and $r = 16$. For Flexibility 8, it can be obtained that

$$D=\begin{pmatrix}
1&0&0&0&0&0&0&0&0&0&0&1&0&0&0&0\\
0&1&0&0&0&0&0&0&0&0&0&0&0&0&0&0\\
0&1&0&1&1&0&0&0&0&0&1&0&0&0&0&1\\
0&0&1&0&0&1&1&0&0&0&0&0&1&0&0&0\\
0&0&0&1&0&1&0&1&1&0&0&0&0&1&0&0\\
0&0&0&0&1&0&1&0&0&1&1&0&0&0&1&1\\
0&0&0&0&0&0&0&1&0&1&0&0&1&0&0&0\\
0&0&0&0&0&0&0&0&1&0&1&0&0&1&1&0
\end{pmatrix}\circ\begin{pmatrix}
1&1&1&1&1&1&1&1&1&1&1&2&1&1&1&1\\
1&1&1&1&1&1&1&1&1&1&1&1&1&1&1&1\\
1&1&1&1&1&1&1&1&1&1&1&2&1&1&1&2\\
1&1&1&1&1&1&1&1&1&1&1&1&\frac{1}{2}&1&1&1\\
1&1&1&1&1&1&1&1&1&1&1&1&1&\frac{1}{2}&1&1\\
1&1&1&1&1&1&1&1&1&1&1&1&1&1&2&2\\
1&1&1&1&1&1&1&1&1&1&1&1&\frac{1}{2}&1&1&1\\
1&1&1&1&1&1&1&1&1&1&1&1&1&\frac{1}{2}&2&1
\end{pmatrix}$$

$$=\begin{pmatrix}
1&0&0&0&0&0&0&0&0&0&0&2&0&0&0&0\\
0&1&0&0&0&0&0&0&0&0&0&0&0&0&0&0\\
0&1&0&1&1&0&0&0&0&0&2&0&0&0&0&2\\
0&0&1&0&0&1&1&0&0&0&0&0&\frac{1}{2}&0&0&0\\
0&0&0&1&0&1&0&1&1&0&0&0&0&\frac{1}{2}&0&0\\
0&0&0&0&1&0&1&0&0&1&1&0&0&0&2&2\\
0&0&0&0&0&0&0&1&0&1&0&0&\frac{1}{2}&0&0&0\\
0&0&0&0&0&0&0&0&1&0&1&0&0&\frac{1}{2}&2&0
\end{pmatrix}$$

then the ability value of candidate resource services S_{11}, S_{21}, S_{31}, S_{32}, S_{41}, S_{42}, S_{51}, S_{52} with correlation is: 3,1,7,7/2,9/2,8,5/2,9/2. By (11.21) and (11.22), the average information quantity of correlation among resource services that RSC having is $(C \& B \& S)_f = 0.8560$.

For Flexibility 9, by the 8 flows of RSC listed in E.g. 11.4.2,

$$Flow1: s_{11} \to s_{21} \to s_{31} \to s_{41} \to s_{51},$$
$$Flow2: s_{11} \to s_{21} \to s_{31} \to s_{42} \to s_{51},$$
$$Flow3: s_{11} \to s_{21} \to s_{32} \to s_{41} \to s_{51},$$
$$Flow4: s_{11} \to s_{21} \to s_{32} \to s_{42} \to s_{52},$$
$$Flow5: s_{11} \to s_{21} \to s_{31} \to s_{42} \to s_{52},$$
$$Flow6: s_{11} \to s_{21} \to s_{32} \to s_{42} \to s_{51},$$
$$Flow7: s_{11} \to s_{21} \to s_{31} \to s_{41} \to s_{52},$$
$$Flow8: s_{11} \to s_{21} \to s_{32} \to s_{41} \to s_{52}$$

it can be obtained that

$$CoCf^{in} = \begin{pmatrix} 1 & 1 & 1 & 1 & 1 & 1 & 1 & 1 \\ 1 & 1 & 1 & 1 & 1 & 1 & 1 & 1 \\ 1 & 1 & 0 & 0 & 1 & 0 & 1 & 0 \\ 0 & 0 & 1 & 1 & 0 & 1 & 0 & 1 \\ 1 & 0 & 1 & 0 & 0 & 0 & 1 & 1 \\ 0 & 1 & 0 & 1 & 1 & 1 & 0 & 0 \\ 1 & 1 & 1 & 0 & 0 & 1 & 0 & 0 \\ 0 & 0 & 0 & 0 & 0 & 0 & 0 & 0 \end{pmatrix},$$

$$CoCf^{out} = \begin{pmatrix} 0 & 0 & 0 & 0 & 0 & 0 & 0 & 0 \\ 1 & 1 & 1 & 1 & 1 & 1 & 1 & 1 \\ 1 & 1 & 0 & 0 & 1 & 0 & 1 & 0 \\ 0 & 0 & 1 & 1 & 0 & 1 & 0 & 1 \\ 1 & 0 & 1 & 0 & 0 & 0 & 1 & 1 \\ 0 & 1 & 0 & 1 & 1 & 1 & 0 & 0 \\ 1 & 1 & 1 & 0 & 0 & 1 & 0 & 0 \\ 0 & 0 & 0 & 1 & 1 & 0 & 1 & 1 \end{pmatrix}$$

then the composable correlation flexibility is $CoC_f = 3.1148$ by (11.23).

Let $\lambda_1 = 0.7, \lambda_2 = 0.3$, then the measurement value of correlation flexibility is $CC_f = 1.5336$.

11.4.5 Flow Flexibility

Flow flexibility refers to the ability of different flow of RSC to execute a certain task. When the task that users submitted does not change, but the original flow of RSC that is selected to execute the task is unavailable due to some reason (e.g. a certain resource service is unavailable, network is overloaded), the different flow of RSC needs to be reconfigured to complete the task. In this case, if RSC can be reconfigured to the new flow of RSC, then RSC has flow flexibility.

Compared with the other types of flexibility in RSC, flow flexibility is a global view on RSC flexibility, while other types are all local view. When changes occur in the life-cycle of RSC, flow flexibility can make the reconfiguration of RSC easier. Not only the rate of success can be enhanced, but also time can be shortened and cost can be reduced, QoS of RSC can be maintained as well.

In order to measure the flow flexibility, which factors are needed to be measured is analyzed first.

The more candidate resource service flows available in a RSC, then the more candidate resource service flows could be selected for executing a task when certain part in the selected flow of the RSC occur abnormality while executing the task. It means that the grater flow flexibility the RSC has.

However, for certain tasks, more numbers of alternative flows does not go far enough. Though there are many alternative flows, if the difference among these alternative flows is not much, then these flows deal with the "uncertain" factors during RSC in a similar manner. When Chang (2007) measures routing flexibility of a manufacturing system, it is thought that the greater the difference between two routes, the greater the flexibility of the manufacturing system is. Therefore, the difference function is taken as the measuring criterion by comparing the difference of the number of machines in two routes during the manufacturing process. This idea can also be applied to the measurement of flow flexibility, that is, the more the differences between two flows is, the greater the flow flexibility is.

However, it is not completely the same when this idea is applied to the measurement of flow flexibility. This is because in different flows of RSC, not only are there differences for the resource services, but also there are differences for the control logical relation owing to the input/output logical relation among resource services. For example, in Figure 11.5 it is assumed that there are two flows of RSC that can execute the same task. The same resource services between two flows are $\{RS_1, RS_2, RS_5, RS_6\}$, and the different resource services between two flows are $\{RS_3, RS_4, RS_7, RS_8\}$. The same logical relations between two flows are denoted by a green arrowhead, and the different ones are denoted by a red arrowhead. Therefore, when comparing the difference between two flows, two aspects should be considered, i.e. the difference of resource service and the logical relation.

In addition, in order to ensure satisfaction with the execution results, if the flow of RSC needs to be substituted to execute the original task due to a certain reason, the degree to maintain the flow of QoS of RSC also needs to be considered.

In view of previous analysis, three attributes of flow flexibility should be considered and measured (Guo *et al.*, 2011):

Flexibility 10: The total number of flows that RSC owns.

Flexibility 11: The difference between flows of RSC, which mainly refers to the difference in resource services and the controlled logical relation between two flows.

Flexibility 12: The ability of RSC to maintain the QoS performance level after the flow of RSC is substituted.

In order to measure flow flexibility of RSC, in a similar manner to the measurement of task flexibility, first take h same sampling time

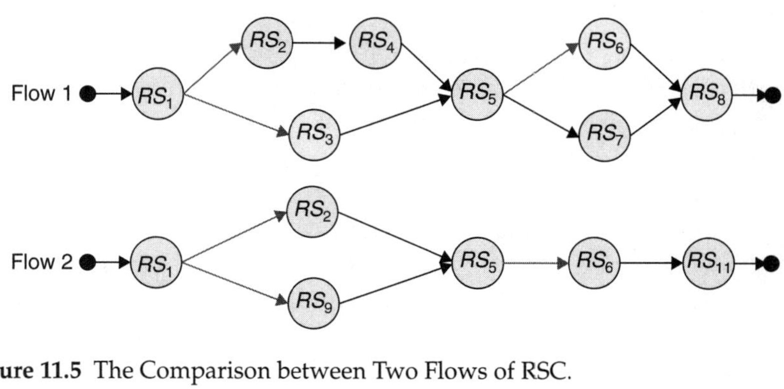

Figure 11.5 The Comparison between Two Flows of RSC.

segments from t_1 to t_h. It is assumed that the tasks set that flows of RSC can cope with is $T=\{T_1,T_2,\ldots,T_m\}$ in sampling time segments, where the number of tasks is m. In the meantime, record the number of flows of RSC to execute task $T_k(k=1,\ldots,m)$ in each time segment $i(i=1,\cdots,h)$ as $N_{C_{ki}}$, where k denotes the kth task, i denotes the ith time segment, as follows:

$$
\begin{array}{l}
\quad\quad\; t_1 \quad\;\; t_2 \quad \cdots \quad t_h \\
T_1 : N_{C_{11}}, N_{C_{12}}, \cdots, N_{C_{1h}} \\
T_2 : N_{C_{21}}, N_{C_{22}}, \cdots, N_{C_{2h}} \\
\cdots \\
T_m : N_{C_{l1}}, N_{C_{l2}}, \cdots, N_{C_{lh}}
\end{array}
$$

For Flexibility 10, the total number of flows of RSC that can execute task T_k in a certain time segment is:

$$\bar{N}_{C_k} = \sum_{i=1}^{h} N_{C_{ki}} \tag{11.25}$$

Normalize (11.25):

$$\bar{N}'_{C_k} = \frac{\bar{N}_{C_k}}{\sum_{k=1}^{m} \bar{N}_{C_k}} \tag{11.26}$$

Because the larger the number of flows of RSC that can execute task T_k is, the greater the flow flexibility of RSC is. Since $0 \le N'_{C_k} \le 1$, use entropy to reflect the average number of flows of RSC that RSC owns as follows:

$$C_{f_1} = -\sum_{k=1}^{m} e^{-\bar{N}'_{C_k}} \log e^{-\bar{N}'_{C_k}} \tag{11.27}$$

For Flexibility 11, it is assumed that in h sampling time segment, the number of the flows of RSC that can be used to execute tasks set

T is l, thus the difference function of resource services between flow C_i and flow C_j is:

$$d_{ij} = 1 - \frac{|SC_i \cap SC_j|}{|SC_i \cup SC_j|} \tag{11.28}$$

where $Sf_i \cap Sf_j$ denotes the set of the same resource services of flow C_i and flow C_j, $Sf_i \cup Sf_j$ denotes the set of all resource services included in flow C_i and flow C_j. The difference function of control logical relation between flow C_i and flow C_j is:

$$d'_{ij} = 1 - \frac{|CC_i \cap CC_j|}{|CC_i \cup CC_j|} \tag{11.29}$$

where $Cf_i \cap Cf_j$ denotes the set of the same control logical relation of flow C_i and flow C_j, $Cf_i \cup Cf_j$ denotes the set of all control logical relation included in flow C_i and flow C_j.

Because there are $C_l^2 = l(l-1)/2$ kinds of comparisons among flows of RSC, the total difference of flows of RSC is:

$$D_{C_f} = \frac{1}{l(l-1)} \sum_{i=1}^{l} \sum_{j=1}^{l} \sqrt{d_{ij} \cdot d'_{ij}} \tag{11.30}$$

For Flexibility 12, the measurement method is similar to (11.13) in Section 11.4.3.

By (11.13), (11.27) and (11.30), the mathematical model of measuring flow flexibility is:

$$C_f = \omega_1 C_{f_1} + \omega_2 D_{C_f} + \omega_3 \Delta QoS_C \tag{11.31}$$

where $\omega_1, \omega_2, \omega_3 \in [0,1]$ are the weight, and $\omega_1 + \omega_2 + \omega_3 = 1$.

E.g. 11.4.5 Compute flow flexibility of RSC shown in Figure 9.1. Take $h = 1$, $m = 4$, the 8 flows in E.g. 11.4.2 are:

$$Flow1: s_{11} \to s_{21} \to s_{31} \to s_{41} \to s_{51},$$
$$Flow2: s_{11} \to s_{21} \to s_{31} \to s_{42} \to s_{51},$$
$$Flow3: s_{11} \to s_{21} \to s_{32} \to s_{41} \to s_{51},$$
$$Flow4: s_{11} \to s_{21} \to s_{32} \to s_{42} \to s_{52},$$
$$Flow5: s_{11} \to s_{21} \to s_{31} \to s_{42} \to s_{52},$$
$$Flow6: s_{11} \to s_{21} \to s_{32} \to s_{42} \to s_{51},$$
$$Flow7: s_{11} \to s_{21} \to s_{31} \to s_{41} \to s_{52},$$
$$Flow8: s_{11} \to s_{21} \to s_{32} \to s_{41} \to s_{52},$$

Then in the sampling time segment, each of 4 tasks has two flows to execute.

For Flexibility 10, in this sampling time, by (11.26) and (11.27), the normalizing total number of flows of RSC to execute tasks T_1, T_2, T_3, T_4, is $1/4, 1/4, 1/4, 1/4$, respectively. Hence, use the entropy to reflect the average number of flows of RSC that RSC owns is $C_{f_1} = 0.7788$.

For Flexibility 11, by (11.29) and (11.30), the difference of resource services and control logical relation is shown in Table 11.4 (Guo *et al.*, 2011).

Table 11.4 The Difference Function Value of Resource Services and Control Logical Relation.

	Flow1	**Flow2**	**Flow3**	**Flow4**	**Flow5**	**Flow6**	**Flow7**	**Flow8**
Flow1	*	1/3	1/3	3/4	4/7	4/7	1/3	4/7
Flow2	2/3	*	4/7	4/7	1/3	1/3	4/7	3/4
Flow3	2/3	6/7	*	4/7	3/4	1/3	4/7	1/3
Flow4	6/7	6/7	2/3	*	1/3	1/3	4/7	1/3
Flow5	2/3	2/5	6/7	2/3	*	1/3	1/3	4/7
Flow6	6/7	2/3	2/3	2/5	5/6	*	3/4	4/7
Flow7	2/5	2/3	6/7	6/7	2/5	6/7	*	1/3
Flow8	6/7	6/7	2/5	2/3	6/7	2/3	2/3	*

Where right upper triangular of Table 11.4 is the difference function value of resource services, and the left upper triangular is the difference function value of control logical relation.

Since the number of flows of RSC is 8, there are 28 kinds of comparisons among flows of RSC. By (11.30), the difference of total flows of RSC is $D_{C_f} = 1/56 \times 16.1294 = 0.2880$.

Through E.g. 11.4.5, the ability for RSC to maintain the QoS performance level after the flow of RSC is substituted is $\Delta QoS_C = 0.5687$. Take $\omega_1 = \omega_2 = \omega_3 = 1/3$, then flow flexibility is $C_f = 0.5452$.

11.5 Case Study and Experiment Results

In this section, "online purchasing automobiles parts (*OPAP*)" in Figure 9.1 is taken as the example to compare the difference of dealing with dynamic changes during RSC with flexibility and with no flexibility.

As shown in Figure 11.6, it is assumed that there are five elementary resource services in the flow of "*OPAP*", and each elementary resource service has five candidate resource services. The 3125 kinds of different flows of RSC that can execute the task of purchasing automobiles parts will be generated in the flow of "*OPAP*". In this chapter, three QoS properties, time, cost, and reliability are mainly considered. First, defaults of QoS are generated for each candidate resource service. For the sake of simplicity, the defaults of QoS are mapped to [1, 10], [1, 10], and [0, 10] respectively. In the meantime, let 25 pairs of resource services have certain correlations, the existence of which are all beneficial to QoS of the flow of RSC. Then assume that the weight of 3 QoS properties is 0.2, 0.5, and 0.3, respectively. According to the QoS computation model of correlation-aware RSC

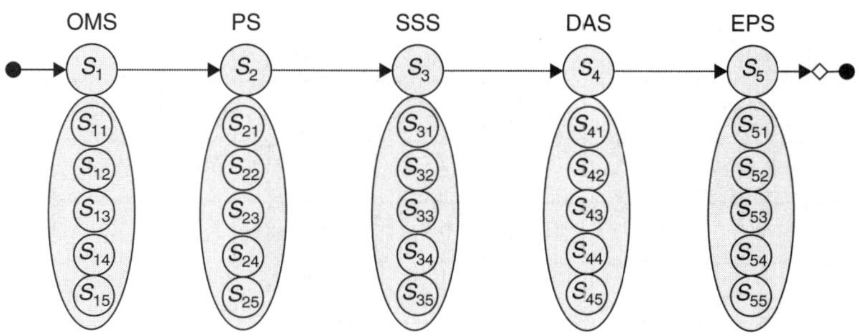

Figure 11.6 Sequence Composite Resource Service Process.

presented in Chapter 9, QoS of 3125 correlation-aware flows of RSC can be calculated. Through calculating, it can be obtained that the range of QoS of flows of RSC is in [9.1953, 31.8001].

In the following simulation experiment, when QoS of tasks satisfies [9.1953, 20], it is thought that the tasks can be executed successfully. It is assumed that in a period of time, users submitted 1000 tasks to be executed by flow of RSC *"OPAP"*, 40% of which can be executed successfully, and another 60% will fail.

The rate of success of RSC with flexibility and without flexibility will be compared when tasks, resource services, QoS, correlation and flow change.

11.5.1 The Change of Tasks

According to the assumption, when tasks do not change, the rate of success to complete the task for RSC is 40%. The comparison of the rate of success of RSC with flexibility and with no flexibility when the QoS of 1000 tasks changes from 10%~90% by 10% as a unit is presented. Test 50 times and take the average value. The final experiment results are shown in Figure 11.7.

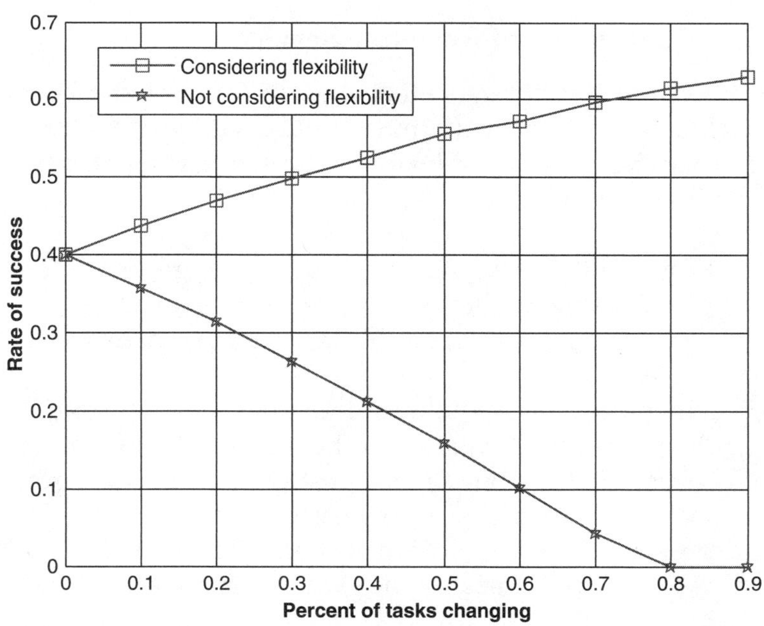

Figure 11.7 The comparison of rate of success when RSC with flexibility and without flexibility if task changes.

It can be seen from Figure 11.7 that:

1. When task changes, the rate of success of RSC with flexibility is higher than that of RSC with no flexibility, and
2. With the increase of the number of changing tasks, the rate of success of RSC with flexibility will also increase, while the rate of success of RSC with no flexibility will be reduced.

This is because only 40% of the original 1000 tasks can be executed successfully. When task changes, if RSC has no flexibility the changed tasks cannot be executed, and the left 60% of the original tasks cannot be executed successfully as well. Therefore, when RSC has no flexibility, the more the number of changing tasks the lower the rate of success that the task is executed. If RSC has flexibility and QoS of the changing task can satisfy the range [9.1953, 20], the new task can be executed successfully. Therefore, when RSC has flexibility, the more the number of changing tasks is, the lower the rate of success that the task is executed.

11.5.2 The Change of Resource Services

In this section, the difference between the rate of success when RSC with flexibility and without flexibility will be compared if the resource service is unavailable due to the exit of resource service.

First of all, it is assumed that:

- When RSC has no flexibility, if there is resource service existing in the flow of RSC that is used to execute a certain task, then this task will be not executed successfully;
- When RSC has flexibility, if there is resource service in the flow of RSC exiting, but there exists other resource service that can be used to substitute this resource service, and the difference between QoS of the substituted flow of RSC and QoS of the original flow of RSC is not more, then this task can be executed successfully.

For example, as shown in Figure 11.8, in flow1: $S_{11} \rightarrow S_{21} \rightarrow S_{32} \rightarrow S_{41} \rightarrow S_{51}$, it is assumed that candidate resource service S_{31} exits due to some reason, if there exists resource service, S_{32} can be used to substitute candidate resource service S_{31}, and the difference between QoS of substituted flow 2: $s_{11} \to s_{21} \to s_{32} \to s_{41} \to s_{51}$ and QoS of original flow 1 is not great, i.e. $\left|QoS_{flow1} - QoS_{flow2}\right| \leq \theta$, where θ is a constant greater than 0, then this task can be executed successfully.

The rate of success of RSC with flexibility and with no flexibility when resource services exit by a certain proportion and $\theta = 1, 5, 10$ will be shown. The average of 50 times implementations is shown in Figure 11. 9.

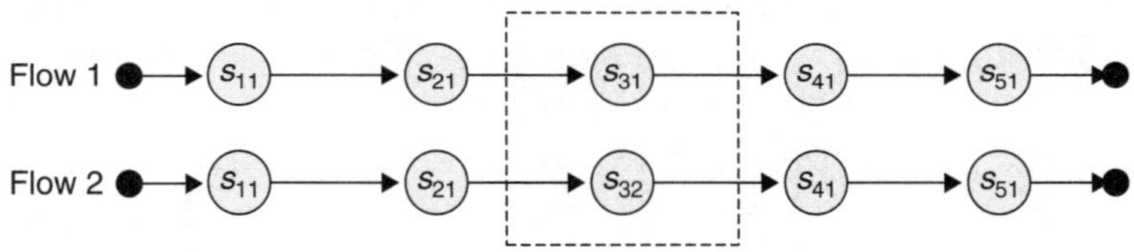

Figure 11.8 The Original Flow of RSC and the Substituted Flow of RSC.

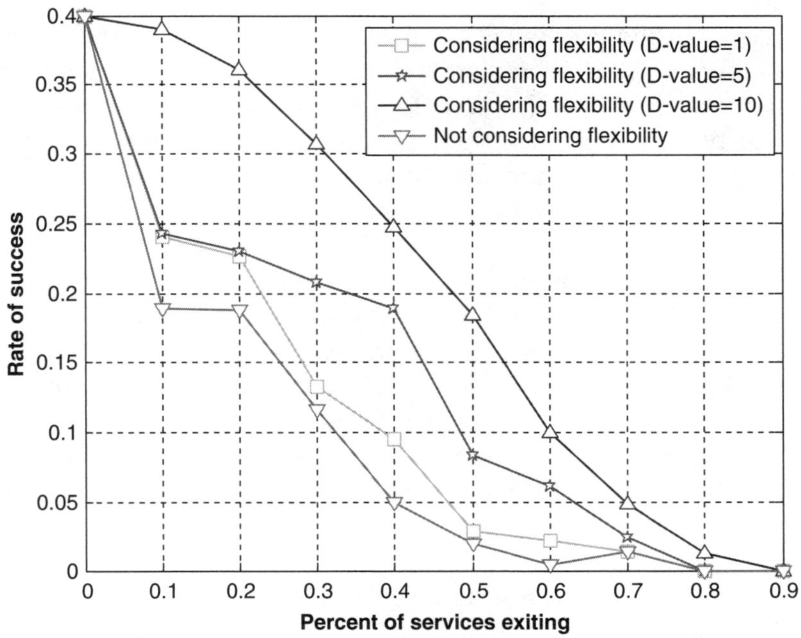

Figure 11.9 The difference of rate of success of RSC with flexibility and with no flexibility when resource services exit by a certain proportion and $\theta = 1, 5, 10$.

It can be seen from Figure 11.9 that:

1. When resource services exit by a certain proportion from 10%~90%, in whichever case, the rate of success will reduce. Moreover, when the number of exiting resource services arrive at a certain level, the rate of success of RSC executing tasks will reduce to 0.
2. When θ is fixed and resource services exit by a certain proportion from 10%~90%, the rate of success of RSC with flexibility to execute the tasks is greater than that of RSC with no flexibility.
3. If the proportion of exiting resource service is fixed along with the difference θ of the setting QoS value of the original flow of RSC, and QoS value of the flow of RSC is substituted with increasing resource services, the number of resource services that can be used to substitute will also increase, and then the rate of success of executing tasks will increase.

11.5.3 The Change of QoS of Resource Services

Similar as in the previous two sections, it is assumed that 40% of 1000 tasks can be executed successfully, that is QoS of tasks satisfy the interval [9.1953, 20]. Then QoS of resource services is changed arbitrarily by a certain proportion, where the changing range of time and cost is [6, 9] and the changing range of reliability is [0, 4]. It is assumed that:

- When RSC has no flexibility, if there is QoS of resource services in flow of RSC to execute a certain task changing, then this task cannot be executed successfully;
- When RSC has flexibility, if there is QoS of resource services in flow of RSC to execute a certain task changing, but the changing QoS of flow of RSC still satisfy the interval [9.1953, 20], then this task can be executed successfully.

Thus, when QoS of resource services changes by a certain proportion, the rate of success of RSC with flexibility and with no flexibility will be calculated. The average of 50 times implementations is shown in Figure 11.10.

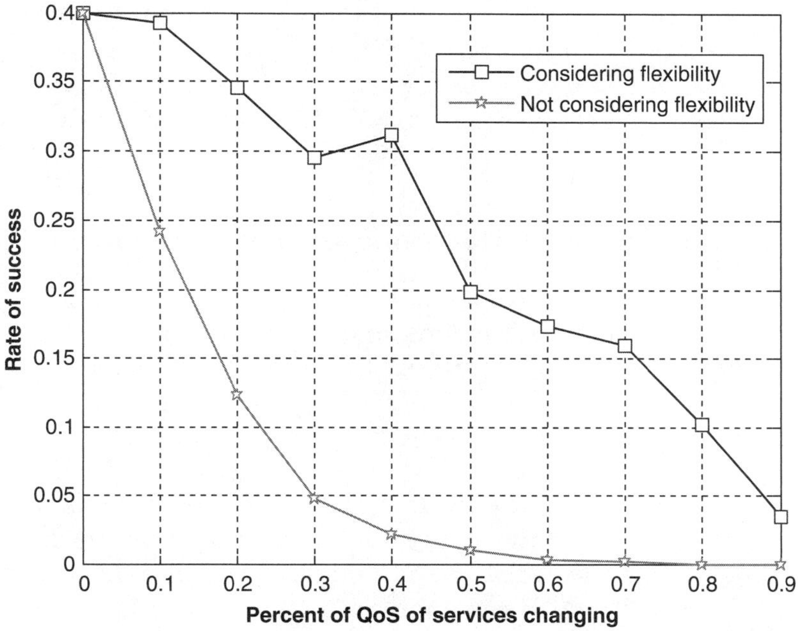

Figure 11.10 The difference of rate of success of RSC with flexibility and with no flexibility when QoS of resource services change.

It can be seen from Figure 11.10 that:

1. When QoS of resource services changes by proportion from 10%~90%, if RSC has no flexibility, along with the increasing of proportion, the number of unavailable resource service will also increase, so the rate of success of RSC executing tasks will reduce.
2. When QoS of resource services changes, the rate of success of RSC with flexibility to execute tasks is greater than that of RSC with no flexibility. This is because when QoS of resource services change randomly, the QoS of the changing resource services still have the possibility to make the QoS of the flow of RSC in interval [9.1953, 20], thereby tasks can be completed successfully.
3. With the increasing of the changing proportion of QoS of resource services, the rate of success to execute tasks for RSC with flexibility will decrease. This is because the setting change of time, cost and reliability is not

beneficial to QoS of RSC. The more the number of the proportion of changing QoS is, the more the number of RSC whose QoS value is beyond the range [9.1953, 20] is, and the smaller the rate of success of RSC executing tasks is.

11.5.4 The Change of Correlations among Resource Services

In this section, when correlations among resource services change, the difference of rate of success of RSC with flexibility and with no flexibility will be considered.

First of all, it is assumed that:

1. When RSC has no flexibility, if there are correlation changing in flows of RSC to execute a certain task, then this task cannot be executed successfully.
2. When RSC has flexibility, if there are correlations changing in flows of RSC to execute a certain task, but the changing QoS of flows of RSC are still in interval [9.1953, 20], then this task can be executed successfully.

Thus, when QoS of resource services changes by a certain proportion, if there are correlations changing in a certain flow of RSC, then QoS of this flow will be set to increase by 5. Test 50 times and take the average value. The final experiment results are shown in Figure 11.11.

It can be seen from Figure 11.11 that:

1. When correlations among resource services change by a certain proportion from 10%~90%, the rate of success of RSC with flexibility to execute tasks is greater than that of RSC with no flexibility.
2. With the increasing of proportion of changing correlations, when RSC has no flexibility according to previous setting, the unavailable number of flows of RSC will also decrease, so the rate of success of RSC executing task will decrease and gradually approaches 0.
3. With the increasing of proportion of changing correlations, when RSC has flexibility, though the rate

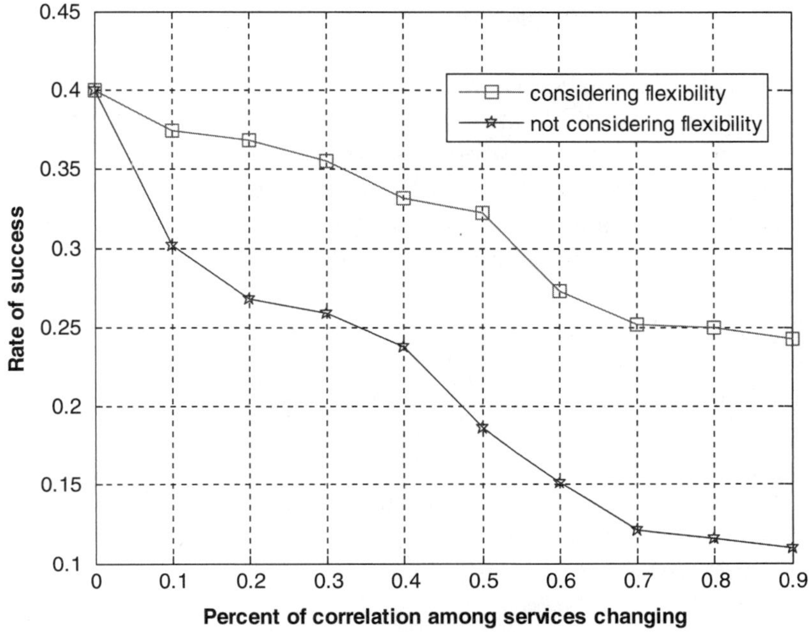

Figure 11.11 The difference of rate of Success of RSC with flexibility and with no flexibility when correlations among resource services change by a certain proportion.

of success of RSC executing tasks also decrease, overall changes are not great. And when the changes achieve a certain proportion (>70%), the rate of success basically does not change. This is because according to the previous assumption, if there are correlations changing in a certain flow of RSC, then QoS of this flow will be set to increase by 5. When the change of correlations arrives at a certain proportion, then the number of QoS of flow of RSC lying in interval [9.1953, 20] will remain stable.

11.5.5 The Change of Flows of RSC

In this section, when flows of RSC change, e.g., the original flow is unavailable and unable to execute the task, the difference of rate of success of RSC with flexibility and with no flexibility executing the task will be considered.

First of all, it is assumed that:

- When RSC has no flexibility, if the flow of RSC executing a certain task is unavailable due to some reasons, then this task cannot be executed successfully;
- When RSC has flexibility, if the flow of RSC executing a certain task is unavailable due to some reasons, but there exist other flows of RSC that can be used to substitute this flows of RSC, and the difference between QoS of substituted flows of RSC and that of original flows of RSC is not great and still in interval [9.1953, 20], then this task can be executed successfully.

The rate of success of RSC with flexibility and with no flexibility when flows of RSC change by a certain proportion and $\theta = 1, 5, 10$ is presented below. Test 50 times and take the average value. The final experiment results are shown in Figure 11.12.

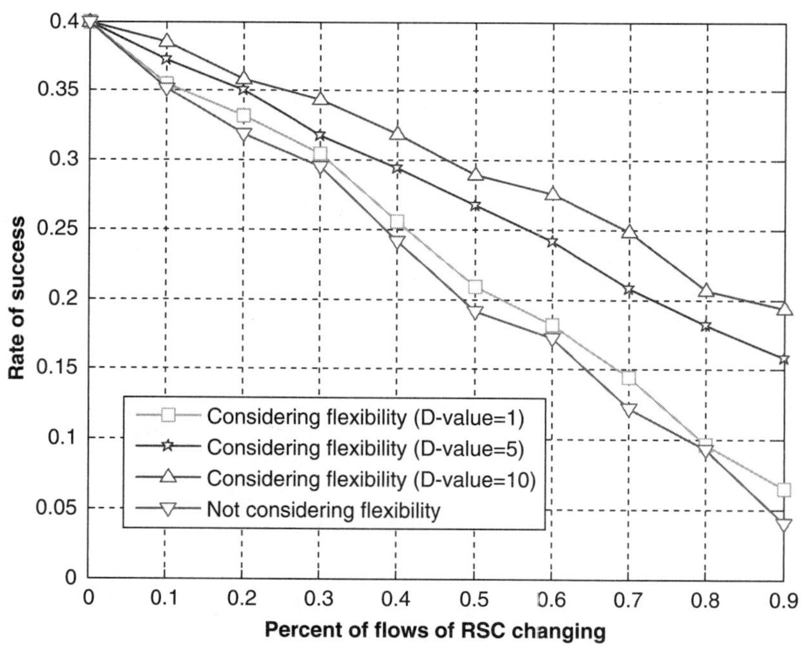

Figure 11.12 The difference of rate of success of RSC with flexibility and with no flexibility when flows of RSC change by a certain proportion.

It can be seen from Figure 11.12 that:

1. Whatever value of θ is, only if $\theta > 0$, when flows of RSC change by a certain proportion from 10%~90%, the rate of success of RSC with flexibility executing task is greater than RSC with no flexibility.
2. If the changing proportion of flows of RSC is fixed, when the difference value θ of the setting QoS value of original flow of RSC, and QoS value of flow of RSC substituted with resource services is very small, the difference of rate of success of RSC with no flexibility and with flexibility executing tasks is not big. Along with the increasing of θ, the number of new flows that can be used to substitute the original flows of RSC will also increase, then the rate of success of RSC executing tasks will increase.

11.6 Summary

With the development of microelectronics, computer technology, information technology, and intelligent machinery and control equipment, the flexible manufacturing technology has become one of the key developing directions for advanced manufacturing technology. For how to make RSC apply to the complex and volatile environment, such as the uncertainty of supply of resources, the change of correlation among resource services and QoS due to the number of resource services increasing, the change of request of enterprise users, etc., the flexibility will be an important and indispensable factor to solve the dynamic change of RSC.

The advantages to introducing flexibility to RSC in the manufacturing grid are as follows (Guo *et al.*, 2011):

1. Shortening the time of RSC. When the dynamic changes in life-cycle of RSC occur, only flow of RSC is needed to be reconfigured while restart RSC is not needed, thus the time of RSC can be effectively reduced.
2. Reduces the cost of RSC. If a certain resource service is unavailable in the executing process of RSC, it only finds the needed related resource services or composite

resource services to be substituted, and the cost loss of invoking resource services again is avoided, thus the cost of RSC can be reduced.

3. Improves QoS of RSC. When RSC is influenced due to dynamic changes, RSC with flexibility can ensure that the difference of QoS of adjusted RSC and QoS of original RSC is not great, thus the satisfied resource service can be provided to users.
4. Improves the satisfaction of users. When the user changes the demand of tasks, cancels or resets tasks, etc., restarting the task every time is not necessary, and better RSC plans can be provided to the user that meet their preferences as closely as possible.
5. Enhances the competitiveness of enterprises. When RSC has flexibility, not only the life-cycle of completing tasks can be shortened by reducing the time of invoking resource services, but also uncertainty in the resource supply can be coped with better, and dynamic intelligent resource sharing and allocation can be realized, which helps make enterprises produce high quality products within a short time.

In view of the importance of RSC, this chapter presented the definition, classification and measurement method of RSC flexibility based on analyzing the factors that influence the life-cycle of RSC. A simulation experiment is also presented that illustrates that the rate of success of RSC with flexibility executing tasks is greater than RSC with no flexibility. Further work will apply the method presented in this chapter to the specific services platform for industrial testing and research on the method of enhancing RSC flexibility.

12

Resource Services Composition Network

12.1 Introduction

Web services are autonomous software systems, programs, and devices that are advertised, located, and accessible via standard Web protocols (e.g., SOAP, WSDL, OWL-S, and UDDI) and transmitted using Internet protocols (Zeng *et al.*, 2004a). As Web services proliferate, it becomes more difficult to find the specific service that can perform the task at hand. Especially for the task with complex flow and multi-service requirements when there is no single service capable of performing that task, but effective combinations of existing services that could (Sirin *et al.*, 2004a). Hence, composition of web services, or service composition, has drawn a great deal of attention recently. Service composition means taking advantage of currently existing web services to provide a new service that does not exist on its own to execute a complex task that a single service cannot do (Hashemian and Mavaddat, 2005). From the software or service developer's perspective, service composition offers reuse possibilities. From the user's perspective, it offers seamless access to a variety of complex services (Milanovic and Malek, 2004). Solving the problem of service composition could decrease the cost of developing software components, in terms of the money, time, and human effort spent (Hashemian and Mavaddat, 2006). Therefore, researchers and engineers, both from the industrial community and academia, have carried out a mass of works on service composition.

In the industrial community, the description languages for service composition are emphasized, and several web service oriented flow descriptive languages have proposed, such as WS-BPEL (2010), OWL-S (2010), WSCI (2010), and WS-CDL (2010). These languages have played very important roles for the development of service composition software and tools. In academia, existing works on service composition concentrate on the service composition framework (Hu *et al.*, 2009b; Al-Nory *et al.*, 2009; Omer and Schill, 2009; Da Silva Santos *et al.*, 2008; Chifu *et al.*, 2008; Gamha *et al.*, 2008), service composition validation (Shiaa and Fladmark, 2009; Bianculli and Ghezzi, 2008; Ait-Sadoune and Ait-Ameur, 2008; Baresi *et al.*, 2007; Zhou and Zeng, 2007), and service composition method, but especially on the service composition method.

Referring to Tao *et al.*, (2010a), existing service composition methods can be primarily summarized into the following five categories: (1) *Business flow-based service composition.* The fundament of this method is business flow. Service selection and binding for each action in business flow is completed via modeling of business flow, as a result a composite service based on workflow is generated. Majority of service composition using this method is based on WS-BPEL and related works can be found in Benatallah *et al.*, 2002; Tosic *et al.*, 2001; Narayanan and Mcllraith, 2002; Fensel and Bussler, 2002; Benatallah *et al.*, 2005; Antonio and Razvan, 2006; Ponnekanti and Fox, 2002; and Benatallah *et al.*, 2003. (2) *AI (artificial intelligence)-based service composition.* In this method, the service composition problem is mapped into a planning problem by representing services as actions, then AI planning techniques are used to address the planning problem (i.e., service composition problem) (Baader *et al.*, 2005; Schuschel and Weske, 2004; Akkiraju *et al.*, 2004; Peer, 2004; Sirin *et al.*, 2004a; Oh *et al.*, 2008). (3) *Graph-based service composition.* The relations among services are represented as a dependency-based graph, and the service composition problem is treated as searching an optimal path from inputs (outputs) to the outputs (inputs) in the graph (Zhang *et al.*, 2003b; Rao *et al.*, 2006a). (4) *QoS (quality of service)-based service composition.* The aggregating QoS evaluation models and algorithms for composition service is completed, then combinational objective functions, or fitness functions, are established based on aggregating QoS, and intelligent computing based optimization algorithms such as genetic algorithms (GA), particle swarm optimization (PSO), and quantum genetic algorithms (QGA) which are employed to find the optimal combination solutions (Lin *et al.*, 2005b; Wang, 2009; Ardagna and Pernici, 2006; Brazier *et al.*, 2003). (5) *Agent-based service composition method* (Maamar *et al.*, 2005). Agent technologies are employed to realize the second model of service in order to make the service more intelligent. Then the service composition is realized based on the service agent.

In addition, different than the above works, Sheng *et al.*, (2009) described the design of a system that provides tools for adaptive service composition and provisioning. In their subsequent works, Ngu *et al.*, (2010) proposed a method to enable progressive composition of non-web-service-based components, such as native widgets, portlets, legacy systems, web applications, and Java Beans. A novel application of semantic annotation with a matching algorithm to find functionally equivalent components is introduced by them. To enable service engineers to design and implement suitable and safe service compositions, Foster *et al.*, (2010b) modeled service orchestration, choreography behavior, and service orchestration deployment through formal semantics applied to service behavior and configuration descriptions. A method for checking these service models is provided in an integrated workbench for the verification and validation of service compositions. Segev and Toch (2008) proposed a two-step, context-based semantic approach to the problem of matching and ranking web services for possible service composition, they presented an analysis of different methods for classifying web services for possible composition, and supply a context-based semantic matching method for ranking these possibilities.

For addressing the problems of how queries are conducted, when to query for revised information, and from which service(s) to obtain information for adapting web service composition, Harney and Doshi (2008) presented a method to selectively query services using the value of changed (VOC) information. In order to determine a subset of Web services to be invoked at runtime so as to successfully orchestrate a composite Web service, Hwang *et al.*, (2008) studied the dynamic web service selection problem in a failure-prone environment. They used a finite state machine (FSM) to model the permitted invocation sequences of web service operations. Next they assigned each state of execution an aggregated reliability to measure the probability that the given state will lead to successful execution in the context where each Web service may fail with some probability. Furthermore, they proposed two strategies to select web services that are likely to successfully complete the execution of a given sequence of operations. Obrenovic and Gasevic (2008) investigated the requirements for spreadsheet-based service composition and present a framework that implements these requirements. The framework can enable spreadsheets to send requests and retrieve results from various local and remote services. In 2010, a special issue on transactional Web services has been published by *IEEE Transactions on Service Computing*. In this special issue, some problems related to service compositions from the aspect of transactional mode are researched (Gaaloul *et al.*, 2010; Liu *et al.*, (2010a); El Haddad *et al.*, 2010).

However, the above research perspective is based on the requirements of service composition, such as describing language supporting service composition, service composition framework, method for service composition, and service composition validation. Few works have been carried out from the perspective of composition service network (CS-Net). Furthermore, the above research falls short of addressing the following issues (Tao *et al.*, 2010a; Tao *et al.*, 2011e):

Issue 1. *The combinable relationship (CoR) among the composite service involved in composition service is not fully considered.* As the authors have pointed out in Che and Gu (2004), the correlation among composite services involved in a composition service can affect the quality of the composition service badly. However, the topics such as: (1) What kinds of CoR exist among composite services, and how to identify them? (2) What's the influence of these CoRs on the *combinable strength*, i.e., the probability of a service being invoked for service composition? (3) How to evaluate the *combinable strength*, and how does it change along with time? There issues are not fully studied and effectively addressed.

Issue 2. *How to establish a dynamic and combinable relationship-based composition service network (CoRCS-Net)?* Traditional works on service composition can be classified into three steps according to a task requirements: (1) searching out the qualified resource service for each functionality requirement of the task, and generating the candidate service set (CSS), (2) selecting one candidate service from each CSS and to assemble them into a composition service according to the flow requirements of the task; and

(3) evaluating the quality of all possible composition services based on their QoS, and selecting the optimal one to execute the task by using intelligent optimization algorithms, such as genetic algorithm (GA) and particle swarm optimization (PSO). However, few works have been carried out from the perspective of composition service network, especially in regards to the probability of a service in the network being invoked for service composition. Hence, the principles and methods for establishing a CoRCS-Net should be investigated.

Issue 3. *How does a CoRCS-Net evolve with time, and how about its dynamic characteristic?* Most of the existing works on service composition consider the services in the network static in the service composition process. The dynamic addition and deletion, and the changing of QoS of the services are not considered. In fact, during the process of service composition, there are services being added into, and deleted from, the network all the time, and the qualities of the services are changing too. Hence, it is very necessary to research the evolving characteristics of service composition network for service composition.

Motivated by addressing these issues, this chapter emphasizes the combinable relationship-based composition service network, i.e., CoRCS-Net. The principles for establishing and modeling CoRCS-Net were studied, and nine class combinable relationships among services in CoRCS-Net are investigated, and fourteen elementary evolving operators for CoRCS-Net dynamic evolution are designed. According to the definition of scale-free network and the investigations on its related research achievements, we have a theoretical hypothesis that **"**CoRCS-Net is a scale-free network**"**. In order to prove the theoretical hypothesis, the concepts of *combinable strength* and *variation of combinable strength* were introduced, and *combinable strength* is used to describe the invoking times of a service being invoked for service composition. The real time *variation of combinable strength* of an arbitrary service in CoRCS-Net is first calculated, and then the corresponding real time *combinable strength* is obtained and its distribution for all services in CoRCS-Net is analyzed. It has been discovered that *"like many nature and social phenomenon, CoRCS-Net is 'scale-free', and it is constructed by few 'active services' and a great deal of 'silent services'. In the process of service composition, the invoking times for majority services are very small,* *while only few services are invoked frequently, and the probability (or invoking times) for the services in a CoRCS-Net to be invoked for service composition decays as a power-law."* (Tao et al., 2010a; Tao et al., 2011e).

12.2 Scale-free Network (SFN)

In a traditional stochastic network, although the connections are randomly set, the numbers of connections for majority nodes are approximately equivalent, i.e., the connection distribution for the nodes is a 'poisson distribution', and its distribution shape is a 'bell'. There exists an 'average' in a stochastic network, and the nodes that have more or less connections than the average are very sparse. At the increasing of the number of connections, the probability is rapidly descending in an exponent way, so the network is called a stochastic network or exponential network (Barabasi and Albert, 1999).

In 1999, Barabasi *et al.*, (Barabasi and Albert, 1999; Barabasi *et al.*, 2000) discovered in their research on WWW that, the distribution shape of connections for WWW is not a stochastic bell. It is connected by a few web pages with high connections. The connections of majority nodes (over 80%) in WWW are less than four, but seldom nodes (less than 1/10,000) have more than 1,000 connections. In other words, they discovered that the 'average' nature of the stochastic network disappeared in WWW. In their subsequent research, they discovered that, in a WWW with k connections, the connections distribution of web pages follows a power-law, i.e., the probability of a connection connecting to the other k connections marked with $P(k)$ decays as a power-law and following $P(k) \sim Ak^{-\gamma}$. The networks with connection distribution following a power-law are called scale-free networks (Barabasi and Albert, 1999; Barabasi *et al.*, 2000).

Definition 12.1 – Scale-free network (SFN) (Barabasi and Albert, 1999; Barabasi *et al.*, 2000): Assume that a network $G(V, E)$ with M vertices and K edges connecting the M vertices. The $G(V, E)$ can be represented by a set of vertices $V = \{m_i \mid i = 1,2,... M\}$ and a set of edges $E = \{l_j \mid j = 1,2 ... K\}$. The *degree* k of a vertex is defined as the number of other vertices the vertex connects to. If the probability of the degree of the vertex, $P(K)$, decays as a power-law and following $P(K) \sim AK^{-\gamma}$, the network is called scale-free network. Studies have discovered that a lot of real networks are SFNs with exponents varying in the range of $2<\gamma<3$.

In their subsequent research, Barabasi *et al.*, (2000) released the reasons why SFN has the characteristic of power-law compared with the stochastic network, and they are as follows: When establishing the ER mode of stochastic network, it is assumed that (1) the node list is known before they are connected, and all nodes are static in the whole process, and (2) the probability that two vertices are connected is random and uniform. In fact, in the forming and evolving process of a network, (1) the nodes in the network are dynamic changing, e.g., new nodes would be added into and some nodes may be quit from the network; and (2) the probabilities for the nodes to be selected as the points of new added edges are not equivalent, and they have the feature of *preferential*, i.e., some nodes will be selected more and more frequently, while some other nodes will be selected less and less. E.g., when we try to find some information on WWW, usually we prefer to view some pages we are familiar with, such as Google, Yahoo, Baidu, and Amazon. Therefore, there will be more and more connections for these web pages.

According to the above discovery, Barabasi *et al.*, (2000), proposed a method named BA model for establishing SFN. The two key operators for generation BA model are as follows:

Growth: Starting with a small number (m_0) of vertices, at every time step a new vertex with $m(\leq m_0)$ edges (that will be connected to the vertices already present in the system) is added into the network.

Preferential attachment: When choosing the vertices to which the new vertex connects, it is assumed that the probability $\prod(k_i)$, that a new vertex will be connected to vertex i depends on the connectivity k_i of that vertex, such that

$$\prod(k_i) = k_i / \sum_j k_j$$

According to the above two operators, they found that when the network evolved into a state that the degree distribution is not changing as the time, the corresponding degree distribution $p(K)$ decays as a power-law, and following $P(k) \sim k^{-3}$.

12.3 The Theoretical Hypothesis: Composition Service Network is a Scale-free Network

According to the definition of scale-free network and the investigations on its related research achievements (Zhang, 2008a; Cai, 2007b; Jovani and Fortuna, 2007; Zhang *et al.*, 2007a; Barabasi *et al.*, 1999; Barabasi *et al.*, 2001), and combining the authors' research works on service composition, the authors have the following theoretical hypothesis (Tao *et al.*, 2010a; Tao *et al.*, 2011e):

"In the process of service composition, the probability (or invoking times) for the services in the network to be invoked for service composition are not equivalent (or do not obey a poisson distribution), the invoking times for the majority of services are very small, while only few services are invoked frequently. It supposes that the composition service network is constructed by few 'active services' and a great deal of 'silent services', it is a scale-free network, and the distribution of the invoking times for its services to be invoked for service composition decays as a power-law".

In order to prove the above theoretical hypothesis, according to the definition of SFN, we just need to prove the distribution of the invoking times for the services in CoRCS-Net being invoked for service composition decays as a power-law. However, it is very difficult to measure the invoking times for each service due to the time limitation, detection tool and method, and the dynamic characteristics of CoRCS-Net. Consequently, it is impossible to research its distribution. Hence, we will introduce the concepts of *combinable strength* and *variation of combinable strength*, and use *combinable strength* to describe the invoking times of a service being invoked for service composition. First we calculate the real-time *variation of the combinable strength* of an arbitrary service in CoRCS-Net, and then obtain the corresponding *combinable strength* and investigate its distribution for all services in CoRCS-Net.

In order to realize the above theoretical proof design, this chapter studies the possible *combinable relationships* among services first, including their definitions and expressions, and then establishes a CoRCS-Net model based on the CoRs. The evolving operators and algorithms for CoRCS-Net are investigated for the sake of calculating the *variation of combinable strength*. Then we prove the hypothesis in theory that CoRCS-Net has the dynamic characteristic of scale-free through analyzing the distribution of *combinable strength* for the services in CoRCS-Net.

12.4 Concepts and Definition in CoRCS-Net

12.4.1 Web Service, Composition Service, and Composition Service Network

Definition 12.2 – Web service: Web services are defined by W3C as software systems, programs, and devices designed to support interoperable machine-to-machine interaction over a network. They are advertised, located, and accessible via standard Web protocols (e.g., SOAP, WSDL, OWL-S, and UDDI). They have an interface described in a format (e.g., WSDL) that a machine can process. Other systems interact with the web service in a manner prescribed by its description using SOAP message (Zeng *et al.*, 2004a; Hashemian and Mavaddat, 2006).

In this chapter, the web services considered are all stateless, i.e., they have a 2-step behavior: receiving some inputs and returning some outputs as the result as shown in Figure 12.1. In this chapter, we use 'RS' to denote a web service for simplicity. The key figures, symbols, and notations used in this article are explained in Table 12. 1.

Definition 12.3 – Composition service or Resource Services Chain (RSC): It refers to the sequence of services of which, in each connected couple-services, the behavior of the first service is the necessary condition to generate the behavior of the second service (as shown in Figure12.2[a]). The edge connects the first service, and the second service denotes the combinable relationships of the two services (as shown in Figure 12.2[b]) (Tao *et al.*, 2010a; Tao *et al.*, 2011e).

Definition 12.4 – Composition service network (CS-Net): A network consisting of a series of service nodes and RSCs that are connected by their combinable relationships is defined as CS-Net. The elements of CS-Net are RS, RSC, and edge (i.e., CoRs). The definitions and meanings of different edges are described in detail in Section 12.4.2. A simple abstract CS-Net example is shown in Figure12.3.

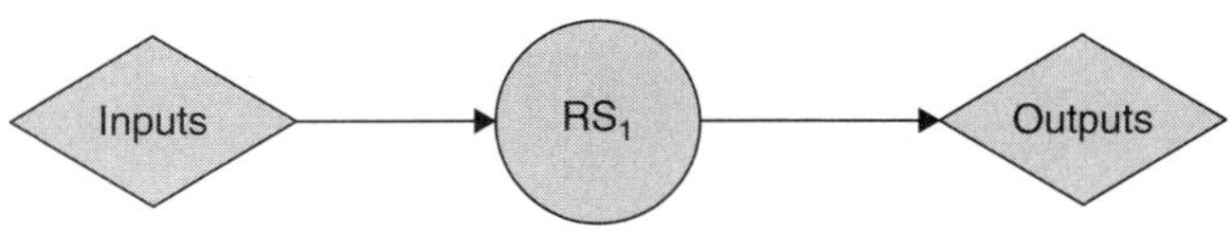

Figure 12.1 Stateless web services.

Table 12.1 The meanings of key figures and symbols used in this article (Tao *et al.*, 2010a; Tao *et al.*, 2011e)

Figures and Symbols	**Meaning**
	Denote the *inputs* of a service
	Denote the *outputs* of a service
	Denote the service, *RS*
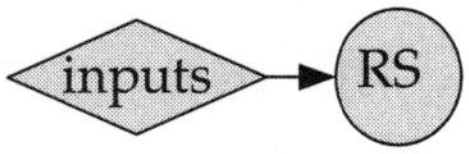	 is the inputs of 
	 is the outputs of 
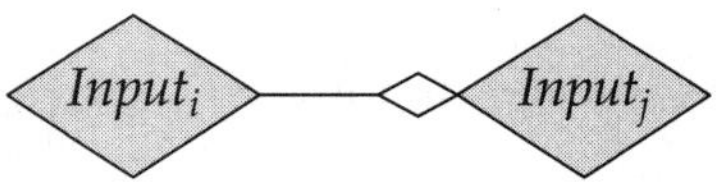	$Input_i$ is a subclass of $Input_j$, i.e., $Input_i \subset Input_j$
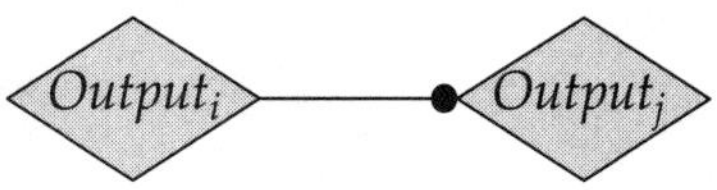	$Output_i$ is a subclass of $Output_j$ i.e., $Output_i \subset Output_j$
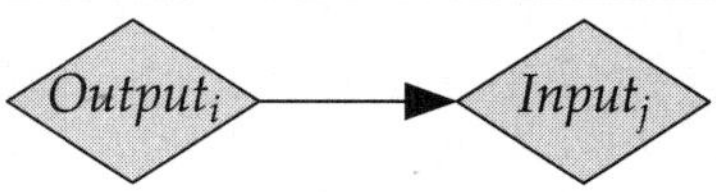	$Output_i$ is the input of RS_j
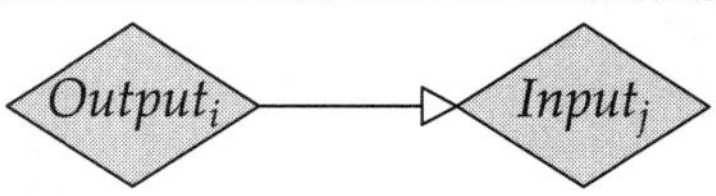	RS_i's output ($Output_i$) is a part of RS_j's inputs ($Input_j$) requirements
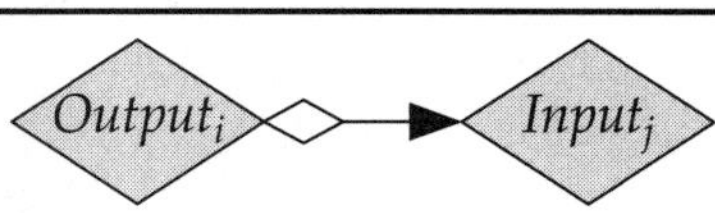	RS_j's inputs ($Input_j$) requirements is a part of RS_i's output ($Output_i$)

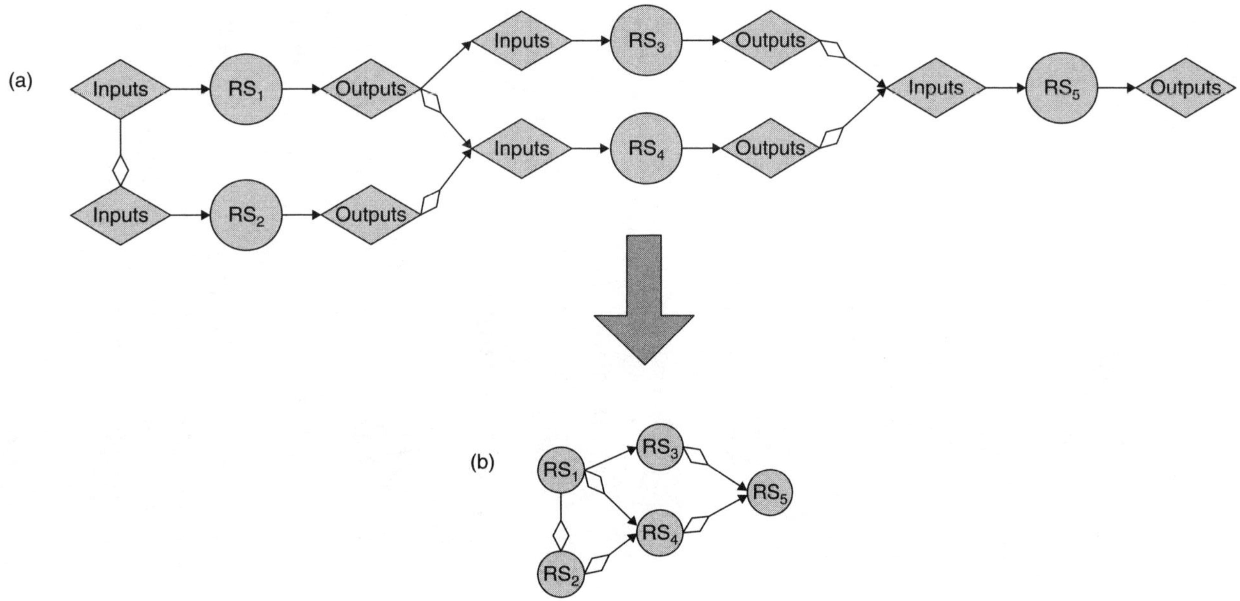

Figure 12.2 A simple example of RSC [the meanings of the edge between service nodes in [b] are shown in Table 12. 3] (Tao *et al.*, 2010a; Tao *et al.*, 2011e).

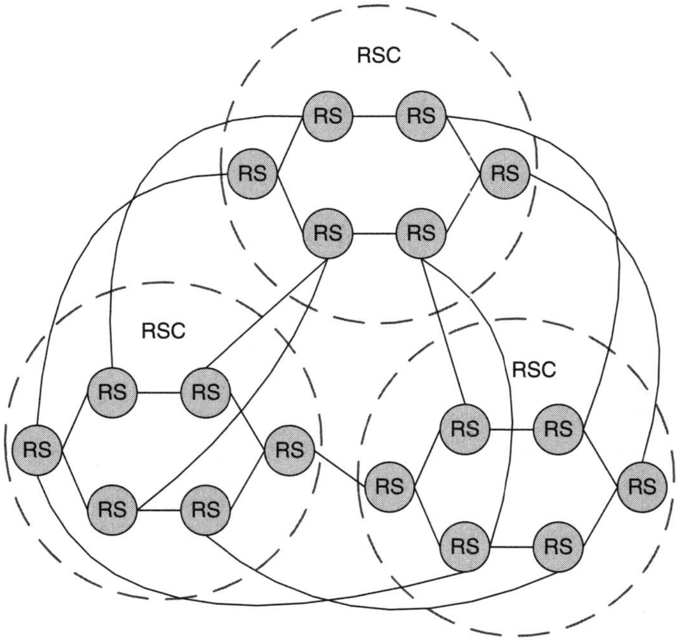

Figure 12.3 A simple abstract CS-Net example.
Note: The edge between each pair of service nodes is a not simple line, the practical line will be designed and explained in Table 12.3 according to different combinable relationships. (Tao *et al.*, 2010a; Tao *et al.*, 2011e)

12.4.2 Combinable Relationship (CoR)

Let the input set and the output set of service RS_i and RS_j in a CS-Net be $Input_i$ and $Output_i$, $Input_j$ and $Output_j$, respectivelyj According to the relationship between input sets and the output sets of RS_i and RS_j, the CoRs of RS_i and RS_j, denoted as $CoR(i, j)$, can be classified into nine categories, they are (Tao *et al.*, 2010a; Tao *et al.*, 2011e):

- *equivalent or competition combinable relationship* (eqCoR),
- *equivalent input combinable relationship* (eiCoR),
- *equivalent output combinable relationship* (eoCoR),
- *input replaceable combinable relationship* (irCoR),
- *output replaceable combinable relationship* (orCoR),
- *partial replaceable combinable relationship* (prCoR),
- *exact combinable relationship* (exCoR),
- *partial* pre-order *combinable relationship* (ppCoR), and
- *partial successor combinable relationship* (psCoR).

(1) Equivalent or competition combinable relationship (eqCoR)

If $Input_i = Input_j$ && $Output_i = Output_j$ && $Output_i \cap Input_j = \Phi$, then the relationship between RS_i and RS_j is defined as *equivalent or competition combinable relationship*, and denoted as $RS_i \equiv RS_j$. As shown in Figure 12.4(a), the inputs and outputs of RS_i and RS_j are exactly the same. During service composition, when input and output are fixed, RS_i and RS_j can be replaced by each other directly, so

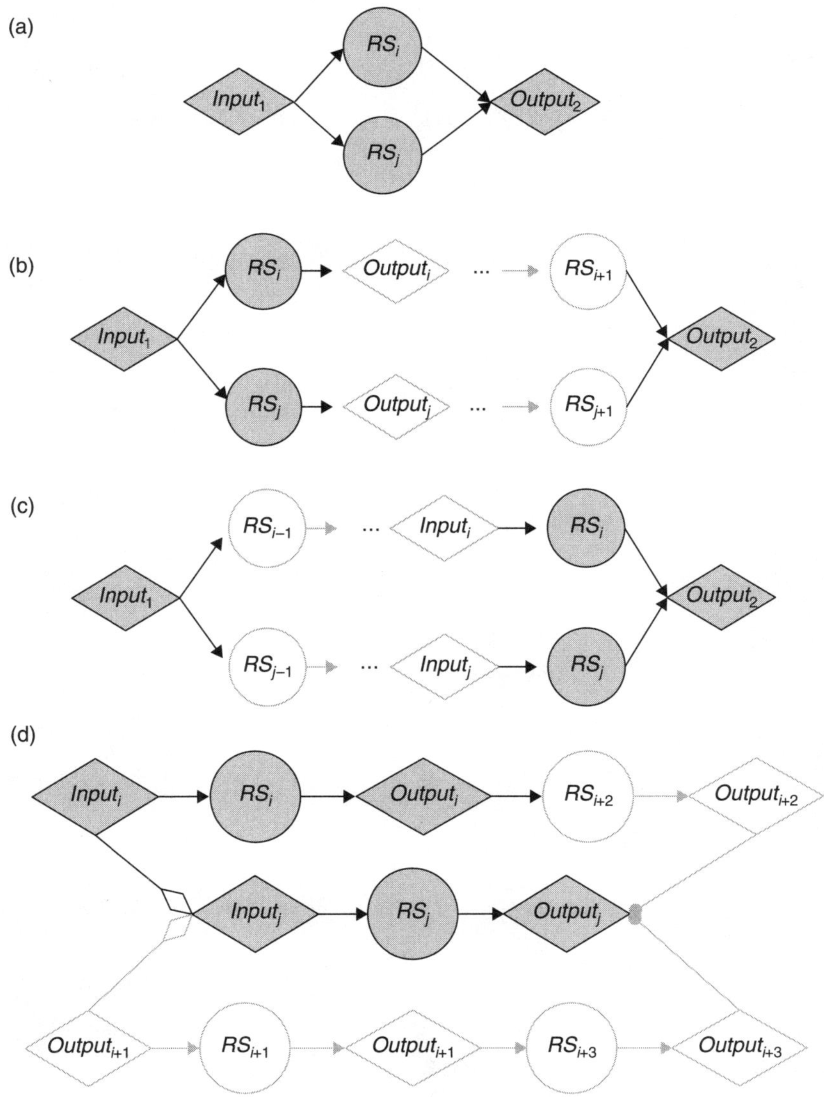

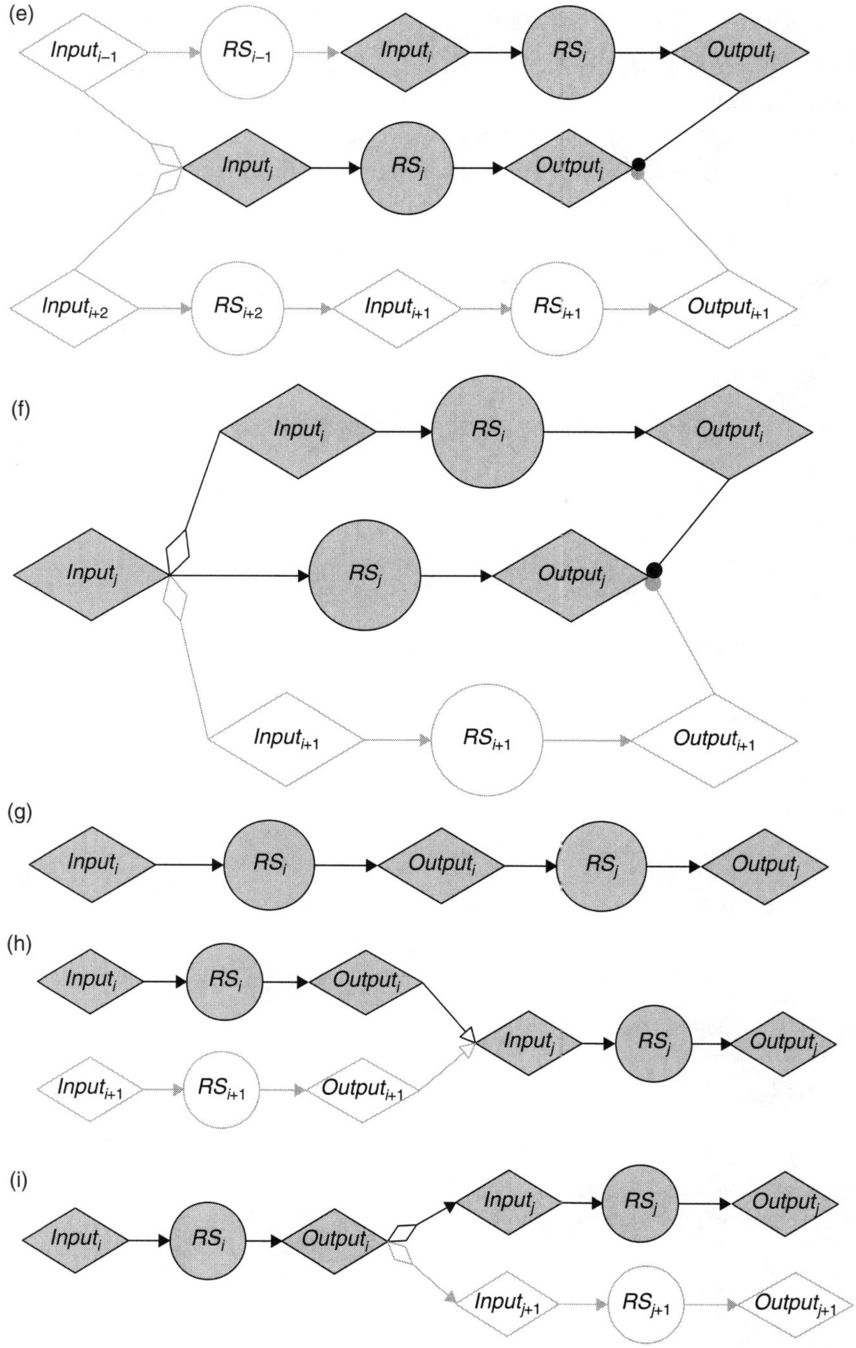

Figure 12.4 The main nine combinable relationship (Tao *et al.*, 2010a; Tao *et al.*, 2011e).

the relationship between them can also be defined as competition combinable relationship.

(2) Equivalent input combinable relationship (eiCoR)

If $Input_i$ = $Input_j$ && $Output_i \cap Input_j$ = Φ, then the relationship between RS_i and RS_j is defined as *equivalent input combinable relationship*, and denoted as $RS_i =\mid RS_j$. As shown in Figure 12.4(b), the inputs of RS_i and RS_j are the same, i.e. $Input_1$, but their outputs are different, which are $Output_i$ and $Output_j$, respectively. During services composition, when the inputs requirements are fixed, but are unable to find an accurate RS_j, the corresponding function of RS_j can be replaced by a local composition of RS_i and its successor services. E.g., <RS_j,..., RS_{j+1}> can be replaced by the composition plan <RS_i,..., RS_{i+1}> as shown in Figure 12.4(b).

(3) Equivalent output combinable relationship (eoCoR)

If $Output_i$ = $Output_j$ && $Output_i \cap Input_j$ = Φ, then the relationship between RS_i and RS_j is defined as *equivalent output combinable relationship*, and denoted as $RS_i \mid= RS_j$. As shown in Figure 12.4(c), the outputs of RS_i and RS_j are the same, i.e., $Output_2$, but their input requirements are different, which are $Input_i$ and $Input_j$, respectively. During services composition, when the required output is fixed but is unable to find an accurate RS_j, the corresponding function of RS_j can be replaced by a local composition RS_i and its pre-order services. E.g., $< RS_{j-1},\cdots,RS_j >$ can be replaced by the composition plan $< RS_{i-1},\cdots,RS_i >$ as shown in Figure 12.4(c).

(4) Input replaceable combinable relationship (irCoR)

If $Input_i \subset Input_j$ && $Output_i \cap Input_j = \Phi$, then the relationship between RS_i and RS_j is defined as *input replaceable combinable relationship*, and denoted as $RS_i \lhd\mid RS_j$. As shown in Figure 12.4(d), the input of RS_j (i.e., $Input_j$) is the union ($Input_i \cup Input_{i+1}$) of the inputs of RS_i and RS_{i+1}. During services composition, when inputs is fixed but be unable to find an accurate RS_j, the corresponding function of RS_j can be replaced by a composition of the services with *input replaceable combinable relationship* to RS_j. E.g., RS_j can be replaced by the composition plan $<< RS_i, RS_{i+2} >, < RS_{i+1}, RS_{i+3} >>$ as shown in Figure 12.4(d).

(5) Output replaceable combinable relationship (orCoR)

If $Output_i \subset Output_j$ && $Output_i \cap Input_j = \Phi$,then the relationship between RS_i and RS_j is defined as *output replaceable combinable relationship*, and denoted as $RS_i \Vdash\!< RS_j$. As shown in Figure 12.4(e), the output of (i.e., $Output_j$) RS_j is the union of the outputs of RS_i and RS_{i+1} (i.e., $Output_i \cup Output_{i+1}$). During services composition, when output is fixed but be unable to find an accurate RS_j, the corresponding function of RS_j can be replaced by a composition of the services with *output replaceable combinable relationship* to RS_j. E.g., RS_j can be replaced by the composition plan $<< RS_{i-1}, RS_i >, < RS_{i+2}, RS_{i+1} >>$ as shown in Figure 12.4(e).

(6) Partial replaceable combinable relationship (prCoR)

If $Input_i \subset Input_j \;\&\&\; Output_i \subset Output_j \;\&\&\; Output_i \cap Input_j = \Phi$, then the relationship between RS_i and RS_j is defined as *partial replaceable combinable relationship*, and denoted as $RS_i \Vdash\!\dashv RS_j$. As shown in Figure 12.4(f), the union ($Input_i \cup Input_{i+1}$) of the inputs of RS_i and RS_{i+1} is the input ($Input_j$) of RS_j, and the union ($Output_i \cup Output_{i+1}$) of the outputs of RS_i and RS_{i+1} is the output ($Output_j$) of RS_j. During services composition, when input and output are fixed but are unable to find an accurate RS_j, the corresponding function of RS_j can be replaced by the services with *partial replaceable combinable relationship* to RS_j. E.g., RS_j can be replaced by the composition plan $< RS_i, RS_{i+1} >$ as shown in Figure 12.4(f).

(7) Exact combinable relationship (exCoR)

If $Output_i = Input_j$, then the relationship between RS_i and RS_j is defined as *exact combinable relationship*, and denoted as $RS_i \rightarrow RS_j$. As shown in Figure 12.4(g), the output $Output_i$ of RS_i is directly the input of RS_j. During resource services composition, RS_i can be regarded as the preorder service of RS_j directly. Similarly, RS_j can be regarded as the successor service of RS_i directly.

(8) Partial pre-order combinable relationship (ppCoR)

If $Output_i \subset Input_j$, then the relationship between RS_i and RS_j is defined as *partial preorder combinable relationship*, and denoted as $RS_i \lhd RS_j$. As shown in Figure 12.4(h), the output of RS_i (i.e., $Output_i$) is the part of the input requirements of RS_j (i.e., $Input_j$). During resource services composition, RS_j can be regarded as the successor service of RS_i directly, while RS_i can be only regarded as the partial preorder service of RS_j.

(9) Partial successor combinable relationship (psCoR)

If $Input_j \subset Output_i$, then the relationship between RS_i and RS_j is defined as *partial successor combinable relationship*, which is denoted by $RS_i \lhd RS_j$. As shown in Figure 12.4(i), the input requirements of

RS_j (i.e., $Input_j$) is a part of the output of RS_i (i.e., $Output_i$). Hence, during resource services composition RS_i can be regarded as the preorder service of RS_j directly, and RS_j can be only regarded as the partial successor service of RS_i.

Definition 12.5 – Combinable relationship set (Tao *et al.*, 2010a; Tao *et al.*, 2011e): Define CoR={Rel_1, Rel_2, …, Rel_k, …, Rel_k} ($0 < k \leq K$ and K=1,2,3…) as *combinable relationship set* among services in a CS-Net, where Rel_k is defined as *combinable relationship unit.* Apparently, there are nine different combinable relationship units. The number of combinable relationship units in combinable relationship set is defined as the module of the combinable relationship set, denoted as $\|CoR\|$.

According to the above definitions of combinable relationship set and nine combinable relationships in CS-Net, combinable relationship set among resource services can be denoted as CoR = {eqCoR, eiCoR, eoCoR, irCoR, orCoR, prCoR, exCoR, ppCoR, psCoR}= $\{\equiv, |=, =|, <|, |<, |<|, \rightarrow, \lhd, \rhd\}$, and $\|CoR\|$ =9.

Definition12.6–Combinable relationship degree (Tao *et al.*, 2010a; Tao *et al.*, 2011e): Let CoR={Rel_1, Rel_2, …, Rel_k, …, Rel_K}, $0 < k \leq K$, be the combinable relationship set for a CS-Net, if there exists combinable relationship Rel_k ($Rel_k \in CoR$) between two resource services RS_i and RS_j in the CRS-Net, then $Rel_k(i,j)=1$; If there doesn't exist combinable relationship Rel_k between RS_i and RS_j, then $Rel_k(i,j)=0$. The *combinable relationship degree* with regard to Rel_k between RS_i and RS_j is defined as l_k.

12.4.3 Decision Rule of Combinable Relationship

Rule 12.1–Decision rule of combinable relationship: Let CoR={Rel_1, Rel_2, …, Rel_k, …, Rel_k} be the combinable relationship set among services in a CS-Net. If there exists combinable relationship between RS_i and RS_j ($\forall$ RS_i, RS_j ∈ CS-Net), then

$$\sum_{k}^{\|CoR_{i,j}\|} Rel_k(i,j) > 0,\ \forall Rel_k \in CoR_{i,j} \subseteq CoR \tag{12.1}$$

and vice versa (Tao *et al.*, 2010a; Tao *et al.*, 2011e).

Definition 12.7–Combinable weight: Let CoR={Rel_1, Rel_2, …, Rel_k, …, Rel_K} be the combinable relationship set among services in a CS-Net, and $CoR_{i,j}$ be the combinable relationship set between

RS_i and RS_j ($\forall\ RS_i, RS_j \in$ CS-Net and $CoR_{i,j} \subseteq CoR$), the *combinable weight* between RS_i and RS_j is defined as $\omega_{i,j}$, and

$$\omega_{i,j} = \sum_{k}^{\|CoR_{i,j}\|} Rel_k(i,j)l_k,\ \forall Rel_k \in CoR_{i,j} \subseteq CoR \qquad (12.2)$$

The *combinable weight* between RS_i and RS_j is also called the *edge weight* of edge $l_{i,j}$ connecting RS_i and RS_j. Let CoR={Rel_1, Rel_2, ..., Rel_k, ..., Rel_K} be the combinable relationship set among services in a CS-Net, if there exists combinable relationship between RS_i and RS_j ($\forall\ RS_i, RS_j \in$ CS-Net), then $\omega_{i,j} \geq 1$; And vice versa.

12.4.4 Combinable Strength

Definition 12.8 – Combinable neighborhood: In a CS-Net, the combinable range of a service node (i.e., the other nodes the service node connects to) is defined as the *combinable neighborhood* of the vertex. E.g., the combinable neighborhood φ_i of service node ($\forall\ RS_i \in$CS-Net) is defined as (Tao *et al.*, 2010a; Tao *et al.*, 2011e):

$$\varphi_i = \{RS_j \left| \sum_{k}^{\|CoR_{i,j}\|} Rel_k(i,j)l_k > 0, \forall Rel_k \in CoR_{i,j} \subseteq CoR\right.\} \qquad (12.3)$$

Definition 12.9 – Combinable strength: In a CS-Net, the combinable strength d_i of a service node RS_i ($\forall\ RS_i \in$CS-Net) is defined as:

$$d_i = \sum_{RS_j \in \varphi_i} \omega_{i,j} + QoS_i \qquad (12.4)$$

where $\sum_{RS_j \in \varphi_i} \omega_{i,j}$ is the sum of the *combinable weight* of RS_i and the nodes in its combinable neighborhood (i.e., φ_i), QoS_i is the comprehensive value of non-functionality properties of RS_i, which can be regarded as the ability of RS_i attracting the users. In general, the higher QoS_i is, the higher the probability the service being invoked or selected is.

12.4.5 Definition and Establishment of CoRCS-Net

Definition 12.10 – Combinable relationship-based composition service network (CoRCS-Net): A network that consists of a series of service nodes and RSCs that are connected by their combinable relationships is defined as CoRCS-Net. The elements of CoRCS-Net are RS, RSC, and edge (i.e., CoRs), the definitions and meanings of different edges are described in detail in Section 12.4.2.

According to the definitions of CoRCS-Net and the nine combinable relationships, the elementary rules of establishing a CoRCS-Net are as follows (Tao *et al.*, 2010a; Tao *et al.*, 2011e):

Rule1. CoRCS-Net consists of service, composition service, and combinable relationships. Of which services are the 'vertices' of the CoRCS-Net and combinable relationships are the edges of the CoRCS-Net;

Rule2. 'Vertices' represents service and is denoted by dot;

Rule3. The combinable relationship among the vertexes is denoted by 'edge' connecting the vertexes. According to the nine kinds of combinable relationship defined in Section 12.4.2, there are nine kinds of edges between corresponding vertexes, as shown in Table 12.3;

Rule4. The "labels" on the edges are denoted by the relational symbols among the vertexes. E.g., "$\equiv$" is used to denote the edge between two vertexes with *equivalent or competition combinable relationship.*

According to the above establishing rules, the relationship of the vertexes in Figures 12.4(a)–12.4(i) in Section 12.4.2 can be expressed by (a)–(i) in Figure 12.5. According to the above four establishing rules and nine elements in Table 12.3 and Figure 12.5, a complex CoRCS-Net based combinable relationship can be established as shown in Figure 12.6.

(a) $RS_i \equiv RS_j$	(b) $RS_i {=}| RS_j$	(c) $RS_i |{=} RS_j$
(d) $RS_i <| RS_j$	(e) $RS_i |< RS_j$	(f) $RS_i |<| RS_j$
(g) $RS_i \rightarrow RS_j$	(h) $RS_i \lhd RS_j$	(i) $RS_i \rhd RS_j$

Table 12.2 The definitions of edges in CoRCS-Net (Tao *et al.*, 2010a; Tao *et al.*, 2011e)

Relationship	Combinable relationship unit (Rel_k)	Expression	Edge	Label of Edge
Equivalent or competition combinable relationship (eqCoR)	Rel_1	$RS_i \equiv RS_j$	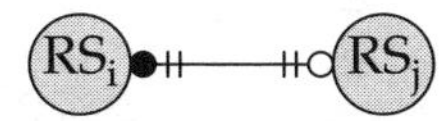	≡
Equivalent input combinable relationship (eiCoR)	Rel_2	$RS_i \vert{=} RS_j$	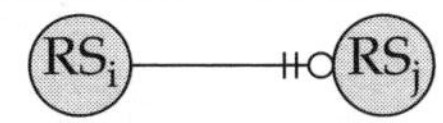	|=
Equivalent output combinable relationship (eoCoR)	Rel_3	$RS_i {=}\vert RS_j$	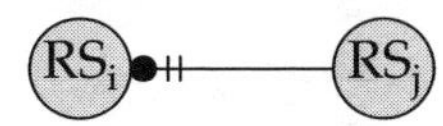	=|
Input replaceable Combinable relationship (irCoR)	Rel_4	$RS_i {<}\vert RS_j$	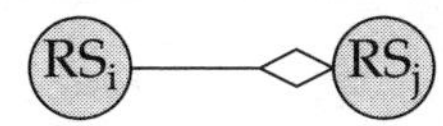	<|
Output replaceable combinable (orCoR)	Rel_5	$RS_i \vert{<} RS_j$		|<
Partial replaced combinable relationship (eqCoR)	Rel_6	$RS_i \vert{<}\vert RS_j$		|<|
Exact combinable relationship (exCoR)	Rel_7	$RS_i \rightarrow RS_j$	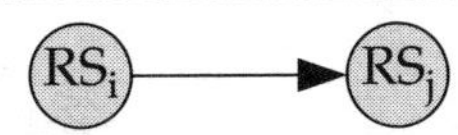	→
Partial preorder combinable relationship (ppCoR)	Rel_8	$RS_i \lhd RS_j$	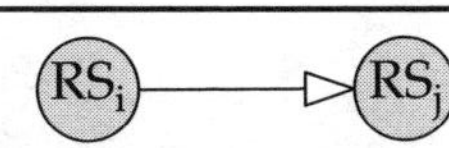	◁
Partial successor relationship combinable (psCoR)	Rel_9	$RS_i \rhd RS_j$	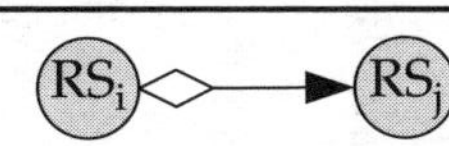	▷

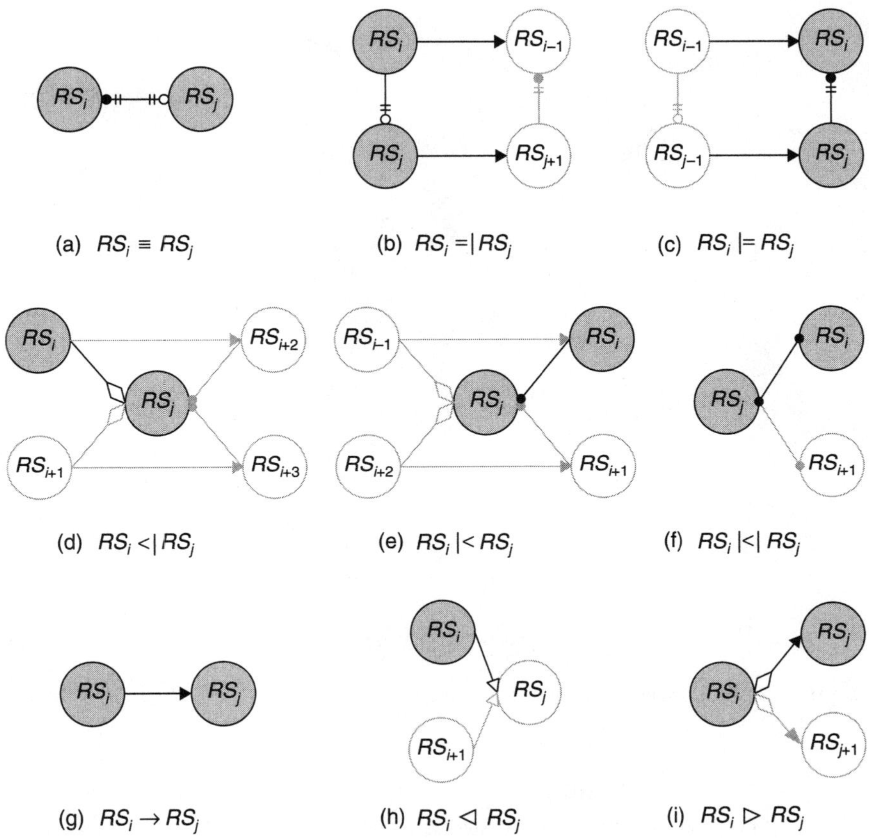

Figure 12.5 Nine elements for CoRCS-Net (Tao *et al.*, 2010a; Tao *et al.*, 2011e).

12.5 The Evolving Behavior of CoRCS-Net

12.5.1 The Elementary Evolving Operators of CoRCS-Net

In the evolving process of CoRCS-Net, the following changes may occur (Tao *et al.*, 2010a; Tao *et al.*, 2011e):

1. New RSC (including the independent and dependent) are added into the CoRCS-Net.
2. New vertexes, i.e., service nodes (including the independent and dependent) are added into the CoRCS-Net.

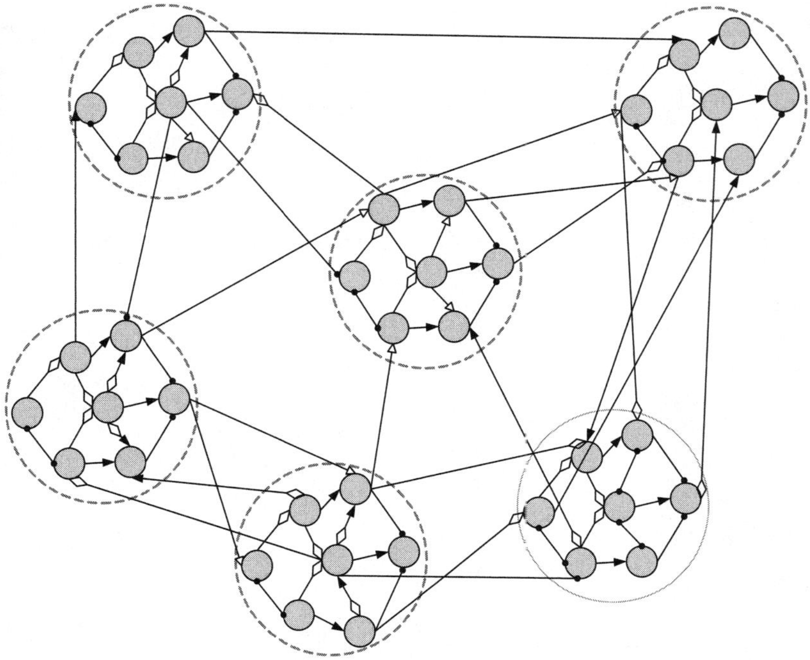

Figure 12.6 A simple example of a CoRCS-Net.

3. New combinable relationships that are added between two vertexes belong to an identical RSC or two different RSCs in the CoRCS-Net.
4. The original combinable relationship between two resource vertexes belong to an identical RSC or two different RSC in the CoRCS-Net changes when the parameters of their inputs and outputs changed, e.g., changed from *equivalent input combinable relationship* into *equivalent combinable relationship*, or from a kind of combinable relationship into no combinable relationship.
5. Some vertexes (including the independent or correlative) is unavailable (e.g., some services cannot be used any more) or deleted.
6. Some RSC (including the independent or correlative) is unavailable or deleted.

In order to describe the above changes during the evolving process of a CoRCS-Net and generate the corresponding real-time and dynamic CoRCS-Net model, 14 basic evolving operators are designed in this chapter. The definitions of the 14 operators and their influences on combinable strength of service vertexes are described in Table 12.3, as well as the examples and interfaces of these operators.

Definition 12.11 – Elementary evolving operators set: The set of the evolving operators whose implementing effects can't be replaced by the other operators or the composition of the other operators is defined as *elementary evolving operation set*. The set is marked with EOS and EOS={*Add_Indep_RSC, Add_Indep_RS, Add_Dep_RS, Add_Intra_Edge, Add_Inter_Edge, Del_Intra_Edge, Del_Inter_Edge, Del_Indep_RS, Del_Indep_RSC, Change_QoS*}.

Definition 12.12 – Combinable evolving operators set: The set of the evolving operators whose implementing effects can be achieved by a certain composition of the operators in EOS is defined as *combinable evolving operation set*, which is denoted by COS and COS={*Add_Dep_RSC, Change_Edge, Del_Dep_RS, Del_Dep_RSC*}.

12.5.2 The Influence of the Evolving Operators on the Combinable Strength in CoRCS-Net

At any time, the possible changes and evolutions of CoRCS-Net can be realized by implementing the elementary evolving operators in EOS or their combinations. Therefore, if the effect of each operator in the EOS can be identified on the *combinable strength* of an arbitrary service vertex (e.g., RS_i) in the CoRCS-Net, then the related real time *combinable strength* for the service vertex can be obtained.

It is assumed that the evolution the CoRCS-Net starts at time $t_\circ$, and there are n_0 RSC and each RSC has m_0^u $(u=1,2,\cdots n_0)$ service vertices and k_0^u combinable relationship edges at $t_\circ$ in the CoRCS-Net. Apparently, there are n_0 RSC, $\sum_{u=1}^{n_0} m_0^u$ service vertices, and $\sum_{u=1}^{n_0} k_0^u$ in CoRCS-Net at $t_\circ$.

It assumed that during an enough long and continuous $t_\circ$ time steps, the CoRCS-Net has the following evolving activities from $t_\circ$:

(1) Add_Indep_RSC: n_1 independent RSC are added into the CoRCS-Net with the probability p_1

During the t time steps, it is assumed that n_1 independent RSC are add into the CoRCS-Net with the probability p_1 at each time step, and each added RSC has m_1^v $(v=1,2,\cdots n_1)$ service vertices

Table 12.3 The 14 evolving operators of CoRCS-Net (Tao *et al.*, 2010a; Tao *et al.*, 2011e)

Evolving operations	**Interfaces**	**The influence to combinable strength of resource vertexes**	**Examples**
Add an independent RSC (e.g.,RSC_m) into the CoRCS-Net	*Add_Indep_RSC*	Because the added RSC_m is independent, and there temporarily are no connections between RSC_m and the vertices in existing CoRCS-Net, so the combinable strength of each vertex in CoRCS-Net does not change over time.	A new independent RSC is added into the existing CoRCS-Net or an unavailable independent RSC is available.
Add a dependent RSC, (e.g., RSC_m) into the CoRCS-Net	*Add_Dep_RSC*	Because the added RSC_m is dependent, there exists combinable relationship among the vertices in RSC_m and existing CoRCS-Net, so the combinable strength of some related vertexes in existing CoRCS-Net will be reinforced. The effect of this operator can be realized by sequentially executing the operators of *Add_Indep_RSC*, *Add_Inter_Edge*, and *Add_Intra_Edge* in turn.	A new dependent RSC is added into the existing CoRCS-Net or an unavailable dependent RSC is available.

Add an independent service vertex (e.g.RS_i) into the CoRCS-Net	*Add_Indep_RS*	Because the added RS_i is independent, and there temporarily are no connections between RS_i and the vertices in existing CoRCS-Net, so the combinable strength of the vertices in CoRCS-Net does not change over time.	A new independent service is added into the existing CoRCS-Net or an unavailable independent service is available.
Add a dependent service vertex (e.g.,RS_i) into the CoRCS-Net	*Add_Dep_RS*	Because the added RS_i is dependent, there exists combinable relationship among the vertices in RS_i and existing CoRCS-Net, so the combinable strength of some related vertexes in existing CoRCS-Net will be reinforced over time.	A new dependent service is added into the existing CoRCS-Net or an unavailable dependent service is available.
Add a combinable relationship edge (e.g., $l_{i,j}$) into a RSC in the CoRCS-Net	*Add_ Intra_Edge*	The combinable strength of the two vertexes (e.g., RS_i and RS_j) belong to an identical RSC connected by $l_{i,j}$ will be reinforced.	The parameters of the input and output of two vertexes (e.g., RS_i and RS_j) belong to an identical RSC in the CoRCS-Net changes, and some combinable relationships are generated between the two vertices.

Table 12.3 (cont.) The 14 evolving operators of CoRCS-Net (Tao *et al.*, 2010a; Tao *et al.*, 2011e)

Evolving operations	Interfaces	The influence to combinable strength of resource vertexes	Examples
Add a combinable relationship edge (e.g., $l_{i,j}$) between two RSCs in the CoRCS-Net	*Add_Inter_Edge*	The combinable strength of the two vertexes (e.g., RS_i and RS_j) belong to two RSCs connected by $l_{i,j}$ will be reinforced.	The parameters of the input and output of two vertexes (e.g., RS_i and RS_j) belong to two RSCs in the CoRCS-Net changes, and some combinable relationships are generated between the two vertices.
Delete a combinable relationship edge (e.g., $l_{i,j}$) from a RSC in the CoRCS-Net	*Del_ Intra_Edge*	The combinable strength of the two vertexes (e.g., RS_i and RS_j) belong to an identical RSC connected by $l_{i,j}$ will be weaken.	The parameters of the input and output of two vertexes (e.g., RS_i and RS_j) belong to an identical RSC in the CoRCS-Net changes, and some combinable relationships are disappeared between the two vertices.

Delete a combinable correlation edge $l_{i,j}$ among the service chain	*Del_Inter_Edge*	The combinable strength of the two vertexes (e.g., RS_i and RS_j) belong to two RSCs connected by $l_{i,j}$ will be weaken.	The parameters of the input and output of two vertexes (e.g., RS_i and RS_j) belong to two RSCs in the CoRCS-Net changes, and some combinable relationships are disappeared between the two vertices.
Change the combinable relationship between two service vertices	*Change_Edge*	The effect of this operator can be realized by sequentially executing the operators of *Del_Inter_Edge* (or *Del_Intra_Edge*), and *Add_Inter_Edge* (or *Add_Intra_Edge*) in turn.	The original combinable relationship between two service vertices belong changes when the parameters of theirs inputs and outputs changed, e.g., form *equivalent input combinable relationship* to *equivalent combinable relationship*.
Delete an independent service vertex (e.g., RS_i) from the CoRCS-Net	*Del_Indep_RS*	Because the added RS_i is independent, and there temporarily are no connections between RS_i and the vertices in existing CoRCS-Net, so the combinable strength of the vertices in CoRCS-Net does not change over time.	An independent service is deleted from the existing CoRCS-Net or an available independent service is unavailable.

Table 12.3 (cont.) The 14 evolving operators of CoRCS-Net (Tao *et al.*, 2010a; Tao *et al.*, 2011e)

Evolving operations	Interfaces	The influence to combinable strength of resource vertexes	Examples
Delete a dependent service vertex (e.g., RS_i) from the CoRCS-Net	*Del_Dep_RS*	The "edge" between RS_i and a vertex in its combinable neighborhood (i.e., φ_i) will disappeared automatically as the deletion of RS_i, as a result, the corresponding combinable strength of related vertices in the CoRCS-Net will be weakened. The effect of this operator can be realized by sequentially executing the operators of *Del_Inter_Edge*, *Del_Intra_Edge*), and *Del_Inde_RS* in turn.	A dependent service is deleted from the existing CoRCS-Net or an available dependent service is unavailable.
Delete an independent RSC (i.e., RSC_m) from the CoRCS-Net	*Del_Indep_RSC*	Because the deleted RSC_m is independent, and there temporarily are no connections between RSC_m and the vertices in existing CoRCS-Net, so the combinable strength of each vertex in CoRCS-Net does not change over time.	An independent RSC is deleted from the existing CoRCS-Net or an independent RSC service is unavailable.

Delete a dependent RSC (e.g., RSC_m) from the CoRCS-Net	*Del_Dep_RSC*	Because the deleted RSC_m is dependent, there exists combinable relationship among the vertices in RSC_m and existing CoRCS-Net, so the combinable strength of some related vertexes in existing CoRCS-Net will be weaken. The effect of this operator can be realized by sequentially executing the operators of *Del_Inter_Edge*, *Del_Intra_Edge*, and *Del_Dep_RS* (or *Del_Indep_RSC*) in turn.	A dependent RSC is deleted from the existing CoRCS-Net or a dependent RSC service is unavailable.
Change The QoS of a service vertex (e.g., RS_i) in the CoRCS-Net	*Change_QoS*	The combinable strength of RS_i will be reinforced (weakened) as the improving (descending) of its QoS over time.	QoS of a resource service RS in CoRCS-Net changes.

and k_1^v combinable relationship edges. After the implementation, there are totally $n_0 + p_1 n_1 t$ RSC, $\sum_{u=1}^{n_0} m_0^u + \sum_{v=1}^{n_1} p_1 m_1^v t$ service vertices, and $\sum_{u=1}^{n_0} k_0^u + \sum_{v=1}^{n_1} p_1 k_1^v t$ combinable relationships edges in the CoRCS-Net. The combinable relationship edges among service vertices in the CoRCS-Net and their corresponding strength keep unchanged.

(2) *Del_Indep_RSC*: n_2 independent RSC are deleted from the CoRCS-Net with the probability p_2

During the t time steps, it is assumed that n_2 independent RSC are deleted from the CoRCS-Net with the probability p_2, and each RSC has m_2^w $(w = 1, 2, \cdots n_2)$ service vertices and k_2^w combinable relationship edges. After the implementation, there totally are $n_0 + n_1 p_1 t - n_2 p_2 t$ RSC, $\sum_{u=1}^{n_0} m_0^u + \sum_{v=1}^{n_1} p_1 m_1^v t - \sum_{w=1}^{n_2} p_2 m_2^w t$ service vertices, and $\sum_{u=1}^{n_0} k_0^u + \sum_{v=1}^{n_1} p_1 k_1^v t - \sum_{w=1}^{n_2} p_2 k_2^w t$ combinable relationship edges in the CoRCS-Net. The combinable relationship edges among service vertices in the CoRCS-Net and their corresponding strength will keep unchanged.

(3) *Add_Dep_RS*: m_3 dependent service vertices are added into the CoRCS-Net with the probability p_3

During the t time steps, it is assumed that m_3 new dependent service vertices are added into a RSC in the existing CoRCS-Net with the probability p_3. Let the selected RSC be RSC°, and the probability of RSC° being selected is $1/(n_0 + n_1 p_1 t - n_2 p_2 t)$. It is assumed that there exist k_3^x $(x = 1, 2, \cdots m_3)$ combinable relationship edges among a new added service vertexes and the vertices in RSC°, and the probability the added vertexes has a combinable relationship with RS_i $(\forall RS_i \in RSC^\circ)$ is $\lambda_i^\circ (0 \le \lambda_i^\circ \le 1)$. After the implementation, there are totally $n_0 + n_1 p_1 t - n_2 p_2 t$ RSC, $\sum_{u=1}^{n_0} m_0^u + \sum_{v=1}^{n_1} p_1 m_1^v t - \sum_{w=1}^{n_2} p_2 m_2^w t + p_3 m_3 t$ service vertices, and $\sum_{u=1}^{n_0} k_0^u + \sum_{v=1}^{n_1} p_1 k_1^v t - \sum_{w=1}^{n_2} p_2 k_2^w t + \sum_{x=1}^{m_3} p_3 k_3^x t + p_5 k_5 t$ combinable relationship edges in the CoRCS-Net. The number of added combinable relationship edges to an arbitrary RS_i in the CoRCS-Net is

$$\sum_{x=1}^{m_3} \frac{p_3 k_3^x \lambda_i^\circ}{n_0 + n_1 p_1 t - n_2 p_2 t} \tag{12.5}$$

It is assumed that the probability for each added combinable relationship unit s ($s \in CoR$) is p_s, and the combinable relationship degree of s is l_s (the value of l_s is defined in Table 12.2), then the enhanced combinable strength for RS_i is:

$$\sum_{x=1}^{m_3} \sum_{s}^{\|CoR\|} \frac{p_3 k_3^x \lambda_i^\circ p_s l_s}{n_0 + n_1 p_1 t - n_2 p_2 t} \tag{12.6}$$

(4) ***Add_Indep_RS***: m_4 independent service vertices are added into the CoRCS-Net with the probability p_4

During the t time steps, it is assumed that m_4 independent service vertices are added into the CoRCS-Net with the probability p_4, the selected RSC is marked with RSC°, and the probability for RSC° being selected is $1/(n_0 + n_1 p_1 t - n_2 p_2 t)$. After the implementation, there are $n_0 + n_1 p_1 t - n_2 p_2 t$ RSC, $\sum_{u=1}^{n_0} m_0^u + \sum_{v=1}^{n_1} p_1 m_1^v t - \sum_{w=1}^{n_2} p_2 m_2^w t + p_3 m_3 t + p_4 m_4 t$ service vertices, and $\sum_{u=1}^{n_0} k_0^u + \sum_{v=1}^{n_1} p_1 k_1^v t - \sum_{w=1}^{n_2} p_2 k_2^w t + \sum_{x=1}^{m_3} p_3 k_3^x t + p_5 k_5 t$ combinable relationship edges in the CoRCS-Net. The combinable relationship edges among service vertices in the CoRCS-Net and their corresponding strength will keep unchanged.

(5) ***Add_Intra_Edge***: k_5 combinable relationship edges are added to an identical RSC in the CoRCS-Net with the probability p_5

During the t time steps, it is assumed that k_5 combinable relationship edges are added to an identical RSC in the CoRCS-Net with the probability p_5. The selected RSC is marked with RSC°, and the probability for RSC° being selected is $1/(n_0 + n_1 p_1 t - n_2 p_2 t)$. Let the two service vertexes connected by a new added combinable relationship edge, l, be RS_i and RS_j ($\forall RS_i, RS_j \in RSC^\circ$), respectively. The probability that RS_i and RS_j are selected are λ_i° and λ_j° ($0 \le \lambda_i^\circ, \lambda_j^\circ \le 1$ and $\lambda_i^\circ + \lambda_j^\circ = 1$), respectively. After the implementation, there are $n_0 + n_1 p_1 t - n_2 p_2 t$ RSC, $\sum_{u=1}^{n_0} m_0^u + \sum_{v=1}^{n_1} p_1 m_1^v t - \sum_{w=1}^{n_2} p_2 m_2^w t + p_3 m_3 t + p_4 m_4 t$ service vertices, and $\sum_{u=1}^{n_0} k_0^u + \sum_{v=1}^{n_1} p_1 k_1^v t - \sum_{w=1}^{n_2} p_2 k_2^w t + \sum_{x=1}^{m_3} p_3 k_3^x t + p_5 k_5 t$ combinable relationship edges in the CoRCS-Net. The number of added combinable relationship edges to an arbitrary RS_i in the CoRCS-Net is

$$\sum_{j}^{NRSC^{\circ}-1} \frac{k_5 p_5 \lambda_i^{\circ} \lambda_j^{\circ}}{n_0 + n_1 p_1 t - n_2 p_2 t} = \sum_{j}^{NRSC^{\circ}-1} \frac{k_5 p_5 \lambda_i^{\circ} (1-\lambda_i^{\circ})}{n_0 + n_1 p_1 t - n_2 p_2 t} \tag{12.7}$$

where $NRSC^{\circ}$ is the number of the resource vertices in RSC°. It is assumed that the probability for each added combinable relationship unit s($s \in CoR$)is p_s, and the combinable relationship degree of s is l_s (the value of l_s is defined in Table 12.2), then the enhanced combinable strength for RS_i is:

$$\sum_{s}^{\|CoR\|} \sum_{j}^{NRSC^{\circ}-1} \frac{k_5 p_5 \lambda_i^{\circ} (1-\lambda_i^{\circ}) p_s l_s}{n_0 + n_1 p_1 t - n_2 p_2 t} \tag{12.8}$$

(6) *Add_Inter_Edge*: k_6 combinable relationship edges are added between different RSC with the probability p_6

During the t time steps, it is assumed that k_6 combinable relationship edges are added between the service vertices belong to two RSC in the CoRCS-Net with the probability p_6. The selected two RSC are marked with RSC° and RSC^{*}, and the probabilities that RSC° and RSC^{*} being selected are $1/(n_0 + n_1 p_1 t - n_2 p_2 t)$ and $1/(n_0 + n_1 p_1 t - n_2 p_2 t - 1)$, respectively. Let the two service vertexes connected by a new added combinable relationship edge, l, be RS_i ($\forall RS_i \in RSC^{\circ}$) and RS_j($\forall RS_j \in RSC^{*}$), respectively. The probability that RS_i and RS_j being selected are λ_i° and λ_j^{*} ($0 \le \lambda_i^{\circ}, \lambda_j^{*} \le 1$), respectively. After the implementation, there are $n_0 + n_1 p_1 t - n_2 p_2 t$ RSC, $\sum_{u=1}^{n_0} m_0^u + \sum_{v=1}^{n_1} p_1 m_1^v t - \sum_{w=1}^{n_2} p_2 m_2^w t + p_3 m_3 t + p_4 m_4 t$ service vertices, and $\sum_{u=1}^{n_0} k_0^u + \sum_{v=1}^{n_1} p_1 k_1^v t - \sum_{w=1}^{n_2} p_2 k_2^w t + \sum_{x=1}^{m_3} p_3 k_3^x t + p_5 k_5 t + p_6 k_6 t$ combinable relationship edges in the CoRCS-Net. The number of added combinable relationship edges to an arbitrary RS_i in the CoRCS-Net is

$$\sum_{j}^{NRSC^{*}} \sum_{1}^{N_t - 1} \frac{k_6 p_6 \lambda_i^{\circ} \lambda_j^{*}}{(n_0 + n_1 p_1 t - n_2 p_2 t)(n_0 + n_1 p_1 t - n_2 p_2 t - 1)} \tag{12.9}$$

where $N_t = n_0 + n_1 p_1 t - n_2 p_2 t$, $NRSC^{*}$ is the number of the service vertices in RSC^{*}. It is assumed that the probability for each added combinable relationship unit s ($s \in CoR$)is p_s, and the combinable relationship degree of s is l_s (the value of l_s is defined in Table 12.2), then the enhanced combinable strength for RS_i is:

$$\sum_{s}^{\|CoR\|}\sum_{j}^{NRSC^*}\sum_{1}^{N_t-1}\frac{p_6k_6\lambda_i^\circ\lambda_j^*p_sl_s}{(n_0+n_1p_1t-n_2p_2t)(n_0+n_1p_1t-n_2p_2t-1)} = \sum_{s}^{\|CoR\|}\sum_{j}^{NRSC^*}\frac{p_6k_6\lambda_i^\circ\lambda_j^*p_sl_s}{n_0+n_1p_1t-n_2p_2t} \quad (12.10)$$

(7) *Del_Intra_Edge*: k_7 combinable relationship edges within an identical RSC are deleted with the probability p_7.

During the t time steps, it is assumed that k_7 combinable relationship edges within an identical RSC are deleted with the probability p_7. The selected RSC is marked with RSC°, and the probability for RSC° being selected is $1/(n_0+n_1p_1t-n_2p_2t)$. Let the two service vertexes connected by a new added combinable relationship edge, l, be RS_i and RS_j ($\forall RS_i, RS_j \in RSC^\circ$), and the probability that RS_i and RS_j are selected is λ_i° and λ_j° ($0 \le \lambda_i^\circ, \lambda_j^\circ \le 1$ and $\lambda_i^\circ + \lambda_j^\circ = 1$), respectively. After the implementation, there are $n_0+n_1p_1t-n_2p_2t$ RSC, $\sum_{u=1}^{n_0} m_0^u + \sum_{v=1}^{n_1} p_1m_1^vt - \sum_{w=1}^{n_2} p_2m_2^wt + p_3m_3t + p_4m_4t$ service vertices, and $\sum_{u=1}^{n_0} k_0^u + \sum_{v=1}^{n_1} p_1k_1^vt - \sum_{w=1}^{n_2} p_2k_2^wt + \sum_{x=1}^{m_3} p_3k_3^xt + p_5k_5t + p_6k_6t - p_7k_7t$ combinable relationship edges in the CoRCS-Net. The number of added combinable relationship edges to an arbitrary RS_i in the CoRCS-Net is:

$$\sum_{j}^{NRSC^\circ-1}\frac{k_7p_7\lambda_i^\circ\lambda_j^\circ}{n_0+n_1p_1t-n_2p_2t} = \sum_{j}^{NRSC^\circ-1}\frac{k_7p_7\lambda_i^\circ(1-\lambda_i^\circ)}{n_0+n_1p_1t-n_2p_2t} \quad (12.11)$$

where $NRSC^\circ$ is the number of the service vertices in RSC°. It is assumed that the probability for each added combinable relationship unit s ($s \in CoR$) is p_s, and the combinable relationship degree of s is l_s (the value of l_s is defined in Table 12.2), then the enhanced combinable strength for RS_i is:

$$\sum_{s}^{\|CoR\|}\sum_{j}^{NRSC^\circ-1}\frac{p_7k_7\lambda_i^\circ(1-\lambda_i^\circ)p_sl_s}{n_0+n_1p_1t-n_2p_2t} \quad (12.12)$$

(8) *Del_Inter_Edge*: k_8 combinable relationship edges between different RSC are deleted with the probability p_8.

During the t time steps, it is assumed that k_6 combinable relationship edges are deleted among the service vertices belong to two different RSC in the CoRCS-Net with the probability p_6. The selected two RSCs are marked with RSC° and RSC^{*}, and the probabilities that RSC° and RSC^{*} being selected are $1/(n_0 + n_1 p_1 t - n_2 p_2 t)$ and $1/(n_0 + n_1 p_1 t - n_2 p_2 t - 1)$, respectively. Let the two service vertexes connected by a new added combinable relationship edge, l, be RS_i ($\forall RS_i \in RSC^{\circ}$) and RS_j ($\forall RS_j \in RSC^{*}$), respectively. The probability that RS_i and RS_j are selected are λ_i° and λ_j^{*} ($0 \le \lambda_i^{\circ}, \lambda_j^{*} \le 1$), respectively. After the implementation, there are $n_0 + n_1 p_1 t - n_2 p_2 t$ RSC, $\sum_{u=1}^{n_0} m_0^u + \sum_{v=1}^{n_1} p_1 m_1^v t - \sum_{w=1}^{n_2} p_2 m_2^w t + p_3 m_3 t + p_4 m_4 t$ service vertices, and

$$\sum_{u=1}^{n_0} k_0^u + \sum_{v=1}^{n_1} p_1 k_1^v t - \sum_{w=1}^{n_2} p_2 k_2^w t + \sum_{x=1}^{m_3} p_3 k_3^x t + p_5 k_5 t + p_6 k_6 t - p_7 k_7 t - p_8 k_8 t$$

combinable relationship edges in the CoRCS-Net. The number of added combinable relationship edges to an arbitrary RS_i in the CoRCS-Net is:

$$\sum_{j}^{NRSC^*} \sum_{1}^{N_t - 1} \frac{k_8 p_8 \lambda_i^{\circ} \lambda_j^{*}}{(n_0 + n_1 p_1 t - n_2 p_2 t)(n_0 + n_1 p_1 t - n_2 p_2 t - 1)} \tag{12.13}$$

where $NRSC^{*}$ is the number of service vertices in RSC^{*}. It is assumed that the probability for each added combinable relationship unit s($s \in CoR$)is p_s, and the combinable relationship degree of s is l_s (the value of l_s is defined in Table 12.2), then the enhanced combinable strength for RS_i is:

$$\begin{aligned} &\sum_{s}^{\|CoR\|} \sum_{j}^{NRSC^*} \sum_{1}^{N_t - 1} \frac{k_8 p_8 \lambda_i^{\circ} \lambda_j^{*} p_s l_s}{(n_0 + n_1 p_1 t - n_2 p_2 t)(n_0 + n_1 p_1 t - n_2 p_2 t - 1)} \\ &= \sum_{s}^{\|CoR\|} \sum_{j}^{NRSC^*} \frac{p_8 k_8 \lambda_i^{\circ} \lambda_j^{*} p_s l_s}{n_0 + n_1 p_1 t - n_2 p_2 t} \end{aligned} \tag{12.14}$$

(9) *Del_Indep_RS*: m_9 independent service vertices are deleted from the CoRCS-Net with the probability p_9.

During the t time steps, it is assumed that m_9 independent services vertexes are deleted from the CoRCS-Net with the probability p_9. After the implementation, there are $n_0 + n_1p_1t - n_2p_2t$ RSC, $\sum_{u=1}^{n_0} m_0^u + \sum_{v=1}^{n_1} p_1m_1^v t - \sum_{w=1}^{n_2} p_2m_2^w t + p_3m_3t + p_4m_4t - p_9m_9t$ service vertices, and $\sum_{u=1}^{n_0} k_0^u + \sum_{v=1}^{n_1} p_1k_1^v t - \sum_{w=1}^{n_2} p_2k_2^w t + \sum_{x=1}^{m_3} p_3k_3^x t + p_5k_5t + p_6k_6t - p_7k_7t - p_8k_8t$ combinable relationship edges in the CoRCS-Net. The combinable relationship edges among service vertices in the CoRCS-Net and their corresponding strength will keep unchanged.

(10) *Change_QoS*: QoS of service vertex RS_i in the CoRCS-Net is changed.

During the t time steps, it is assumed that QoS of service vertex RS_i in the CoRCS-Net are changing. Let the variation function of the QoS for RS_i over time t be $QoS_if(t)$, then the change of combinable strength of an arbitrary RS_i in the CoRCS-Net is:

$$\Delta QoS_if(t) \tag{12.15}$$

Formulas 12.5–12.15 satisfy $\forall p_i \in [0,1]$ and $\sum_{i=1}^{9} p_i = 1 (i = 1,2,3\cdots 9)$.

During the evolving process of the CoRCS-Net, the above 10 steps are implemented in accordance with the order. After the above implementations, we have (Tao et al., 2010a; Tao *et al.*, 2011e):

1. The total number of RSC in the CoRCS-Net is:

$$N_t^{RSC} = n_0 + n_1p_1t - n_2p_2t$$

2. The total number of service vertexes in the CoRCS-Net is:

$$N_t^{RS} = \sum_{u=1}^{n_0} m_0^u + \sum_{v=1}^{n_1} p_1m_1^v t - \sum_{w=1}^{n_2} p_2m_2^w t + p_3m_3t + p_4m_4t - p_9m_9t$$

3. The total number of combinable relationship edges in the CoRCS-Net is:

$$N_t^l = \sum_{u=1}^{n_0} k_0^u + \sum_{v=1}^{n_1} p_1k_1^v t - \sum_{w=1}^{n_2} p_2k_2^w t + \sum_{x=1}^{m_3} p_3k_3^x t + p_5k_5t + p_6k_6t - p_7k_7t - p_8k_8t$$

4. The variation of combinable strength d_i, for an arbitrary service vertex RS_i in the CoRCS-Net is:

$$\left(\begin{array}{l}\sum_{x=1}^{m_3}\sum_{s}^{\|CoR\|}\frac{p_3 k_3^x \lambda_i^\circ p_s l_s}{n_0+n_1p_1t-n_2p_2t}+\sum_{s}^{\|CoR\|}\sum_{j}^{NRSC^\circ-1}\frac{k_5p_5\lambda_i^\circ(1-\lambda_i^\circ)p_sl_s}{n_0+n_1p_1t-n_2p_2t}+\\ \sum_{s}^{\|CoR\|}\sum_{j}^{NRSC^*}\frac{p_6k_6\lambda_i^\circ\lambda_j^*p_sl_s}{n_0+n_1p_1t-n_2p_2t}-\sum_{s}^{\|CoR\|}\sum_{j}^{NRSC^\circ-1}\frac{p_7k_7\lambda_i^\circ(1-\lambda_i^\circ)p_sl_s}{n_0+n_1p_1t-n_2p_2t}-\\ \sum_{s}^{\|CoR\|}\sum_{j}^{NRSC^*}\frac{p_8k_8\lambda_i^\circ\lambda_j^*p_sl_s}{n_0+n_1p_1t-n_2p_2t}+\Delta QoS_i f(t)\end{array}\right)$$

12.6 Theoretical Proof of the Scale-free Characteristics of CoRCS-Net

12.6.1 The Variation Rate of Combinable Strength for Each Service Vertexes in the CoRCS-Net

The following assumptions are made before the proof (Tao *et al.*, 2010a; Tao *et al.*, 2011e):

1. The evolution of CoRCS-Net is continuous, and the t time steps are also continuous.
2. Every RS_i before joined into the CoRCS-Net has a certain combinable strength, $d_i(t_0)$, and $d_i(t_0) = B$
3. The evolution of the CoRCS-Net starts at t_o and there are n_0 RSC and each RSC has m_0^u $(u=1,2,\cdots n_0)$ service vertices (denoted as RS) and k_0^u combinable relationship edges at t_o in the CoRCS-Net.
4. The sum D_o of the combinable strength of all service vertices in the CoRCS-Net at t_o is known. The sum of the combinable strength of all service vertices in the CoRCS-Net at t, $D(t)$, changed along with time.
5. According to the definition of *Preferential attachment* in scale-free network in Section 12.2, the probability for RS_i being selected in the operators of *Add_Dep_RS, Add_Intra_Edge, Add_Inter_Edge, Del_Intra_Edge, Del_Inter_Edge,* is the ratio between the combinable strength of RS_i (i.e., d_i) and the sum d_{RSC° of the

combinablestrengthofallverticesin RSC° ($RS_i \in RSC^{\circ}$), i.e., $\lambda_i^{\circ} = d_i / d_{RSC^{\circ}}$. Let the total number of RSC in CoRCS-Net at time t be N_t, and $d_{RSC^{\circ}} = D(t)/N_t$.

6. The comprehensive value of non-functionality properties (i.e., QoS) of RS_i, doesn't change along with t, i.e., QoS_i keeps unchanged.

After the implementations of the operators in Section 12.6.1, the sum of the combinable strength for all the service vertices in the CoRCS-Net at t is as follows (Tao *et al.*, 2010a; Tao *et al.*, 2011e):

$$D(t) = 2N_t^l \sum_s^{\|CoR\|} p_s l_s + \sum_i^{N_t^{RS}} QoS_i f(t)$$

$$= 2\sum_s^{\|CoR\|}\sum_u^{n_0} k_0^u p_s l_s + 2t\begin{pmatrix} \sum_s^{\|CoR\|}\sum_v^{n_1} p_1 k_1^v p_s l_s - \sum_s^{\|CoR\|}\sum_w^{n_2} p_2 k_2^w p_s l_s + \\ \sum_s^{\|CoR\|}\sum_x^{m_3} p_3 k_3^x p_s l_s + \sum_s^{\|CoR\|} p_s l_s p_5 k_5 + \\ \sum_s^{\|CoR\|} p_6 k_6 p_s l_s - \sum_s^{\|CoR\|} p_s l_s p_7 k_7 - \sum_s^{\|CoR\|} p_8 k_8 p_s l_s \end{pmatrix} + \sum_i^{N_t^{RS}} QoS_i f(t) \tag{12.6}$$

$$= 2\sum_s^{\|CoR\|}\sum_{u=0}^{n_0} k_0^u p_s l_s + 2t\begin{pmatrix} \sum_{v=1}^{n_1} p_1 k_1^v - \sum_{w=1}^{n_2} p_2 k_2^w + \sum_{x=1}^{m_3} p_3 k_3^x \\ + p_5 k_5 + p_6 k_6 - p_7 k_7 - p_8 k_8 \end{pmatrix} \sum_s^{\|CoR\|} p_s l_s + \sum_i^{N_t^{RS}} QoS_i f(t)$$

According to the **Assumption 6**

$$QoS_i f(t) = 0, \sum_i^{N_t^{RS}} QoS_i f(t) = 0 \tag{12.17}$$

Set

$$A_1 = 2\Big(\sum_{v=1}^{n_1} p_1 k_1^v - \sum_{w=1}^{n_2} p_2 k_2^w + \sum_{x=1}^{m_3} p_3 k_3^x \\ + p_5 k_5 + p_6 k_6 - p_7 k_7 - p_8 k_8\Big) \sum_s^{\|CoR\|} p_s l_s \tag{12.18}$$

when $t \to \infty$,

$$D(t) = 2\sum_s^{\|CoR\|}\sum_{u=1}^{n_0} k_0^u p_s l_s + A_1 t + \sum_i^{N_t^{RS}} QoS_i f(t) \approx A_1 t \tag{12.19}$$

Therefore, the variation rate of combinable strength, $\frac{dd_i}{dt}$, for an arbitrary service vertex RS_i in the CoRCS-Net in time period t can be expressed as follows (Tao *et al.*, 2010a; Tao *et al.*, 2011e):

$$\frac{dd_i}{dt}=\left(\begin{array}{l}\sum_{x=1}^{m_3}\sum_{s}^{\|CoR\|}\frac{p_3k_3^x\lambda_i^\circ p_sl_s}{n_0+n_1p_1t-n_2p_2t}+\sum_{s}^{\|CoR\|}\sum_{j}^{NRSC^\circ-1}\frac{k_5p_5\lambda_i^\circ(1-\lambda_i^\circ)p_sl_s}{n_0+n_1p_1t-n_2p_2t}+\\ \sum_{s}^{\|CoR\|}\sum_{j}^{NRSC^*}\frac{p_6k_6\lambda_i^\circ\lambda_j^*p_sl_s}{n_0+n_1p_1t-n_2p_2t}-\sum_{s}^{\|CoR\|}\sum_{j}^{NRSC^\circ-1}\frac{p_7k_7\lambda_i^\circ(1-\lambda_i^\circ)p_sl_s}{n_0+n_1p_1t-n_2p_2t}-\\ \sum_{s}^{\|CoR\|}\sum_{j}^{NRSC^*}\frac{p_8k_8\lambda_i^\circ\lambda_j^*p_sl_s}{n_0+n_1p_1t-n_2p_2t}+\Delta QoS_if(t)\end{array}\right)$$

$$=\frac{\left(\sum_{x=1}^{m_3}p_3k_3^x-\sum_{j}^{NRSC^\circ-1}p_5k_5\lambda_i^\circ+\sum_{j}^{NRSC^*}p_6k_6\lambda_i^\circ+\sum_{j}^{NRSC^\circ-1}p_7k_7\lambda_i^\circ-\sum_{j}^{NRSC^*}p_8k_8\lambda_i^*\right)\sum_{s}^{\|CoR\|}p_sl_s}{n_0+n_1p_1t-n_2p_2t}\lambda_i^\circ+\frac{(p_5k_5-p_7k_7)\sum_{j}^{NRSC^\circ-1}\lambda_i^\circ\sum_{s}^{\|CoR\|}p_sl_s}{n_0+n_1p_1t-n_2p_2t}+\Delta QoS_if(t) \quad (12.20)$$

According to the **Assumption 5**

$$\lambda_i^\circ=d_iN_t^{RSC}/D(t)=\frac{d_iN_t^{RSC}}{D(t)}\approx\frac{d_i(n+n_1p_1t-n_2p_2t)}{A_1t} \quad (12.21)$$

$$\sum^{NRSC^\circ}\lambda_i^\circ=1,\ \sum_{j}^{NRSC^*}\lambda_j^*=1,\ \sum^{NRSC^\circ-1}\lambda_i^\circ\approx 1 \quad (12.22)$$

By Equation 12.17, Equation 12.21, and Equation 12.22, Equation 12.20 is transformed as

$$\frac{dd_i}{dt}=\frac{\left(\sum_{x=1}^{m_3}p_3k_3^x-p_5k_5+p_6k_6+p_7k_7-p_8k_8\right)\sum_{s}^{\|CoR\|}p_sl_s}{n_0+n_1p_1t-n_2p_2t}\lambda_i^\circ-\frac{(p_5k_5-p_7k_7)\sum_{s}^{\|CoR\|}p_sl_s}{n_0+n_1p_1t-n_2p_2t} \quad (12.23)$$

Set

$$A_2 = \left(\sum_{x=1}^{m_3} p_3 k_3^x - p_5 k_5 + p_6 k_6 + p_7 k_7 - p_8 k_8\right) \sum_{s}^{\|CoR\|} p_s l_s \quad (12.24)$$

$$A_3 = (p_5 k_5 - p_7 k_7) \sum_{s}^{\|CoR\|} p_s l_s \quad (12.25)$$

Then

$$\begin{aligned} \frac{dd_i}{dt} &= \frac{A_2}{n_0 + n_1 p_1 t - n_2 p_2 t} \lambda_i^{\circ} + \frac{A_3}{n_0 + n_1 p_1 t - n_2 p_2 t} \\ &= \frac{A_2}{n_0 + n_1 p_1 t - n_2 p_2 t} \frac{d_i (n + n_1 p_1 t - n_2 p_2 t)}{A_1 t} \\ &\quad + \frac{A_3}{n_0 + n_1 p_1 t - n_2 p_2 t} \\ &= \frac{A_2 d_i}{A_1 t} + \frac{A_3}{n_0 + n_1 p_1 t - n_2 p_2 t} \end{aligned} \quad (12.26)$$

When $t \to \infty$,

$$\frac{dd_i}{dt} \approx \frac{A_2 d_i}{A_1 t} + \frac{A_3}{(n_1 p_1 - n_2 p_2) t} \quad (12.27)$$

By solving the above differential equation Equation 12.29, we have

$$d_i(t) = \frac{A_1}{A_2} e^{\left(\frac{A_2 C}{A_1}\right)} t^{\left(\frac{A_2}{A_1}\right)} + \frac{A_1 A_3}{A_2 (n_1 p_1 - n_2 p_2)} \quad (12.28)$$

where C is a constant. According to the **Assumption 5**

$$\text{when } t = t_0, \text{ then } d_i(t_0) = B \quad (12.29)$$

Combine Equation 12.29 into Equation 12.28, then

$$\frac{A_1}{A_2} e^{(\frac{A_2 C}{A_1})} = \left(B - \frac{A_1 A_3}{A_2(n_1 p_1 - n_2 p_2)} \right) t_0^{(-\frac{A_2}{A_1})} \tag{12.30}$$

Combine Equation 12.30 into Equation 12.28, then

$$\begin{aligned} d_i(t) &= \left(B - \frac{A_1 A_3}{A_2(n_1 p_1 - n_2 p_2)} \right) t_0^{(-\frac{A_2}{A_1})} t^{(\frac{A_2}{A_1})} + \frac{A_1 A_3}{A_2(n_1 p_1 - n_2 p_2)} \\ &= \left(B - \frac{A_1 A_3}{A_2(n_1 p_1 - n_2 p_2)} \right) \left(\frac{t}{t_\circ} \right)^{(\frac{A_2}{A_1})} + \frac{A_1 A_3}{A_2(n_1 p_1 - n_2 p_2)} \end{aligned} \tag{12.31}$$

Conclusion 1: The variation relation of combinable strength of an arbitrary service vertex RS_i in CoRCS-Net over time is as follows (Tao *et al.*, 2010a; Tao *et al.*, 2011e):

$$d_i(t) = \left(B - \frac{A_1 A_3}{A_2(n_1 p_1 - n_2 p_2)} \right) \left(\frac{t}{t_\circ} \right)^{(\frac{A_2}{A_1})} + \frac{A_1 A_3}{A_2(n_1 p_1 - n_2 p_2)}$$

12.6.2 The Proof of Scale-free Characteristic of CoRCS-Net

Let d $(0 < d)$ be an arbitrary combinable strength value in CoRCS-Net. According to the definition of scale-free network 'when the network evolves into the state that the distribution of degree doesn't change along with time, the corresponding distribution of degree follows a power-law distribution', if the distribution of combinable strength decays as power-law, i.e., following $P(d) \sim Ad^{-\varepsilon}$, then it can be concluded that CoRCS-Net is a scale-free network. Let $d_i(t) < d$, then

$$\begin{aligned} d_i(t) < d &\Leftrightarrow \left(B - \frac{A_1 A_3}{A_2(n_1 p_1 - n_2 p_2)} \right) \left(\frac{t}{t_\circ} \right)^{(\frac{A_2}{A_1})} + \frac{A_1 A_3}{A_2(n_1 p_1 - n_2 p_2)} < d \\ &\Leftrightarrow t_\circ > \left(\left(B - \frac{A_1 A_3}{A_2(n_1 p_1 - n_2 p_2)} \right) \Big/ \left(d - \frac{A_1 A_3}{A_2(n_1 p_1 - n_2 p_2)} \right) \right)^{(\frac{A_1}{A_2})} t \end{aligned} \tag{12.32}$$

Let

$$Y=\left(\left(B-\frac{A_1A_3}{A_2(n_1p_1-n_2p_2)}\right)\Big/\left(d-\frac{A_1A_3}{A_2(n_1p_1-n_2p_2)}\right)\right)^{(\frac{A_1}{A_2})}t \tag{12.33}$$

Since the probability of $t_\circ < Y$ (i.e., $P(t_\circ < Y)$) is

$$\begin{aligned} P(t_\circ < Y) &= \int_0^Y \frac{1}{N_t^{RS}}dt_\circ \\ &= Y/N_t^{RS} \\ &= Y/(\sum\nolimits_{u=1}^{n_0} m_0^u + \sum\nolimits_{v=1}^{n_1} p_1 m_1^v t - \sum\nolimits_{w=1}^{n_2} p_2 m_2^w t \\ &\quad + p_3 m_3 t + p_4 m_4 t - p_9 m_9 t) \end{aligned} \tag{12.34}$$

Let

$$A_4 = \sum\nolimits_{v=1}^{n_1} p_1 m_1^v - \sum\nolimits_{w=1}^{n_2} p_2 m_2^w + p_3 n_3 + p_4 n_4 - p_9 n_9 \tag{12.35}$$

When $t \to \infty$

$$\begin{aligned} P(t_\circ < Y) \approx Y/(&\sum\nolimits_{v=1}^{n_1} p_1 m_1^v - \sum\nolimits_{w=1}^{n_2} p_2 m_2^w \\ &+ p_3 n_3 + p_4 n_4 - p_9 n_9)t \end{aligned} \tag{12.36}$$

And

$$P(t_\circ < Y) = Y/A_4 t \tag{12.37}$$

Then

$$\begin{aligned} P[d_i(t) < d] &= P(t_\circ > Y) \\ &= 1 - P(t_\circ < Y) \\ &= 1 - Y/A_4 t \\ &= 1 - \frac{1}{A_4}\left(\left(B-\frac{A_1A_3}{A_2(n_1p_1-n_2p_2)}\right)\Big/\right. \\ &\quad \left.\left(d-\frac{A_1A_3}{A_2(n_1p_1-n_2p_2)}\right)\right)^{(\frac{A_1}{A_2})} \end{aligned} \tag{12.38}$$

Because

$$P[d_i(t)=d]=\frac{\partial(P[d_i(t)<d])}{\partial d}$$

$$=\frac{\partial\left(1-\frac{1}{A_4}\left(\left(B-\frac{A_1A_3}{A_2(n_1p_1-n_2p_2)}\right)\Big/\left(d-\frac{A_1A_3}{A_2(n_1p_1-n_2p_2)}\right)\right)^{(\frac{A_1}{A_2})}\right)}{\partial d}$$

$$=\frac{1}{A_4}\left(B-\frac{A_1A_3}{A_2(n_1p_1-n_2p_2)}\right)^{(\frac{A_1}{A_2})}\left(d-\frac{A_1A_3}{A_2(n_1p_1-n_2p_2)}\right)^{-(\frac{A_1}{A_2}+1)}\frac{A_1}{A_2} \quad (12.39)$$

Let

$$\varepsilon=\frac{A_1}{A_2}+1 \quad (12.40)$$

$$A_5=\frac{A_1}{A_4A_2}\left(B-\frac{A_1A_3}{A_2(n_1p_1-n_2p_2)}\right)^{(\frac{A_1}{A_2})} \quad (12.41)$$

$$A_6=\frac{A_1A_3}{A_2(n_1p_1-n_2p_2)} \quad (12.42)$$

Then

$$P[d_i(t)=d]=A_5(d-A_6)^{-\varepsilon} \quad (12.43)$$

Equation 12.43 demonstrates that the distribution of the combinable strength of the services in a CoRCS-Net decays a power-law with the exponent $\varepsilon(\varepsilon=A_1/A_2+1)$, and following $P(d)=A_5(d-A_6)^{-\varepsilon}$. Therefore, according to the definition of scale-free network, the following conclusions can be obtained (Tao *et al.*, 2010a; Tao *et al.*, 2011e):

"The combinable relationship-based composition service network (i.e., CoRCS-Net) is a scale-free network, and it is constructed by few 'active services' and a great deal of 'silent services'. In the process of service composition, the invoking times for majority services are very small, while only few services are invoked frequently, and the probability (or invoking times) for the services in a CoRCS-Net to be invoked for service composition decays as a power-law, following $P(d)=A_5(d-A_6)^{-\varepsilon}$ *and* $\varepsilon=\frac{A_1}{A_2}+1$*."*

It demonstrates that our theoretical hypothesis in Section 12.3 is right in theory.

12.6.3 Significance

The service composition network has been proven to be a scale-free network. Therefore, the related theories and methods of scale-free network and complex network can be applied to a service-oriented system. The results of this chapter can be used to guide the service providers in the information industry to deploy their services in an optimal way for commercial purposes. For example, the modeling method of scale-free network (i.e., the BA model)) can be used to model the service network in the service-oriented manufacturing system (e.g., manufacturing grid system, cloud manufacturing system (Tao *et al.*, 2011b) and e-commerce system. Then the dynamic characteristics of service node and service network in service-oriented manufacturing, such as invoking frequency, robustness, security, reliability, and so on can be further studied. This can further provide decision supports and theoretical guidance for service developers or providers on the follow issues: (a) what kinds of service they should develop or provide, (b) how to enhance their services' utilization, (c) how to realize their services added value, and (d) how to protect their core service.

In addition, as mentioned before, existing works on service composition are primarily based on the requirements of service composition, and focus on describing language supporting service composition, service composition framework, method for service composition, and service composition validation. Few works have been carried out from the perspective of combinable relationship among composite services and composition service network. However, this work studies the service composition from a totally new point, and we believe it will inspire more related people to study service from a networked and systematic view. Maybe some new research topic will be generated.

12.7 Summary

Existing works on service composition are based on the requirements of service composition, such as describing language supporting service composition, service composition framework, method for service composition, and service composition validation. Few works have been carried out from the perspective of combinable relationship among composite services and composition service network. This chapter puts the emphasis on composition service network based on combinable relationship, i.e., CoRCS-Net. According to the definition of scale-free network and the investigations on its related research achievements, we supposed that that 'CoRCS-Net is a scale-free network'.

In order to prove the above theoretical hypothesis, the possible combinable relationships among services in CoRCS-Net were studied. A CoRCS-Net modeling method, and the elementary evolving operators and algorithms for CoRCS-Net dynamic evolution were proposed. We also introduced the concepts of *combinable strength* and *variation of combinable strength*, and used *combinable strength* to describe the invoking times of a service being invoked for service composition. First we calculated the real time *variation of combinable strength* of an arbitrary service in CoRCS-Net, and then obtained the corresponding *combinable strength* and researched its distribution for all services in CoRCS-Net. The main creative works and contributions of this chapter are as follows:

1. Nine kinds of combinable relationships in CoRCS-Net are investigated, including their definitions, and graphical expression. The concept of *combinable strength* for calculating the degree of the combinable relationship among services is introduced. A rule for identifying whether there are combinable relationships existing among services in CoRCS-Net is proposed.
2. Principles for establishing a composition service network based on the proposed combinable relationships, i.e., CoRCS-Net, are established, as well as its nine elements. Fourteen evolving operators for CoRCS-Net dynamic evolution are designed, and their influences on the combinable strength for the resource vertices in CoRCS-Net are analyzed.
3. By studying the distribution of the combinable strength of service vertices in CoRCS-Net, we have proven in theory that: CoRCS-Net is a scale-free network and is constructed by few 'active services' and a great deal of 'silent services'. In the process of service composition, the invoking times for majority services are very small, while only few services are invoked frequently, and the probability (or invoking times) for the services in a CoRCS-Net to be invoked for service composition

decays as a power-law, following $P(d) = A_5(d - A_6)^{-\varepsilon}$ and $\varepsilon = \frac{A_1}{A_2} + 1$.

However, the CoRCS-Net is a very complex network, and the research on it is just in its infancy. Although we have proven that CoRSC-Net is a scale-free network, the conclusion is only proofed in theory and under many assumptions. The modeling for a real time and practice CoRCS-Net is very complex, as well as its dynamic evolution. The dynamic characteristics of CoRCS-Net such as invoking frequency, robustness, security, reliability, and so on, are recommended for further study.

13

Failure Detection and Recovery in Resource Service Optimal-Allocation

13.1 Introduction

In Figure 6.1 in Chapter 6, when RSD and RSP have a resource service transaction, the transaction will be stopped or will fail if any one of the following phenomena occurs (Tao *et al.*, 2010a; 2010b):

1. The virtual link between two entities (i.e., RSD and RSP) is disconnected, or the bandwidth between them decreases suddenly due to some reason.
2. Resource service has failures during a transaction process, e.g., resource service quits suddenly, resource service is overloaded, mismatch during resource service composition, etc.
3. Submitted task has failures during a transaction process, e.g., task cancellation, task suspension, task requirement cannot be satisfied, task and resource service are mismatched, etc.
4. Wrong access permissions; full disk or memory outages, memory leak, numerical exception, etc.

The above phenomena are all defined as failure during MGrid resource service scheduling. In order to enable MGrid resource service scheduling to be successfully performed, the following three issues must be effectively addressed:

1. What failures will be generated during MGrid resource service scheduling?
2. How does the system learn that a failure has generated? Namely failures detection.
3. How to deal with the detected failures? Namely, failure recovery.

Motivated by addressing the above issues and realizing MGrid resource service scheduling, we will carry out investigations into failure management in the MGrid system in this chapter, including failure classification and definition, failure management system and its architecture, failure detection, and failure recovery.

The main works and contributions of this chapter are:

1. Classification and definition of failures in MGrid are presented. MGrid resource service failures are classified into four categories, including virtual link related failures, resource service related failures, task related failures, and application related failures; and 13 failures are defined in detail.
2. A failure management system is proposed. Corresponding failure detection mechanisms and methods according to each defined failure are proposed.
3. An ECA (event–condition–action) rules based failure recovery mechanism is put forward, and the corresponding failures recovery methods are presented.

13.2 Related Works

13.2.1 Failure Management in Manufacturing System

To the best of our knowledge, MGrid resource service failure management, including failure detection and recovery, which is the core part and one of the key technologies to realize a dynamic MGrid executing platform, has not been adequately addressed. Some works about failure management in the manufacturing system are as follows:

Alcaraz-Mejia *et al.*, (2006, 2007) studied fault recovery of discrete manufacturing systems through the reconfiguration of the controller, and proposed a technique for fault recovery that exploits the redundancies included in the Petri net model. They assumed that the system is controlled using a Petri net regulation controller scheme, and presented a novel control reconfiguration technique for fault recovery. Their main contribution is an incidence matrix based technique and an abelian group of Petri net modules to recompute the controller each time that a resource fails. Fries *et al.*, (2007) researched multi-agent fault diagnosis in manufacturing systems using multiple fuzzy intelligent agents that examine the problem domain from a variety of perspectives. Du *et al.*, (2006) discussed the application of a distributed sensor system for fault detection and isolation in multistage manufacturing systems. Fletcher *et al.*, (2001) researched a model of fault-tolerant holonic manufacturing systems (HMS) where each holon's activities are controlled by an intelligent software agent. Multiple agents schedule actions, resolve conflicts and manage information to produce, transport, assemble, inspect, and store customized products. The model provides robustness and distribution transparency across a shop floor where unpredictable failures occur with machines, control software, and communication networks. Ulieru *et al.*, (2000) introduced a novel distributed fault recovery method that could be useful in the design of re-configurable holonic manufacturing systems. It can accomplish fault-recovery by redistribution of tasks in case a resource fails. It also can reconfigure production by dynamic reallocation of resources for newly received high-priority tasks.

Byrne and Sheahan (2005, 2006) used integrated risk minimization (IRM) methodology to reduce potential risks of failures in high volume manufacturing, and the proposed methodology was implemented and tested in an electroplating process. They also researched the technology and method to automate the inspection of surface quality through the application of inline vision systems (i.e., a high-speed greyscale system, a standard color system and a high-speed color system), and applied them to a high-volume connector manufacturer. The overall benefit of the method as shown in their study was an improvement in the defect rate, and a reduction in risk priority number on the failure, modes and effects analysis of the process.

Mannar and Ceglarek et al., (2004) proposed a machine learning approach using rough sets to realize rapid diagnosis of dimensional failures. In their later work (Mannar and Ceglarek et al., 2006), they proposed a novel fault region localization methodology to link warranty failures to manufacturing measurements (hence, to design and process parameters) for diagnosing warranty failures and to perform tolerance revaluation. The methodology consists of identifying relations between warranty failures and design/process variables using rough sets-based analysis on training data consisting of warranty information and manufacturing measurements. The methodology expands the rough set-based analysis by introducing parameters for inclusion of noise and uncertainty of warranty data classes. Based on the identified parameters related to the failure, a revaluation of the original tolerances can be performed to improve product robustness.

The failures considered in the above studies primarily result from physical components of the manufacturing system. Although these studies proposed efficient failure-tolerance (including both failure detection and recovery) in a traditional manufacturing system, their schemes are not suited for a SOA (service-oriented architecture) manufacturing system such as MGrid. In an MGrid system, all resources or manufacturing capabilities are encapsulated into service, and during an MGrid resource service scheduling process, the system does not consider the physical state (such as temperature, pressure) of resources. It only considers the availability of the encapsulated services, such as accessibility and load.

13.2.2 Failure Management in Distributed System

In the current grid application system, tools such as Globus (Foster *et al.*, 1997; 1998), Condor-G (Frey *et al.*, 2002), Nimnod-G (Buyya *et al.*, 2000), Ninf-G (Tanaka *et al.*, 2003) and others (Hwang *et al.*, 2003; Iamnitchi *et al.*, 2000; Dialani *et al.*, 2002), have been addressing either failure tolerance issues and do not provide a general mechanism for resolving failures, or different applications have been adopting *ad hoc* fault tolerance mechanisms which cannot be reused, nor shared among them. Furthermore, several related works about failure management in distributed computing system can be discovered in literature: Fang *et al.*, (2007) proposed a fault tolerant web service SOAP (Simple Object Access Protocol) called fault tolerant SOAP (or FT-SOAP) through which web services can be built with higher resilience to failure; Luckow *et al.*, (2008) presented a fault-tolerant service framework for MPI application in a grid system, namely Migol; Domingues *et al.*, (2007) investigated sabotage-tolerance and trust management in desktop grid computing, and presented a mechanism to detect sabotage attempts based on checkpoint comparison; Lee *at al.*, (2005c) studied the fault tolerance service in grid computing, and classified failures into process failure, processor failure, and network failure. A fault detector and fault manager components were also proposed by them; Khanli *et al.*, (2008) researched grid resource fault management based on ECA (event-condition-action) rules. Huedo *et al.*, (2006) investigated the evaluation method of the reliability of computational grid from the end user's point of view, and proposed six kinds of failures.

The resources considered in the above research are primarily computational resources, such as processors, memories, storages, networks, processes, which are inadequate in an MGrid system. In addition, the detailed fault detection implementing methods have not been studied, and they do not provide a detailed and executable mechanism or method for handling the detected failures either.

Therefore, in this chapter, based on the above work, we investigated all potential failures during MGrid resource service scheduling, and defined four categories and 13 failures in detail, they are (Tao *et al.*, 2010b):

1. Virtual link related failures,
2. resource service related failures,
3. task related failures, and
4. application related failures.

A failure management system is proposed for providing failure tolerance service. Detailed detecting mechanisms and methods for each failure are put forward, as well as handling methods for each detected failure.

13.3 Define and Classification of MGrid Failure

It is a failure if and only if, one of the following two conditions is satisfied (but not both) (Lee *at al.*, 2005c):

- A resource service stops due to a resource crash.
- The availability of a resource does not meet the minimum levels of a task.

As discussed in the introduction, all the factors such as virtual link, resource service, task, and application involved in MGrid resource service scheduling could cause failures. Therefore, we investigated and defined the possible failures from the above four categories, namely:

- Virtual link related failure.
- Resource service related failure.
- Task related failure.
- Application related failure.

13.3.1 Virtual Link Related Failure

Virtual link (VL) is a general connection between two entities (e.g., resource service provider (RSP) and resource service demander (RSD), resource and resource, user and user) in an MGrid. VL can be made up of various elements, such as routers, computers, internet/intranet, wireless communication channels, and so forth. In this Chapter, we primarily consider the following two failures resulting from virtual link (Tao *et al.*, 2010b):

1. Virtual link disconnected failure (**VL_Disconnect_Failure**): VL_Disconnect_Failure is defined as the failure caused by disconnected virtual link. The connection of virtual link could unpredictably fail due to power, hardware, base software, and so forth. Without the connection, any service transaction in MGird has to be stopped.
2. Inadequate Bandwidth failure **(Bandwidth_Failure):** MGrid is a Web-based manufacturing mode, so the network bandwidth critical to the realizing of MGrid due to network traffic varies dynamically. We define the failure which causes invalid resource or task connection, or date transmission and accessing due to inadequate bandwidth in MGrid resource service scheduling process as a Bandwidth_Failure.

13.3.2 Resource Service Related Failure

In MGrid resource service scheduling process, resource services are the scheduling objects and the executor of tasks. Therefore, the changing of resource service state, such as quit from MGrid system, overloading, QoS ability decreasing, would result in failures of MGrid resource service scheduling. There are many failures caused by resource service, taking into account their importance in MGrid resource service scheduling, we will primarily consider the following four kinds of failures resulting from resource services:

1. Resource service quit failure (**RS_Quit_Failure**): In an MGrid, RSP can add or move a resource service from it freely at any time; moreover, a resource service can be shut down by system administrators freely due to a certain reason which will cause the executing task (or the task allocated to the resource service) to not be completed in time. Therefore, these failures caused by resource service quit are defined as an RS_Quit_Failure.
2. Resource service overload or saturation failure (**RS_Overload_Failure**): Resource services are shared by all users in an MGrid; hence resource load changes dynamically. We only consider the failure resulting from resource overload if it causes a task to not be completed in accordance with its quality requirements in time, and define these failures as RS_Overload_Failure.
3. Resource service composition failure (**RS_Composition_Failure**): Due to the complexity of the manufacturing process, a submitted manufacturing task (or resource service request) in MGrid usually cannot be completed by invoking only one resource service. It is usually decomposed into several subtasks, and completed by invoking several resource services in sequence, which is defined as resource service composition in Chapter 7. Assume that RS_i and RS_j are arbitrary two neighboring resource services in a composition path, and RS_j is the direct successor (Zeng *et al.*, 2004a) of RS_i. During MGrid resource services composition process, the following four kinds of failures would be generated between RS_i and RS_j:

(I) **Conceptual mismatch between RS_i and RS_j**. For example:

- RS_i provides "cutting" service, but RS_j requires "planning" service, so there exist a gap between RS_i and

RS_j. We define this kind of concept relationship as $RS_i \perp RS_j$.

- RS_i is a superclass of RS_j. We define this kind of concept relationship as $RS_i \supset RS_j$.
- RS_i is a subclass of RS_j. We define this kind of concept relationship as $RS_i \subset RS_j$.
- RS_i is a part of RS_j. We define this kind of concept relationship as $RS_i \in RS_j$.

(II) **Data mismatch between RS_i and RS_j.** For example:

- RS_i and RS_j have the same data type associated with a parameter (such as input parameter, output parameter, QoS parameter), but they have different units, e.g., RS_i uses feet, but RS_j requires meter. We define this kind of relationship as $RS_i \approx RS_j$.
- RS_i and RS_j have the same parameter concepts, but different data type. E.g., RS_j employs "string" data type to define parameters, but RS_i requires "character" data type for the corresponding parameters. We define this kind of relationship between RS_i and RS_j as $RS_i \neq RS_j$.

(III) **Attributes mismatch between RS_i and RS_j:** RS_j has more attribute requirements than RS_i can offer. E.g., RS_j requires five attributes, but RS_i only have two. We define this kind of relationship between RS_i and RS_j as $RS_i < RS_j$.

(IV) **QoS inconsistent:** In resource service composition, the QoS consistency between two neighboring resource services must be taken into account. If the consistency between two neighboring resource services is bad, it would result in the failure of the whole composition.

4. Resource service ability changed failure (**RS_AbilityChange_Failure**): During the execution, a resource service's ability may change suddenly for several reasons, such as hardware failure, etc., which would cause the task it served having to be migrated or rescheduled. We define the failure that results in the

task allocated to it that cannot be accomplished successfully due to the ability change of resource service as an RS_AbilityChange_Failure.

13.3.3 Task Related Failure

Task is the executed objects in MGrid, and a task could be canceled, suspended, or mismatched in a resource service due to some reason, which will result in failures of MGrid resource service scheduling. There are many failures caused by task, taking into account their importance in MGrid resource service scheduling, we will primarily consider the following four kinds of failures resulting from tasks (Tao *et al.*, 2010b):

1. Task cancellation failure (**Task_Cancel_Failure**): A submitted or executing task could be cancelled for several reasons, e.g., stopped by user when it exceeds the wall time limit, or by the system administrator to preserve system performance, and so forth. We define the failure in MGrid resource service scheduling caused by task cancellation as Task_Cancel_Failure.
2. Task suspension failure (**Task_Suspension_Failure**). A task could be suspended for several reasons, e.g., (a) the requirements of the task are too onerous to be satisfied with the current available time; (b) higher priority tasks continuously entering the system could prevent the executing of the task; and (c) the system administrator could freely hold a waiting task for system performance, and so forth.
3. Mismatch failure between a task and resource service (**Task_RS_Mimatch_Failure**): If an assigned resource service cannot satisfy the requirements of a task due to some reason (e.g., mismatch, the ability of the resource service changed suddenly, etc.), then the system has to search and assign a new resource service for the task, or the task has to be migrated. Failures caused by mismatch between a task and resource service during MGrid resource service scheduling are defined as Task_RS_Mimatch_Failure.

4. Changed task requirements failure (**Task_Require-Change_Failure):** The requirement of a task could be changed by the user if the requirements of the task are changed into a higher level, such as higher precision, shorter execution time, lower price, and so forth. This will cause a rescheduling for the task. Failures resulting from the changes of task requirements are defined as Task_RequireChange_Failure.

13.3.4 Application Related Failure

1. Trust Failure **(Trust Failure)**: In an MGrid, there are two main kinds of user: (a) resources enterprise or resource service provider (RSP), and (b) user enterprise or resource service demander (RSD). The former, RSP publishes its idle resource service to meet users' requirements in MGrid. The later, RSD searches the optimal manufacturing resource and service it requires, and selects the corresponding partner to establish the virtual collaboration manufacturing net. When RSD and RSP have resource service transaction with each other, the following problems will be generated:

   From the RSD point of view, before submitting its manufacturing task, there are three problems as described in Chapter 6:

- RSD must know whether the RSP (to be visited) who is providing the required resource service is trustworthy, or has the ability to faithfully accomplish the task;
- If the answer to the first problem is yes, then RSD must determine if the RSP will trust in it and be willing to accept and execute the submitting manufacturing task;
- RSD must confirm that the RSP will not destroy the submitting manufacturing task, and the latter will not provide unsuitable results and other uncertainties.

   From the RSP point of view, before the submitted manufacturing task is executed, there exist two problems:

- RSP must check the validity of the manufacturing task submitted by RSD (to be served), and make sure the RSD will not submit unsuitable or destructive tasks which would destroy the MGrid system and similar things;
- RSP must be certain that the task was not destroyed and tampered with by the third sides before submission.

Therefore, the undesired actions or results during resource service transaction resulting from trust problems about RSDs and RSPs will cut down the QoS of MGrid, such as the quality of resources scheduling, the quality of resources management, etc.

2. Application design or coding failures (**App_DesignCode_Failure**), such as memory leak, numerical exception, and so forth.
3. Accessing right failures (**App_AccessRight_Failure**), such as wrong access permissions, full disk or memory outages.

13.4 Architecture of MGrid Failure Management System

Figure 13.1 shows the architecture of the failure management system (FMS) in MGrid (Tao *et al.*, 2010b). The proposed FMS can provide failure detecting service and failure recovery service during MGrid resource service scheduling. The main components of the proposed FMS include an *information service model* (ISM), an *MGrid resource service optimal-selection system* (MGRSOSS), a *failure detector* (FD), and a *failure recovery* (FR).

(1) *MGrid resource service optimal-selection system* (MGRSOSS)

MGRSOSS is the source of failure generation. As presented by the authors in Chapter 2, the key technology to realize MGrid resource service scheduling is the resource service optimal-selection, including operations of resource service math and search, resource service QoS synthetically processing (QoS extraction, QoS evaluation, QoS comparison), resource service composition, etc. Almost all the failures are generated in these operations. The detailed model,

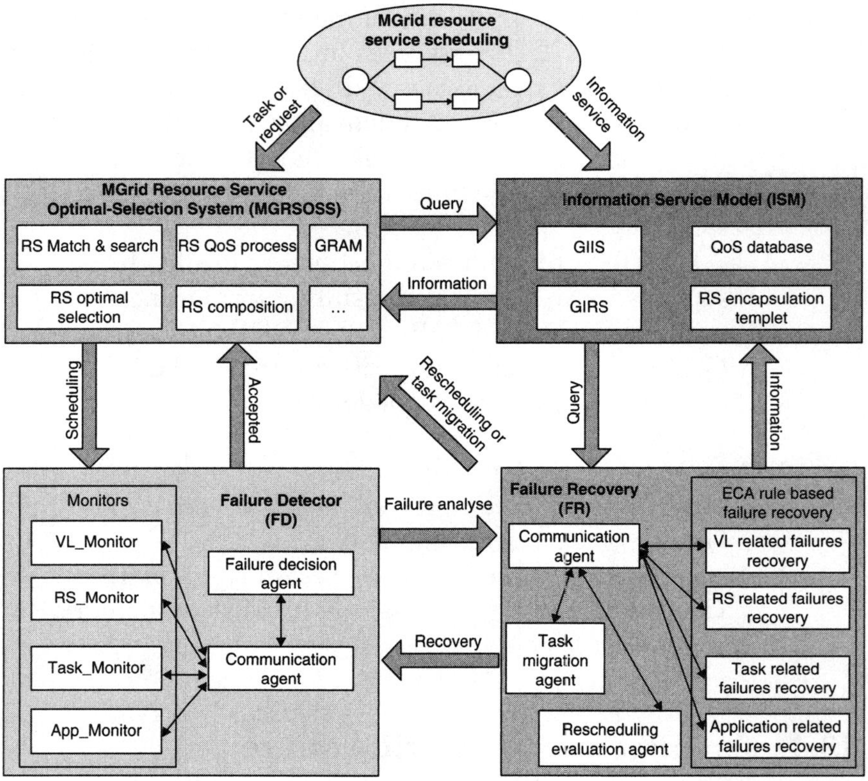

Figure 13.1 Architecture of failure management system in MGrid.

workflow, and algorithms of resource service optimal-selection in MGrid have been studied in detail in Chapter 6–Chapter 10.

(2) *Information service model (ISM)*
ISM provides FD, FR, and MGRSOSS with data and information. It consists of GIIS (Grid Information Index Service) and GRIS (Grid Resource Information Service) which is provided by Globus (Foster *et al.*, 2001a; Du *et al.*, 2002), and the nine kinds of resource service encapsulation templets presented in Chapter 3, and a QoS database for depositing QoS information and datum.

(3) *Failure detector (FD)*
It monitors the state of virtual link, resource service, task, and application involved in MGrid resource service optimal-allocation. It consists of: (a) four *monitors*, i.e., VL_Monitor, RS_Monitor,

Task_Monitor, and App_Monitor, which are responsible for monitoring the state of virtual link, resource service, task, and application involved in MGrid resource service optimal-allocation process; (b) *failure decision agent,* which decides the occurrence of a failure by analyzing states of information of virtual link, resource service, task, and application; it identifies the related failures such as a VL related failure, a RS related failure, a task related failure, or an application related failure; if a failure occurred, it reports the failure information to failure recovery model, and (c) *communication agent,* which manages the communication between detectors and agents.

(4) *Failure recovery* (FR)

It is responsible for analyzing the detected failure, and provides failure recovery service. It consists of: (a) *ECA rule base failure recovery service,* which provides failure recovery service based on ECA rules, detailed methods of ECA rule based failure recovery are presented in Section 13.6. (b) *Communication agent,* which manages the communication between detectors and agents. (c) *Rescheduling evaluation agent,* if the detected failure cannot be recovered based on ECA rules, then the rescheduling evaluation agent evaluates the total system benefit that can be obtained from task migration and decides whether a task migration occurs or not. (d) *Task migration agent,* if it receives a rescheduling result from rescheduling evaluation agent, it requests a reallocation to resource service to the task or restarts the task.

13.5 Detection of MGrid Failure

13.5.1 Detection of Virtual Link Related Failures

(1) Detection of VL_Disconnect_Failure

A **VL_Disconnect_Failure** usually can be detected by the security policy or system middleware embedded in an MGrid. In this chapter, we employ the Globus GRAM (Grid Resource Allocation and Management) service, i.e., GRAM callbacks, to detect the virtual link disconnection failure (Czajkowski *et al.*, 1998). GRAM callbacks notify of submission failures that include connection, authentication and authorization, RSL (Resource Specification Language) parsing and executable or input staging (Huedo *et al.*, 2006).

(2) Detection of Bandwidth_Failure

For **Bandwidth_Failure** we employ two criteria: *communication time* and *probability of success communication* to judge whether a **Bandwidth_Failure** was generated between two entities. Suppose the virtual link between entities A and B is denoted as $VL(A,B)$, the total size of information exchanged through $VL(A,B)$ is $SumInfor(A,B)$, and the speed (bandwidth) of $VL(A,B)$ is $VL(A,B)$, then the estimated *communication time*, $T_c(A,B)$, between A and can be calculated as follows (Tao *et al.*, 2010b):

$$T_c(A,B) = \frac{SumInfor(A,B)}{V(A,B)} + Waite(A,B) \qquad (13.1)$$

$Waite(A,B)$ is the waiting time. Assume the failure intensities of $VL(A,B)$ is $a\,(A,B)$, and the failure intensities of entities A and B are $a\,(A)$ and $a\,(B)$, respectively. Therefore, the *probability of success communication*, $S_C(A,B)$, between A and B can be expressed as:

$$S_C(A,B) = \exp\{-(a(A) + a(B) + a(A,B))T_c(A,B)\} \qquad (13.2)$$

Set the limit of requested communication time between A and B is $T_c^{\circ}(A,B)$, and the lowest requested probability of success communication is $S_C^{\circ}(A,B)$, if $Tc(A,B) > T_C^{\circ}(A,B)$ or $S_C(A,B) < S_C^{\circ}(A,B)$, then the system determines that a **Bandwidth_Failure** occurred.

13.5.2 Detection of Resource Service Related Failures

(1) Detection of RS_Quit_Failure.

In order to detect an **RS_Quit_Failure**, the resource service monitor (i.e., RS_Monitor in Figure 13.1) checks on a regular basis, that each resource service associated with a task is alive. If the resource service fails to respond to RS_Monitor, then the system determines that the resource service has failed or quit, i.e., an **RS_Quit_Failure** occurred. The main advantage of this detection mechanism is that, as the RS_Monitor periodically polls resource services in MGrid, resource services may be found to be faulty prior to the moment when an application or a task would wish to use them; therefore they can be replaced with zero replacement waiting time.

(2) De tection of RS_Overload_Failure.
Let RS_i be an arbitrary resource service, and RS_i is modeled as a vector with four parameters: $RS_i = (B_i, M_i, \Gamma_i, D_i)$,where B_i is the bandwidth of the storage subsystem of RS_i; M_i is the memory capacity; Γ_i is the set of tasks allocated to RS_i, assume there are k tasks allocated to RS_i, then $\Gamma_i = \{Task_1, Task_2, \cdots, Task_j, \cdots, Task_k\}$; D_i is an array of datasets accessed and generated by an application.

Let an arbitrary task be expressed by $Task_j = (n_j^{RS}, d_j, et_j, td_j, m_j)$, where n_j^{RS} is the number of requested resource services; d_j is the total size of information that has to be exchanged; et_j is the required execution time; td_j is the required time spent on data communication, $td_j = \sum_{i=1}^{n} td_{i,j}$ and $td_{i,j} = \frac{d_{i,j}}{V(i,j)}$, of which $d_{i,j}$ is the amount of accessed data and $V(i,j)$ is the bandwidth of the jth dedicated storage subsystem (i.e.,RS_i); m_j is the memory requirements of $Task_j$.

We make use of the following three equations to quantify the communication, computational, and data accessing load of RS_i (Tao *et al.*, 2010b):

$$Communication_{RS_i} = \sum_{Task_j \in \Gamma_i} \left(\frac{d_{i,j}}{V(i,j)} \right) \tag{13.3}$$

$$Computational_{RS_i} = \sum_{Task_j \in \Gamma_i} \left(n_j^{RS} \cdot et_j \right) \tag{13.4}$$

$$Data_{RS_i} = \sum_{Task_j \in \Gamma_i} d_{i,j} \tag{13.5}$$

Set the limits of communication time, execution time, and data capability of RS_i are $Lim_{RS_i}^{\mathrm{CT}}$, $Lim_{RS_i}^{\mathrm{ET}}$, and $Lim_{RS_i}^{\mathrm{DC}}$ respectively. Therefore, if any one of the following three conditions is satisfied,

$$Computational_{RS_i} = \sum_{Task_j \in \Gamma_i} \left(n_j^{RS} \cdot tc_j \right) > Lim_{RS_i}^{\mathrm{ET}} \tag{13.6}$$

$$Communication_{RS_i} = \sum_{Task_j \in \Gamma_i} \left(\frac{d_{i,j}}{V(i,j)} \right) > Lim_{RS_i}^{\mathrm{CT}} \tag{13.7}$$

$$Data_{RS_i} = \sum_{Task_j \in \Gamma_i} d_{i,j} > Lim_{RS_i}^{\mathrm{DC}} \tag{13.8}$$

Then the system determines that an **RS_Overload_Failure** occurred.

(3) Detection of RS_Composition_Failure.

In an MGrid, RS_i and RS_j are annotated with OWL-S and the match and search mechanisms described in Chapter 4 are used in order to identify a list of possible web services. Identification rules of the above RS_Composition_Failure are as follows:

(a) Conceptual mismatch detection rules:

- If $RS_i \subset RS_k \,\&\&\, RS_k \not\subset RS_j \Rightarrow RS_i \perp RS_j$, then system determines that a conceptual mismatch (concept gap) occurred between RS_i and RS_j.
- If $RS_i \subset RS_k \,\&\&\, RS_k \subset RS_j \Rightarrow RS_i \subset RS_j$, then system determines that a conceptual mismatch (RS_i is a super-class of RS_i, RS_i cannot satisfy RS_j's requirements) occurred between RS_i and RS_j.

(b) Data mismatch detection rules:

- If $\mathrm{DU_{nit}T_{ransfer}}(RS_i) = RS_j$ then $RS_i \approx RS_j$, then system determines that a data mismatch (same data type but different unit) occurred. $\mathrm{DU_{nit}T_{ransfer}}()$ is a data unit transformation function.
- If $\mathrm{DT_{ype}T_{ransfer}}(RS_i) = RS_j$ then $RS_i \neq RS_j$, then system determines that a data mismatch (same concept but different data type) occurred. $\mathrm{DT_{ype}T_{ransfer}}()$ is a data type transformation function.

(c) Attributes mismatch detection rules

- If $RS_i < RS_j \,\&\&\, RS_i \cap \varsigma(RS_j) \neq \Phi$, then it demonstrates that RS_i's attributes cannot satisfy RS_j's requirements, and system determines that an attributes_Failure occurred. $\varsigma()$ is a split function.
- The above detection rules to **RS_Composition_Failure** are examples, but not an exhaustive list.

(d) QoS inconsistent detection rules

- Assume $Dim(Q_j^{inputs})$ and $Dim(Q_i^{outputs})$ are the number of the parameters of RS_i and RS_j, respectively, for

$\forall 1 \leq m \leq Dim(Q_j^{inputs}), \exists 1 \leq n \leq Dim(Q_i^{outputs})$ if RS_i and RS_j subject to

$$\begin{cases} q_{j.m}^{inputs} \leq q_{i.n}^{inputs} & (q_{j.m}^{inputs} \text{ is a single and negative parameter}) \\ q_{j.m}^{inputs} \geq q_{i.n}^{inputs} & (q_{j.m}^{inputs} \text{ is a signal and positive parameter}) \\ q_{j.m}^{inputs} \subseteq q_{i.n}^{inputs} & (q_{j.m}^{inputs} \text{ is an interval number parameter}) \end{cases} \tag{13.9}$$

It demonstrates that RS_i and RS_j are QoS consistent, and the composition path they located is effective. Otherwise, the system determines that an **RS_Composition_Failure** occurred between RS_i and RS_j.

13.5.3 Detection of Task Related Failures

(1) Detection of T ancel_Failure and Task_Suspension_Failure

The task exit conditions can be checked through the drama_wif*() routines included in the DRMAA interface (Rajic *et al.*, 2003), which is fully implemented by GridWay (Huedo *et al.*, 2004), so that the appropriate corrective action can be undertaken if application-related failures occur.

(2) Detection of Task_Suspension_Failure

In order to detect a Task_Suspension_Failure, the Task_Monitor in Figure 13.1 monitors the task locally, and checks on a regular basis the state of each task. A Task_Suspension_Failure is detected when a task remained in its *SUSPEND* state longer than the given threshold specified by the user.

(3) Detection of Task_RS_Mismatch_Failure

Assume $Task_j$ is the requirement description of a subtask or a task that cannot be decomposed (i.e., the requested resource service), and RS_i is a candidate resource service for $Task_j$, the matching model between $Task_j$ and RS_i are defined as follows:

$$\text{Match}(RS_i, Task_j) = \prod \begin{pmatrix} \text{Match}_{bas}(RS_i, Task_j), \text{Match}_{i/o} \\ (RS_i, Task_j), \text{Match}_{QoS}(RS_i, Task_j) \end{pmatrix} \tag{13.10}$$

Where

- $$\text{Match}_{bas}(RS_i, Task_j) = \begin{pmatrix} \omega_1 \times \text{Match}_\text{w}(RS_i.ServiceName, Task_j. \\ ServiceName) + \\ \omega_2 \times \text{Match}_\text{T}(RS_i.ServiceDescription, \\ Task_j.ServiceDescription) \end{pmatrix} \quad (13.11)$$

- $$\text{Match}_{i/o}(RS_i, Task_j) = \begin{pmatrix} \sum_{i_1=1}^{\text{n}_\text{d}} \omega_{d_{i_1}} \text{Match}_\text{D}(RS_i.d_{i_1}, RS^j.d_{i_1}) + \\ \sum_{i_2=1}^{\text{n}_\text{c}} \omega_{\text{c}_{i_2}} \text{Match}_\text{c}(RS_i.c_{i_2}, Task_j.c_{i_2}) + \\ \sum_{i_3=1}^{\text{n}_\text{w}} \omega_{\text{w}_{i_3}} \text{Match}_\text{w}(RS_i.w_{i_3}, Task_j.w_{i_3}) \end{pmatrix} \quad (13.12)$$

- $$\text{Match}_{QoS}(RS_i, Task_j) = \sum_{i=1}^{\text{m}} \omega_{QoS_{i_4}} \text{Match}_\text{d}(RS_i.\text{QoS}_{i_4}, Task_j.\text{QoS}_{i} \quad (13.13)$$

- $\prod()$ is the integrated match processing function. It integrates the matching results of $\text{Match}_{bas}(RS_i, Task_j)$, $\text{Match}_{QoS}(RS_i, Task_j)$, $\text{Match}_{i/o}(RS_i, Task_j)$, and *QoS matching*, and generates an integrated matching result.

The detailed definitions and explanations of Equations 13.10–13.13 are described in Chapter 6. It assumes that the declared threshold values are ζ_{bas} (i.e., the basic matching threshold), $\zeta_{i/o}$ (i.e., the I/O matching threshold), ζ_{QoS} (i.e., the QoS matching threshold) and ζ (i.e., the total matching threshold), therefore:

(a) If $\text{Match}_{bas}(RS_i, Task_j) < \zeta_{bas}$, then system determines that a basic mismatch failure occurred between RS_i and $Task_j$.
(b) If $\text{Match}_{i/o}(RS_i, Task_j) < \zeta_{i/o}$, then system determines that an I/O mismatch failure occurred between RS_i and $Task_j$.
(c) If $\text{Match}_{QoS}(RS_i, Task_j) < \zeta_{QoS}$, then system determines that a QoS mismatch failure occurred between RS_i and $Task_j$.
(d) If $\text{Match}(RS_i, Task_j) < \zeta$, then system determines that an integrated mismatch failure occurred between RS_i and $Task_j$.

RS_AbilityChange_FailureandTask_RequireChange_Failuredetectionmethodsarethesameasthoseof Task_Resource_Mismatch_Failure.

13.5.4 Detection of Application Related Failures

(1) Detection of Trust_Failure

Trust is the belief or faith that a trusting entity (e.g., RSD) has in a trusted entity's ability (e.g., RSP) in a given context and time slot, that gives it confidence to carry out a transaction with the trusted entity. Trust is realized by the concept of *trust-QoS relationship* as described in Chapter 6. Each *trust-QoS relationship* denotes a trust value from a trusting entity to a trusted entity, for a particular service content and time slot. The *trust-QoS relationship* is represented as a tuple: (A,C,Γ,R,E,B,T) in Chapter 6.

Provided that x and y stand for two entities that have resource service transaction in an MGrid, and x is the trusting entity while y is the trusted entity. According to the works in Chapter 6, the trust value between x and y, $T_{x\to y}$, can be evaluated as follows:

(a) If x and y belong to a same resource domain, then

$$T_{x\to y} = \alpha \times d_{td} + \beta \times r_{td}$$

Where:

- d_{td} is the direct trust-QoS evaluation value between x and y.
- r_{td} is the comprehensive combined recommended trust-QoS evaluation value of each recommender's individual recommended trust with y.
- α ,β are the weights of d_{td} and r_{td} respectively.

(b) If x and y belong to two different resource domains A and B respectively, then

$$T^*_{x\to y} = W^B_y \times (\alpha^* \times d_{td}^* + \beta^* \times r_{td}^*)$$

Where

- d_{td}^* is the direct trust evaluation value of A about B.
- r_{td}^* is the comprehensive combined recommended trust evaluation value of each recommender's individual recommended trust with B.

- α^* and β^* are the weights of d_{td}^* and r_{td}^* respectively.
- W_y^B is the trust weight of entity y occupied in resource domain B.

Detailed algorithms for evaluating d_{td} and r_{td} can be found in Chapter 6.

Set $T_{x \to y}^\circ$ is the lowest trust degree of x's requirement to y during the resource service transaction, if

$$T_{x \to y} < T_{x \to y}^\circ$$

Then system determines that a **Trust_Failure** occurred.

(2) Detection of App_DesignCode_Failure and App_AccessRight_Failure

We employed the Globus GRAM (Grid Resource Allocation and Management) service, i.e., GRAM callbacks, to detect App_DesignCode_Failure and App_AccessRight_Failure. GRAM callbacks notify submission failures that include connection, authentication and authorization, RSL (Resource Specification Language) parsing, and executable or input staging (Huedo et al., 2006).

13.6 MGrid Failure Recovery Based on ECA Rules

In this chapter, we will introduce a novel idea in which ECA (Event-Condition-Action) are used to provide failure recovery service. ECA rules are one way of implementing this kind of functionality (Papamarkos *et al.*, 2006). The reason we use ECA rules is due to the following advantages (Khanli *et al.*, 2008):

1. ECA rules allow functionality to be specified and managed within a rule base rather than being dispersed in diverse programs, thus enhancing the modularity, maintainability, and extensibility of applications.
2. ECA rules have a high-level, declarative syntax, and are thus amenable to analysis and optimization techniques which cannot be applied if the same functionality is expressed directly in application code.

3. ECA rules are a generic mechanism that can abstract a wide variety of reactive behaviors, in contrast to an application code that is typically specialized to a particular kind of reactive scenario.

Reference to the structure of traditional ECA rules, and the proposed architecture for processing failure during resource service scheduling based on ECA rules, is shown in Figure 13.2 (Tao *et al.*, 2010b).

Its working flow is described as follows. During MGrid resource service scheduling process, the failure detector will detect if the failure occurred according to the mechanisms and algorithms described in Section 13.5; the event detector embedded in the scheduling system waits for an event (i.e., detected failure), then evaluates and checks if the conditions of the triggered rule is satisfied. When the condition is satisfied and the rule has not been marked for dynamic conflict, the corresponding action is executed immediately.

An ECA rule has the general syntax of:

On *event* **If** *Condition* **Do** *Action*.

- An *event* is an incident that triggers a rule. In this chapter it is classified into four primitive events, namely VL event (VL_Event), resource service event (RS_Event), task event (Task_Event), and application event (App_Event). VL_Event is generated from VL related failures, such as VL_Disconnect_Failure and Bandwidth_Failure. RS_Event is generated from resource service related failures, such as RS_Quit_Failure, RS_Overload_Failure, and RS_Composition_Failure. Task_Event is generated from task related failures, such as Task_Cancel_Failure, Task_Suspension_Failure, and Task_RS_Mismatch_Failure. App_Event is generated from application related failures, such as Trust_Failure, App_DesignCode_Failure, and App_AccessRight_Failure.
- The *condition* of a rule is a Boolean statement that must be satisfied in order to activate a rule. It is described in terms of an XPath expression (Jung *et al.*, 2007), and the expression in the condition may refer to values from the event definition.
- The *action* part contains the instruction that is executed when a triggered rule is activated.

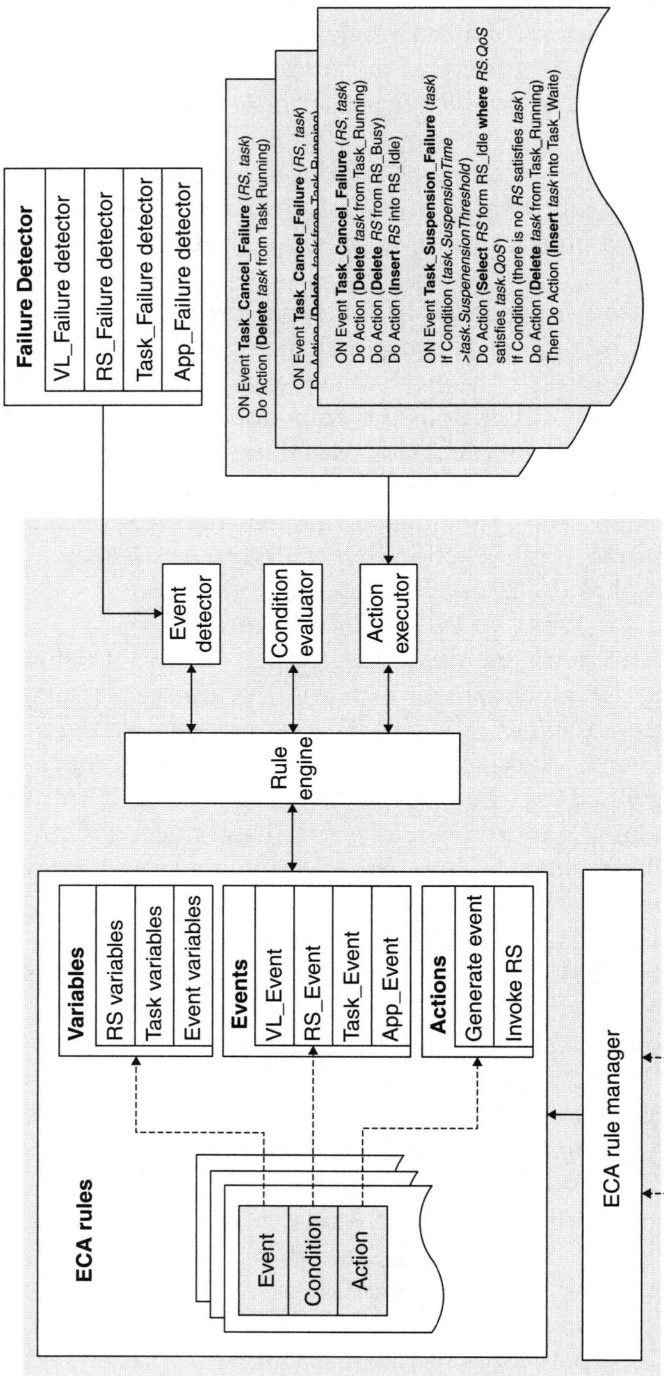

Figure13.2 ECA based MGrid resource service scheduling failure recovery.

In our proposed work, ECA rules are used directly to provide failure recovery service during MGrid resource service scheduling.

Before introducing the corresponding failure recovery ECA rules, we will first investigate the life-cycle of the two main factors that cause failure in the MGrid scheduling process: resource service and task.

During the whole process of MGrid resource service scheduling, the state of a resource service can be classified into three categories: (a) *ARRIVAL*: a resource service is published and inserted into RS_Advertise queue; (b) *IDLE*, a resource service is ready to execute a task at any time and it is inserted into RS_Idle queue; and (c) *BUSY*, a resource service is executing a task and it is inserted into RS_Busy queue. If a resource service has completed its allocated tasks, or the resource service is improper to the allocated task, the resource service then will be deleted from RS_Busy queue and inserted into RS_Idle queue. The whole life-cycle of a resource service and its state transformation in the MGrid resource service scheduling process are shown in Figure 13.3.

In the same way, the state of a task can be classified into four categories: (a) *ARRIVAL*: a task is submitted by user and inserted into Task_Arrival queue; (b) *WAIT*, a task is ready to be executed at any time when a proper resource service is free, and any task on *WAIT* state is inserted into Task_Wait queue; (c) *RUNNING*: a task is being executed on a resource service, and any task on *WAIT* state is inserted into Task_Running queue; and (d) *FINISHED*: a task has been completed and it is inserted into Task_Finished queue. The whole life-cycle of a task and its state transformation during MGrid resource service scheduling are shown in Figure 13.4.

Based on the above works, we designed the following ECA rules for the recovery of the above mentioned failures in the MGrid resource service scheduling process:

ON Event **RS_Quit_Failure (***RS, task***)**
  Do Action **(**delete *RS* from RS_Busy**)**
  If Condition (Select *RS* from RS_Idle where *RS.QoS* satisfies *task.QoS*)
  Do Action Match *RS* and *task*
  If Condition (there is no *RS* in RS_Idle satisfies *task*)
  Do Action (Delete *task* from Task_Running)
  Then Do Action (Insert *task* into Task_Wait)

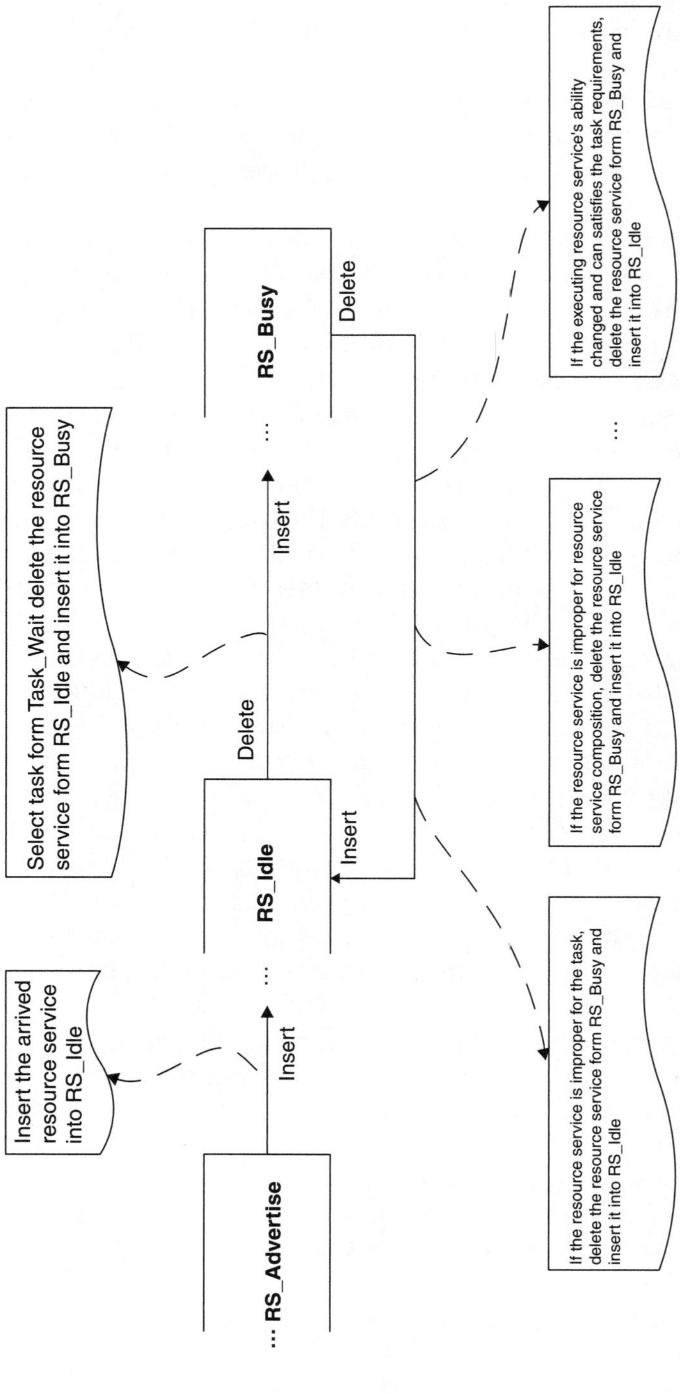

Figure 13.3 Life-cycle of a resource service in MGrid resource service scheduling.

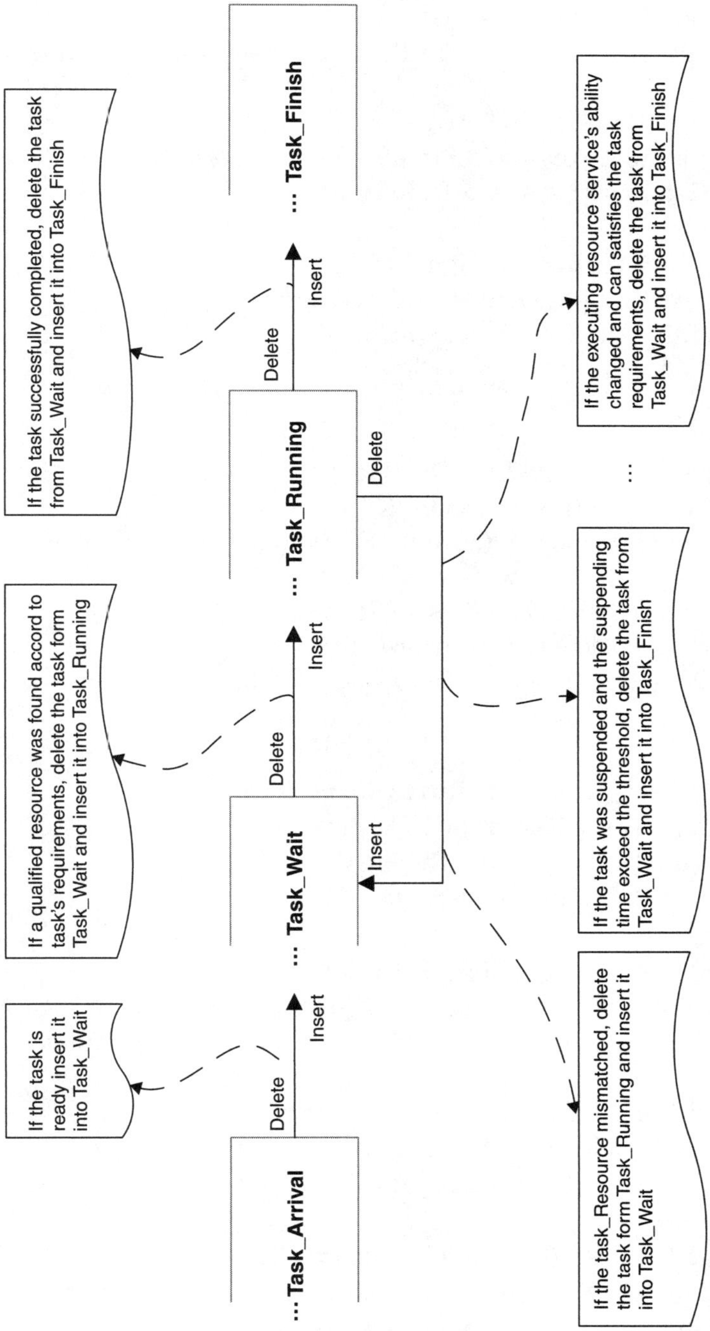

Figure 13.4 Life-cycle of task in MGrid resource service scheduling.

ON Event **RS_Overload_Failure (*task*)**
Do Action {Insert *task* into Task_Wait}
If Condition (Select *RS* from RS_Idle where *RS.QoS* satisfies *task.QoS*)
Do Action Match *RS* and *task*
If Condition (there is no *RS* in RS_Idle satisfies *task k*
Do Action {Insert *task* into Task_Wait}

ON Event **RS_AbilityChange_Failure (***RS, task*)
Do Action (compare *RS.QoS* and *task.QoS*)
If Condition (*RS.QoS* satisfies *task.QoS*)
Do Action (Exit)
Else
{
Do Action (Delete *RS* from RS_Busy)
Then Do Action (Insert *RS* into RS_Idle)
And Do Action (Select *RS* from RS_Idle where *RS.QoS* satisfies *task.QoS*)
If Condition (there is no *RS* satisfies *task*)
{Do Action (Delete *task* from Task_Running)
Then Do Action (Insert *task* into Task_Wait)}
}

ON Event **Task_Cancel_Failure (***RS, task***)**
Do Action (Delete *task* from Task_Running)
Do Action (Delete *RS* from RS_Busy)
Do Action (Insert *RS* into RS_Idle)

ON Event **Task_Suspension_Failure (***task***)**
If Condition (*task.SuspensionTime >task.SuspenensionThreshold*)
Do Action (Select *RS* form RS_Idle where *RS.QoS* satisfies *task.QoS*)
If Condition (there is no *RS* satisfies *task*)
Do Action (Delete *task* from Task_Running)
Then Do Action (Insert *task* into Task_Wait)

ON Event **Task_RS_Mismatch_Failure (***RS, task*)
Do Action (Delete *RS* from RS_Busy)
Then Do Action (Insert *RS* into RS_Idle)
Do Action (Select *RS* form RS_Idle where *RS.QoS* satisfies *task.QoS*)
If Condition (there is no *RS* satisfies *task*)

Do Action (Delete *task* from Task_Running)
Then Do Action (Insert *task* into Task_Wait)

ON Event **Task_RequireChange_Failure (***RS, task***)**
Do Action (compare *RS.QoS* and *task.QoS*)
If Condition (*RS.QoS* satisfies *task.QoS*)
Do Action (Exit)
Else Do Action (Delete *RS* from RS_Busy)
Then Do Action (Insert *RS* into RS_Idle)
And Do Action (Select *RS* from RS_Idle where RS.QoS satisfies task. QoS)
If Condition (there is no *RS* satisfies *task*)
Do Action (Delete *task* from Task_Running)
Then Do Action (Insert *task* into Task_Wait)

ON Event **Trust_Failure (***x, y, task, RS***)**
// here *x* denote the resource service demander (RSD) or the entity who submit
//the *task*, and *y* denoted the resource service provider (RSP) who provides *RS*
If condition ($T_{x \to y} < T_{x \to y}^{\circ}$)
Do Action (Delete *RS* from RS_Busy)
Then Do Action (Insert *RS* into RS_Idle)
Do Action (Select a new *RS* belong to RSP *z* form RS_Idle where $T_{x \to z} \geq T_{x \to z}^{\circ}$)
If condition (there is no *RS* satisfies the *task* of *x*)
Do Action (Delete *task* from Task_Running)
Then Do Action (Insert *task* into Task_Wait)

The above ECA rules to failure recovery are examples, but not an exhaustive list.

13.7 Implementation and Simulation

In order to verify the rationality and usability of our proposed methods, we combined our failure detection and recovery method with *Min-min* (Tracy *et al.*, 2001) scheduling method, and named it as *F-Min-min* scheduling method. *Min-min* scheduling method is described as follows:

- The *Min-min* heuristic begins with the set of tasks. Then the set of minimum completion times for each task is found. Next, the task with the overall minimum completion time from the task set is selected and assigned to the corresponding resource services. Last, the newly assigned task is removed from the task set, and the process repeats until all tasks are assigned. *Min-min* is based on the minimum completion time (Tracy *et al.*, 2001).

We compared our methods (i.e., *F-Min-min*) with *Min-min* from the aspects of the number of task failures, success rate, and total execution time. We first take 100–500 magnetic bearing parameterized design request as the tasks respectively, the requested parameters are generated randomly by computer. We select 100 resource services available in the experimental.

13.7.1 Comparison of Success Rate

In these experiments, the four kinds of failures proposed in Section 13.3 were artificially generated. E.g., during the execution of tasks, some resource service providers or demanders were disconnected from the network by cutting off the network wire; several running tasks were suspended by using stop command on task manager; resource service was overloaded with an artificially generated load. The distributions of each kind of failure in these experiments are shown in Figure 13.5.

During resource service scheduling, the comparisons of the total failure numbers and the detected failure numbers in regards to the *F-Min-min* method are shown in Figure 13.6. In Figure 13.5, we found that more than 80 percent of the total failures can be detected using the proposed failure detection methods. It proved that the proposed failure detection methods are effective.

The comparison of task failure number and success rate between *F-Min-min* and *Min-min* are shown in Figures 13.7 and 13.8 respectively. From these figures we can see that *F-Min-min* scheduling method evidently has a smaller failure task number than *Min-min* scheduling method, and has a higher success rate than *Min-min*. The reason is that *F-Min-min* scheduling method takes into account the failure that would be generated during the scheduling process, and has failure detection and recovery service. *F-Min-min* method detected that 80 percent of the failures occurred during the scheduling process, and most of the detected failures can be recovered by using the methods proposed in Section 13.5. Hence, *F-Min-min has* a higher success rate and smaller failure task number than the

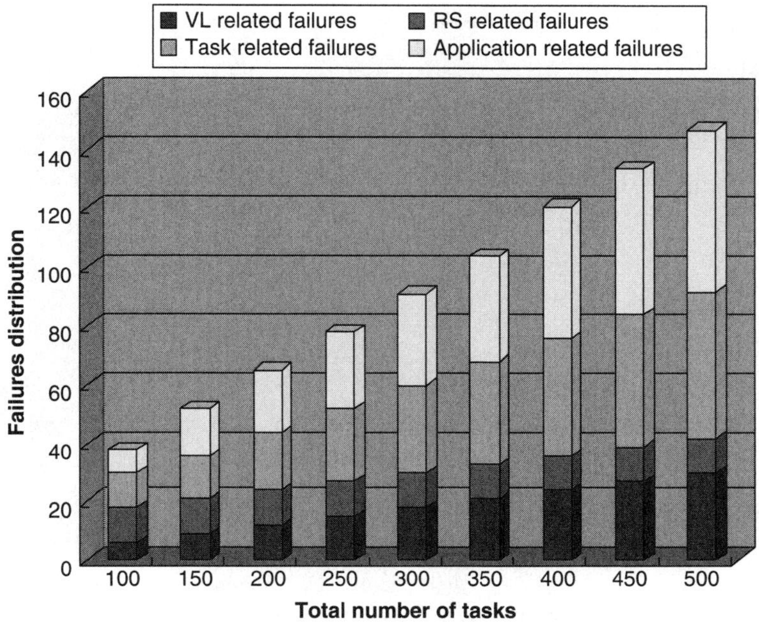

Figure 13.5 Distribution of failures.

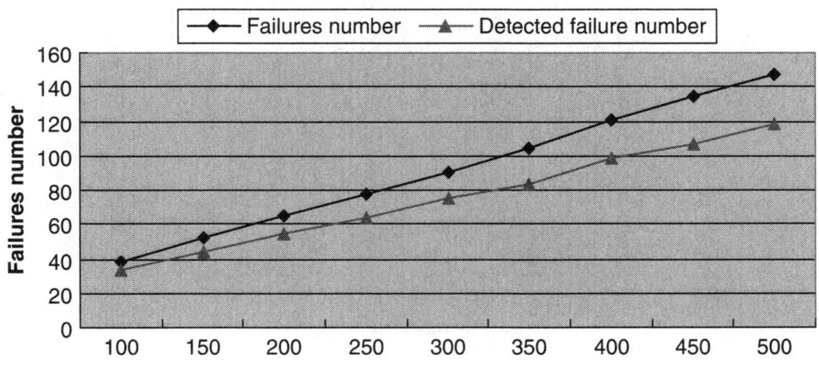

Figure 13.6 Detected failures under different task number.

simple *Min-min* method. It demonstrates that our proposed failure detection and recovery method is effective and useful.

13.7.2 Comparison of Execution Time

In order to further test the effectiveness of our proposed method under different task and resource service, in addition to the above simple task with a single resource service request, we considered

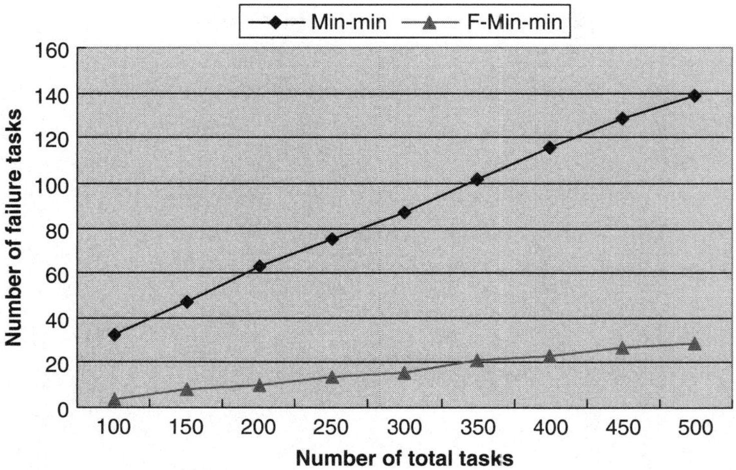

Figure 13.7 Comparison of failure number of tasks.

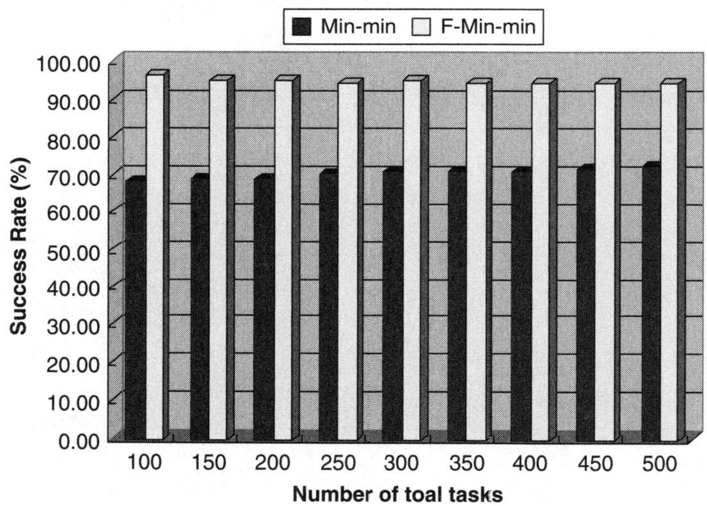

Figure13.8 Comparison of success rate.

some more complicated tasks in the developing process of the magnetic bearing system as experimental objects. We selected 10 tasks of which the decomposed subtasks vary from 1–10 with increment of 1 as experimental objects, such as online analyze and simulation of axial magnetic bearing system (marked with $T_{magnetic}$), which can be decomposed into five subtasks (Tao *et al.* 2008c): (a) axial magnetic bearing structure parameterized design service, (b) mechanics analyze and simulation service, (c) temperature field simulation and analyze service, (d) magnetic field simulation and analyze service, and (e) controlling system simulation and analyze service. The system has to invoke five different resource services in a certain sequence to execute $T_{magnetic}$. During the scheduling process of these tasks, failures proposed in Section 13.3 are artificially generated in the same way as mentioned above. We selected the average execution time of these tasks as the index to evaluate the effectiveness of our proposed method.

Figure 13.9 shows the comparison of the execution time between using only *Min-min* algorithms and performing failure recovery through task scheduling when failures happen (namely *F-Min-min*). Each result is an average of 20 successful test cases. In this chapter, the total execution time of a task depends on the longest execution time of its subtasks, It consists of resource service selecting time, composing time, QoS information processing time, task executing time, etc., during the scheduling process. In Figure 13.9, we found that the total execution time decreases largely due to failure recovery using the *F-Min-min* method. This is because the simple *Min-min* method has no additional failure detection and recovery services or functions. Therefore, subtasks (or tasks) are stopped or suspended and have to wait to restart until the system assigns a new resource service to it. Hence, the total execution time becomes longer due to the unforeseen waiting time. However, there are additional failure

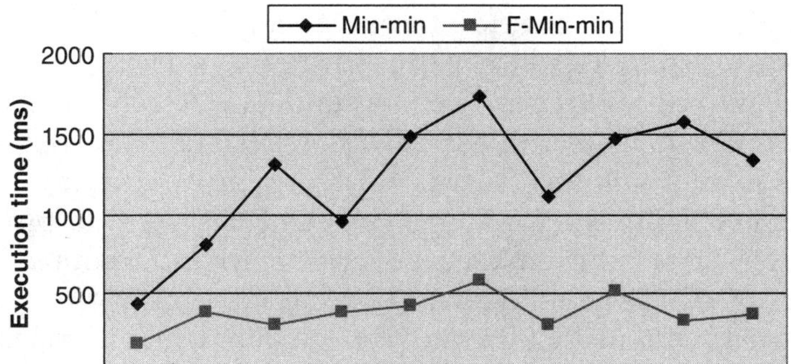

Figure 13.9 Comparison of the execution time between using only Min-min algorithms and performing failure recovery through task scheduling when failures happen (namely F-Min-min). The total execution time of a task depends on the longest execution time of sub-tasks.

detection and recovery services in *F-Min-min*, and when a failure occurs on a subtask (or task), the system can detect and recover it as soon as possible, and the corresponding waiting time is much shorter than that in *Min-min*. Therefore the total execution time of *F-Min-min* is obviously shorter than *Min-min*. This demonstrates that the proposed failure detection and recovery methods are effective and useful. They are also useful in shortening the execution time of a complicated task and improving the quality of service of scheduling in MGrid.

13.8 Conclusion

MGrid system is different from conventional distributed computing systems because of its focus on larger-scale distributed and heterogeneous manufacturing resources and services sharing, including equipment resources, application system resources, material resources, technical resources, public service resources, etc., and because remote resource service scheduling (e.g., remote accessing, information communication) has a great influence on its quality of service (QoS). The probability of failures is higher in MGrid resource service scheduling, and the failures fatally affect the task execution and quality of service of MGrid. However, failure detection and recovery, the core part in realizing dynamic MGrid resource service scheduling, are mostly absent and have not been effectively addressed.

Therefore, in this chapter, the potential failures that could be generated during MGrid resource service scheduling are investigated. Thirteen failures are defined in detail, which are classified into four classifications: (a) virtual link related failures, (b) resource service related failures, (c) task related failures, and (d) application related failures. A failure management system is proposed, which provides failure-tolerance service in MGrid resource service scheduling. Corresponding detection mechanisms and methods to address each defined failure are presented in detail, in conjunction with the corresponding failure recovery methods. The implementation and simulation results indicate that the proposed approaches are sound when it comes to the promotion of the MGrid resource service scheduling success rate.

14

Summary of the Application of Grid Technology in Manufacturing

14.1 Introduction

In this chapter, the related works on the application of grid technology in manufacturing are investigated, including research on manufacturing grid (MGrid) theories and applications. Also, several key future research issues of MGrid are discussed.

14.2 Review of MGrid Theories

In this section, some related MGrid theories are presented in a tabular format. Current research on MGrid can be classified into the following 11 categories (Tao *et al.*, 2011d):

1. Architecture of MGrid,
2. Resource management system,
3. Modeling and encapsulation of MGrid resource,
4. Resource service discovery and scheduling,
5. QoS management,

6. Service composition,
7. Workflow management,
8. Job management,
9. Reliability management,
10. ecurity and trust management, and
11. Others such as MGrid portal design, semantic support, and performance evaluation.

A brief summary of current research on MGrid theories is illustrated in Table 14.1.

Table 14.1 Study on MGrid theories (Tao *et al.*, 2011d)

Classification	Methods or Specific Contents	References
Concept	The concepts and connotation of MGrid have been studied	Tao *et al.* (2008a, 2008b, 2006b), Qiu (2004), Fan *et al.* (2003a), McFarlan (2004), Hu *et al.*(2006)
	A five-layered MGrid architecture based on five-layered grid architecture was proposed and a simple prototype application of MGrid was presented in this article	Qiu (2004)
Architecture	Nine-layered MGrid architecture, including network layer, unit and protocol layer, resource encapsulation layer, grid middleware layer, enabling layer, application layer, portal layer and enterprise cooperation layer, and MGrid operations and management system layer	Fan *et al.*(2003a)
	Service-oriented MGrid architecture are investigated	Tao *et al.*(2008a), Liu 2004, Liu 2006, Sun *et al.*(2008b)

(Continued)

Table 14.1 (cont.) Study on MGrid theories (Tao *et al.*, 2011d)

Classification	Methods or Specific Contents	References
Architecture	A MGrid architecture driven by manufacturing process service is investigated	Zhang (2008)
	Ontology-based and with semantics supported MGrid architecture is studied	Zhang *et al.*, (2008a, 2008b)
	MGrid architecture based on multi-view (e.g., resource view, function view, information view, process view) is proposed	Lv *et al.* (2007), Shi (2007)
	Collaborative manufacturing supported MGrid platform architecture	Tao *et al.* (2006b)
	Organization model framework for virtual enterprise based on MGrid application is researched	Wang *et al.* (2005a)
	MGrid architecture based on ASP (Application Service Provider)	Wu *et al.* (2006c)
	SMT (Surface Mount Technology)-oriented MGrid architecture	Li (2007)
	MGrid architecture based on economy theory	Zhou *et al.* (2007b)
	MGrid architecture based on service control center for general enterprises	Tan *et al.* 2009
Portal design	GridSphere based MGrid portal design is investigated, which considers user management, resource management, service management, and security management.	Wang (2008)

(Continued)

Table 14.1 (cont.) Study on MGrid theories (Tao *et al.*, 2011d)

Classification	Methods or Specific Contents	References
Resource management system or framework	Service-oriented MGrid resource service optimal allocation system, including resource service encapsulation model, resource service publication and discovery model, QoS evaluation mode, resource service composition and optimal selection model, failure tolerance management model.	Tao *et al.* (2008a)
	MGrid resource knowledge and information abstraction framework based on agents is studied	Zhang *et al.* (2009a)
	A fuzzy resource reservation framework based on SLA (Service Level Agreement) is proposed	Xie *et al.* (2008b)
	Mining model of MGrid Resource service	Li *et al.* (2008b)
	Resource management system based on SLA is studied	Shen *et al.* (2006b)
	Resource management system based on WRSF is studied	He *et al.* (2006b, 2006d)
	A MGrid data management framework is proposed	Ma *et al.* (2006)
	Resource management framework based on GRIS (Grid Resource Information Service), GIIS (Grid Index Information Service), and GRAM (Globus Resource Allocation Manager) is put forward	Liu 2004
	MGrid resource management mode based on transactions	Wu *et al.* (2005a)

(Continued)

Table 14.1 (cont.) Study won MGrid theories (Tao *et al.*, 2011d)

Classification	Methods or Specific Contents	References
Modeling and encapsulating of resource services	Resource service description method based on MGrid ontology and OWL-S is reserached	Hu *et al.* (2009), Li *et al.* (2007b)
	Modeling method for machining equipment based on web service library and standard parts library is studied	Shi (2007)
	Resource service modeling method based on STEP, XML, and WSDL	Shi (2007), Tao *et al.* (2006c)
	Resource service modeling and encapsulating methods based on WRSF are studied	Wu *et al.* (2007b, 2007c)
	Resource service matching and searching based on matching algorithms of resource service description information, such as word concept information matching algorithms, number interval and fuzzy number information matching algorithms, sentence information matching algorithms, and entity class information matching algorithms.	Tao *et al.* (2009b)
Resource service discovery	Hierarchy-searching (mapping layer, clustering layer, physical resource layer) method	Fu (2007)
	Service semantic discovery based on ontology and OWL-S	Hu *et al.* (2008), He *et al.* (2006c)
	Agent-based MGrid search model is proposed	Wen (2005)

(Continued)

Table 14.1 (cont.) Study on MGrid theories (Tao *et al.*, 2011d)

Classification	Methods or Specific Contents	References
Resource service discovery	TQPS (Time, Quality, Price, Service) based MGrid resource discovery	He *et al.* (2006e)
	Resource organization and discovery based on distributed hash tables	Liu (2006), Meng *et al.* (2006)
	A stratified resource discovery model in MGrid	Wang *et al.* (2009)
Resource service scheduling	Establish scheduling objective function based on QoS, and then apply optimization algorithms (e.g., particle swarm optimization (PSO), genetic algorithm (GA), AHP, wasp-based, simulated annealing algorithm, ant colony algorithm) to attain optimal results.	Liu (2004) Bao *et al.* (2008), Yuan (2006), Zhang *et al.* (2006a)
	Manufacturing ability and utility driven MGrid resource scheduling strategy	Lu and Liu (2008)
	Resource allocation method based on competitiveness equilibrium for MGrid	Shi *et al.* (2009), Liu *et al.* (2008a)
	Evaluation index system is established and a resource optimal-selection model based on agents is proposed	Tao *et al.* (2005a), Wen (2005)
	Workflow scheduling based on Petri net is investigated	Hu and Li (2009)
	Extended enterprise oriented MGrid resource optimal allocation strategies based on Bayesian theory are studied	Zhou *et al.* (2007a)

(Continued)

Table 14.1 (cont.) Study on MGrid theories (Tao *et al.*, 2011d)

Classification	Methods or Specific Contents	References
Resource service scheduling	Scheduling strategies and mechanisms based on trust and economy model are studied	Xiang (2008), Tao *et al.*(2009a)
QoS management	QoS whole-lifecycle management mode, QoS evaluation model, and QoS-based resource service management framework are studied.	Tao *et al.* (2009c)
	MGrid QoS management framework, relationship model among QoS and workflow, resource management, project management, and job decomposition	Liu (2004), Sun *et al.* (2006), Shi (2007)
	SLA management (e.g., interest allocation, protocol management, dynamic negotiation) research in MGrid QoS management system	Liu (2004)
Service composition	Basic structural model of composite service, aggregation model evaluation, generation of composition path, PSO based composition path optimal-selection are researched.	Tao *et al.* (2008c, 2010e)
	Service composition frameworks and methods based on MGrid ontology, which is established based on TOVE, STEP, and PSL, are investigated	Zhang *et al.* (2003b, 2005b, 2006d).
	A service composition framework based on BPEL4WS is studied	Duan *et al.* (2006)

(Continued)

Table 14.1 (cont.) Study on MGrid theories (Tao *et al.*, 2011d)

Classification	Methods or Specific Contents	References
Semantic support	A five-layered MGrid architecture with semantic support based on MGrid ontology is proposed. MGrid ontology is established based on grid service ontology, manufacturing domain ontology, SLA ontology depends on QoS negotiation	Zhang (2006a)
Workflow management	Workflow management based on service, including modeling of service, service discovery, service bidding, and service invoking.	Yuan 2006
	Workflow system encapsulation method under the MGrid environment	Yuan 2006
	Agent and knowledge driven workflow management system framework in MGrid	He *et al.* (2006a), Yu *et al.* (2005)
	MGrid workflow architecture, workflow model (process model, organization model, resource model, and service model)	Yuan 2006
Job management	MGrid job management system model (definition of job, job planning, job and resource matching, monitoring of job state)	Shi (2007), Mo (2006)
	MGrid job management based on composite service scheduling engine, web service scheduling engine, and OGSA-DAI service scheduling engine	Zhang *et al.* (2006e)

(Continued)

Table 14.1 (cont.) Study on MGrid theories (Tao *et al.*, 2011d)

Classification	Methods or Specific Contents	References
Job management	Job decomposition based on QoS in MGrid	He *et al.* (2005a)
Reliability management	Thirteen failures defined in detail and are classified into four categories: (a) virtual-link-related failures, (b) resource service-related failures, (c) task-related failures, and (d) application-related failures. A failure management system of the MGrid system is presented together with its architecture. Corresponding detection mechanisms and methods for each defined failure are presented in detail, as well as the corresponding failure recovery methods.	Tao *et al.* (2010b)
	Definition, evaluation index system, and evaluation model of reliability in MGrid system are investigated, as well as a reliability management framework.	Zou *et al.* (2006), Li *et al.* (2005b)
	A method for collecting reliability data in MGrid based on mobile agents is proposed	Zhou and Yu (2006)
Security	Identity-based designated multi-receiver sign-encryption scheme for MGrid is studied	Wang (2007)
	MGrid security management architecture, security ontology, and role-based access control model and strategies are researched	Cai (2007a)

(Continued)

Table 14.1 (cont.) Study on MGrid theories (Tao *et al.*, 2011d)

Classification	Methods or Specific Contents	References
Security	MGrid security infrastructure based on GSI, distributed authorization, and real time authentication are investigated.	Shu (2006)
	RBAC-based access control strategy and authorization service strategy based on XACML (eXtensible Access Control Markup Language) are proposed	Chen (2006)
	An enterprise qualification authentication system and method in an MGrid system is studied.	Sun *et al.* (2008a)
	Trust evaluation of MGrid node based on AHP, identity authentication and password authentication based multipart key negotiation agreement, and security management methods based on identity/certification authorization and proxy digital signature.	Cai (2007a), Wang (2007)
	Single sign-on and intranet-based security infrastructure is studied in MGrid	Cheng and Zhang (2008)
Trust	Trust-QoS relationship mode, inter-domain and intra-domain trust-QoS evaluation model, and dynamic updating mode are proposed.	Tao *et al.* (2009a)

(Continued)

Table 14.1 (cont.) Study on MGrid theories (Tao *et al.*, 2011d)

Classification	Methods or Specific Contents	References
Trust	Simple evaluation model of resource trust via auditing resource qualification information is studied	Zhang *et* al. (2009b)
Performance evaluation	Performance evaluation of MGrid system based on decision support is investigated	Xie *et al.* (2008a)

14.2.1 Architecture of MGrid

The functions and basic modules of an MGrid system are decided by the architecture of MGrid, which primarily studies the components of MGrid, the relationships of components, the way to integrate these components, and the implementing mechanisms of the MGrid system. Many researchers have proposed different forms of MGrid architecture. According to grid architecture, these proposed MGrid architectures can be classified into two categories:

1. *Layered MGrid architecture based on the hourglass model,* and
2. *Service-oriented MGrid architecture.*

The layered MGrid architecture is proposed according to the grid architecture based on the hourglass model, and is shown in Figure 14.1. The hourglass model based grid architecture is a protocol-centric architecture. In this architecture, the neck of the hourglass consists of *resource* and *connectivity* protocols, which facilitates the sharing of individual resources. Protocols at these layers are designed such that they can be implemented on top of a diverse range of resource types, defined at the *fabric* layer. They can in turn be used to construct a wide range of global services and application-specific behaviors at the *collective* layer – so called because they involve the coordinated ("collective") use of multiple resources (Foster *et al.*, 2001a). In accordance with the hourglass model based grid architecture, several layered MGrid architectures have been proposed, e.g., a five-layered MGrid architecture and a nine-layered MGrid architecture were proposed by Qiu (2004)

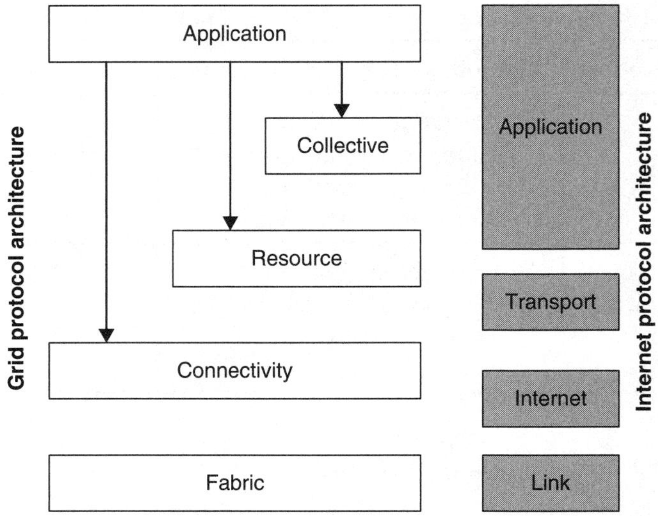

Figure 14.1 The hourglass model based grid architecture and its relationship to the internet protocol architecture (Foster *et al.*, 2001a).

and Fan *et al.*, (2003a), respectively. Liu (2004) proposed a four-layered rapid MGrid architecture.

Service-oriented MGrid architecture is based on OGSA, a new grid architecture proposed by Foster *et al.*, (2001a) based on the hourglass model. OGSA is a service-centric architecture based on OGSI. In OGSA, all the resources are encapsulated into services, and these services are described using WSDL, and can be accessed using SOAP. Service-oriented MGrid architecture has been studied by many researchers recently (Tao *et al.*, 2008a; Liu, 2004; Sun *et al.*, 2008b). The third row in Table 14.1 summarizes some of the proposed MGrid architectures.

Although a number of MGrid architectures have been proposed, they are not very useful for supporting the semantics in an organization. In addition, these proposed MGrid architectures did not consider the commercial operations of MGrid. Without the consideration of commercial operations, the resource service providers and demanders will lose confidence in its benefits. As a result, resource service providers will not like to publish their manufacturing resources or abilities in MGrid. Additionally, resource service demanders will not want to use MGrid because not enough high quality services are available in the MGrid system. Then the development and application of MGrid will be hindered.

14.2.2 Resource Management System (RMS)

In a grid system, 'resource management' refers to the operations used to control how capabilities are provided by the grid resources and services made available to the other entities such as users, applications, and services providers (Foster and Kesselman, 2004a). Referring to Krauter *et al.*, (2002), RMS in MGrid is defined as a service that is provided by a distributed manufacturing system (*i.e.*, the MGrid system) that manages a pool of named resources that are available to a distributed manufacturing system, such that the system or the job-centric performance metric is optimized. RMS is therefore the central component of an MGrid system.

The specific services and functions involved in RMS in an MGrid system primarily involve resource organization and encapsulation, resource discovery, resource scheduling, QoS management, service composition, workflow management, job management, etc. In this section, the emphasis is not on the realization and implementation of these services and functions, as the authors primarily focus on the related studies of the existing RMS frameworks that organize these services and functions.

Existing MGrid RMS frameworks can be classified into four categories:

1. MGrid RMS based on a specific standard or protocol, such as RMS based on SLA (Xie *et al.*, 2008b; Shen *et al.*, 2006b), RMS based on WRSF (He *et al.*, 2006b, 2006d), RMS based on GRIS, GIIS, and GRAM of Globus (Liu, 2004),
2. RMS based on agent (Zhang *et al.*, 2009a),
3. RMS based on economic theories (Wu *et al.*, 2005a), and
4. Service oriented MGrid RMS.

In Chapter 2 a service-oriented MGrid resource service optimal-selection system (RSOSS) has been proposed and discussed. RSOSS includes the key services of RMS, such as resource service encapsulation, resource service publication and discovery, QoS evaluation, resource service composition and optimal-selection, and failure tolerance analysis. The fourth row in Table 14.1 summarizes some proposed RMS of MGrid.

However, the above studies on MGrid RMS lack integrality, as only one or two aspects of performance are emphasized, e.g., some of them only emphasize the scheduling performance, some pay attention to QoS management, and some highlight the workflow management ability. In an effective and applicable RMS, many factors have to be considered, such as security, scheduling ability, reliability, scalability, etc. Therefore, a more reasonable RMS based on the requirements of MGrid implementation has to be studied. In addition, the existing research studies on MGrid RMS are too conceptual, and the majority of these studies emphasize qualitative research, while quantitative research on specific implementing operations is inadequate. Hence, in Section 14.6 the authors point out that quantitative research on MGrid is a key research to focus on in the future.

14.2.3 Modeling and Encapsulating of MGrid Resource Services

The aim of MGrid resource modeling and encapsulation is to encapsulate the various heterogeneous manufacturing resources or manufacturing capabilities into services using some web service descriptive languages or standards. This is done in order to enable MGrid resource information to be published and registered in an MGrid system. Such information can be stored, delivered, accessed, and invoked via specific protocols or languages, such as WSDL, UDDI, SOAP, and HTTP. Without digital descriptive information of resource services, the implementing operations of MGrid would lose data and information support. Current methods for modeling and encapsulating MGrid resources are primarily based on several descriptive languages such as OWL-S (Hu *et al.*, 2009; Li *et al.*, 2007b), STEP, XML, WSDL (Shi, 2007; Tao *et al.*, 2006c), and WRSF (Wu *et al.*, 2007c; 2007b) as shown in the fifth row in Table 14.1.

The above methods can realize the digital description of the MGrid resources to some extent, but there are still some shortcomings:

1. The hidden knowledge in the manufacturing processes and manufacturing activities has not been successfully captured,
2. The digital descriptive information of resource service lacks semantics, and
3. Correlations among manufacturing processes are not well described.

Therefore, a more comprehensive and effective method for MGrid resource service modeling and encapsulating should be studied in the future.

14.2.4 Resource Service Discovery and Scheduling

Existing studies on MGrid resource discovery, primarily emphasize search mechanisms or frameworks at the abstract level, such as the hierarchy-searching method (Fu, 2007), ontology, OWL-S based searching method (Hu *et al.*, 2008; He *et al.*, 2006c), agent-based searching method (Wen, 2005), TQPS (time, quality, price, and service) based searching method (He *et al.*, 2006e), and searching methods based on distributed hash tables (Meng *et al.*, 2006; Liu 2006).

Different from the above research methodologies, the authors (Tao *et al.*, 2009b) investigated the method for resource service match and search from the descriptive information of resource service, as described in Chapter 4. After searching for the qualified resource services for each task, the next step is to realize the mapping between resource services and tasks, namely resource service scheduling. Current studies on resource service scheduling are briefly summarized as follows:

1. QoS and intelligent computing based resource scheduling. This method first establishes the scheduling objective function based on QoS, and then applies the optimization algorithms based on intelligent computing (e.g., particle swarm optimization (PSO), genetic algorithm (GA), AHP, wasp-based, simulated annealing algorithm, ant colony algorithm, etc.) to arrive at an optimal scheduling solution (Liu, 2004; Bao *et al.*, 2008; Yuan 2006).
2. Economic theories based resource scheduling methods, such as competitiveness equilibrium based resource scheduling, trade-based resource scheduling, and market-based resource scheduling (Shi *et al.*, 2009; Liu *et al.*, 2008a).
3. Agent-based resource optimal selection and scheduling methods (Tao *et al.*, 2005a; Wen, 2005).
4. Workflow-based resource scheduling (Hu and Li, 2009), and
5. Trust-based resource scheduling methods (Xiang 2008, Tao *et al.*, 2009a).

In addition, Deng *et al.*, (2006a) and Chen *et al.*, (2004a) proposed a resource characteristics-based scheduling strategy and applied this to the development of a customized artificial joint.

The abovementioned service scheduling methods can address the resource scheduling problem effectively to some extent, but they lack dynamism. During the implementation of these scheduling methods, the resource services existing in the MGrid system are treated as stable and unchanged. As it is known, any grid system is a dynamic system, and its nodes or services are changing all the time. For example, new nodes with better quality will be added while some nodes will be deleted, and the quality of the nodes will also change with time. Hence, when designing a resource scheduling mechanism, these dynamic factors should be considered in the future. In addition, the degree of intelligence and autonomy should be emphasized.

14.2.5 QoS Management

Providing extraordinary quality of service is one of the three judging criteria of a grid system. QoS management has been studied in various contexts, particularly for computer networks and multimedia applications, network QoS, and end-system QoS. However, little work has been carried out in an MGrid system. In Chapter 5, the classification and modeling of MGrid QoS have been investigated from three different perspectives: (1) QoS whole life-cycle management, (2) MGrid architecture view, and (3) QoS attribute parameters. The measurement and evaluation models of QoS parameters in MGrid system have been presented. In addition, Tao *et al.*, (2009d) first put forward the concept of resource service trust-QoS in MGrid, and a trust-QoS relationship model which is capable of capturing a comprehensive range of trust relationships which exist in the MGrid system is presented. Liu (2004), Sun *et al.*, (2006) and Shi (2007) proposed an MGrid QoS management framework, and they investigated the relationship model among QoS and workflow, resource management, project management, job decomposition, and SLA management.

As has been described in Section 14.3, the contents of QoS management involve the establishment of the QoS index system, description of QoS, modeling of QoS, evaluation of QoS, QoS ontology, life-cycle management of QoS, QoS information storage and extraction, etc. Apparently, existing studies on QoS management research is insufficient, and more studies are needed that emphasize QoS management in the future.

14.2.6 Service Composition

As describe in Chapter 7, in an MGrid system, there are primarily two kinds of manufacturing tasks (or resource service requests): (1) single resource service request task (SRSRTask), and (2) multi-resource service request task (MRSRTask) (Tao *et al.*, 2008c). For a SRSRTask, the system searches the resource services that are qualified for its functional requirements and chooses the optimal one for execution. For an MRSRTask, in addition to the search for all qualified resource services according to each subtask, the system selects one candidate resource service for each subtask. Then the system generates a new composite resource service (CRS) and selects the optimal resource service composite path from all possible paths to execute the task under a given multi-objective (e.g., time minimization, cost minimization, and reliability maximization) and constraints. The above is defined as a multi-objective MGrid resource service composition and optimal-selection (MO-MRSCOS) problem (Tao *et al.*, 2008c).

A few studies have been carried out for addressing the MO-MRSCOS problem as depicted in Table 14.1. In Chapter 10, the formulation for the MO-MRSCOS problem to minimize execution time and cost, and maximize the reliability, has been presented. Four basic resource service composite modes (RSCM), i.e., *sequence model*, *parallel model*, *selective model*, and *circular model*, for CRS are described. The principles for transforming a complicated RSCM into a simple sequence RSCM, and the aggregating QoS evaluation method for CRS are presented. Particle swarm optimization (PSO) combined with non-dominant sorting technique is used to realize the optimal selection of service composition paths. In addition, Zhang *et al.*, (2005; 2006d) proposed a service composition framework based on MGrid ontology, which was established based on TOVE, STEP, and PSL. Furthermore, an MGrid service composition framework based on BPEL4WS was proposed by Duan *et al.*, (2006).

It is apparent that existing studies on service composition are highly insufficient. Problems such as, (1) validation and verification of CRS, (2) correlation relationship among element services in a CRS, (3) measurement and evaluation of the invoking times for a service to be selected for composition, (4) establishing a composition service network, and (5) determination of the dynamic characteristics of the composition service network, have not been investigated. Therefore, more studies are needed on service composition in the future, and the authors have listed this as one of the key research issues for future MGrid research in Section 14.6.5.

14.2.7 Workflow Management

MGrid workflow can be viewed as the manufacturing activities which can be executed in heterogeneous, distributed, and job-dependent resources or services in the MGrid environment according to business processes and rules. On the other hand, workflow management for the computing grid has been widely investigated to enable users to compose and execute complex grid applications with distributed heterogeneous and unreliable computing resources.

Existing studies on MGrid workflow management primarily concentrate on two aspects: workflow management framework and workflow scheduling. For the first aspect, Yuan (2006) studied the workflow system encapsulation method under an MGrid environment. He *et al.*, (2006a) and Yu *et al.*, (2005) investigated agent and knowledge driven workflow management framework, and a service-based workflow management framework was proposed in Yuan (2006). In addition, Yuan (2006) studied an MGrid workflow model (including process mode, organization mode, resource model, and service model) and architecture, and Hu *et al.*, (2009) presented a modeling and scheduling method for MGrid workflow using timed Petri nets. A brief summary of the related studies on workflow management is listed in Table 14.1.

The proposed workflow management and scheduling methods primarily evolve from the research achievements in distributed computing systems. Workflow in MGrid resembles traditional workflow management in distributed computing systems but with the following distinctive characteristics:

1. *Service-centered.* All manufacturing resources or manufacturing capabilities must be encapsulated into services before being published and registered in an MGrid system. Hence, workflow management

architecture and workflow applications in MGrid must be service-oriented.

2. *Heterogeneity and diversity of resource.* In a traditional computing grid system, resources are limited to computing resources, communication resources, and storage resources. However, in MGrid, additional resources such as equipment, software, technical, material, and public service are also involved. These resources can be added on or removed from an MGrid dynamically.
3. *Complexity and dynamic characteristics of business processes.* In MGrid, business processes are the primary manufacturing activities or processes, and they can be very complex.

Therefore, when designing a workflow management framework for MGrid, the above issues should be considered.

14.2.8 Job Management

In this study, jobs are defined as objects that are executed in order to solve a problem in MGrid. The objects are executed by invoking a single resource service or several resource services in a certain sequence in an MGrid system. MGrid jobs are anything that needs resource services. It can be a simple request (e.g., bandwidth, a simple calculation, a query, etc.), an application request (e.g., a structural design, a simulation, etc.), or an aggregate of applications (e.g., large-scale computing, and remote immersion). It can also be a product or a part for design or manufacturing, etc. Job management is primarily responsible for the whole life-cycle management of jobs in an MGrid, including job creation, job definition, job description, job decomposition, job planning, and job deployment and scheduling.

Existing research studies on MGrid job management concentrate on frameworks and job scheduling strategies. A job management system was proposed by Shi (2007) and Mo (2006), which consists of five modules, *i.e.*, job modeling, job planning, job-resource matching, job scheduling, and job state monitoring. The workflow of each module has been described. Zhang *et al.*, (2006e) presented dynamic manufacturing job management, which was designed to invoke various types of services. In order to realize dynamic service invoking, they divided the invoking engine into three categories: composite service invoking engine, web service invoking engine, and OGSA-DAI service invoking engine. The architecture and workflow for these three kinds of invoking engines were presented by them in detail. In addition, a QoS-based task decomposing method was researched in He *et al.*, (2005a).

Undoubtedly, job scheduling strategies are very important in a grid system. However, in the MGrid system, manufacturing jobs are more complicated than the jobs in a computing grid system. The description and decomposition of manufacturing jobs are the two basic issues for realizing job scheduling. Without job description, the system will not know the detailed requirements of the job. If the job decomposition cannot be realized, the system would not be able to interpret the kind of task (SRSRTask or MRSRTask) the submitted job belongs to, and it will be impossible for the implementation of the subsequent operations such as service discovery, matching, and composition. Therefore, more research is required on these two aspects in future.

14.2.9 Reliability Management

Existing research studies on the reliability of an MGrid system can be classified into two categories: (1) failure detection and recovery, and (2) MGrid system reliability management, such as the establishment of a reliability evaluations index, a reliability evaluation model, and a collecting method of reliability data.

Referring to Lee *et al.*, (2005c), the failure in an MGrid system was defined by Tao *et al.*, (2010b) as follows: "It is a failure if and only if one of the following two conditions is satisfied (but not both): (a) a resource service stops due to a resource crash, and (b) the availability of a resource does not meet the minimum levels of a task." In Chapter 14, failure detection and recovery during MGrid resource service scheduling were investigated. In addition, Zou *et al.*, (2006) studied MGrid reliability from three aspects: (a) system operational reliability, (b) system inherent reliability, and (c) system use reliability. A hierarchical reliability evaluation mode for an MGrid system was proposed by them. Li *et al.*, (2005b) investigated a data mining-based MGrid reliability system, and a reliability data collecting method based on agents was reported in Zhou and Yu (2006).

An MGrid system is different from a conventional manufacturing system due to its focus on larger-scale, distributed and heterogeneous manufacturing resources and services sharing. These resources may be geographically distributed and registered, can be moved dynamically, and are owned by different enterprises or organizations with their own policies. Therefore, remote resource service scheduling (e.g., remote accessing and information communication) has a great influence on MGrid quality of service (Tao *et al.*, 2010b). The probability of failure is higher in MGrid resource service scheduling, and the failure can critically affect task execution and the quality of service of an MGrid. Therefore, reliability is a significant and complex issue in an MGrid system. As described in the above paragraphs, few studies have been reported on MGrid reliability. Other aspects such as manufacturing service reliability, result reliability, data reliability, etc., should be studied further in the future.

14.2.10 Security and Trust Management

Successful execution of any information system cannot be maintained without a security guarantee, especially for the MGrid system. Existing research studies on MGrid security can be classified into the following four categories.

1. *Security architecture of MGrid system*, such as an ontology-based MGrid security management architecture that was investigated by Cai (2007a), which involved security ontology, a role-based access control model, and strategies; An MGrid security architecture based on GSI, distributed authorization, and real-time authentication was proposed by Shu (2006); A single sign-on and intranet-based security architecture was put forward by Cheng and Zhang (2008).
2. *Access control models and strategies*, such as role-based access control (RBAC) strategy (Chen, 2006), proxy-based access control strategy, and CAS (community authorization service) based access control strategy.
3. *Authentication and authorization mechanism*, such as XACML-based authorization strategy (Chen, 2006), enterprise qualification authentication method (Sun *et al.*, 2008a), identity authentication and password authentication (Cai, 2007a; Wang, 2007).

4. *Trust management*, such as trust evaluation of the MGrid node based on AHP (Wang, 2007), and resource trust and trust-QoS evaluation model reported in Tao *et al.*, (2009d) and Zhang *et al.*, (2009b).

The tenth row in Table 14.1 summarizes the related studies of security and trust management in an MGrid system.

One of the key bottlenecks that hinder the development and application is the security problem, including system security, network communication security, enterprise core data security, etc. The same problem has been encountered when implementing MGrid. Due to the lack of a commercial operation model, the return space for both service provider and demander enterprises using NM system is limited. As a result, few special commercial security products for NM are developed by IT companies. The research on MGrid security is primarily carried out by researchers in universities and research institutions, and the related studies primarily are either conceptual or experimental. In the future, more studies on the practical application aspects should be emphasized, e.g., research an MGrid security service model based on the concept of clouds security.

14.2.11 Others

In addition to the above ten main research points, some researchers have carried out studies on portal design (Wang, 2008), semantic support (Zhang, 2006a), performance evaluation of MGrid systems (Xie *et al.*, 2008a), and resource reservation, as shown in Table 14.1. Performance evaluation is listed as one of the key research issues for future MGrid research in Section 14.6.7.

14.3 Investigation of Application Research on MGrid

The first project that introduced grid technology into manufacturing is the information power grid (IPG) (Johnston *et al.*, 2000; IPG, 2009), supported by the American National Aeronautics and Space Administration (NASA) and Natural Science Foundation (NSF). The aim of IPG is to establish a distributed computing and data management environment supporting remote collaboration, so as to solve large-scale scientific and engineering problems. Typical applications of grid technology in manufacturing can be classified into the following six aspects, as shown in Figure 14.2 (Tao *et al.*, 2011d):

1. *Product design:* Apply grid technology to the development of a new product (especially for complex products), including large-scale computing, simulation, optimization, and visualization, so as to enhance the development of the new product.
2. *Manufacturing resource sharing and integration*: Grid technology is used to enable distributed and heterogeneous manufacturing resources to be fully connected, so as to realize the integration and exchange of manufacturing data and manufacturing capability, and realize the sharing and interoperation of manufacturing resources.
3. *Enterprise management:* Grid technology is used in enterprise management activities, such as supply chain management, purchasing management, customer relations management, workflow management, and resource management.
4. *Service encapsulation of manufacturing resource*: Encapsulates manufacturing resources and manufacturing capabilities into services under the support of grid technology, web service, WRSF, and ontology, so as to publish them on the Internet to be invoked by other users.
6. *Optimal-allocation and scheduling of manufacturing resources*: Grid technology is used to realize optimal

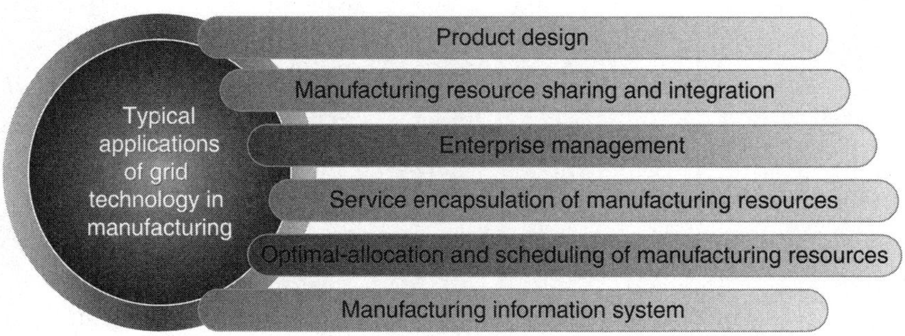

Figure 14.2 Six typical applications of grid technologies in manufacturing.

allocation and scheduling of global manufacturing resources based on economic theory, intelligent computation, etc.

7. *Manufacturing information system*: Grid technology is introduced in the construction of enterprise information systems.

The specific contents and instances for each typical application are shown in Table 14.2. In addition, many scholars and research institutes have studied the application of grid technology in specific manufacturing fields, such as virtual enterprise, aeronautical manufacturing, mould and die manufacturing, equipment support, engineering simulation, modern logistics, and rapid manufacturing. A detailed study can be found in Tao *et al.*, (2008a).

Table 14.2 Typical application areas of grid technology in manufacturing

Application Areas	Specific Content	Examples
Product design	• Denseness data computing • Design, simulation, and optimization of complex products • Distributed cooperative work • Solving for large-scale scientific and engineering problems • Visualization	• IPG (Johnston *et al.*, 2000; IPG 2009) (Information power grid), supported by the American National Aeronautics and Space Administration (NASA) and Natural Science Foundation (NSF) of America, has realized the connection of hardware and software in many research institutions in North America, and established a distributed computing and data management environment supporting remote collaboration for addressing large-scale scientific and engineering problems. Integrated simulation of airplane can be realized using IPG.

(Continued)

Table 14.2 (cont.) Typical application areas of grid technology in manufacturing

Application Areas	Specific Content	Examples
Product design	• Multi-disciplinary collaborative design	• Boeing (Zhang, 2004) simulated the CFD corporately in the design process of F-18E/F by their multi-research center linked with grid technology. • For promoting cooperative design between engineers and customers, Honeywell, General Motors have connected their expensive equipments and design centers (Ccidnet, 2010) • Jaguar's automotive products research information center used GT3.0 and BPEL4WS as an intermediate language to define and manage grid services in engineering workflow, and it applied grid services to its automobile industry (Mahon *et al.*, 2004). • Wind tunnel experiments and circuit experiments have been replaced by simulation and testing based on grid technologies (Paventhan *et al.*, 2006) in some automobile companies. • Semantic Grid and ontology were employed for multidisciplinary collaborative design of metal stamping over the internet (Zhang and Yin, 2009)

(Continued)

Table 14.2 (cont.) Typical application areas of grid technology in manufacturing

Application Areas	Specific Content	Examples
Resource sharing and integration	• Data integration and exchange • Resource connection, sharing, and co-operation • Resource sharing platform in specific manufacturing fields • Distributed collaborative manufacturing	• The "Enterprise information integration" software of Avaki aims to realize the seamless integration of distributed data sources. The Avaki Grid (Avaki, 2002) was developed from the Legion (Legion, 2009) project based on grid technology. • In the application protocol of STEP-NC, a project of STEP Tool Inc., the manufacturing grid was proposed to eliminate the islands of automation. The aim is to realize integration and exchange among different databases in ISO (Steptools 2009). • MGrid application platforms include rapid MGrid (Liu, 2004), collaborative MGrid (He et al., 2005b), textiles-oriented MGrid (Zhang et al. 2005b), mould and die MGrid (Chen et al., 2003), logistics resource grid system (Huang et al., 2005b), and AmGrid (Liang, 2004). • An agent-based semantic grid for distributed manufacturing collaboration across ubiquitous virtual enterprises was studied by Yin et al.,(2009)

(Continued)

Table 14.2 (cont.) Typical application areas of grid technology in manufacturing

Application Areas	Specific Content	Examples
Enterprise management	• Supply chain management • Purchasing management • Customer relationship management (CRM) • Workflow management • Resource management	• Ford (Ccidnet, 2010) has built an online market and linked its 30,000 suppliers together using grid technology, which not only decreases the operating cost and reduces the procurement costs, but also shortens procurement time. Ford estimates that it can save approximately US$80 billion in procurement costs alone. • Oracle's Database 11g (Oracle, 2009) connects all its computing resources using grid technology, and uses them effectively through virtual technology. • Sun Grid Engine (Sun, 2009) of Sun Microsystems. Inc., adopts strategies management to share its computing resources easily among different groups, and realizes re-allocation of computing resources according to requirements. • Allstate Insurance Company has connected its 15,000 business offices and 220 claims offices, and 60,000 personal computers via internet using grid technology (Larry, 2002). • Distributed supply chain (DSC) based on grid technology (Tan *et al.*, 2003). • Workflow management based on grid technology (Neubauer *et al.*, 2006)

(Continued)

Table 14.2 (cont.) Typical application areas of grid technology in manufacturing

Application Areas	Specific Content	Examples
Service encapsulation of manufacturing resources	• Modeling of resources • Resource encapsulation • Semantic description of resources	• Manufacturing resource connection, integration, and co-operation based on XML (Kazuo 2003) • Modeling and encapsulation of resource encapsulation based on WSDL/XML (Shi, 2007, Tao *et al.*, 2006c) • Modeling and encapsulation of resource encapsulation based on WRSF (Wu *et al.*, 2007b; 2007c) • Modeling and encapsulation of resource encapsulation based on ontology (Hu *et al.*, 2009, Li *et al.*, 2007b)
Optimal allocation and scheduling of manufacturing resource	• Manufacturing resource optimal allocation and scheduling method using grid technology	• QoS and intelligent computing based resource scheduling (Bao *et al.*, 2008, Zhang *et al.*, 2006a) • Economy theory based resource scheduling (Shi *et al.*, 2009, Liu *et al.*, 2008a) • Agent-based resource optimal selection and scheduling method (Tao *et al.*, 2005a) • Workflow-based resource scheduling (Hu and Li, 2009) • Trust-based resource scheduling (Xiang 2008, Tao *et al.*, 2009a) • In addition, Deng *et al.*, (2006a) and Chen *et al.*, (2004a) proposed a resource characteristics-based scheduling strategy and have applied this to the development of customized artificial joints.

(Continued)

Table 14.2 (cont.) Typical application areas of grid technology in manufacturing

Application Areas	Specific Content	Examples
Manu-facturing information system	• Application of grid technology in enterprise information systems	• Application of grid technology in enterprise information systems, such as MIS, ERP, SCM, CAD, CAM, CAPP MRP (material requirement planning) (Huang *et al.* 2005b, Lee *et al.*, 2009).

Although many researchers have endeavored to make practical application of MGrid, and many companies have used grid technologies in their manufacturing activities and product design, the outcome is not satisfactory and the current profit margin for the enterprises is limited. No commercial entities have taken the role in developing the MGrid platform for a special industrial domain, and the manufacturing services are not well collated and operated as products. Therefore, from the service provider's perspective, they lack motivation to share and provide services because there are no entities available to help them sell the service, and they do not have the time to devote to doing business. As a result, services available for the service demander are inadequate, and the quality and reliability of the available services are not very high. Hence, from the service demander's perspective, they do not want to use MGrid widely.

Therefore, in order to enhance the application of MGrid, the commercial operation model should be considered in the future. That is the reason the authors list the commercial operation model, cost, and price management as key issues in the future for MGrid in Section 14.6.

14.4 Key Future Research Issues

After investigating the existing research studies on MGrid, the following MGrid research issues are recommended for further study.

14.4.1 Basic Protocols, Standards, and Criteria

One of the main differences between MGrid and NM is that MGrid has the support of unified technologies (e.g., Web service, grid service), unified platform (e.g., Globus Toolkit), and unified standards and protocols (e.g., OGSA, WSRF, OGSI), while NM does not have such support. However, these standards and protocols are inadequate for the successful execution of an MGrid system. Specific and basic standards, protocols, and criteria are required to be developed, so as to support key operations involved in MGrid, such as resource service encapsulation, matching and searching, scheduling, composition, remote access, security management, data mining, and data delivery and communication.

14.4.2 Service-oriented MGrid Architecture with Semantic Support

Although a number of MGrid architectures have been proposed, they are not very useful for supporting the semantics in an organization. Self-organization, SOA, and semantic net are mainstream developments for the future of information technology. Hence, the service-oriented MGrid architecture with the characteristics of semantic support such as self-organization, intelligence, and dynamism, is still a key research issue. In addition, adopting the theories of the ecosystem, such as self-organization, self-construction, and self-evolvement in order to organize and manage resources in MGrid is recommended for future research.

14.4.3 MGrid Ontology

In order to support the research of service-oriented MGrid architecture, the ontology of the MGrid has to be studied and established. Therefore, the method for establishing MGrid ontology is also recommended as a future research issue. MGrid domain ontologies for manufacturing, resources, services, organization, business flow, and QoS would need to be established.

14.4.4 Quantitative Research on MGrid Resource Service

Existing research on MGrid primarily focuses on its concept, architecture, resource management and scheduling framework, and application prototype in a specific manufacturing environment.

A majority of the studies put the emphasis on qualitative research, while quantitative research on specific implementing operations in the MGrid system is inadequate. The authors have conducted some quantitative research on MGrid resource services, such as the QoS evaluation of a single resource and the aggregation of the QoS evaluation of composite resource services, but this is not sufficient. Hence, quantitative research is one of the key issues in the future, including quantitative evaluation of MGrid performance, quantitative matching algorithms for service descriptive information, optimization algorithms for service composition and optimal-selection, and cost estimates.

14.4.5 Resource Service Composition and Its Optimal-selection

As seen in Section 14.4.6, few studies have been conducted on MGrid resource service composition and optimal selection (RSCOS). Existing research on service composition focuses on web service composition in distributed service systems, and e-commerce systems. In the industrial community, several web service oriented flow descriptive languages have been proposed, such as WS-BPEL, OWL-S, WSCI, and WS-CDL. These languages have played very important roles in the development of service composition software. In academia, existing service composition methods can be summarized in the following five classifications according to the type of web service composition: (1) business flow based service composition method (Narayanan and Mcllraith, 2002; Antonio *et al.*, 2006; Ponnekanti *et al.*, 2002; Benatallah *et al.*, 2003), (2) AI-based (artificial intelligence) service composition method (Baader *et al.*, 2005; Schuschel and Weske, 2004; Akkiraju *et al.*, 2004; Sirin, [2004a]), (3) graph-search-based service composition method (Rao *et al.*, 2006a), (4) QoS-based service composition method (Lin *et al.*, 2005b; Ardagna and Pernici, 2006), and (5) agent-based service composition method (Brazier *et al.*, [2003; Maamar *et al.*, 2005). Methods to solve RSCOS problems in MGrid are recommended in addition to new methods for addressing RSCOS problems.

14.4.6 Cost and Price Management, and Electronic Payment of MGrid

Investigation of existing studies on MGrid reveals that there is no research on cost and price management, as well as electronic payment. Apparently, the successful implementation of an MGrid system cannot be realized without economic operations Therefore, cost and price management, and electronic payment are key issues to be addressed. Specific research areas should include:

1. Cost factors and its measurement from the perspectives of resource service provider, resource service demander, and manager of the MGrid system,
2. Pricing and cost management strategies, and
3. Electronic accounting and electronic payment.

14.4.7 Performance Measurement and Evaluation of an MGrid System

The evolvement of any system is based on performance measurement and evaluation, especially for the information systems. Performance measurement and evaluation of the MGrid system are classified into two parts: establishment of evaluation index system, and quantitative evaluation or qualitative evaluation. The former aims to establish the performance evaluation index, while the latter, to collect performance data, realize its statistics and analysis, and quantitative or qualitative evaluation. It consists of three subparts:

1. MGrid system performance evaluation, such as resource organization performance, system utility performance, and throughput;
2. Resource service performance evaluation, such as quality of service, load, and usability, and
3. Performance evaluation of core subsystem or middleware, such as an efficiency evaluation of the resource service publication and discovery system, a performance evaluation of the service composition system, an executing performance evaluation of composite services, and a performance evaluation of the failure tolerance management systems.

14.4.8 Commercial Operation Model of MGrid

One issue that hinders the practical application of MGrid is the lack of a commercial operation model. Economic and market theories are recommended while researching for the commercial operation model for implementing MGrid in the future. In addition, service and operation models based on cloud computing can be adopted when carrying out MGrid commercial operations. The authors envisage research in cloud manufacturing to be a hot topic in the future.

14.4.9 Embedded Connect Technology for Physical Equipment

The resources encapsulated and published in MGrid, networked manufacturing systems, and virtual manufacturing systems, are primarily information and soft resources, such as software application systems, various data, computational equipment, a few simple physical equipment, etc. Most physical manufacturing equipment may not be connected by a network, and their manufacturing abilities cannot be shared currently. Hence, the embedded connect technologies for physical manufacturing facilities are recommended for further research.

14.5 Summary

Many research studies on the application of grid technologies in manufacturing have been carried out, and a large number of articles have been published during the first six years since the concept of a manufacturing grid (MGrid) was first introduced. Existing research on MGrid theories (including eleven categories) and applications (including six aspects) were reviewed. The existing problems of the current research on MGrid were analyzed, and nine future key issues involving MGrid were highlighted.

15

Cloud Manufacturing: Development and Commerce Realization of MGrid

15.1 Introduction

The manufacturing grid (MGrid) has experienced rapid development, and many researchers have studied its theories and applications. However, there are many limitations for the wider practical application of MGrid, such as lack of (a) effective models, standards, and criteria supporting free circulation and concurrency, and on-demand use of manufacturing resources, capabilities, and services, (b) manufacturing knowledge management and innovative utilization, (c) intelligent perception and connection into the Internet of various physical manufacturing resources, (d) virtualization and service encapsulation of manufacturing resources and capabilities based on knowledge, and (e) reliable security solutions and technologies under a wide Internet environment, etc. These bottlenecks have hindered the further practical commerce application and development of MGrid.

At the same time, some new technologies have emerged and been widely applied in various fields, such as advanced computing models and technologies (e.g., cloud computing (Marston *et al.*, 2011; Rosenthal *et al.*, 2009), high performance computing (HPC), ubiquitous computing, service-oriented technologies (e.g., service-oriented architecture [Bosin *et al.*, 2011]), Web service, and semantic Web ontologies), internet of thing (IoT) (Atzoria *et al.*, 2010), virtualization technologies, intelligent embedded system and technologies, etc. These technologies are the enabling force for effectively addressing the above bottlenecks in manufacturing.

In this chapter, combing the above-mentioned new technologies and existing advanced manufacturing technologies, a computing and service-oriented manufacturing model, i.e., cloud manufacturing (CMfg) is discussed based on the previous work of this book. CMfg is the development and future commerce realization of MGrid. The concept, architecture, core enabling technologies, and typical characteristics of CMfg are to be abstractly studied in this chapter. Four typical CMfg service platforms, (i.e., public, private, community, and hybrid CMfg service platforms) are to be investigated. The key advantages and challenges for implementing CMfg are analyzed, as well as the key technologies and main research contents.

15.2 Concept and Architecture of Cloud Manufacturing

15.2.1 Concept of Cloud Manufacturing

Cloud manufacturing is a computing and service-oriented manufacturing model developed from existing advanced manufacturing models (e.g., ASP, AM, NM, MGrid) and enterprise information technologies under the support of cloud computing, IoT (Internet of Things), virtualization and service-oriented technologies, and advanced computing technologies. It aims to realize the full sharing and circulation, high utilization, and on-demand use of various manufacturing resources and capabilities. It does this by providing safe and reliable, high quality, inexpensive, on-demand use of manufacturing services for the whole life-cycle of manufacturing (Tao *et al.*, 2011b).

The abstract running principle for CMfg is shown in Figure 15.1.

In CMfg system, various manufacturing resources and capabilities can be intelligently sensed and connected into a wider internet, and automatically managed and controlled using IoT technologies (e.g., RFID, wired and wireless sensor network, embedded system).

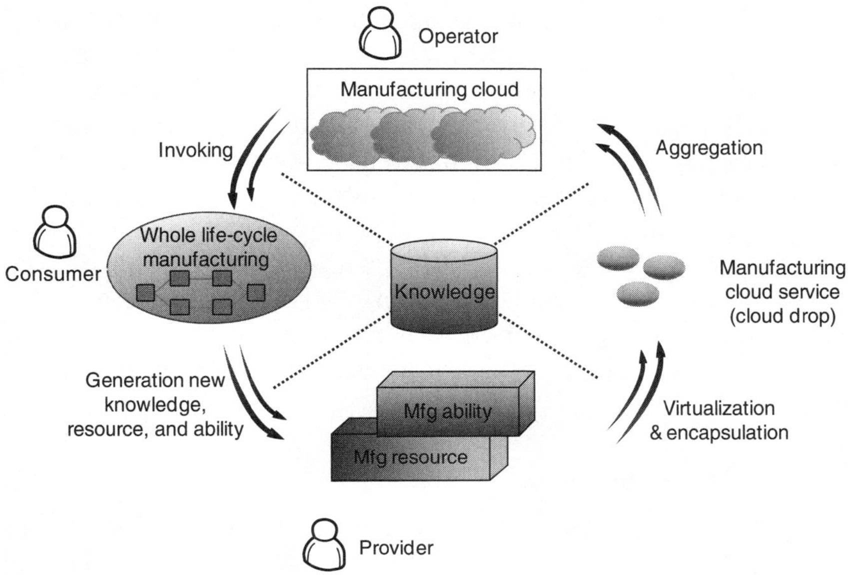

Figure 15.1 Abstract running principle for cloud manufacturing system.

Then the manufacturing resources and capabilities are virtualized and encapsulated into different manufacturing cloud services (MCSs) that can be accessed, invoked, and deployed based on knowledge by using virtualization technologies, service-oriented technologies, and cloud computing technologies. The MCSs are classified and aggregated according to specific rules and algorithms, and different kinds of manufacturing clouds are constructed. Different users can search and invoke the qualified MCSs from a related manufacturing cloud according to their needs. Next, they are assembled as a virtual manufacturing environment or solution in order to complete the manufacturing task involved in the whole life-cycle of the manufacturing process under the support of cloud computing, service-oriented technologies, and advanced computing technologies. As illustrated in Figure 15.1, there are primarily three categories of users in a CMfg system, briefly described as follows:

1. *Provider:* The providers own and provide the manufacturing resources and capabilities involved in the whole life-cycle of the manufacturing process of which consumers were in charge. They can be a person, an organization, an enterprise, or a third party.

2. *Operator:* The operators operate the CMfg platform to deliver services and functions to providers, consumers, and third parties. They deal with the organization, sales, licensing, and consulting of the MCSs, and provide, update, and maintain the technologies and services involved in the operations to MCSs and the platform.
3. *Consumer:* The consumers are the subscribers of the MCSs available in a CMfg service platform. They purchase the use of the MCSs from the operator on an operational expense basis according to their needs.

15.2.2 Architecture of Cloud Manufacturing

The proposed architecture of the CMfg system is a hierarchical structure as illustrated in Figure 15.2. It consists of the following ten layers (Tao *et al.*, 2011b).

1. *Resource layer*: It provides the manufacturing resource and capability involved in the whole life-cycle of manufacturing that can be encapsulated into service which can be invoked by user or consumer.
2. *Perception layer:* It is responsible for sensing the physical manufacturing resources and capabilities, enabling them to be connected into the wider network, and processing the related data and information.
3. *Resource virtualization layer*: It is responsible for the virtualization of manufacturing resources and capabilities, and encapsulating them into manufacturing cloud service.
4. *Cloud service layer*: Cloud service layer primarily provides two categories of service: manufacturing cloud service (MCS) and CMfg core service. MCS is the result of service encapsulation of manufacturing resources and capability which can be invoked by end users. CMfg core services are the main services provided by the CMfg platform for the three categories of users (i.e., the provider, operator, and consumer) to manage, access, and invoke MCSs, including description,

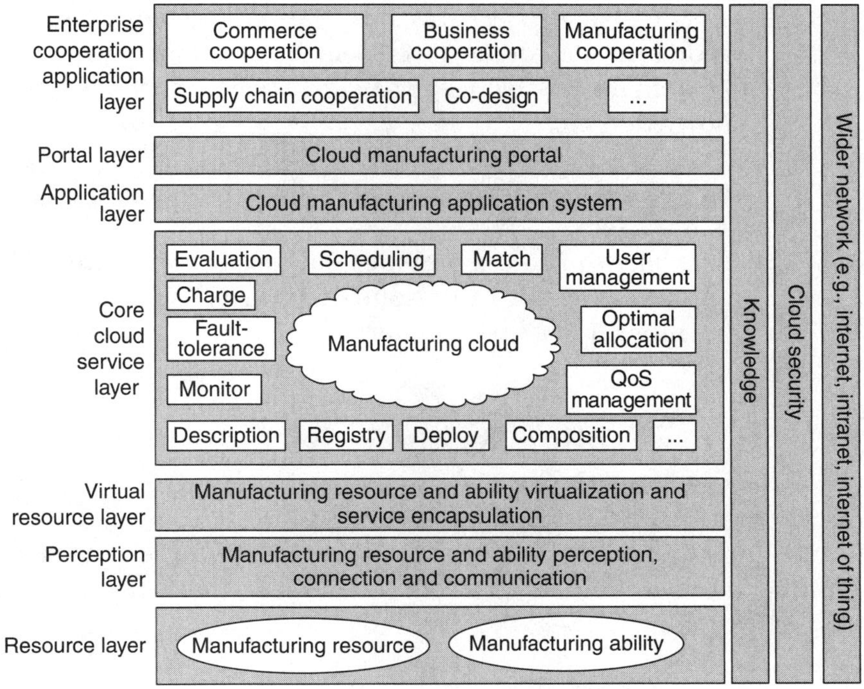

Figure 15.2 Architecture of cloud manufacturing system.

registry, publication, match and search, scheduling, charge, evaluation, and so forth.

5. *Application layer:* Based on the integration of the existing manufacturing system in an enterprise with CMfg, application layer develops a special manufacturing application system according to specific demands, such as cooperative supply chain management system, CMfg-based ERP, and so forth.
6. *Portal layer:* It provides various man-machine interaction interfaces for users accessing and invoking MCSs in CMfg.
7. *Enterprise cooperation application layer:* Supported by the MCSs and CMfg core services, different kinds of cooperation application can be realized, including commerce cooperation, business cooperation, collaborative design, manufacturing cooperation, and so forth.

8. *Knowledge layer:* It provides various knowledge needed in the other layers, such as manufacturing domain knowledge, process knowledge, model knowledge, etc.
9. *Cloud security layer*: It provides different security architecture, mechanisms, and strategies for the CMfg system.
10. *Wider internet layer:* It provides the base communication environment to all resources, service, users, and operations in the CMfg system.

15.3 Core Enabling Technologies for Cloud Manufacturing

While the evolution of CMfg will take several years or an even longer time to fully unfold, the six core technologies that will enable it are rapidly taking shape. They are (Tao *et al.*, 2011b):

1. *Existing manufacturing informationization technologies* are the foundation of CMfg;
2. *Cloud computing* is the enabling technology for intelligent information processing and decision making in manufacturing activities;
3. *Internet of thing technologies* can provide enabling technologies for the realization of intelligent perception and effective connection among different things (including men-to-men, machine-to-machine, and men-to-machine);
4. *Virtualization* is the technology that hides the physical characteristics of the manufacturing resource and capability of a CMfg platform from users;
5. *Service-oriented technologies*, such as SOA, web service, and semantic web, can provide enabling technologies for intelligently constructing a virtual manufacturing and service environment; and,
6. *Advanced computing technologies* provide enabling technologies for solving large-scale and complex manufacturing problems and carrying out large-scale cooperative manufacturing and work, such as high performance computing (HPC) technology.

15.4 Typical Characteristics of Cloud Manufacturing

Compared with existing manufacturing models, CMfg has the following typical characteristics (Tao *et al.*, 2011b):

1. *Service and requirement-oriented:* Most manufacturing models are resource or order-oriented, while CMfg is a service and requirement-oriented manufacturing model. In CMfg, both manufacturing resource and capability are virtualized and encapsulated into manufacturing cloud service, and can be used by all users according to their requirements via a wide internet. The core thought of CMfg is manufacturing as a service. Its aim is to realize the free currency and share of manufacturing resource and capability, and enable manufacturing resource and capability to be used according to requirements.
2. *Dynamic and uncertainty:* The resources and services in CMfg are various and dynamic, and the solutions for addressing a manufacturing task is dynamic too.
3. *Knowledge-based:* The whole life-cycle of CMfg needs knowledge supports, including knowledge based (a) manufacturing resource and capability perception, connection and virtualization, (b) cloud service description, match and search, aggregation, and composition, (c) optimal allocation and scheduling, and (d) business workflow management, and so on.
4. *Initiative:* In a CMfg system, both manufacturing cloud service and manufacturing task (or requirements) are active, they can automatically find and match with each other under the supports of semantic reasoning based on knowledge.
5. *Physical distributed and logical centralized:* The physical manufacturing resource and capability in CMfg are located in different places and are controlled by different persons or organizations. However, they are all virtualized and encapsulated into MCSs that are managed, controlled, and used centrally in logic.
6. *Wikipedia style and group innovation based manufacturing:* Any person, institute, or enterprise can participate

in and contribute their manufacturing resources, capabilities, and knowledge to CMfg service platform. Meanwhile, any enterprise can carry out their manufacturing actions based on these manufacturing resources, capabilities, and knowledge. Similar to Wikipedia, CMfg is a group innovation-based manufacturing model.

7. *Lower threshold and outsourcing:* Under a CMfg environment, an enterprise does not need to possess all the hardware manufacturing environment (such as workshop, equipments, IT infrastructures, and people) and software manufacturing capability (such as design, manufacturing, management, and sales capability). It can rent or invoke the resources, capabilities, and services in CMfg platform according to its requirements after payment. It reduces the entry threshold for a new enterprise to carry out manufacturing actions.

15.5 Difference and Relationship between Cloud Computing and Cloud Manufacturing

CMfg is proposed after cloud computing, and cloud computing is a core enabling technology for CMfg as described in Section 15.3. However, the resources involved in cloud computing primarily are computational resources (e.g., server, storage, network, software), and they are primarily provided as services for the user in the following three models:

1. *Infrastructure as a service* (IaaS): The storage and compute capabilities are provided as a service; examples are Amazon's S3 storage service, Rackspace Cloud Servers, and EC2 computing platform.
2. *Platform as a service* (PaaS): The platform is provided as a service which enables the development and deployment of applications without the cost and complexity of buying and managing the underlying hardware and software layers; examples are Microsoft's Azure Services Platform, Google App Engine, and Amazon's Relational Database Services.

3. *Software as a service* (SaaS): The application and software are offered as a service in which the application runs on the cloud. Therefore, the need to install and run the application on the client's computer are eliminated; examples are Salesforce, Gmail, Facebook, and Twitter (Marston *et al.*, 2011).

While in CMfg, in addition to the IT resources, all manufacturing resources and capabilities involved in the whole life-cycle of manufacturing are aimed to be provided for the user in different service models based on IaaS, PaaS, and SaaS, including the following models:

1. *Design as a service* (DaaS): The design resource and capability are provided as a service.
2. *Manufacturing as a service* (MFGaaS): The manufacturing resource and capability are offered as a service.
3. *Experimentation as a service* (EaaS): The experimentation resource and capability are provided as a service.
4. *Simulation as a service* (SIMaaS): The simulation resource and capability are provided as a service.
5. *Management as a service* (MaaS): The management resource and capability are provided as a service.
6. *Maintain as a service* (MAaaS): The maintain resource and capability are provided as a service.
7. *Integration as a service* (INTaaS): The integrated resource and capability, information system, and platform are provided as a service.

The difference and relationships between cloud computing and CMfg are illustrated in Figure 15.3.

15.6 Classification of Cloud Manufacturing Service Platform

Because of security and safety concerns, and scope of utilization, there are primarily four kinds of CMfg service platform, they are:

a. Public CMfg service platform,
b. Private CMfg service platform,

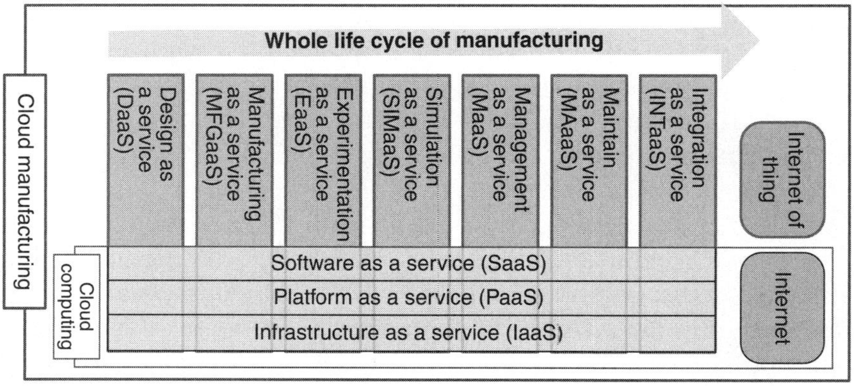

Figure 15.3 Relationship between cloud computing and CMfg.

c. Community CMfg service platform, and
d. Hybrid CMfg service platform.

A public CMfg service platform is characterized as being available from a third party service provider and demander, and is used to realize the sharing and optimal-allocation of the entire manufacturing resources and capabilities distributed in, and owned by, different enterprises and organizations, especially for small and medium sized enterprises (SMEs).

A private CMfg service platform offers many of the benefits and services of a public CMfg service platform environment, but is managed within an organization or enterprise. It provides greater control over its resources and services, and is usually established and used to realize the sharing and optimal-allocation of a group enterprise's in-house manufacturing resources and capabilities distributed in its branch companies or subsidiary companies, research center, different departments, etc. This promotes utilization and reduces cost.

A community CMfg service platform is controlled and used by a group of organizations that have shared interests, such as specific security requirements or a common mission.

A hybrid CMfg service platform is a combination of a public and private CMfg service platform. Noncritical service or information is outsourced to a public CMfg service platform, while business-critical service and data are kept within the control of the organization.

15.7 Key Technologies and Main Research Contents of Cloud Manufacturing

The key technologies and primary research content that need to be investigated for use in implementing CMfg are as follows:

1. Architecture, models, standards and criteria for CMfg;
2. Intelligent perception, connection technologies, and equipments for various physical manufacturing resource and capability, including hardware and software resource;
3. Knowledge-based resource virtualization and virtual resource management technologies;
4. Cloud service encapsulation technologies of manufacturing resource and capability, including cloud service modeling, description, deployment, and so forth.
5. Cloud service comprehensive management technologies, such as aggregation and classification, cost management, QoS (quality of service) management, integrated evaluation, composition, trust and reliability management, scheduling, and optimal-allocation of cloud service;
6. CMfg task management, such as task definition, description, decomposition, evaluation, and scheduling;
7. Manufacturing capability and cloud service transaction process management;
8. CMfg user business management and ubiquitous interface development;
9. CMfg service platform development and management; and
10. CMfg safety and security technologies and so forth.

15.8 Key Advantages and Challenges of Cloud Manufacturing

Specifically, CMfg can offer the following key advantages:

1. It can reduce resource idle capacity and increase utilization. In conventional manufacturing models, manufacturing resource utilization is low, e.g., estimated at

15–20% for IT resources, and the estimates for other manufacturing resources are lower. In contrast, a CMfg platform can increase the utilization across many customers, e.g., the IT resource utilization attains at least 40% (Rosenthal *et al.*, 2009).

2. It can reduce the upfront investments and lower the cost of entry for SMEs trying to benefit from high-value manufacturing resources (e.g., high-end equipment, expensive application systems) and specific manufacturing capability that were until now available only to the larger corporations. These manufacturing resources or capabilities involve large amount investments for a relatively short amount of time. CMfg makes such dynamic provisioning of resources possible, and provides an almost immediate access to these resources and capabilities without upfront capital investments.
3. It can lead to reduced infrastructure and administrative costs, energy savings, and reduced upgrades and maintenance cost according to the above advantage (Marston *et al.*, 2011).
4. CMfg makes it easier for manufacturing enterprises (both smaller and larger) to scale their production and business according to client demand. Through the resource and service can be easily obtained, so they can be invoked, and assembled very fast as new requirements arise by an enterprise. The successful application of cloud computing in IT enterprise is a very convictive example (Marston *et al.*, 2010; Rosenthal *et al.*, 2009).
5. CMfg also can generate new types and classes of manufacturing/business models or processes and deliver manufacturing services that were not possible before, such as MFG.COM (www.mfg.com) and Local-Motors.com (www.local-motors.com).
6. It can optimize the industrial distribution, and speed the transformation from a distributed and high energy consumption manufacturing model to a centralized resource and environmentally friendly manufacturing model, and from production-oriented manufacturing to service-oriented manufacturing. Furthermore, it can

promote the specialization division of manufacturing, because under cloud manufacturing environment, an enterprise does not need to carry out all activities involved in the whole life-cycle of manufacturing, they only need to emphasize their core business and service, even if it is a very small part.

Meanwhile, there are many challenges that need to be addressed before CMfg can be accepted as a viable choice in manufacturing, such as:

1. The possibility of backlash from entrenched ideas, manufacturing processes, and models,
2. Lack of standards and criteria,
3. Regulation at the personal, enterprise, local, national, and international level, and
4. Safety and security problems, and so on.

15.9 Summary

CMfg enhances resource utilization and reduces resource and energy consumption by realizing free transaction and currency, and the sharing and cooperation of manufacturing resources and capabilities. In order to speed up the transformation from production-oriented manufacturing to service-oriented manufacturing, and promote the power of service competition and innovation based on the knowledge of manufacturing enterprise, a computing and service-oriented manufacturing model – cloud manufacturing (CMfg) – has been proposed under the supports of cloud computing, Internet of things, advanced computing, and virtualization and service-oriented technologies. In this chapter the concept, architecture, core enabling technologies, and typical characteristics of CMfg were discussed and investigated, as well as the differences and relationship between cloud computing and CMfg. Four typical CMfg service platforms, i.e., public, private, community, and hybrid, were introduced. The key advantages and challenges for implementing CMfg were analyzed, as well as the key technologies and main research contents.

Bibliography

Abdul-Rahman, A., Hailes, S. (2000). Supporting trust in virtual communities. *Proceedings of the 33rd Hawaii International Conference on System Sciences.* January 4–7, Maui, Hawaii, USA pp.1–9.

Abeni, L., Cucinotta, T., Lipari, G., Marzario, L., Palopoli, L. (2005). QoS management through adaptive reservations. *Real-Time Systems,* 29(2–3):131–155.

Aggarwal, R., Verma, K., Miller, J., Milnor, W. (2004). Constraint driven web service composition in METEOR-S. *Proceedings of IEEE SCC2004,* September 15–18, Shanghai, China, pp.23–30.

Ait-Sadoune, I., Ait-Ameur, Y. (2008). Verification and validation of web service composition using event B method. *Lecture Notes in Computer Science,* 5238:339–340.

Akkermans, H.A., Van der Horst, H. (2002). Managing IT infrastructure standardization in the networked manufacturing firm. *International Journal of Production Economics,* 75(1–2):213–228.

Akkiraju, R., Verma, K., Goodwin, R., Doshi, P., Lee, J., (2004). Executing Abstract Web Process Flows. *International Conference on Automated Planning and Scheduling (ICAP2004),* 3–7 June, Whistler, British Columbia, Canada, http://www.isi.edu/ikcap/icaps04-workshop/final/akkiraju.pdf.

Al-Ali, R., Amin, K., Laszewski, G., (2003a). An OGSA-based quality of service framework. *Proceedings of the second international Workshop on grid and cooperative computing (GCC2003),* December 7–10, Shanghai, China, pp.361–378.

Al-Ali, R., Hafid, A., Rana, W.D. (2004a). An approach for quality of service adaptation in service-oriented grids. *Concurrent and computation: practice and experience,* 16:401–412.

Alberto, F., Matteo V., Cesar, C., Sascha, O., (2007). A role-based support mechanism for service description and discovery. *Lecture Notes in Computer Science,* 4504:132–146.

Alcaraz-Mejia, M., Lopez-Mellado, E., Ramirez-Treviio, A., (2007). Redundancy Based Controller Reconfiguration for Fault Recovery of Manufacturing Systems. *The 4th annual IEEE Conference on Automation Science and Engineering,* September 22–25, Scottsdale, Arizona, USA, pp.128–133.

Alcaraz-Mejia, M., Lopez-Mellado, E., Ramirez-Trevino, A. (2006). Fault recovery of manufacturing systems based on controller reconfiguration. *2006 IEEE/SMC International Conference on System of Systems Engineering*, April 24–26, Los Angeles, California, USA, p 6.

Alexopoulos, K., Mourtzis, D., Papakostas, N., Chryssolouris, G. (2007a). DESYMA – Assessing flexibility for the lifecycle of manufacturing systems. *International Journal of Production Research*, 45 (7):1683–1694.

Alexopoulos, K., Papakostas, N., Mourtzis, D., Chryssolouris, G. (2010). A method for comparing flexibility performance for the lifecycle of manufacturing systems under capacity planning constraints. *International Journal of Production Research* (DOI: 10.1080/00207543.2010.482566) [published online first].

Alexopoulos, K., Papakostas, N., Mourtzis, D., Gogos, P., Chryssolouris, G. (2007b). Quantifying the flexibility of a manufacturing system by applying the transfer function. *International Journal of Computer Integrated Manufacturing*, 20(6):538–547.

Allen, D.K., Bryan, J. (1986). Architecture for Computer-Integrated Manufacturing. *CIRP Annals - Manufacturing Technology*, 35(1): 351–354.

Al-Nory, M., Brodsky, A., Nash, H. (2009). A service composition framework for decision making under uncertainty. *Lecture Notes in Business Information Processing*, 24:363–75.

Antonio, B., Razvan, P. (2006). Automated generation of BPEL adapters. *Proceedings of International Conference on Service Oriented Computing (ICSOC 2006)*, December 4–7, Chicago, Illinois, USA, pp.27–39.

Ardagna, D., Pernici, B. (2006). Dynamic web service composition with QoS constraints. *International Journal of Business Process Integration and Management*, 1(4): 233–243.

Atanassov, K.T., (1986). Intuitionistic fuzzy sets. *Fuzzy Sets and Systems*, 20:87–96.

Atzoria, L., Ierab, A., Morabito, G. (2010). The Internet of things: A survey. *Computer Networks*, 54(5):2787–2805.

Avaki (2002). Avaki grid software: concepts and architecture, using comprehensive grid software from AVAKI to provide wide-area access to processing power, applications, and data. *White paper of Avaki Corporation*, March, http://www.avaki.com/

Azzedin, F., Maheswaran, M. (2002). Integrating trust into grid resource management system. *International Conference on Parallel Processing (ICPP'02)*, August 20–23, Vancouver, BC, Canada, pp.47–54.

Baader, F., Lutz, C., Miličić, M., Sattler, U., Wolter, F. (2005). A Description Logic Based Approach to Reasoning about Web Services. *Proceedings of the WWW 2005 Workshop on Web Service Semantics (WSS2005)*, May 10–14, Chiba City, Japan.

Balken, R., Haukrogh, J., Jensen, J.L., Jensen, M.N., Roost, L.J., Toft, P.N., Olsen, R.L., Schwefel, H.P. (2007). Context-Sensitive Service Discovery Experimental Prototype and Evaluation, *Wireless Personal Communications*, 40(3):417–431.

Bao, Z.Q., Li, X.Q., Zhang, D., Wang, P., Gao, K.Z. (2008). Wasp Algorithm for Dynamic Scheduling Based on Manufacturing Grid. *2008 International Symposiums on Information Processing (ISIP 2008)*, May 23–25, Mosco, pp.83–87.

Barabasi, A.-L., Albert, R. (1999). Emergence of scaling in random network. *Science*, 286(5439):509–512.

Barabasi, A.-L., Albert, R., Jeong, H. (2009). Mean-field theory for scale-free random networks. *Physica A: Statistical Mechanics and its Applications*, 272(1–2):73–187.

Barabasi, A.-L., Jeong, H., Albert, R. (2000). Scale-free characteristics of random networks: the topology of the world-wide web. *Physica A: Statistical Mechanics and its Applications*, 281(1–4):69–77.

Barabasi, A.-L., Ravasz, E., Vicsek, T. (2001). Deterministic scale-free networks. *Physica A: Statistical Mechanics and its Applications*, 299(3–4):559–564.

Barash, M.M. (1980). Computer Integrated manufacturing system. American Society of Mechanical Engineers. *Production Engineering Division, (Publication) PED*, 1:37–50.

Baresi, L., Bianculli, D., Ghezzi, C., Guinea, S., Spoletini, P. (2007). Validation of web service compositions. *IET Software*, 1(6):219–232.

Bedi, P., Kaur, K., Marwaha, S. (2007). Trust based recommender system for semantic web. *Proceedings of 20th International Joint Conferences on Artificial Intelligence (IJCAI'07)*, January 6–12, Hyderabad, India, pp.2677–2682.

Benatallah, B., Casati, F., Grigori, D., Nezhad, M., Toumani, F. (2005). Developing Adapters for Web Services Integration. *Proceedings of the International Conference on Advanced Information Systems Engineering (CAiSE)*, June 13–17, Porto, Portugal, pp.415–429.

Benatallah, B., Dumas, M., Sheng, Q.Z., Ngu, A.H.H. (2002). Declarative composition and peer-to-peer provisioning of dynamic Web services. *Proceedings of IEEE International Conference Data Engineering*, February 16–March 1, San Jose, California, USA pp.297–308.

Benatallah, B., Sheng, Q.Z., Dumas, M. (2003). The self-serv environment for web services composition. *IEEE Internet Computing*, 7(1):40–48.

Berardi, D., Calvanese, D., De Giacomo, G., Lenzerini, M., Mecella, M. (2003). Automatic composition of e-services that export their behavior. *Lecture Notes In Computer Science*, 2910:43–58.

Berardi, D., Calvanese, D., De Giacomo, G., Mecella, M. (2005). Composition of services with nondeterministic observable behavior. *Lecture Notes in Computer Science*, 3826:520–526.

Bernet, Y., Ford, P., Yavatkar, R. (2000). A framework for integrated operation over DiffServ. Networks, Internet request for comments RFC 2998, Internet engineering task force, 2000–11.

Bianchini, D., DeAntonellis, V., Melchiori, M. (2005). An ontology-based architecture for service discovery and advice system. *Proceedings of the 16th International Workshop on Database and Expert Systems Applications (DEXA'05)*, August 22–26, Copenhagen, Denmark, pp.551–556.

Bianculli, D., Ghezzi, C. (2008). Towards a methodology for lifelong validation of service compositions. *30th International Conference on Software Engineering (ICSE 2008)*, May 10–18, Leipzig, Germany, pp.7–12.

Blake, S., Black, D., Carlson, M. (1998). An architecture for differentiated services. Internet request for comments RFC 2475, Internet engineering task force, 1998–12.

Bosin, A., Dessì, N., Pes, B. (2011). Extending the SOA paradigm to e-Science environments. *Future Generation Computer Systems*, 27(1):20–31.

Bozsak Erol, Ehrig Marcc, Handschuh Siegfried, et al. (2002). KAON Towards a large scale Semantic Web. EC-Web 2002, LECTURE NOTES IN COMPUTER SCIENCE 2455, 2002:304–313.

Braden, B., Clark, D., Shenker, S. (1994). Integrated services in the internet architecture: an overview. Internet request for comments RFC 1633, Internet engineering task force, 1994–06

Brazier, F.M.T., Richards, D., van Splunter, S., Brazier, E. (2003). Composing Web Services Using an Agent Factory. *The 1st International Workshop on Web Services and Agent-based Engineering*, July 14, Sydney, Australia, pp.1–10, http://www.iids.org/publications/richards.pdf.

Brill, P.H., Mandelbaum, M. (1989). On measures of flexibility in manufacturing systems. *International Journal of Production Research*, 27:747–756.

Buyya, R., Abramson, D., Giddy, J. (2000). Nimrod/G: An architecture of a resource management and scheduling system in a global computational grid. *Proceedings. The Fourth International Conference/Exhibition on High Performance Computing in the Asia-Pacific Region (HPC Asia 2000)*, May 14–17, Beijing, China, pp.283–289.

Byrne, G., Sheahan, C. (2005). Comparative analysis of vision systems for electroplating surface quality inspection. *International Journal of Production Research*, 43(18):3787–3801.

Byrne, G,. Sheahan, C. (2006). Integrated risk minimization methodology for high volume manufacture. *Journal of Manufacturing Technology Management*, 18(5):538–548.

Cai, H. (2007b). Scale-Free Web Services. *IEEE International Conference on Web Services (ICWS 2007)*, 9–13 July, Salt Lake City, Utah, USA, pp.288–295.

Cai, H.X., (2007a). "Research and realization on the technologies of the manufacturing grid security". Thesis (PhD), Shanghai University (in Chinese).

Cardoso, J., Sheth, A. (2003). Semantic e-Workflow Composition. Journal of Intelligent Information Systems, 21(3):191–225.

Ccidnet. http://news.ccidnet.com/art/964/20031117/71759_1.html. Dec. 25, 2010.

Chang, A.Y. (2001). An approach to the measurement of single-machine flexibility. *International Journal of Production Research*, 39(8):1589–1601.

Chang, A.Y. (2004). On the measurement of labor flexibility. *Proceedings of International Engineering Management Conference*, August 12–13, Austin, Texas, USA, pp.163–167.

Chang, A.Y. (2007). On the measurement of routing flexibility: A multiple attribute approach. *International Journal of Production Economics*, 109(1–2):122–136.

Chang, A.Y. (2009). An attribute approach to the measurement of machine-group flexibility. *European Journal of Operational Research*, 194(3):774–786.

Che, H., Gu, J. (2004). Scale-free networks and their significance for systems science. *System Engineering-Theory & Practice*, 24(4):82–88 (in Chinese).

Chen, L., Deng, H., Deng, Q.N., Wu, Z.Y. (2004a). A research of grid manufacturing and its application in custom artificial joint. *Lecture Notes in Computer Science*, 3036:507–510.

Chen, L.J. (2006). "The study of access control problem based on manufacturing grid". Thesis (Master), Wuhan University of Technology (in Chinese).

Chen, X., Tian, W., Chen, X., Deng, B., Sun, J. (2003). Framework of grid-manufacturing system for mould and die making industry franchise business mode. *Computer Integrated Manufacturing Systems*, 9(7):595–600 (in Chinese).

Cheng, N., Zhang, Y. (2008). Intranet Based Security Infrastructure for Manufacturing Grid. *Second International Conference on Genetic and Evolutionary Computing (WGEC'08)*, September 25–26, Jingzhou, China, pp.247–250.

Chi, Y.L., Lee, H.M. (2007). A formal modeling platform for composing web services. *Expert Systems with Applications*, 34(2):1500–1507.

Chifu, V.R., Salomie, I., Chifu, E.S., Pârtac, C. (2008). Harmony-A framework for automatic web service composition. *4th International Conference on Web Information Systems and Technologies*, May 4–7, Funchal, Madeira, Portugal, pp.240–243.

Chryssolouris, G. (1996). Flexibility and its measurement. *CIRP Annals*, 45(2):581–587.

Chryssolouris, G., Lee, M. (1992). An assessment of flexibility in manufacturing systems. *Manufacturing Review*, 5(2):105–116.

Coello, C.A., Pulido, G.T., Lechuga, M.S. (2004). Handling Multiple Objectives With Particle Swarm Optimization, *IEEE Transactions On Evolutionary Computation*, 3:256–279.

COHSE (2007). http://cohse.semanticweb.org/, 2007.10.

Cristani, M., Cuel, R. (2004). A survey on ontology creation methodologies. *International Journal on Semantic Web and Information Systems*, 1(2):49–69.

Czajkowski, K., Foster, I., Karonis, N., Kesselman, C., Martin, S., Smith, W., Tuecke, S. (1998). A resource management architecture for metacomputing systems. *Lecture Notes in Computer Science*, 1459:62–82.

Czajkowski, K., Fitzgerald, S. (2001). Grid Information Services for Distributed Resource Sharing. *Proceedings of the 10th IEEE International Symposium on High Performance Distributed Computing (HPDC-10)*, August 7–9, San Francisco, California, USA pp.181–194.

da Silva Santos L.O.B., Pires, L.F., Van, S.M.. (2008). A goal-based framework for dynamic service discovery and composition. *Proceedings of the 2nd International Workshop on Architectures, Concepts and Technologies for Service Oriented Computing, ACT4SOC 2008 – In Conjunction with ICSOFT 2008*, July 5–8, Porto, Portugal, pp.67–78.

Dai, Y., Yang, L., Zhang, B., Gao, Y. (2006a). QoS for composite Web services and optimizing. *Chinese Journal of Computers*, 29(7):1167–1178.

Dai, Y.S., Wang, X.L. (2007). Optimal resource allocation on grid systems for maximizing service reliability using a genetic algorithm. *Reliability Engineering and System Safety*, 91(9):1071–1082.

DAML Ontology (2008). http://www.daml.org/ontologies/, 2008-July-14.

D'Amours, S., Montreuil, B., Lefrancois, P., Soumis, F. (1999). Networked manufacturing: The impact of information sharing. *International Journal of Production Economics*, 58(1):63–79.

Dasgupta, G.B., Dasgupta, K., Purohit, A.T., Viswanathan, B. (2006). QoS-GRAF: A framework for QoS based Grid Resource Allocation with Failure provisioning, *14th IEEE International Workshop on Quality of Service (IWQoS 2006)*, June 19–21, New Haven, Connecticut, USA pp.281–283.

Deb, K., Pratap, A., Agarwal, S., Meyarivan, T. (2002). A Fast and Elitist Multiobjective Genetic Algorithm: NSGA-II. *IEEE Transactions On Evolutionary Computation*, 2 (2002):182–197.

Deng, H., Chen, L., Deng, Q.N. (2006b). Organization changing of virtual enterprise and scheduling of manufacturing resources based on manufacturing grid. *Computer integrated manufacturing systems*, 12(8):1211–1219.

Deng, H., Chen, L., Wang, C.T., Deng, Q.N. (2006a). A grid-based scheduling system of manufacturing resources for a virtual enterprise. *International Journal of Advanced Manufacturing Technology*, 28(1–2):137–141.

Deng, S., Wu, Z., Kuang, L. (2004a). Management of service flow in a flexible way. *Proceedings of the 5th International Conference on Web Information Systems Engineering*, November 22–24, Brisbane, Australia, pp.428–438.

Deora, V., Shao, J., Gray, W.A., Fiddian, N.J. (2006). Expectation-based quality of service assessment, *International Journal on Digital Libraries*, 6(3): 260–269.

Dialani, V., Miles, S., Moreau, L., Roure, D.D., Luck, M. (2002). Transparent fault tolerance for web services based architectures. *Eighth International Europar Conference (EURO-PAR'02)*, August 27–30, Padeborn, Germany, pp.889–898.

Diamadopolou, V., Makris, C., Panagis, Y., Sakkopoulos, E. (2008). Techniques to support Web Service selection and consumption with QoS characteristics. *Journal of Network and Computer Application*, 31(2):108–130.

Dickinson, D.A.., Draper, C.W., Saminathan, M., Sohn, J.E., Williams, G. (1995). Green product manufacturing. *AT&T technical journal*, 74(6):26–35.

Ding, Y.F., Tao, F., Buyun, S., Zude, Z. (2008a). Modeling and application of optimal-selection evaluation for manufacturing grid resource. *International Journal of Computer Integrated Manufacturing*, 21(1):62–72.

Ding, Y.F., Tao, F., Hu, Y.F., Wu, B., Tian, J. (2005a). Research on data exchange between PDM and ER. *The Proceedings of the 11th International Conference on Industrial Engineering and Engineering Managemen (IEEM'2005)*, April 23–25, Shenyang, China, pp.1099–1133.

Domingues, P., Sousa, B., Silva, L.M. (2007). Sabotage-tolerance and trust management in desktop grid computing. *Future Generation Computer System*, 23(7):904–912.

Doulkeridis, C., Zafeiris, V., Norvag, K., Vazirgiannis, M., Giakoumakis, E.A. (2007). Context-based caching and routing for P2P Web service discovery. *Distributed and Parallel Database*, 21(1):59–84.

Du, S.C., Xi, L.F., Shi, J.J. (2006). Distributed sensor system for fault detection and isolation in multistage manufacturing systems. *International Journal of Computer Applications in Technology*, 25(4):182–191.

Duan, L., Liu, S.J., Tu, C.H., Meng, X.X. (2006). A BPEL4WS-based Composite Service Modeling Solution in Manufacturing Grid. *IEEE Asia-Pacific Conference on Services Computing (APSCC'06)*, December 12–15, Guangzhou, China, pp.405–410.

El Haddad, J., Manouvrier, M., Rukoz, M. (2010). TQoS: Transactional and QoS-aware selection algorithm for automatic Web service composition. *IEEE Transactions on Service Computing*, 3(1):73–85.

Eberhart, R.C., Shi, Y. (1998). Evolving artificial neural networks. *Proceedings of International Conference on Neural Networks and Brain*, October 27–30, Beijing, China, pp.5–13.

Elkins, D.A., Huang, N., Alden, J.M. (2004). Agile manufacturing systems in the automotive industry. *International Journal of Production Economics*, 91(3):201–214.

Fan, Y., Zhao, D., Zhang, L., Huang, S., Liu, B. (2003a). Manufacturing grid needs, concept, and architecture. *International Workshop on Grid and Cooperative Computing (GCC 2003)*, December 7–10, Shanghai, China, pp.653–656.

Fang, C.L., Liang, D., Lin, F., Lin, C.C. (2007). Fault tolerant web service. *Journal of Systems Architecture*, 53(1):21–38.

Fensel, D., Bussler, C. (2002). The Web Service Modeling Framework WSMF. *Electronic Commerce Research and Applications*, 1(2):113–137.

Fernández, A., Vasirani, M., Cáceres, C., Ossowski, S, (2007), A role-based support mechanism for service description and discovery. *AAMAS 2007 International Workshop on Service-Oriented Computing: Agents, Semantics, and Engineering (SOCASE 2007)*, May 14, Honolulu, Hawaii, USA, pp.132–146.

Fletcher, M., Deen, S.M. (2001). Fault-tolerant holonic manufacturing systems. *Concurrency Computation Practice and Experience*, 13(1):43–70.

Forte, M., Souza, W.L., Prado, A.F. (2008). Using ontologies and Web services for content adaptation in Ubiquitous Computing. *The Journal of Systems and Software*, 81(3):368–381.

Foster, H., Uchitel, S., Magee, J., Kramer, J. (2010b). An integrated workbench for model-based engineering of service compositions. *IEEE Transactions on Service Computing*, 3(2):131–144.

Foster, I. (2002a). What is the grid? A three point checklist. *GRID Today*. July 20, http://www.chinagrid.net/grid/paperppt/GlobusPaper/WhatIsTheGrid.pdf 2009.6.

Foster, I., Kesselman, C. (1997). Globus: A metacomputing infrastructure toolkit. *International Journal of Supercomputer Applications*, 11(2):115–128.

Foster, I., Kesselman, C. (1998). *The Grid: Blueprint for a New Computing Infrastructure*. Los Altos, CA: Morgan Kaufmann Publishers, 1998.

Foster, I., Kesselman, C. (1999). A distributed resource management architecture that supports advance reservation and co-allocation. *Proceedings of the 7th International Workshop on Quality of Service (IWQOS 1999)*, June 1–4, London, pp.27–36.

Foster, I., Kesselman, C. (2004a). *The Grid 2: Blueprint for a New Computing Infrastructure*. Los Altos, CA: Morgan Kaufmann Publishers, 2004.

Foster, I., Kesselman, C., Tuecke, S. (2001a). The anatomy of the grid: Enabling scalable virtual organizations. *International Journal of High Performance Computing Applications*, 15(3):200–222.

Foster, I., Roy, A., Sander, V. (2000a). A quality of service architecture that combines resource reservation and application adaptation. *Proceedings of the 8th International Workshop on Quality of Service (IWQOS 2000)*, June 5–7, Pittsburgh, Pennsylvania, USA, pp.81–188.

Frayret, J.M., D'Amours, S., Montreuil, B., Cloutier, L. (2001). A network approach to operate agile manufacturing systems. *International Journal of Production Economics*, 74(1–3):239–259.

Frey, J., Tannenbaum, T., Livny, M., Foster, I., Tuecke, S. (2002). Condor-G: A computation management agent for multi-institutional grids. *Journal of Cluster Computing*, 5(3): 237–246.

Fries, T.P. (2007). Multi-agent fault diagnosis in manufacturing systems using soft computing. *International Conference on Integration of Knowledge Intensive Multi-Agent Systems (KIMAS 2007)*, April 30–May 3, Waltham, MA, USA, pp.168–173.

Fu, J.Z. (2007). "Research on architecture and resource management for manufacturing grid system". Thesis (PhD), Nanjing University of Science and Technology, (in Chinese).

Fujii, K., Suda, T. (2005). Semantics-based dynamic service composition. *IEEE Journal on Selected Areas in Communications*, 12 (2005):2361–2372.

Gaaloul, W., Bhiri, S,, Rouached, M. (2010). Event-based design and runtime verification of composite service transactional behavior. *IEEE Transactions on Service Computing*, 3(1):32–45.

Gamha, Y., Bennacer, N., Naquet, G.V., Ayeb, B. (2008). A framework for the semantic composition of web services handling user constraints. *Proceedings of the IEEE International Conference on Web Services (ICWS 2008)*, September 23–26, Beijing, China, pp.228–237.

Gangemi, A., Catenacci, C., Ciaramita, M., Lehmann, J. (2006). Modeling ontology valuation and validation. ESWC, *Lecture Notes in Computer Science*, 4011:140–154.

Gholamian, M.R., Fatemi Ghomi, S.M.T. (2007). Meta knowledge of intelligent manufacturing: An overview of state-of-the-art. *Applied Soft Computing*, 7(1):1–16.

Globus (2009). http://www.globus.org, 2009.1.

Gu, X., Nahrstedt, K. (2006). Distributed multimedia service composition with statistical QoS assurances. *IEEE Transactions on Multimedia*, 1(2006):141–151.

Guo, H., Tao, F., Zhang, L., Su, S.Y., Si, N. (2010). Correlation-aware web services composition and QoS computation model in virtual enterprise. *International Journal of Advanced Manufacturing Technology*, 51:817–827.

Guo, H., Tao, F., Zhang, L., Laili, Y. J., and Liu, D. K. (2011) Research on measurement method of flexibility of Resource service composition in service–oriented manufacturing. *International Journal of Computer Integrated Manufacturing*, DOI:10.1080/0951192X.2011.596572

Haase, P., Siebes, R., Harmelen, F. (2008). Expertise-based peer selection in peer-to-peer networks. *Knowledge and Information Systems*, 15(1):75–107.

Harney, J., Doshi, P. (2008). Selective querying for adapting Web service compositions using the value of changed information. *IEEE Transactions on Service Computing*, 1(3):169–185.

Hashemian, S.V., Mavaddat, F. (2005). A graph-based approach to Web services composition. *Proceedings of the 2005 Symposium on Applications and the Internet (SAINT'05)*, January 31–February 4 2005, Trento, Italy, pp.183–189.

Hashemian, S.V., Mavaddat, F. (2006). A Graph-Based framework for composition of stateless Web services. *Proceedings of the European Conference on Web Services (ECOWS'06)*, December 4–6, Zurich, Switzerland, pp.75–86.

He, X.S., Sun, X.H. (2003). QoS guided task scheduling algorithm for grid computing. *Journal of Computer Science and Technology*, 18(4):442–451.

He, Y., He, W., Yang, H., Hao, G., Zhao, K. (2006a). Towards an agent and knowledge enacted dynamic workflow Management System for Intelligent Manufacturing Grid. *Lecture Notes in Computer Science*, 4101:75–82.

He, Y., Yang, H., He, W., Mo, R., Jing, S.K. (2005b). Framework of cross-enterprise collaborative manufacturing based on grid technology. *Computer Integrated Manufacturing System*, 11(5):636–641 (in Chinese).

He, Y., Yu, T., Liu, L., Shen, B., Sun, H. (2005a). A QoS-based task decomposing middleware research for Manufacturing Grid. *First International Conference on Semantics, Knowledge and Grid (SKG'05)*, November 27–29, Beijing, China, pp.50–50.

He, Y., Yu, T., Liu, L., Shen, B., Sun, H.Y. (2006d). A WSRF-based Resource Management System of Manufacturing Grid. *Proceedings of the Sixth IEEE International Symposium on Cluster Computing and the Grid (CCGRID'06)*, May 16–19, Singapore, pp.174–177.

He, Y., Yu, T., Liu, L., Sun, H. (2006c). Research on manufacturing resource discovery based on ontology and QoS in Manufacturing Grid. *International Conference on Cyberworlds (CW '06)*, November 28–30, Lausanne, Switzerland, pp.209–215.

He, Y., Yu, T., Shen, B. (2006b). Research on WSRF-based resource management model and encapsulation in Manufacturing Grid. *IFIP International Federation for Information Processing, Knowledge Enterprise: Intelligent Strategies in Product Design, Manufacturing, and Management.* June 15–17, Shanghai, China, pp.908–913.

He, Y., Yu, T., Shen, B., Liu, L., Sun, Y. (2006e). Research on manufacturing resource discovery and scheduling in Manufacturing Grid. *IFIP International Federation for Information Processing, Knowledge Enterprise: Intelligent Strategies in Product Design, Manufacturing, and Management*, June 15–17, Shanghai, China, pp.902–907.

Hu, H., Li, Z. (2009). Modeling and scheduling for manufacturing grid workflows using timed Petri nets. *International Journal of Advanced Manufacturing Technology*, 42(5–6):553–568.

Hu, J., Guo, C., Wang, H., Zou, P. (2005a). Quality driven web service selection. *Proceedings of the 2005 IEEE international conference on e-business engineering (ICEBE'05)*, October 12–18, Beijing, China, pp.681–688.

Hu, J., Li, J., Zeng, Z. (2009b). SWSCF: A semantic-based Web service composition framework. *Journal of Networks*, 4(4): 290–297.

Hu, Y.F., Tao, F., Ding, Y.F., Sheng, B.Y., Zhou, Z.D. (2006). Study on framework of Manufacturing Grid supporting collaborative manufacturing. *China Mechanical Engineering*, 17(18):1903–1908 (In Chinese).

Hu, Y.F., Tao, F., Zhao, D., Zhou, Z.D. (2009a). Manufacturing Grid resource and resource service digital description. *International Journal of Advanced Manufacturing Technology*, 44(9–10):1024–1035.

Hu, Y.F., Tao, F., Zhou, Z.D. (2006b). The connotation of Manufacturing Grid and its key technology. *4th International IEEE Conference on Industrial Informatics (INDIN'06)*, August 16–18, Singapore, pp.293–298.

Hu, Y.F., Zhang, H.J., Tao, F., Zhou, Z.D. (2008). Research on OWL-S based Manufacturing Grid service discover. *Chinese Mechanical Engineering*, 19(21):2594–2599 (in Chinese).

Huang, C., Huang, B.Q., Li, C.P. (2005b). Research on resources integration framework of logistics resource grid. *Computer Integrated Manufacturing Systems*, 11(5), 630–635 (in Chinese).

Huang, C.L., Lo, C.C., Li, Y., Chao, K.M., Chung, J.Y., Huang, Y. (2005b). Service discovery through multi-agent consensus. *2005 IEEE international workshop on service-oriented system engineering (SOSE2005)*, October 20–21, Beijing, China, pp.37–44.

Huedo, E., Montero, R.S., Llorente, I.M. (2004). A framework for adaptive execution on girds. *Software-practice and experience*, 34(7):631–651.

Huedo, E., Montero, R.S., Llorente, I.M., (2006). Evaluating the reliability of computational grids from the end user's point of view. *Journal of System Architecture*, 52(12):727–736.

Hung, W.L., Yang, M.S. (2008). On similarity measures between intuitionistic fuzzy sets. *International Journal of Intelligent Systems*, 23(3):364–383.

Hwang, S., Kesselman, C. (2003). A generic failure detection service for the grid. Technical Report ISI-TR-568, USC Information Sciences Institute, 2003.

Hwang, S.Y., Lim, E.P., Lee, C.H., Chen, C.H. (2008). Dynamic Web service selection for reliable Web service composition. *IEEE Transactions on Service Computing*, 1(2):104–116.

Iamnitchi, A., Foster, I. (2000). A problem-specific fault-tolerance mechanism for asynchronous, distributed systems. *Proceedings of the 2000 International Conference on Parallel Processing*, August 21–24, Toronto, Canada, pp.4–13.

IPG (2009). http://www.gloriad.org/gloriad/projects/project000053.html 2009.1.

Iravani, S.M., Oyen, M.P.V., Sims, K.Y. (2005). Structural flexibility: A new perspective on the design of manufacturing and service operations. *Management Science*, 51(2):151–166.

Irvine, C.E., Levin, T. (1999). An approach characterizing resource usage and user preferences in benefit functions. Technical Report, NPS-CS-99-005, NPS, 1999.

Irvine, C.E., Levin, T. (2000). Toward quality of security service in a resource management system benefit function. *The 9th Proceedings of the Heterogeneous Computing Workshop (HCW2000)*, May 1, Cancun, Mexico, pp.133–139.

Jaeger, M.C., Rojec-Goldmann, G., Mühl, G. (2004). QoS aggregation for service composition using workflow patterns. *Proceedings of the 8th International Enterprise Distributed Object Computing Conference (EDOC 2004)*, September 20–24, Monterey, CA, USA, pp.149–159.

Jaeger, M.C., Rojec-Goldmann, G., Mühl, G. (2005). QoS aggregation in web service compositions. *Proceedings of the 2005 IEEE International Conference on e-Technology, e-Commerce and e-Service (EEE-05)*, March 29–April 1, Hong Kong, pages 181–185.

Howe, J. (2008). *Why the power of the crowd is driving the future of business.* Beijing: China CITIC Press, 2009.

Jena http://jena.sourceforge.net,2007.10.

Jia, Y., Srikumar, V., Rajlumar, B. (2006a). A market-oriented grid directory service for publication and discovery of grid service providers and their services. *The Journal of Supercomputing*, 36(1):17–31.

Jiang, J.J., Conrath, D.W. (1997). Semantic similarity based on corpus statistics and lexical taxonomy. *International Conference on Research in Computational Linguistics (ROCLING X 1997)*, August 22–24, Taiwan, pp.19–33.

Jiang, Z., Zhang, H., Luo, X. (2008a). Study on the cutting technology based on green manufacturing. *Key Engineering Materials*, 375–376:158–162.

Johnston, W., Gannon, S., Nitzberg, B., Woo, A., Thigpen, B., Tanner, L. (2000). Computing and data grids for science and engineering. *ACM/IEEE SC 2000 Conference (SC 2000)*, November 4–10, Dallas, Texas, USA, p 52.

Jordan, W.C., Graves, S.C. (1995). Principles on the benefits of manufacturing process flexibility. *Management Science*, 41(4):577–594.

Jovani, R., Fortuna, M.A. (2007). The shape of the past in the World Wide Web: Scale-free patterns and dynamics. *Physica A: Statistical Mechanics and its Applications*, 385(2):683–688.

Jung, J.Y., Park, J., Han, S.K., Lee, K. (2007). An ECA-based framework for decentralized coordination of ubiquitous web services. *Information and Software Technology*, 49(11–12):1141–1161.

Juric, M., Mathew, B., Sarang, P. (2004). Business Process Execution Language for Web Services. Packt Publishing, 2004.

Kalasapur, S., Kumar, M., Shirazi, B.A. (2007). Dynamic service composition in pervasive computing. *IEEE Transactions on Parallel and Distributed Systems*, 7 (2007):907–918.

Kazuo, M. (2003). Advanced technology for manufacturing engineering development: XML technology on a system that enables user to view required information from the workshop through Web browser. *JSAE Review*, 24(7):303–312.

Kennedy, J., Eberhart, R. (1995). Particle swarm optimization. *Proceedings of the IEEE International Conference on Neural Networks*, November 27–December 1, Perth, Australia, 4:1942–1948.

Kennedy, J., Eberhart, R., Shi, Y. (2001). *Swarm Intelligence*. Morgan Kaufmann, San Mateo, CA, 2001.

Khanli, L.M., Analoui, M. (2008). An approach to grid resource selection and fault management based on ECA rules. *Future Generation Computer System*, 24(4):296–316.

Kokash, N., Birukou, A., D'Andrea, V. (2007). Web service discovery based on past user experience. *10th International Conference on Business Information Systems (BIS 2007)*, April 25–27, Poznan, pp. 95–107.

Kokash, N., Birukou, A., D'Andrea, V. (2007). Web service discovery based on past user experience. *Lecture Notes in Computer Science*, 4439:95–107.

Koufteros, X., Vonderembse, M., Doll, W. (2001). Concurrent engineering and its consequences. *Journal of Operations Management*, 19(1):97–115.

Krauter, K., Buyya, B., Maheswaran, M. (2002). A taxonomy and survey of grid resource management systems for distributed computing. *Software-Practice and Experience*, 32(2):135–164.

Kumar, D.S., Biswas, R., Roy, A.R. (2000). Some operations on intuitionistic fuzzy sets. *Fuzzy Sets and Systems*, 114(3):477–484.

Kumar, V. (1987). Entropic measurement of manufacturing flexibility. *International Journal of Production Research*, 25(7):957–966.

Lee, C., Helal, S. (2003). Context attributes: an approach to enable context-awareness for service discovery. *Proceedings of the 2003 symposium on application and the Internet (SAINT'03)*, January 27–31, Orlando, Florida, pp.22–30.

Lee, H.G., Parl, N., Park, J. (2009). A high performance finite capacitated MRP process using a computational grid. *International Journal of Production Research*, 47(11):2109–2123.

Lee, H.M., Chung, K.S., Chin, S.H., Lee, J.H., Lee, D.W., Park, S., Yu, H.C. (2005c). A resource management and fault tolerance services in grid computing. *Journal of Parallel and Distributed Computing*, 65(11):1305–1317.

Lee, K.H. (2005b). Fuzzy number, first course on fuzzy theory and applications. *Advance in Soft Computing*, Springer Berlin/Heidelberg, 27:129–151, ISBN:978-3-540-22988-9, http://dx.doi.org/10.1007/3-540-32366-X

Lee, K.M., Chio, K.H., Her, S.P., Shin, D.R. (2005a). Matchmaking algorithms to improve dynamitic service matching in ubiquitous environments. *Proceeding of fourth annual ACIS international conference on computer and information science (ICIS'05)*, July 14–16, Jeju Island, South Korea, pp.239–244.

Legion. http://www.legion.com/ Dec. 1, 2009.

Lei, D. (2008). A Pareto archive particle swarm optimization for multi-objective job shop scheduling. *Computers & Industrial Engineering*, 4(2008):960–971.

Li, B., Tsai, W.T. (2005). Ontology and service oriented programming. *International Journal of Web Services Research*, 2(3):33–68.

Li, C, (2007). "Research on key technology of SMG product manufacturing grid". Thesis (PhD), Shanghai University (in Chinese).

Li, J., Meng, X., Liu, S., Yuan, R. (2007b). Ontology-based resource description in Manufacturing Grid. *11th International Conference on Computer Supported Cooperative Work in Design (CSCWD 2007)*, April 26–28, Melbourne, Australia, pp.646–650.

Li, L., Horrock, I. (2004). A software framework for matchmaking based on semantic web technology. *International Journal of Electronic Commerce*, 8(4):39–60.

Li, M., Yu, B., Rana, O., Wang, Z. (2008a). Grid service discovery with rough sets. *IEEE Transactions on Knowledge and Data Engineering*, 20(6):851–862.

Li, R., Yu, T., Fang, M. (2005b). DM-based reliability management for manufacturing grid. *Proceedings of the Ninth International Conference on Computer Supported Cooperative Work in Design (CSCWD2005)*, May 24–26, Coventry, UK, pp.1117–1122.

Li, T.Y., Zhu, H.F., Lam, K.Y. (2003a). A novel two-level trust model for grid. *4th International Conference on Information, Communications & Signal (ICICS 2003)*, Singapore, December 15–18, pp.214–225.

Li, Y.F., Sheng, B.Y., Tao, F. (2008b). Research on the structure model of Manufacturing Grid resources services Mining. *Journal of Wuhan University (Information & Management Engineering)*, 30(2):252–255 (in Chinese).

Li, Y.H., Bandar, Z.A., McLean, D. (2003). An approach for measuring semantic similarity between words using multiple information sources. *IEEE transactions on knowledge and data engineering*, 15(4):871–882.

Li, Y.H., McLean, D., Bandar, Z.A., O'Shea, J.D., Crockett, K. (2006a). Sentence similarity based on semantic nets and corpus statistics. *IEEE transactions on knowledge and data engineering*, 18(8):1138–1150.

Li, Z., Jin, X., Cao, Y., Zhang, X., Li, Y. (2007). Conception and implementation of a collaborative manufacturing grid. *International Journal of Advanced Manufacturing Technology*, 34(11–12):1224–1235.

Li, Z., Yang, F.C., Su, S. (2009a). Fuzzy multi-attribute decision making-based algorithm for semantic Web services composition. *Journal of Software*, 20(3):583–596.

Liang, W.Y., Huang, C.C., Chuang, H.F. (2007). The design with object (DwO) approach to Web services composition. *Computer Standards & Interfaces*, 1(2007):54–68.

Liang, Y. (2004). Study on the architecture and key technology of grid application platform in network manufacture. *Computer Research and Development*, 41(12):2060–2065 (in Chinese).

Lima, S.R., Carvalho, P., Freitas, V. (2004). Distributed admission control for QoS and SLS. *Journal of Network and Systems Management*, 12(3):397–426.

Lin, C., Varadharajan, V., Wang, Y., Pruthi, V. (2004a). Enhancing grid security with trust management. *Proceedings of 2004 IEEE International Conference on Services Computing (SCC2004)*, September 15–18, Shanghai, China, pp.303–310.

Lin, L.F., Gao, P., Cai, M., Dong, J.X. (2005a). A knowledge service-based model of collaborative manufacturing process planning for networked manufacturing. *Journal of Computer-aided Design & computer Graphics*, 17(9):2085–2091.

Lin, M., Xie, J., Guo, H., Wang, H. (2005b). Solving QoS-driven Web service dynamic composition as fuzzy constraint satisfaction. *2005 IEEE International conference on e-technology, e-commerce and e-service (EEE-05)*, May 29–April 1, Hong Kong, China, pp.9–14.

Lin, W.L., Lo, C.C., Chao, K.M., Younas, M. (2006a). Fuzzy consensus on QoS in web services discovery *20th International Conference on Advanced Information Networking and Applications*, April 18–20, Vienna, Austria, pp.791–798.

Lin, W.L., Lo, C.C., Chao, K.M., Younas, M. (2008a). Consumer-centric QoS-aware selection of web services. *Journal of Computer and System Sciences*, 74(2):211–231.

Liu, A., Li, Q., Huang, L., Xiao, M. (2010a). FACTS: A framework for fault-tolerant composition of transactional Web services. *IEEE Transactions on Service Computing*, 3(1):46–59.

Liu, J., Zhou, H., Liao, H., Liang, Z., Sun, X. (2008a). Study on resources location and trading techniques for Manufacturing Grid. *4th International Conference on Wireless Communications Networking and Mobile Computing (WiCOM '08)*, Oct 12–14, Dalian, pp.1–6.

Liu, L. (2004). Research on manufacturing grid and its QoS-based resource management system. Thesis (PhD), Shanghai University (in Chinese).

Liu, S.J. (2006). "Study on manufacturing architecture and manufacturing resource cooperative management technology". Thesis (PhD), Shandong University (in Chinese).

Liu, Y., Ngu, A.H.H., Zeng, L. (2004b). QoS computation and policing in dynamic web service selection. *Thirteenth International World Wide Web Conference Proceedings (WWW2004)*, May 17–22, New York, New York, USA, pp.798–805.

Lu, S., Liu, T. (2008). Study on the scheduling policy for Manufacturing Grid resource based on comprehensive manufacturing ability. *Modular Machine Tool & Automatic Manufacturing Technique*, 10:106–109.

Luckow, A., Schnor, B. (2008). Migol: a fault-tolerant service framework for MPI application in the grid. *Future Generation Computer System*, 24:142–152.

Lv, B., Shi, S., Yang, H., Mo, R., Chang, Z. (2007). Multi-view model for manufacturing grid architecture. *Computer Integrated Manufacturing Systems*, 13(1):38–43 (in Chinese).

Lv, B.S., Shi, S. Y., Mo, R., Chang, Z.Y., Yang, H.C. (2006a). Market equilibrium based resource optimal allocation for manufacturing grid. *Computer Integrated Manufacturing System*, 12(12):2011–2016.

Ma, R., Meng, X., Liu, S. (2006). Data management system in Manufacturing Grid. *IEEE International Conference on Services Computing (SCC '06)*, September 18–22, Chicago, Illinois, USA, pp. 335–340.

Maamar, Z., Benslimane, D., Thiran, P., Ghedira, C., Dustdar, S., Sattanathan, S. (2007). Towards a context-based multi-type policy approach for Web services composition. *Data & Knowledge Engineering*, 2:327–351.

Maamar, Z., Mostéfaoui, S.K., Yahyaoui, H. (2005). Towards an agent-based and context-oriented approach for web services composition. *IEEE Transactions on Knowledge and Data Engineering*, 17 (5):686–697.

Madhusudan, T., Uttamsingh, N. (2006). A declarative approach to composing web services in dynamic environments, *Decision Support Systems*, 2:325–357.

Maedche, A., Staab, S. (2000). The TEXT-TO-ONTO Ontology Learning Environment.

Mahon, R., Chao, K.M., Younas, M. (2004). Cooperation design in grid services. *Proceedings of the 8th International Conference on Computer Supported Cooperative Work in Design (CSCWD 2004)*, May 26–28, Xiamen, China, pp. 406–412.

Mannar, K., Ceglarek, D. (2004). Continuous failure diagnosis for assembly systems using rough set approach. *Annals of the CIRP*, 53(1):39–42.

Mannar, K., Ceglarek, D., Niu, F., Abifaraj, B. (2006). Fault region localization: Product and process improvement based on field performance and manufacturing measurements. *IEEE Transactions on Automation Science and Engineering*, 3(4):423–439.

Márkus, A. (1987), AI tools for intelligent manufacturing. *Robotics and Computer-Integrated Manufacturing*, 3(2):263–268.

Marston, S., Li, Z., Bandyopadhyay, S., Zhang, J., Ghalsasi, A. (2011). Cloud computing – the business perspective. *Decision Support Systems*, 51(1):176–189.

McFarlan, D. (2004). Cambridge supply networks activities. *The strategic and operational management of manufacturing and technology*. May 2004. http://www.ifm.eng.cam.ac.uk/imrc/grandchallenges/cambridge.ppt, 2009.5.

Menasce, D.A. (2004). Composing web services: A QoS view. *IEEE Internet Computing*, 8(6):88–90.

Menasce, D.A., Ruan, H.L., Gomaa, H. (2007). QoS management in service-oriented architectures. *Performance Evaluation*, 64(7–8):646–663.

Meng, X., Tong, Y., Gong, B., Liu, S., Wu, L. (2006). A solution for resource sharing in Manufacturing Grid. *Lecture Notes in Computer Science*, 3865:235–243.

Milanovic, N., Malek, M. (2004). Current solutions for Web service composition. *IEEE Internet Computing*, 8(6):51–59.

MnM (2007). http://kmi.open.ac.uk/projects/akt/MnM/ , 2007.10.

Mo, R. (2006). Research on manufacturing task management system in manufacturing grid environment. Thesis (Master). Northwestern Polytechnical University (in Chinese).

Nabrzyski, J., Schope, J.M., Weglarz, J. (2003). *Grid Resource Management: State of Art and Future Trends*. Spring-Verlag: Kluwer Academic Publishers, 2003.

Nanda, M.G., Karnik, N. (2003). Synchronization analysis for decentralizing composite web services. *Proceedings of ACM Symposium on Applied Computing*, March 9–12, Melbourne, Florida, USA.

Narayanan, S., Mcllraith, S.A. (2002). Simulation, verification and automated composition of Web services. *Proceedings of International WWW Conference (WWW 2002)*, May 5–7 2002, Honolulu, Hawaii..

Neubauer, F., Hoheisel, A., Geiler, J. (2006). Workflow-based grid applications. *Future Generation Computer System*, 22(1–2): 6–15.

Ngu, A.H.H., Carlson, M.P., Sheng, Q.Z., Paik, H.Y. (2010). Semantic-based mashup of composite applications. *IEEE Transactions on Service Computing*, 3(1): 2–15.

Ni, W.C., Liu, L.C., Wu, C., Liu, W. (2007a). Conceptual correlation-based method for grid services composition. *Journal of TSINGHUA University (Science and Technology)*, 47(10):1581–1585.

Ni, Y.H. (2006a). "The research of knowledge integration technology based on ontology for manufacturing enterprise". Thesis (Ph.D): Zhejiang University, Hangzhou, 2005.

OASIS. Web Services Business Process Execution Language Version 2.0. [EB/OL]. http://docs.oasis-open.org/wsbpel/2.0/CS01/wsbpel-v2.0-CS01.html.

Obrenovic, Z., Gasevic, D. (2008). End-user service computing: Spreadsheets as a service composition tool. *IEEE Transactions on Service Computing*, 1(4):229–242.

OGSA, http://www.globus.org/ogsa/, 2009.1.

OGSA, http://www.ogsadai.org.uk/, 2009.1.

OGSI, http://en.wikipedia.org/wiki/Open_Grid_Services_Infrastructure, 2009.1.

OGSI, http://www.ggf.org/ogsi-wg/, 2009.1.

OGSI, http://www.gridforum.org/ogsi-wg/, 2009.1.

Oh, S.C., Lee, D., Kumara, S.R.T. (2008). Effective Web service composition in diverse and large-scale service networks. *IEEE Transactions on Service Computing*, 1(1):15–32.

OilEd (2007) http://oiled.man.ac.uk/building/ , 2007.10.

Omer, M., Schill, A.A. (2009). Framework for dependency based automatic service composition. *Lecture Notes in Business Information Processing*, 17: 535–541.

Onosato, M., Iwata, K. (1993). Development of a virtual manufacturing system by integrating product models and factory models. *CIRP Annals-Manufacturing Technology*, 42 (1): 475–478.

Ontoedit (2007) http://www.ontoknowledge.org/tools/ontoedit.shtml 2007 .10.

Ontolingua (2008) http://www.ksl.stanford.edu/software/ontolingua/ , 2008-July-14.

OntoSaurus (2007) http://www.isi.edu/isd/ontosaurus.html, 2007.10.

Opitz, A., König, H., Szamlewska, S. (2008). What does grid computing cost. *Journal of Grid Computing*, 6(4):385–397.

Oracle. http://www.oracle.com/technology/products/database/oracle11g/index.html 2009.1.

OWL-S (2010) OWL-S: Semantic Markup for Web Services [EB/OL] http://www.w3.org/Submission/OWL-S/.

Oztemela, E., Tekezb, E.K. (2009). A general framework of a Reference Model for Intelligent Integrated Manufacturing Systems (REMIMS). *Engineering Applications of Artificial Intelligence.*, 22(6):855–864.

Paolucci, M., Kawamura, T., Payne, T., Sycara, K. (2002). Semantic matching of web services capabilities. *Proceedings of the First International Semantic Web Conference (ISWC 2002)*, June 9–12, Sardinia, Italy, pp. 333–347.

Papakostas, N., Efthymiou, K., Mourtzis, D., Chryssolouris, G. (2009). Modeling the complexity of manufacturing systems using nonlinear dynamics approaches. *CIRP Annals - Manufacturing Technology*, 58: 437–440.

Papamarkos, G., Poulovassilis, A., Wood, P.T. (2006). Event-condition-action rules on RDF metadata in P2P environments. *Computer Networks*, 50:1513–1512.

Paventhan, A., Takeda, K., Cox, S.J., Nicole, D.A. (2006). Workflows for wind tunnel grid applications. *Lecture Notes in Computer Science*, 3993: 928–935.

Peer, J. (2004). A PDDL Based Tool for Automatic Web Service Composition. *Lecture Notes in Computer Science*, 3208: 149–163.

Perryea, C.A., Chuang, S. (2006). Community-based service discovery, The 2006 IEEE International Conference on Web Service (ICWS'06), September 18–22, Chicago, Illinois, USA, pp. 903–906.

Poler, R., Hernandez, J.E., Mula. J., Lario, Francisco C. (2008). Collaborative forecasting in networked manufacturing enterprises. *Journal of Manufacturing Technology Management*, 19(4):514–528.

Ponnekanti, S.R., Fox, A. (2002). SWORD: A developer toolkit for Web service composition. *Proceedings of International WWW Conference (WWW 2002)*, May 5–7, Honolulu, Hawaii, USA, pp. 83–107.

Protégé-2000 (2007) http://protege.stanford.edu/, 2007.10.

Qiu, R.G. (2004). Manufacturing Grid: A next generation manufacturing. *2004 IEEE International Conference on System, Man, and Cybernetics (SMC2004)*, October 10–13, The Hague, Netherlands, pp.4667–4672.

Raje, R.R., Bryant, B.R., Olson, A.M., Auguston, M., Burt, C. (2002). A quality-of-service-based framework for creating distributed heterogeneous software components. *Concurrency Computation Practice and Experience*, 14(12):1009–1034.

Rajic, H., Brobst, R., Chan, W., Ferstl, F., Gardiner, J., Robarts, J.P., Haas, A., Nitzberg, B., Tollefsrud, J. (2003). *Distributed resource management application API specification 1.0*. Tech. Rep., DRMAA Working Group-The Global Grid Forum 2003.

Rao, J., Küngas, P., Matskin, M. (2006a). Composition of semantic web services using linear logic theorem proving. *Information Systems*, 31:340–360.

Rasmussen, L.B., Wang, A. (2007). Work in the virtual enterprise—creating identities, building trust, and sharing knowledge. *AI & Society*, 21:184–199.

Ratnaweera, A., Halgamuge, S.K., Watson, H.C. (2004). Self-organizing hierarchical particle swarm optimizer with time-varying acceleration coefficients. *IEEE Transactions on Evolutionary Computation*, 3: 240–255.

Raverdy, P.G., Issarny, V. (2005). Context-aware service discovery in heterogeneous networks. *Proceedings of the sixth IEEE International Symposium on a World of Wireless Mobile and Multimedia Networks (WoWMoM'05)*, June 13–16, Taormina, Italy, pp.478–480.

Resconi, G., Kovalerchuk, B. (2008). Agents' model of uncertainty. *Knowledge and Information Systems*, 18(2):13–229.

Resnik, P. (1995). Using information content to evaluate semantic similarity in a taxonomy. *Proceedings of the 14th International Joint Conference on Artificial Intelligence*, August 20–25, Montreal, Quebec, Canada, pp.448–453.

Richards, D., Splunter, S.V., Brazier, F.M.T., Sabou, M. (2003). Composing web services using an agent factory. *The 1st International Workshop on Web Services and Agent-based Engineering(AAMAS 2003),* July 14–15, Melbourne, Australia, pp.57–66.

Rodriguez, M.A., Egenhofer, M.J. (2003). Determining semantic similarity among entity classes from different ontologies. *IEEE Transactions on Knowledge and Data Engineering,* 15(2): 442–456.

Rosenthal, M., Mork, P., Li, M.H., Stanford, J., Koester, D., Reynolds, P. (2009). Cloud computing: a new business paradigm for biomedical information sharing. *Journal of Biomedical Informatics,* 43(2): 342–353.

Sabata, B., Chatterjee, S., Davis, M., Sydir, J., Lawrence, T. (1997). Taxonomy for QoS specifications. *Proceedings of the IEEE Computer Society 3rd International Workshop on Object-oriented Real-time Dependable Systems (WORDS 97),* February 5–7, Newport Beach, California, USA, pp.100–107.

Sabbaghi, A., Montazemi, A.R. (2004). Computer-integrated manufacturing. *Encyclopedia of Information Systems,* 14:229–242.

Schuschel, H., Weske, M. (2004). Automated planning in a service-oriented architecture. *Proceedings of the 13th IEEE International Workshops on Enabling Technologies: Infrastructure for Collaborative Enterprises (WET ICE'04),* June 14–16, Washington, DC, USA, pp.75–80.

Segev, A., Toch, E. (2008). Context-based matching and ranking of Web services for composition. *IEEE Transactions on Service Computing,* 2(3): 220–222.

Semantic web (2007) http://kaon.semanticweb.org, 2007.12.

Sesame (2007). http://www.openrdf.org, 2007.10.

Sethi, A.K., Sethi, S.P. (1990). Flexibility in manufacturing: A survey. *The International Journal of Flexible Manufacturing Systems,* 2:289–328.

Shannon, C.E. (1948). A mathematical theory of communication. *Bell System Technical Journal,* 27:379–423, 623–656.

Shen, B., Liu, L., Yu, T. (2006b). SLA-based resource management model in manufacturing grid. *Computer Applications,* 26(2): 512–514.

Shen, Z.D., Yan, F., Qiang, W.Z. (2006a). Grid system integrated with trusted computing platform. *Proceedings of the First International Multi-Symposiums on Computer and Computational Sciences (IMSCCS'06),* June 20–24, Hangzhou, China, pp.619–625.

Shen, Z.N., Su, J.W. (2005). Web service discovery based on behavior signatures. *Proceedings of the 2005 IEEE International Conference on Service Computing (SCC'05),* July 11–15, Orlando, Florida, USA, pp.279–286.

Sheng, Q.Z., Benatallah, B., Maamar, Z., Ngu, A.H.H. (2009). Configurable composition and adaptive provisioning of Web services. *IEEE Transactions on Service Computing,* 2(1):34–49.

Shi, S. (2007). "Research on key technologies of resource management and configuration in manufacturing grid". Thesis (PhD). Northwestern Polytechnical University (in Chinese).

Shi, S., Yang, H., Liu, H., Hou, J. (2009). A resource allocation method based on competitiveness equilibrium for manufacturing grid. *International Journal of Advanced Manufacturing Technology,* 41(9–10):997–1002.

Shi, S.Y., Mo, R., Yang, H.C. (2006a). Research on modeling resources in manufacturing grid. *Computer Engineering Design,* 27(16):2925–2927.

Shia, M.M., Fladmark, J.O. (2009). Validation aspects of automatic service composition. *Lecture Notes in Computer Science*, 4907:258–267.

Shu, D. (2006). "The research of security problem based on manufacturing grid". Thesis (Master). Wuhan University of Technology (in Chinese).

Shuiabi, E., Thomson, V., Bhuiyan, N. (2005). Entropy as measure of operational flexibility. *European Journal of Operational Research*, 165:696–707.

Siddiqui, M., Villazon, A., Fahringer, T. (2006). Grid capacity planning with negotiation-based advance reservation for optimized QoS. *Proceedings of the ACM/IEEE Supercomputing 2006 (SC '06)*, November 11–17, Tampa, Florida, USA, pp.103–118.

Sirin, E., Parsia, B., Wu, D., Hendler, J., Nau, D. (2004a). HTN planning for Web service composition using SHOP2. *Web Semantics: Science, Services and Agents*, 1(4):377–396.

Sirin, E., Parsia, B., Hendler, J. (2004). Filtering and selecting semantic web services with interactive composition techniques. *IEEE Intelligent Systems*, 19(4):42–49.

Smarr, L., Visualization needs in science and technology, national computational science alliance, 2002.7.

SOAP (2009). http://www.w3.org/TR/soap/, 2009.1.

Song, S., Kwok, Y.K., Wang, K.H. (2005a). Security-driven heuristics and a fast genetic algorithm for trusted grid job scheduling. *Proceedings of the 19th IEEE International Parallel and Distributed Processing Symposium (IPDPS'05)*, April 4–8, Denver, Colorado, USA, pp.65–74.

Steptools (2009). http://www.steptools.com/ 2009.1.

Stollberg, M., Keller, U., Lausen, H., Heymans, S. (2007). Two-phase web service discovery based on rich functional description. *Proceedings of the 4th European Semantic Web Conference (ESWC 2007)*, June 3–7, Tyrol region of Innsbruck, Austria, pp.99–113.

Sun, H., Yu, T., Liu, L., Cui, M. (2008a). Manufacturing Grid's enterprise qualification authentication system and method. *Proceedings of the 7th World Congress on Intelligent Control and Automation (WCICA 2008)*, June 25–27, Chongqing, China, pp.4901–4905.

Sun, H., Yu, T., Liu, L., He, Y. (2008b). Service-oriented manufacturing system. *Computer Integrated Manufacturing Systems*, 14(1):56–63.

Sun, H.Y., Yu, T., Liu, L., He, Y.A. (2006). QoS management in Manufacturing Grid. *Proceedings of the IFIP International Federation for Information Processing, Knowledge Enterprise: Intelligent Strategies in Product Design, Manufacturing, and Management*, June 15–17, Shanghai, China, pp.831–839.

Sun. http://www.sun.com/software/sge/ 2009.1.

Sycara, K., Klusch, M., Widoff, S., Lu, J. (2002). Larks: Dynamic matchmaking among heterogeneous software agents in cyberspace. *Journal of Autonomous Agents and Multi-Agent Systems*, 5(2):173–203.

Tan, G., Zhao, N., Taylor, S.J.E. (2003). Automobile manufacturing supply chain simulation in the grids environment. *Proceedings of the 2003 Winter Simulation Conference*, December 7–10, New Orleans, Louisiana, USA, pp.1149–1157.

Tan, W., Fan, Y. S. (2005). Research on service matching and composition in networked manufacturing environment, *Computer Integrated Manufacturing Systems*, 11(106):1408–1413.

Tan, W., Yao, X., Liu, X. (2009). Frame of manufacturing grid based on universal enterprise service control center. *Journal of Jinan University (Nature Science)*, 30(1):5–8.

Tanaka, Y., Nakada, H., Sekiguchi, S., Suzumura, T., Matsuoka, S. (2003). Ninf-G: A reference implementation of RPC-based programming middleware for grid computing. *Journal of Grid Computing*, 1 (1):41–51.

Tandon, V. (2000). "Closing the gap between CAD/CAM and optimized CNC end milling". Indianapolis: Master's thesis) Purdue School of Engineering and Technology.

Tao, F., Zhang, L., Nee, A.Y.C. (2011d). A review of the application of grid technology in manufacturing. *International Journal of Production Research*, 49(13):4119–4155.

Tao, F., Hu, Y.F., Zhang, L. (2010a). *Theory and practice: Optimal Resource Service Allocation in Manufacturing Grid*. Beijing: China Machine Press, 2010 (in Chinese).

Tao, F., Hu, Y.F., Ding, Y.F., Shen, B.Y., Zhou, Z.D. (2006a). Resources publication and discovery in manufacturing grid. *Journal of Zhejiang University (Science)*, 7(10):1676–1682.

Tao, F., Hu, Y.F., Ding, Y.F., Sheng, B.Y., Song, C.S., Zhou, Z.D. (2006c). Modeling of manufacturing resource in manufacturing grid based on XML. *Proceedings of the 2006 IEEE International Conference on Industrial Informatics (INDIN'06)*, August 16–18, Singapore, 1261–1266.

Tao, F., Hu, Y.F., Ding, Y.F., Sheng, B.Y., Zhou, Z.D. (2005a). Resource optimization selection evaluation model based on agent in Manufacturing Grid system. *China Mechanical Engineering*, 16(24):2192–2197 (In Chinese).

Tao, F., Hu YF, Zhao D, Zhou ZD (2006b) Study on framework of manufacturing grid supporting collaborative manufacturing. *Chinese Mechanical Engineering*, 17(18):1903–1907. (in Chinese)

Tao F, Hu, Y.F., Zhao, D., Zhou, Z.D. (2009a). An approach to Manufacturing Grid resource service scheduling based on Trust-QoS. *International Journal of Computer Integrated Manufacturing*, 22(2):100–111.

Tao, F., Hu, Y.F., Zhao, D.M., Zhou, Z.D. (2009b). Study on resource service match and search in manufacturing grid system. *International Journal of Advanced Manufacturing Technology*, 43(3–4):379–399.

Tao, F., Hu, Y.F., Zhao, D.M., Zhou, Z.D. (2010b). Study of failure detection and recovery in manufacturing grid resource service scheduling. *International Journal of Production Research*, 48(1):69–94.

Tao, F., Hu, Y.F., Zhao, D.M., Zhou, Z.D., Zhang, H.J., Lei, Z.Z. (2009c). Study on Manufacturing Grid resource service QoS modeling and evaluation. *International Journal of Advanced Manufacturing Technology*, 41(9–10):1034–1042.

Tao, F., Hu, Y.F., Zhou, Z.D. (2008a). Study on Manufacturing Grid and its resources optimal-selection system. *International Journal of Advanced Manufacturing Technology*, 37(9–10):1022–1041.

Tao, F., Hu, Y.F., Zhou, Z.D. (2008b). Study on Manufacturing Grid and its executing platform. *International Journal of Manufacturing Technology and Management*, 14(1–2):35–51.

Tao, F., Hu, Y.F., Zhou, Z.D. (2009d). Application and modeling of resource service trust-QoS evaluation in Manufacturing Grid system. *International Journal of Production Research*, 47(6):1521–1550.

Tao,F., Guo, H., Zhang, L., Cheng, Y. (2011e). Modelling of combinable relationship-based composition service network and theoretical proof of its scale-free characteristics . *Enterprise Information Systems*, (Online first) DOI: 10.1080/17517575.2011.621981

Tao, F., Qiao, K., Zhang, L., Li, Z., Nee, A.Y.C. (2011c). GA-BHTR: An improved genetic algorithm for partner selection in virtual manufacturing. *International Journal of Production Research* (Online first) DOI: 10.1080/00207543.2011.561883 (to be pressed).

Tao, F., Zhang, L., Lu, K., Zhao, D. (2011a). Research on manufacturing grid resource service optimal-selection and composition framework. *Enterprise Information Systems*, (Online first) DOI: 10.1080/17517575.2010.540677. May, 2011.

Tao, F., Zhang, L., Venkatesh, V.C., Luo, Y.L., Cheng, Y. (2011b). Cloud manufacturing: a computing and service-oriented manufacturing model. *Proceedings of the Institution of Mechanical Engineers, Part B, Journal of Engineering Manufacture* (Online first) DOI: 10.1177/0954405411405575 (to be pressed).

Tao, F., Zhang, L., Zhang, Z.H., Nee, A.Y.C. (2010d). A quantum multi-agent evolutionary algorithm for selection of partners in a virtual enterprise. *CIRP Annals-Manufacturing Technology*, 59(1):485–488.

Tao, F., Zhang, L., Zhao, D. (2009e). A framework for MGrid resource service composition and optimal-selection. *IEEE International Conference on Industrial Engineering and Engineering Management (IEEM 2009)*, December 8–11, Hongkong, China, pp.110–114.

Tao, F., Zhao, D., Hu, Y.F., Zhou, Z.D. (2008c). Resource service composition and its optimal selection based on particle swarm optimization in manufacturing grid system. *IEEE Transactions on Industrial Informatics*, 4(4): 315–327.

Tao, F., Zhao, D., Hu, Y.F., Zhou, Z.D. (2010e). Correlation-aware resource service composition and optimal-selection in manufacturing grid. *European Journal of Operational Research*, 201(1):129–143.

Tao, F., Zhao, D., Zhang, L. (2010f). Resource service optimal-selection based on intuitionistic fuzzy set and non-functionality QoS in manufacturing grid system. *Knowledge and Information Systems*, 25(1):185–208.

Thomas, M. (1996). Concurrent engineering: Supporting subsystems. *Computers & Industrial Engineering*, 31(3–4):571–575.

Tomas, V., Maciej, Z., Matthew, M. (2007). Dynamic service discovery through met-interaction with service provider. *Proceedings of the 4th European Semantic Web Conference (ESWC 2007)*, June 3–7, Innsbruck, Austria, pp.84–98.

Tong, Y.X., Meng, X.X., Gong, B. (2005). Resource organizing in the Manufacturing Grid. *Proceedings of the 9th International Conference on Computer Supported Cooperative Work in Design (CSCWD 2005)*, May 24–26, Coventry, UK, pp.1071–1075.

Tosic, V., Mennie, D., Pagurek, B. (2001). On dynamic service composition and its applicability to e-business software systems. *Proceedings of the WOOBS 2001 (Workshop on Object-Oriented Business Solutions) Workshop (at ECOOP 2001)*, June 18–22, Budapest, Hungary, pp.95–108.

Tracy, D.B., Howard, J.S., Noah, B. (2001). A comparison of eleven static heuristics for mapping a class of independent tasks onto heterogeneous distributed computing systems. *Journal of Parallel and Distributed Computing*, 61:810–837.

Traverso, P., Pistore, M. (2004). Automated composition of semantic Web services into executable processes. *Proceedings of the 3rd International Semantic Web Conference-ISWC 2004*, November 7–11, Hiroshima, Japan, pp.380–394.

Tripathi, P.K., Bandyopadhyay, S., Pal, S.K. (2007). Multi-objective particle swarm optimization with time variant inertia and acceleration coefficients. *Information Sciences*, 22:5033–5049.

Tversky, A. (1977). Features of similarity. *Psychological Review*, 84(4):327–352.

UDDI, http://en.wikipedia.org/wiki/Universal_Description_Discovery_and_Integration, 2009.1.

UDDI, http://uddi.xml.org/, 2009.1.

UDDI, http://www.oasis-open.org/committees/tc_home.php?wg_abbrev=uddi-spec 2009.1.

Ulieru, M., Norrie, D. (2000). Fault recovery in distributed manufacturing systems by emergent holonic re-configuration: A fuzzy multi-agent modeling approach. *Information Sciences*, 127(3):101–123.

Ulieru, M., Norrie, D., Kremer, R., Shen, W. (2000). A multi-resolution collaborative architecture for web-centric global manufacturing. *Information Sciences*, 127(1–2):3–21.

van den Bergh, F. (2001). "An analysis of particle swarm optimizers". Thesis (Ph.D), Faculty of Natural and Agricultural Science, University of Pretoria, Pretoria, South Africa, 2001.

van der Aalst, W.M.P. (2003). Don't go with the flow: Web services composition standards exposed. *IEEE Intelligent System*, 18(1):72–76.

Venkatesh, V.C., Izman, S. (2008). *Precision Engineering*, USA: McGraw Hill, 2008.

Vitvar, T., Zaremba, M., Moran, M. (2007). Dynamic service discovery through met-interaction with service provider. *Proceedings of the 4th European Semantic Web Conference (ESWC 2007)*, June 3–7, Innsbruck, Austria, pp.84–98.

W3C OWL-S: Semantic Markup for Web Services [EB/OL]. http://www.w3.org/Submission/OWL-S/.

W3C WSCDL: Web Services Choreography Description Language Version 1.0. [EB/OL]. http://www.w3.org/TR/2004/WD-ws-cdl-10-20041217.

W3C. WSCI: Web Services Choreography Interface (WSCI) 1.0. [EB/OL]. http://www.w3.org/TR/wsci/.

Wadhwa, S., Mishra, M., Chan, F.T.S. (2009). Organizing a virtual manufacturing enterprise: An analytic network process based approach for enterprise flexibility. *International Journal of Production Research*, 47(1):163–186.

Wang, A., Fan, L., Xiao, T., Fan, W. (2005a). Research on application platform and virtual enterprise modeling for manufacturing grid. *Chinese Journal of Mechanical Engineering*, 41(2):176–181 (in Chinese).

Wang, F. (2008). "The design and Implementation of manufacturing grid portal". Thesis (Master). Wuhan University of Technology (in Chinese).

Wang, H., Cao, J.K., Xin, D.X., Shao, Y. (2009). *Communications Technology*, 42(2):130–135.

Wang, P. (2009). QoS-aware web services selection with intuitionistic fuzzy set under consumer's vague perception. *Expert Systems with Applications*, 36(3):4460–4466.

Wang, X.F. (2007). "Research on information security technology of manufacturing grid system". Thesis (PhD). Xi'an University of Technology (in Chinese).

Wang, X.F., Dong, H., Wang, S.P., Zhao, T. (2008). Identity-based designated multi-receiver sign cryption scheme for manufacturing grid. *Computer Integrated Manufacturing Systems*, 14(12):2463–2468.

Wang, Y., Wang, L., Hu, C.M. (2006). A QoS Negotiation protocol for grid workflow. *Proceeding of the 5th International Conference on Grid and Cooperative Computing (GCC'06)*, October 21–23, Hunan, China, pp.295–198.

WebODE (2007). http://webode.dia.fi.upm.es/WebODEWeb/index.html, 2007.10.
WebOnto (2007). http://www.aktors.org/technologies/webonto/, 2007.10.
Wen, H.Y. (2005). "Study on several key technical issues of manufacturing grid". Thesis (PhD). Xidian University (in Chinese).
WRSF. http://en.wikipedia.org/wiki/WRSF, 2009.1.
WS-CDL (2010). Web Services Choreography Description Language Version 1.0. [EB/OL]. http://www.w3.org/TR/2004/WD-ws-cdl-10-20041217.
WSCI (2010). Web Services Choreography Interface (WSCI) 1.0. [EB/OL]. http://www.w3.org/TR/wsci/.
WSDL (2009). http://www.w3.org/TR/wsdl, 2009.1.
WSMO (2011). http://www.wsmo.org/ 2011.4.
WS-PBEL (2010). Web Services Business Process Execution Language Version 2.0. [EB/OL] http://docs.oasis-open.org/wsbpel/2.0/CS01/wsbpel-v2.0-CS01.html.
Wu, J.H., Yin, Z.M., Wang, C.B. (2007a). An OWL-S based framework for quality of semantic Web service. *Journal of Information*, 10: 75–77.
Wu, L., Davari, S. (2002). Multi-protocol label switching (MPLS) support of differentiated services. Internet request for comments RFC 3270, Internet engineering task force, 2002–05.
Wu, L., Davari, S. (2005). RFC 3270: Multi-protocol label switching (MPLS) support of differentiated services. Internet request for comments, Internet engineering task force, 2002–05.
Wu, L., Meng, X., Liu, S. (2007c). A Service-oriented, scalable approach to grid-enabling of manufacturing resources. *Lecture Notes in Computer Science*, 4674:175–183.
Wu, L., Meng, X., Liu, S., Tong, Y., Wei, M. (2005a). A trading supported manufacturing resource sharing model for manufacturing grid. *Proceedings of the 9th International Conference on Computer Supported Cooperative Work in Design (CSCWD 2005)*, May 24–26, Coventry, UK, pp.339–344.
Wu, L., Meng, X.X., Liu, S., Jiao, Y.C. (2007b). A solution of industrial simulation codes sharing in Manufacturing Grid. *Lecture Notes in Computer Science*, 4674:168–179.
Wu, Y., Qi, G., Xie, Q., Gu, X. (2006c). Research on architecture of Manufacturing Grid for application service provider mode. *Proceedings of IEEE International Conference on Networking, Sensing and Control (ICNSC '06)*, April 23–25, Fort Lauderdale, Florida, USA, pp.233–237.
Wu, Y.D., Qi, G.N., Xie, Q.S., Gu, X.J. (2006a). Research on architecture of Manufacturing Grid for application service networking. *Proceedings of the 2006 IEEE International Conference on Networking, Sensing and Control (ICNSC 06)*, April 23–25, Florida, USA, pp.233–237.
Wu, Z.A., Luo, J.Z. (2006b). The measurement of Grid QoS. *Proceedings of the 10th International Conference on Computer Supported Cooperative Work in Design (CSCWD 2006)*, May 3–5, Nanjing, China, pp.1–6.
Xiang, F. (2008). "Algorithms based trust and economic model for manufacturing grid resource scheduling". Thesis (Master). Wuhan University of Technology (in Chinese).
Xie, X.L., Zhou, D.J., Li, C.Q. (2008b). Study on Manufacturing Grid fuzzy resource reservation technology based on SLA. *Chinese Mechanical Engineering*, 19(1): 64–68 (in Chinese).

Xie, X.L., Zhou, D.J., Li, X.X. (2008a). Research on Manufacturing Grid performance evaluations model based on decision support. *Journal of Wuhan University of Technology*, 30(6):125–129.

Xiong, N., Litz, L., Ressom, H. (2002). Learning premises of fuzzy rules for knowledge acquisition in classification problems. *Knowledge and Information Systems*, 4(1):96–111.

XIU, Y.S., CUI, D.G. (2004). Application of grid technology in aeronautical manufacturing. *Computer Research and Development*, 41(12):2073–2078.

Xu, Z. (2007). Some similarity measures of intuitionistic fuzzy sets and their applications to multiple attribute decision making. *Fuzzy Optimization and Decision Making*, 6(2):109–121.

Xu, Z., Yager, R.R. (2007a). Dynamic intuitionistic fuzzy multi-attribute decision making. *International Journal of Approximate Reasoning*, 48(1):246–262.

Yang, D., Qi, G.N. (2005). A cooperative model of VO based on grid and its task scheduling. *China Mechanical Engineering*, 16(9):784–786.

Yang, M.S., Hung, W.L., Chang-Chien, S.J. (2005a). On a similarity measure between LR-type fuzzy numbers and its application to database acquisition. *International Journal of Intelligent Systems*, 20(10):1001–1016.

Yang, X.F., Li, F.X., Zhan, S.Y., Chiai, X.D. (2005). Key technologies of simulation grid resources pool. *Computer Integrated Manufacturing Systems*, 11(8):1174–1177.

Yao, D.D. (1985). Material and information flows in flexible manufacturing systems. *Material Flows*, 2:143–149.

Ye, S.Y., Wei. J., Li. L., Huang, T. (2008a). Service-correlation aware service selection for composite service. *Chinese Journal of Computers*, 31(8):1383–1397.

Ye, Z.L., Gu, X.J., Qian, Y.D., Qi, G.N. (2004). Manufacturing Grid – the applications of grid technology in manufacturing industry. *China Mechanical Engineering*, 15(19):1717–1720.

Yin, J.W., Zhang, W.Y., Cai, M. (2009). Weaving an agent-based Semantic Grid for distributed collaborative manufacturing. *International Journal of Production Research*, 48(7):2109–2126.

Yin, P.Y. (2006). Particle swarm optimization for point pattern matching. *Journal of Visual Communication and Image Representation*, 17:143–162.

Yoshida, H., Kawata, K., Fukuyama, Y., Nakanishi, Y. (1999). A particle swarm optimization for reactive power and voltage control considering voltage stability. *Proceedings of International Conference on Intelligent System Application to Power Systems*, April 4–8, Rio de Janeiro, Brazil, pp.117–121.

Yu, J., Venuqopal, S., Buyya, R. (2006). A market-oriented grid directory service for publication and discovery of grid service providers and their services. *The Journal of Supercomputing*, 36(1):17–31.

Yu, T., Yuan, L.P., Li, J., Xiong, F., Fang, M.L. (2005). A multi-agent based approach for Manufacturing Grid workflow. *Proceedings of the Fourth International Conference on Machine Learning and Cybernetics*, August,18–21, Guangzhou, China, pp.199–204.

Yuan, Y.P. (2006). "Research and development on manufacturing grid workflow". Thesis (PhD). Shanghai University (in Chinese).

Yue, P., Di, L., Yang, W., Yu, G., Zhao, P. (2007). Semantics-based automatic composition of geospatial Web service chains. *Computer & Geosciences*, 5:649–665.

Zadeh, L.A. (1972). Fuzzy-set-theoretic interpretation of linguistic hedges. *Journal of Cybernetics*, 2(3):4–34.

Zeng, L., Benatallah, B., Dumas, M., Kalagnanam, J., Sheng, Q.Z. (2003a). Quality driven web services composition. *Proceedings of the 12th International World Wide Web Conference (WWW2003)*, May 20–24, Budapest, Hungary, pp.411–421.

Zeng, L., Benatallah, B., Lei, H., Ngu, A., Flaxer, D., Chang, H. (2003). Flexible composition of enterprise Web services. *Electronic Markets-Web Services*, 13(2): 141–152.

Zeng, L., Benatallah, B., Ngu, A.H.H., Dumas, M., Kalagnanam, J., Chang, H. (2004a). QoS-aware middleware for Web services composition. *IEEE Transactions on Sofeware Engineering*, 30(5):311–327.

Zhang, C.S., Mo, R., Shi, S.Y., Chang, Z.Y. (2006a). Research on manufacturing grid resource scheduling based on genetic algorithm. *China Mechanical Engineering*, 17(18):1916–1920.

Zhang, G., Shao, X., Li, P., Gao, L. (2009). An effective hybrid particle swarm optimization algorithm for multi-objective flexible job-shop scheduling problem. *Computers & Industrial Engineering*, 56:1309–1318.

Zhang, H., Hu, Y., Tao, F., Zhou, Z. (2008a). Study on semantic-aware Manufacturing Grid architecture. *Proceedings of the 5th International Conference on Fuzzy Systems and Knowledge Discovery (FSKD '08)*, October 18–20, Jinan, China, pp.626–630.

Zhang, H.Y. (2008a). The scale-free nature of semantic Web ontology. *WWW 2008*, April 21–25, Beijing, China, pp.1047–1048.

Zhang, J. (2004). Grid: The third tidal wave of information technology. *China Information Review*, 1:54–57.

Zhang, J. (2008). The service-driven manufacturing grid system. *Systems Engineering*, 26(4):88–93.

Zhang, L. (2006a). "Research on semantic web based resource integration and scheduling for manufacturing grid". Thesis (PhD). Northwestern Polytechnical University (in Chinese).

Zhang, L., Yuan, W.Z., Wang, W. (2006b). An ontology based approach of automated service chaining for manufacturing grid. *China Mechanical Engineering*, 17(14):1484–1488.

Zhang, L., Yuan, W.Z., Wang, W. (2006c). Automotive service composition for manufacturing grid based on domain-specific ontology. *Journal of Computer Application*, 26(1):57–60.

Zhang, L., Yuan, W.Z., Wang, W. (2005). Towards a framework for automatic service composition in manufacturing grid. *Lecture Notes in Computer Science*, 3795:238–243.

Zhang, L., Yuan, W.Z., Wang, W. (2006d). An ontology based approach of automated service chaining for Manufacturing Grid. *Chinese Mechanical Engineering*, 17(14):1484–1488 (in Chinese).

Zhang, Q.Y., Vonderembse, M.A., Lim, J.S. (2003a). Manufacturing flexibility: Defining and analyzing relationships among competence, capability, and customer satisfaction. *Journal of Operations Management*, (21):173–191.

Zhang, R., Arpinar, I.B., Aleman-Meza, B. (2003b). Automatic composition of semantic Web services. *Proceeding of the International Conference on Web Services (ICWS03)*, February 14–16, Amsterdam, Netherlands, pp.38–41.

Zhang, S., Liu, Q., Zhang, P., Chen, X. (2008b). Construction mechanism of manufacturing grid based on ontology semantics. *Application Research of Computers*, 25(8): 2289–2291.

Zhang, S., Meng, X., Liu, S., Ma, R., Wu, L. (2006e). Dynamic manufacturing job management in Manufacturing Grid. *Proceedings of the 10th International Conference on Computer Supported Cooperative Work in Design (CSCWD'06)*, May 3–5, Nanjing, China, pp.1–6.

Zhang, T., He, Y.L., Deng, L., Wu, J. (2009b). Research on credibility model in manufacturing grid contests. *Computer Engineering and Applications*, 45(1): 211–241.

Zhang, T.H., Gu, X.J., Ye, Z.L., Hu, H. (2009a). System framework for knowledge and information acquisition of manufacturing grid resources. *Journal of Zhejiang University*, 43(1):77–82.

Zhang, W.G., Xiao, W.L. (2009). On weighted lower and upper possibilistic means and variances of fuzzy numbers and its application in decision. *Knowledge and Information Systems*, 18(3):311–330.

Zhang, W.Y., Yin, J.W. (2009). Weaving a semantic grid for multidisciplinary collaborative design. *International Journal of Production Research*, 47(11):3079–3095.

Zhang, X.L., Chen, D.R., Lu, L. (2005). Research of architecture for Grid Manufacturing. *Proceedings of the 9th International Conference on Computer Supported Cooperative Work in Design Proceedings*, May 24–26, Coventry, UK, pp.345–49.

Zhang, Z., Rong, L., Zhou, S. (2007a). A general geometric growth model for pseudofractal scale-free web. *Physica A: Statistical Mechanics and its Applications*, 377(1):329–339.

Zhao, S.Y., Lo, V., Dickey, C.G. (2005a). Result verification and trust-based scheduling in peer-to-peer grids. *Proceedings of the Fifth IEEE International Conference on Peer-to-Peer Computing (P2P'05)*, August 31–September 2, Konstanz, Germany, pp.31–38.

Zheng, X.L., Chen, D.R., Lu, L. (2005b). Research of architecture for Grid Manufacturing. *Proceedings of the 9th International Conference on Computer Supported Cooperative Work in Design Proceedings (CSCWD 2005)*, May 24–26, Coventry, UK, pp.345–349.

Zhou J, Li Q, Browne J, Wang Q, Folan P, Xiao T (2007a) Performance-Based Bayesian Learning for Resource Collaboration Optimization in Manufacturing Grid. *Lecture Notes in Computer Science*, 4431: 498–507.

Zhou J, Zeng G (2007) Validity checking on grid service composition. *Proceedings of the 31st Annual International Computer Software and Applications Conference (COMPSAC 2007)*, 24–27 July, Beijing, China, 379–382.

Zhou QM, Yu T (2006) Information Collecting of MG Resources' Reliability Based on Mobile Agent. *Computer Engineering*, 32(22): 119–120.

Zhou, Y.L., Chen, Y.W. (2007b). A new economics architecture for Manufacturing Grid. *Proceedings of the 3rd International Conference on Natural Computation (ICNC 2007)*, August 24–27, Xi'an, China, pp.720–724.

Zhou, Z.D., Tao, F., Hu, Y.F. (2007). Manufacturing grid resource service scheduling based on trust-QoS. *Proceedings of the 3rd Virtual Conference, of the EU-funded FP6 I*PROMS Network of Excellence on Innovative Production Machines and Systems (IPROMS 2007)*, July 2–13, Cardiff, UK.

Zisman, A., Spanoudakis, G. (2006). UML-based service discovery framework. *Proceedings of the 4th International Conference on Service Oriented Computing (ICSOC 2006)*, December 4–7, Chicago, Illinois, USA, pp.402–414.
Zou, Q., Yu, T., Sun, H. (2006). Reliability Analysis for Manufacturing Grid. *Proceedings of the IFIP International Federation for Information Processing, Knowledge Enterprise: Intelligent Strategies in Product Design, Manufacturing, and Management*, June 15–17, Shanghai, China, pp.803–811.

Index

.NET 18

Acceleration Coefficients 298
Accessing right failures (App_AccessRight_Failure) 413
Agile manufacturing (AM) 4
Analytic hierarchy process (AHP) 24
Application design or coding failures (App_DesignCode_Failure) 413
Application related failure 408
Application System Resource (ASR) 28
ARMatcher: ability requirement matcher
Artificial intelligence (AI) 54
ASP (Application Service Provider) 437
ASR-Class 46
ASRD: Application System Resources 151
Automated composition 276

Bandwidth failure (Bandwidth_Failure) 408
Basic algorithms depository (AlgDep) 197
Basic-matching 97
Benign competition correlation (BCC) 264
BPEL 277
Business entity correlation (BuC) 263

CAD (computer-aided design) 3
Candidate resource service set (CRSS) 186
CAPP (computer-aided process planning) 3
CAS (community authorization service) 456
Changed task requirements failure (Task_Require_Change_Failure) 411
Chief Evaluation Agent (CEAgent) 124
Choice model 193
Choice model 193
Classification of Pareto front 290
Cloud manufacturing (CMfg) 470
Cognitive acceleration coefficient 298
Combinable Correlation 258
Combinable correlation degree 258
Combinable evolving operators set 379
Combinable neighborhood 374
Combinable relationship (CoR) 360
Combinable relationship based composition service network (CoRCS-Net) 360
Combinable relationship degree 373
Combinable relationship set 373
Combinable strength 374
Combinable weight 373
Comparison of Pareto solutions 291
Complete Product Design and Manufacturing Service (CPDMS) 280

Component Runtime Environment (CoRE) 277
Component Service Model with Semantics (CoSMoS) 277
Composite resource service (CRS) 40
Composition service network (CS-Net) 365
Composition service or Resource Services Chain (RSC) 365
Comprehensive Combining Recommended Trust-QoS Evaluating Algorithm 169
Computational grid 10
Computational Resource (CR) 29
Computer-integrated manufacturing (CIM) 4
Computing grid (CG) 11
Conceptual Model 31
Concurrent engineering 4
CORBA 18
Correlation flexibility 322
Correlation Relationship 253
Correlation resource service set (CRSS) 282
Cost (Resource Service Execution Cost) 136
CR-Class 46
CRM (customer relationship management) 3
Crowding distance 291
Customer relationship management (CRM) 461

Data grid 10
Decision rule of combinable relationship 373
Delivery arrangement service (DAS) 279
Deploy phase (Dep-phase) 190
Depth of two compared words 85
Description-QoS 131
Design and analysis service (D&AS) 279
Design as a service (DaaS) 477
Design phase (Des-phase) 190
Digital Description of Resource Service (DDoRS) 39
Direct trust 157
Direct trust degree data structure 163
Direct Trust-QoS Evaluating Algorithm 166
Direct trust-QoS evaluation value 157
Discounted cash flow (DCF) 313
Distributed database 22
Domain Node (DN) 151

E-Agent (evaluation agent) 128
ECA (event–condition–action) 404
EIS-Agent (Evaluation Indices Set Agent) 127
Electronic pay service (EPS) 279
Elementary evolving operators set 379
Entity class matching algorithms (ECMAs) 93
Equipment Resource (ER) 28
  equivalent input combinable relationship (eiCoR) 368
  equivalent or competition combinable relationship (eqCoR) 368
  equivalent output combinable relationship (eoCoR) 368
ER-Class 46
ERD: Equipment Resources Domain 151
ERP (enterprise resource planning) 3
Evaluation Expert Agent (EEAgent) 124
  exact combinable relationship (exCoR) 368
Exact match 79
Executable service composition path (ESCP) 191
Execution and Monitor phase (E&M-phase) 190
Experimentation as a service (EaaS) 477

Failure detector (FD) 414
Failure recovery (FR) 415
Failure-tolerance 28
fault tolerant SOAP (FT-SOAP) 406
Finite state machine (FSM) 359
Flexible Manufacturing System (FMS) 313
Flow flexibility 322
FRMatcher : function requirement matcher
Function Similarity 137
Fuzzy number 89
Fuzzy number similarity 90

Genetic algorithms (GA) 75
GIIS (Grid Index Information Service) 438
Globus Toolkit 6
GRAM (Globus Resource Allocation Manager) 438
GRAM (Grid Resource Allocation Management) 127
Green manufacturing (GM) 4
Grid FTP service 30
Grid Security Infrastructure (GSI) 149
Grid service 22
GRIS (Grid Resource Information Service) 438
hasEffect 69
hasInput 69
hasOutput 69
hasPrecondition 69

Hierarchical Task Network (HTN) 278
High performance computing (HPC) 470
HR-Class 46
HRD: Human Resources Domain 151
Human Resource (HR) 28

I/O-matching 98
Individual evaluation indices (IEIs) 116
Inertia weight 297
Information grid 10
Information power grid (IPG) 456
Information technology (IT) 7
Infrastructure as a service (IaaS) 476
input replaceable combinable relationship (irCoR) 368
Inputs, Outputs, Preconditions, and Effects (IOPE) 64
Integrated-matching 101
Integration as a service (INTaaS) 477
Intelligent grid 10
Intelligent manufacturing (IM) 4
Inter-domain trust-QoS 137
Internet of thing (IoT) 470
Intra-domain trust-QoS 137
Intuitionistic fuzzy set (IFS) 40

J2EE 18
Joint production correlation (JCC) 263

Key Words similarity 88
Knowledge grid 10

Liner logic (LL) 278

MADM: multiple attribute decision making
Maintain as a service (MAaaS) 477
Maintainability 136
Management as a service (MaaS) 477
Manufacture service (MfgS) 279
Manufacturing as a service (MFGaaS) 477
Manufacturing cloud service (MCS) 472
Manufacturing cloud services (MCSs) 471
Manufacturing grid (MGrid) 1
Manufacturing plan and arrangement service (MPAS)
Material Resource (MR) 28
Materialized resources (MRs) 44
MDS (Miscomputing Discovery Service) 127

Membership function 91
MfgCapability 67
MfgProcess 65
MfgResource 67
MGrid Collaborative Manufacturing Platform 31
MGrid core middleware 28
MGrid executing process view 131
MGrid function view 131
MGrid job management system (MGJMS) 32
MGrid ontology (MGrid-Ontology) 55
MGrid resource conceptual model (MRCM) 58
MGrid resource layer 28
MGrid resource objective model (MROM) 58
MGrid resource service (MGRS) 68
MGrid resource service classification 45
MGrid resource service composition and optimal-selection (MRSCOS) 276
MGrid resource service contexts 162
MGrid resource service optimal-allocation system (MRSOAS) 32
MGrid Resource Service Optimal-Selection and Composition (RSOSC) 40
MGrid resource service optimal-selection and composition framework (MGrid-RSOSCF) 188
MGrid resource view 131
MGrid service view (or information view) 131
MGrid task decomposition 186
MGrid user layer 28
MGrid user portal layer 28
MGrid user service middleware layer 28
MGridGrounding 63
MGridMode 63
MGrid-Ontology repository (MGOR) 72
MGridProfile 63
MGridService 63
MGridState 67
MGridTask 65
MGRSOA (MGrid resource service optimal allocation) 28
MGRSR (MGrid resource service repository) 71
MIS (manufacturing information system) 3
Mismatch 79
Mismatch failure between a task and resource service (Task_RS_Mimatch_Failure) 411
Mixed composite resource service (MCRS) 284
MR-Class 46
MRD: Material Resources Domain 151
Multi-objectives MGrid resource service composition and optimal-selection (MO-MRSCOS) 276
Multi-resource service request task (MRSRTask) 40

Networked manufacturing (NM) 4
NIS: Negative Idea Solution
Non-dominated solution 290
No-trust QoS request 155
NRS: New Resource Service
Number information 83
Number matching algorithms (NMAs) 89
Numerical interval 89
Numerical interval similarity 89

Objective-based trimming operators (OB Trim) 304
Object-oriented model 58
OGSA (Open Grid Service Architecture) 5

OGSI (Open Grid Service Infrastructure) 6
Online purchasing automobiles parts (OPAP) 255
Ontology 54
OR-Class 46
ORS: Old Resource Service
Other Resource (OR) 29
output replaceable combinable relationship (orCoR) 368
Overlap of two compared words 84
OWL-S (Ontology Web Language for Services) 59
OWL-S (OWL for Services) 313
OWL-S ontology 57

Parallel model 193
Pareto optimal 290
Pareto Solution 290
partial pre-order combinable relationship (ppCoR) 368
partial replaceable combinable relationship (prCoR) 368
partial successor combinable relationship (psCoR) 368
Path length of two compared words 84
PDM (product data management) 3
Performance-QoS 32
Permutation-based trimming operators (PB Trim) 304
Pervasive Information Communities Organization (PICO) 277
PIS: Positive Idea Solution
Platform as a service (PaaS) 476
Plug-in match 79
Poisson distribution 362
Precision 110
Precondition 67
Preferential cooperation correlation (PCC) 263
PSO (partial swarm optimization) 206
PSR-Class 46
Public Service Resource (PSR) 28

QoS (Quality of Service) 32
QoS comparison 139
QoS evaluation 139
QoS extraction 139
QoS flexibility 322
QoS inconsistent 411
QoS monitor 139
QoS search 139
QoS updating 139
QoS: Quality of Service
QoS-based resource service management (QoS-RSM) 138
QoS-matching 100
QPRM: QoS Performance Rating Matrix
QPRMW: Weighted QPRM
Qualifi edRate 248
Quantum genetic algorithms (QGA) 359

Real-time and Dynamical Updating Algorithm of Trust-QoS Degree 170
Recall 110
Recommend entity 154
Recommended trust 158
Recommended trust degree data structure 164
Recommended Trust-QoS Evaluating Algorithm 166
Recommended trust-QoS evaluation value 158
Reliability 136
Requirements of resource service QoS 157
Resource classifier 152
Resource enterprise or resource service provider (RSP) 31
Resource Node (RN) 151

Resource service ability changed failure (RS_AbilityChange_Failure) 410
Resource service composition and optimal-selection (RSCOS) 40
Resource service composition executing path (RSCEP) 205
Resource service composition failure (RS_Composition_Failure) 409
Resource service composition layer (C-Layer) 189
Resource service flexibility 322
Resource service match and search (RSMS) 39
Resource service match and search layer (S-Layer) 189
Resource service match engine (MatchEngine) 197
Resource service matching type 79
Resource Service Optimal-Allocation System 27
Resource service optimal-selection (RSOS) 40
Resource service optimal-selection layer (O-Layer) 189
Resource service overload or saturation failure (RS_Overload_Failure) 409
Resource service QoS synthetically processing layer (Q-Layer) 189
Resource service quit failure (RS_Quit_Failure) 409
Resource service related failure 408
Resource service selection (RSS) 215
Resource service trust-QoS 157
Resource Services Composition Flexibility 311
RMS (Resources Management System) 150
Role-based access control (RBAC) 456
RSC flexibility 318
RSC-Coordinator: Resource service coordinator 211
RSCE-Controller: Resource service executing controller 210
RSC-Engine: Resource service composition engine 205
RSCEP-Generator: Resource service composition executing paths generator 207
RSCEP-Selector: Resource service composition executing paths selector 208
RSC-Monitor: Resource service monitor 210
RSIC (resource service information center) 80
RS-Matcher (resource service matcher) 80

Scale-free Network (SFN) 362
SCM (supply chain management) 3
Semantic Distance 165
Semantic Graph-Based Service Composition (SeGSeC) 277
Semantic grid 10
Sentence information 83
Sentence length similarity 88
Sentence matching Algorithms (SeMAs) 88
Sequence composite resource service (SCRS) 284
Sequence model 192
Service grid 10
Service-oriented MGrid 28
Set of entities in MGrid 156
Set of MGrid resource service context 156
Set of trust degree 156
Similarity matching algorithms (SMAs) 76
Simulation as a service (SIMaaS) 477
Single resource service request task (SRSRTask) 40
SLA(Service Level Agreement) 438
Small and medium-sized enterprises (SMEs) 2

SMAs (similarity matching algorithms) 82
SMT (Surface Mount Technology) 437
SOAP (Simple Object Access Protocol) 5
Social acceleration coefficient 298
Software as a service (SaaS) 477
Space-Time Genetic Algorithm (STGA) 150
Special evaluation indices (SEIs) 116
Strong-trust QoS request 155
Subsume match 79

Task cancellation failure (Task_ Cancel_Failure) 411
Task flexibility 322
Task layer (T-Layer) 188
Task related failure 408
Task suspension failure (Task_ Suspension_Failure) 411
Technological Driving Forces (TDFs) 6
Technology Resource (TR) 28
The statistical cooperate correlation (StC) 264
Time (Resource Service Execution Duration) 134
TQCSEFK (fastest Time-to-market, highest Quality, lowest Cost, best Service, cleanest Environment, greatest Flexibility, and high Knowledge) 4
Trapezoidal fuzzy number 91
Trapezoidal fuzzy numbers matching algorithms 91
TR-Class 46
Triangular fuzzy number 91
Triangular fuzzy numbers matching algorithms (TriFNMAs) 91
Trust decay grade 159
Trust Failure (Trust Failure) 412
Trust type 137
Trusted entity 154
Trusting entity 154
Trust-QoS 137
Trust-QoS benefit 181
Trust-QoS relationship 137
Trust-QoS relationship model 153
Trust-QoS request type 137
Trust-QoS value 137
Trust-time decay function 159

UDDI (Universal Description, Discovery and Integration) 5
UIR-Class 46
User enterprise or resource service demander (RSD) 31
User Information Resource (UIR) 29
UserInfo 68

Verification and validation (VV) 58
Vicious competition correlation (VCC) 264
Virtual link (VL) 409
Virtual link disconnected failure (VL_Disconnect_Failure) 408
Virtual link related failure 408
Virtual manufacturing (VM) 4
Virtual organization (VO) 14

WCS-Agent (Weight Coefficients Set Agent) 129
Weak-trust QoS request 155
Web Service 22
Whole-lifecycle management (WLCM) 130
Word concept information 82
Word matching algorithms (WMAs) 83
Word order similarity 88
WS-BaseFaults, WS-ServiceGroup 9
WS-BPEL (Web Services Business Process Execution Language) 313

WS-CDL (Web Service Choreography Description Language) 313
WSCI (Web Services Choreography Interface) 313
WSDL (Web Service Description Language) 5
WS-Notification 9
WS-RenewableReferences 9
WS-ResourceLifetime 9
WS–ResourceProperties 9
WSRF (Web Service Resource Framework) 5

XACML (eXtensible Access Control Markup Language) 444

Also of Interest

Check out these published and forthcoming related titles from Scrivener Publishing

Evolutionary Computing in Advanced Manufacturing
Edited by Manoj Tiwari and Jenny Harding
Published 2011. ISBN 978-0-470-63924-5

Manufacturing Grid and Resource Service Management
By Fei Tao, Lin Zhang, and Yefa Hu
Published 2011. ISBN 978-1-118-12231-0

Construction Management for Industrial Projects
By Mohamed El-Reedy
Published 2011. ISBN: 978-0-470-87816-3

The Basics of Troubleshooting in Plastics Processing
By Muralisrinivasan Subramanian
Published 2011. ISBN 978-0-470-62606-1

Cost Reduction and Optimization for Manufacturing and Industrial Companies
By Joseph Berk
Published 2010. ISBN 978-0-470-60957-6

Roll-Roll- Vacuum Deposition of Barrier Coatings
By Charles A. Bishop
Published 2010. ISBN 978-0-470-60956-9

Introduction to Industrial Polyethylene
Properties, Catalysts, Processes
By Dennis B. Malpass
Published 2010. ISBN 978-0-470-62598-9